Thermodynamics and Fluid Mechanics of Turbomachinery Volume II

NATO ASI Series

Advanced Science Institutes Series

A Series presenting the results of activities sponsored by the NATO Science Committee, which aims at the dissemination of advanced scientific and technological knowledge, with a view to strengthening links between scientific communities.

The Series is published by an international board of publishers in conjunction with the NATO Scientific Affairs Division

A	**Life Sciences**	Plenum Publishing Corporation
B	**Physics**	London and New York
C	**Mathematical and Physical Sciences**	D. Reidel Publishing Company Dordrecht and Boston
D	**Behavioural and Social Sciences**	Martinus Nijhoff Publishers Dordrecht/Boston/Lancaster
E	**Applied Sciences**	
F	**Computer and Systems Sciences**	Springer-Verlag Berlin/Heidelberg/New York
G	**Ecological Sciences**	

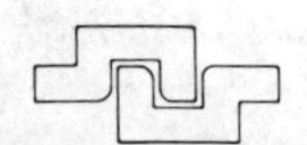

Series E: Applied Sciences – No. 97B

Thermodynamics and Fluid Mechanics of Turbomachinery

Volume II

Edited by

A.Ş. Üçer
Mechanical Engineering Department
Middle East Technical University
(ODTU)
Ankara, Turkey

P. Stow
Theoretical Science Group
Rolls-Royce Limited
Derby, UK

Ch. Hirsch
Free University of Brussels
Brussels, Belgium

1985 **Martinus Nijhoff Publishers**
Dordrecht / Boston / Lancaster
Published in cooperation with NATO Scientific Affairs Division

Proceedings of the NATO Advanced Study Institute on Thermodynamics and Fluid Mechanics of Turbomachinery, Izmir, Turkey, September 17-28, 1984

Library of Congress Cataloging in Publication Data

Main entry under title:

Thermodynamics and fluid mechanics of turbomachinery.

(NATO ASI series. Series E, Applied sciences ; no. 97A-)
"Proceedings of the NATO Advanced Study Institute on "Thermodynamics and Fluid Mechanics of Turbomachinery," Izmir, Turkey, September 17-29, 1984"--V. 1, p. iv.
Bibliography: p.
Includes index.
1. Turbomachines--Congresses. 2. Thermodynamics--Congresses. 3. Fluid mechanics--Congresses. I. Üçer, A. Ş. II. Stow, P. III. Hirsch, Ch. IV. NATO Advanced Study Institute on "Thermodynamics and Fluid Mechanics of Turbomachinery" (1984 : Izmir, Turkey) V. Series: NATO ASI series. Series E, Applied sciences ; no. 97A, etc.
TJ267.T57 1985 621.406 85-18778
ISBN 90-247-3223-9 (set)
ISBN 90-247-3221-2 (v. 1)
ISBN 90-247-3222-0 (v. 2)

ISBN 90-247-3222-0 (this volume)
ISBN 90-247-2689-1 (series)
ISBN 90-247-3223-9 (set)

Distributors for the United States and Canada: Kluwer Boston, Inc., 190 Old Derby Street, Hingham, MA 02043, USA

Distributors for the UK and Ireland: Kluwer Academic Publishers, MTP Press Ltd, Falcon House, Queen Square, Lancaster LA1 1RN, UK

Distributors for all other countries: Kluwer Academic Publishers Group, Distribution Center, P.O. Box 322, 3300 AH Dordrecht, The Netherlands

Printed in The Netherlands

PREFACE

During the last decade, rapid advances have been made in the area of flow analysis in the components of gas turbine engines. Improving the design methods of turbomachine blade rows and understanding of the flow phenomena through them, has become one of the major research topics for aerodynamists. This increase of research efforts is due to the need of reducing the weight and fuel consumption of turbojet engines for the same thrust levels. One way of achieving this is to design more efficient components working at high local velocities. Design efforts can lead to desired results only if the details of flow through the blade rows are understood.

It is also known that for aircraft propulsion systems development, time and cost can be reduced significantly if the performance can be predicted with confidence and enough precision. This generally needs sophisticated two or three dimensional computer codes that can give enough information for design and performance prediction. In the recent years, designers also started to use these sophisticated codes more and more with confidence, in connection with computer aided design and manufacturing techniques. On the other hand, the modelling and solution of flow and the measurement techniques inside the blade rows are difficult, and present challenging problems of fluid dynamics.

It was with these facts in mind that the organizers of the NATO Advanced Study Institute, held in İzmir (Çeşme), Turkey from 17 to 28 September 1984, invited lecturers to review the current state of knowledge in the advanced design and performance prediction of turbomachines. The lecturers were: H.A. Akay, Purdue University at Indianapolis, Indiana, U.S.A.; R.V.D. Braembussche, von Karman Institute, Belgium; T. Cebeci, Douglas Aircraft Company

Long Beach, California, U.S.A.; J. Chauvin, Institute de Mécanique des Fluides, Marseille, France; J. D. Denton, Whittle Lab., Cambridge University, Cambridge, U.K.; E.M. Greitzer, Massachusetts Institute of Technology, Cambridge, Mass., U.S.A.; Ch. Hirsch, Vrije Universiteit Brussel, Brussels, Belgium; E. Macchi, Politecnico di Milano, Milano, Italy; G. Meauze, ONERA, Chatillon, France; R. Schodl, DFVLR-Institut für Antriebstechnik, Köln, W. Germany; G. K. Serovy, Iowa State University, Ames, Iowa, U.S.A.; C. Sieverding, von Karman Institute, Belgium; P. Stow, Rolls-Royce Limited, Derby, U.K.; A.Ş. Üçer, Middle East Technical University, Ankara, Turkey; H. Weyer, Deutsch-Niederlaendischer Windkanal, the Netherlands. This book is compiled from the lectures of the above scientists and their co-workers, and it is aimed to review the experimental and theoretical developments in the field of aerothermodynamics and design aspects of turbomachinery. Naturally the coverage is not complete. However, lecturers have been very conscientious in adopting their work to the aims set down by the organizers. As an outcome of this, the product has exceeded, both in breadth and coherence, the expectations of the initiators of the Institute. The topics are arranged so that the material compiled will serve as an advanced reference book to the turbomachine designers and research workers.

The material is grouped into seven parts. The first part gives the fundamentals of compressible flow in turbomachines. In the second part, solution methods for inviscid and viscous representations are included. The third part presents existing test facilities and the measurement techniques for turbomachinery research and development. In the fourth part, problems associated with secondary flows in radial and axial flow machines and end wall boundary layer modelling and solution methods are discussed. The fifth part covers performance prediction methods and recent design techniques for advanced turbomachinery. In the sixth part unsteady effects and flow instabilities are discussed. In the final part, the round table discussion which covers suggestions for future research in the turbomachinery aerothermodynamics is presented.

The editors would like to thank the NATO Scientific Affairs Division, the Middle East Technical University and the Scientific and Technical Research Council of Turkey for making the Advanced Study Institute possible, and to Martinus Nijhoff Publishers for publishing the edited proceedings. We wish to extend a word of appreciation to DFVLR, General Dynamics Corporation, MTU, Nouvo Pignone, Rolls Royce Limited, U.S. Army Research, Development and Standardization Group, Dokuz Eylül University, and to Turkish firms, GAMA, LAMAŞ, Teknik Komandit Şirketi, TURBOSAN and Yazar Pompa for their support. We are also grateful to the authors for their contributions and their cooperativeness when revisions were requested. Our special gratitute goes to Professor G.K. Serovy, a member of the organizing committee, because of his invaluable contributions at all stages of this endeavour. We also extend our gratitute to the members of the programme committee, session chairmen, co-chairmen, and scientific secretaries in programming, organizing, and executing the technical sessions. The participants of the Institute are acknowledged for their attendance and for their questions and comments which made the sessions lively and convinced everybody that the undertaking had been worthwhile.

Finally we want to express our special appreciation to Drs. H. Aksel, N. Alemdaroğlu, C. Eralp, E. Paykoç, and to Messrs. H. Arıtürk, O. Boratav, M. Çetin, Ü. Enginsoy, O. Oğuz, E. Hatip for their valuable contributions and assistance and to Mrs. G. Beyaz for her efficient secretarial work.

A.Ş. ÜÇER
P. STOW
Ch. HIRSCH

CONTENTS

SECONDARY FLOWS AND END WALL EFFECTS

SECONDARY FLOWS IN AXIAL-FLOW COMPRESSORS

George K. Serovy

Department of Mechanical Engineering
Iowa State University
Ames, IA 50011 U.S.A.

INTRODUCTION

The term "secondary flow" has been present in the key word list of the literature of turbomachinery for many years. There have been, however, substantial changes in the nature and scope of the material discussed in the papers and reports identified with the subject of secondary flow. Although this lecture is principally concerned with secondary flows and secondary losses in axial-flow compressors, it is appropriate to refer to the excellent summary of the topic as related to axial-flow turbines by Sieverding (1984) as well as the lecture in this ASI by Sieverding on the same subject. While there may be differences in quantitative significance, the phenomena involved are fundamentally the same. Key experiments and key analyses tend to overlap any boundary which might be suggested between the two areas of emphasis.

DEFINITION AND CLASSIFICATION

To permit a closer association with current compressor design systems, secondary flow is defined as the "difference between the real flow (excluding small-scale turbulent fluctuations) and the idealized axisymmetric flow." This definition is due to L. H. Smith, Jr. (1955), and its principal shortcoming might be that the "idealized axisymmetric flow" has been a target which moves with time. Its major advantage is that it spotlights the fact that in all axisymmetric flow computation systems used in design and analysis of axial-flow compressors, major discrepancies occur due to flow turning and irreversibility effects caused by secondary effects.

For purposes of reference the following is a partial list of terms included in papers dealing with secondary flow phenomena:

Cascade passage vortex
Distributed secondary vortex
Trailing filament vorticity
Trailing shed vorticity
Corner vortex
Tip clearance vortex
Scraping vortex
Horseshoe vortex
Annulus end wall boundary layer
Blade wake mixing

These terms do not refer to flow field occurrences which are independent mechanisms. As can be observed in experiments as well as in the set of figures accompanying this lecture, they overlap and interact. Terminology and interpretation change with time.

USING THE REFERENCE LIST AND FIGURES

The references include a selection from the review and survey papers published during the last thirty years as well as from the original analytical and experimental efforts.

Any investigator entering the secondary flow and loss area for the first time should study Lakshminarayana and Horlock (1963), Horlock and Lakshminarayana (1970), Woods (1971), Horlock and Perkins (1974), Salvage (1974), Horlock (1977), Chauvin (1981), Hirsch (1981), and Sitaram and Lakshminarayana (1983).

To estimate the influences of secondary flow on compressor design in terms of losses consideration should be given to Vavra (1960), Smith (1970), Salvage (1974), Grieb, Schill and Gumucio (1975), Koch and Smith (1976), Chauvin (1981), and Hirsch (1981).

To develop and evaluate means for control of secondary flows and losses look at the work of Andrews, Jeffs and Hartley (1956), Smith (1974), Wennerstrom (1984), Wisler (1984), Wagner, Dring and Joslyn (1983 and 1984), and Joslyn and Dring (1984).

All of the figures are based on the reference works indicated. An attempt has been made to show the changing patterns in identification and evaluation of secondary flow phenomena.

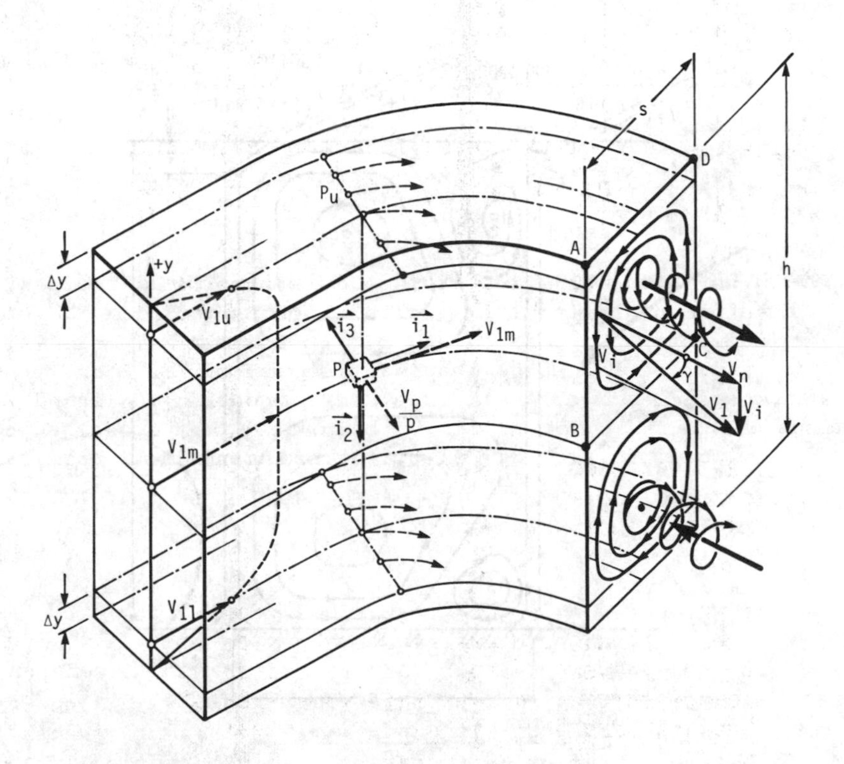

Figure 1. Conceptual sketch of passage vortex resulting from velocity profile at entrance to curved duct (after Vavra (1960)).

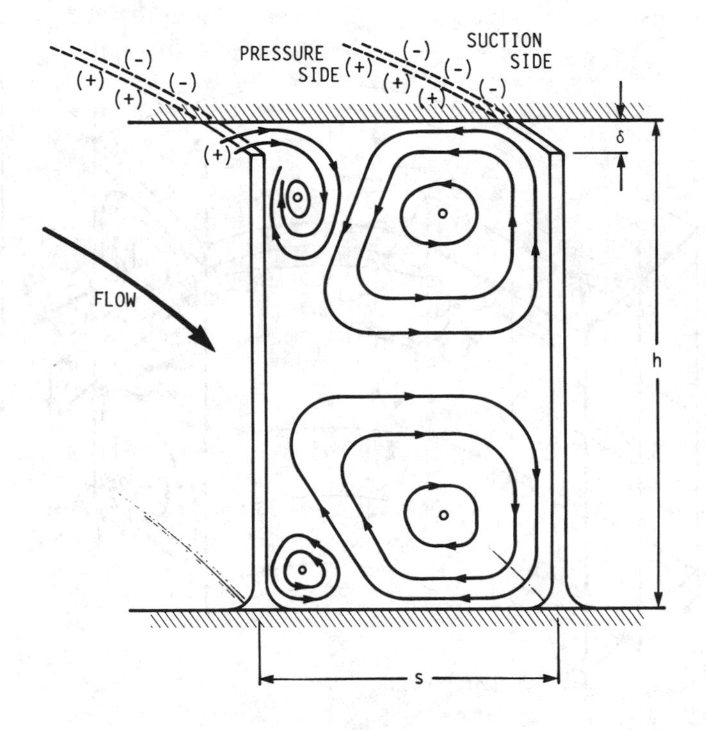

Figure 2. Conceptual sketch of vortex formation in a cascade with tip clearance (after Vavra (1960)).

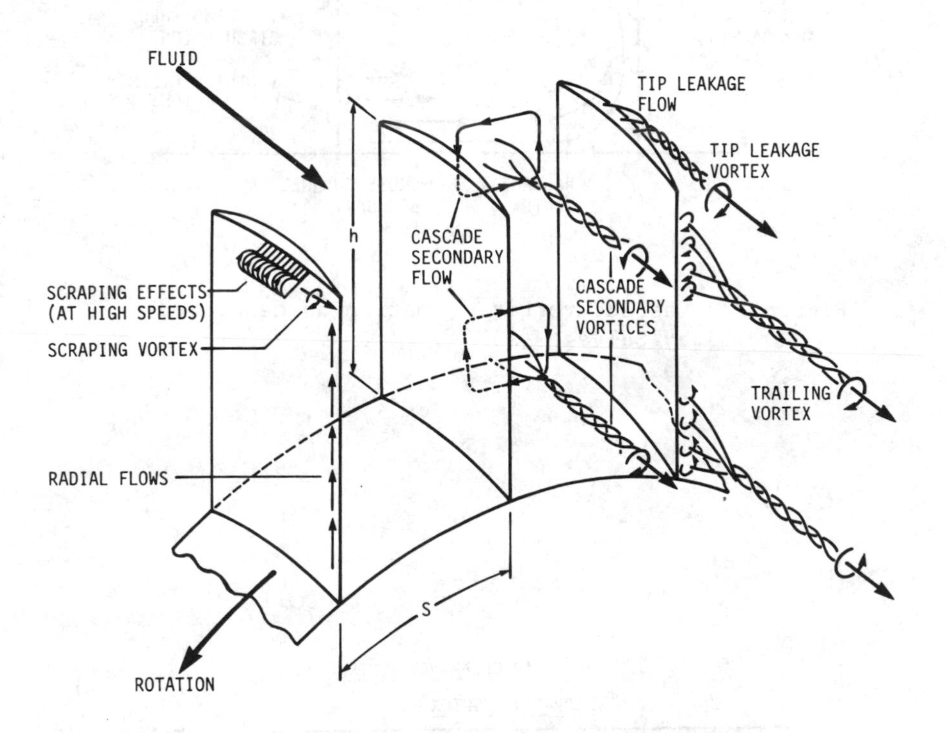

Figure 3. Secondary flow patterns in axial-flow compressor rotor as described by Lakshminarayana and Horlock (1963).

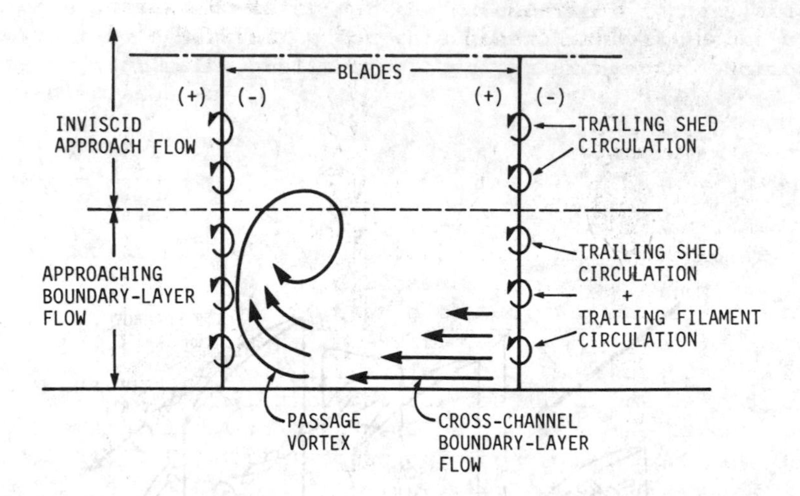

Figure 4. Cascade vortex formation as described by Salvage (1974).

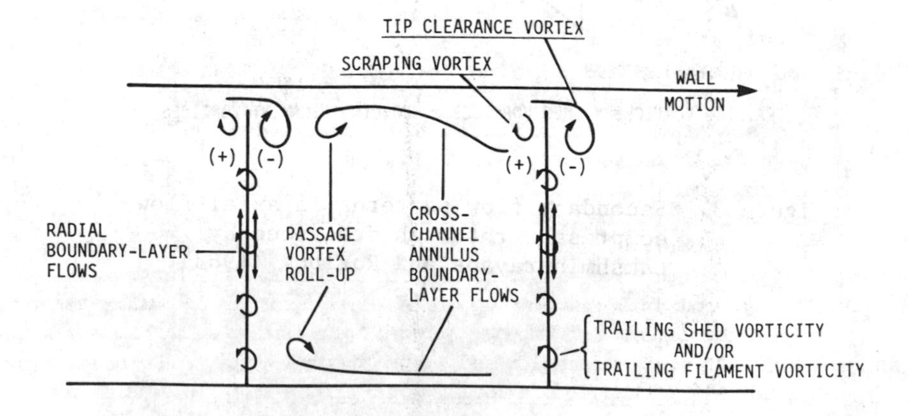

Figure 5. Secondary flow in axial-flow compressor rotor trailing-edge region (after Salvage (1974)).

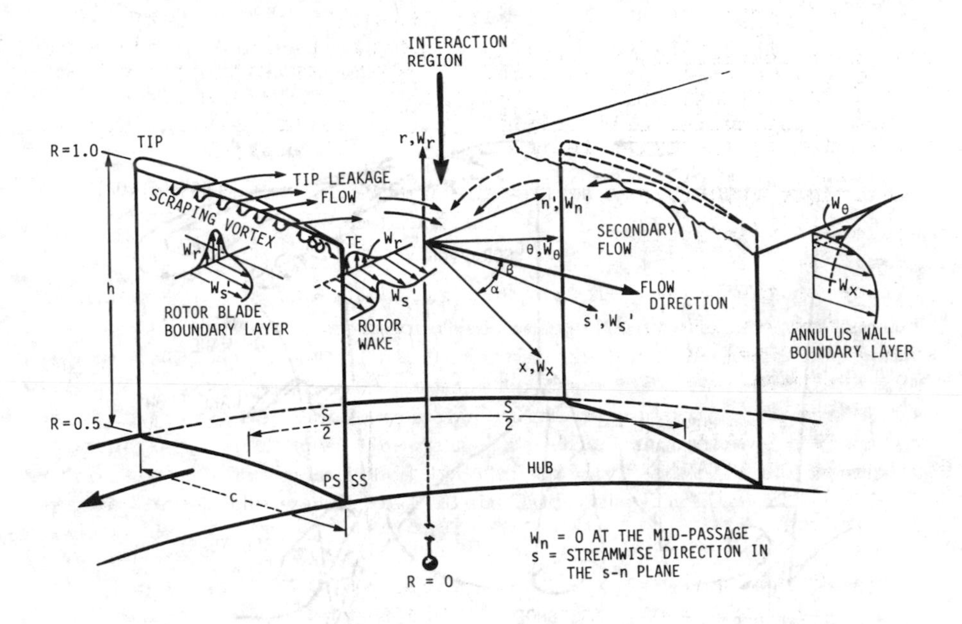

Figure 6. Secondary flows interacting in tip endwall region of axial-flow compressor rotor (Sitaram and Lakshminarayana (1983)).

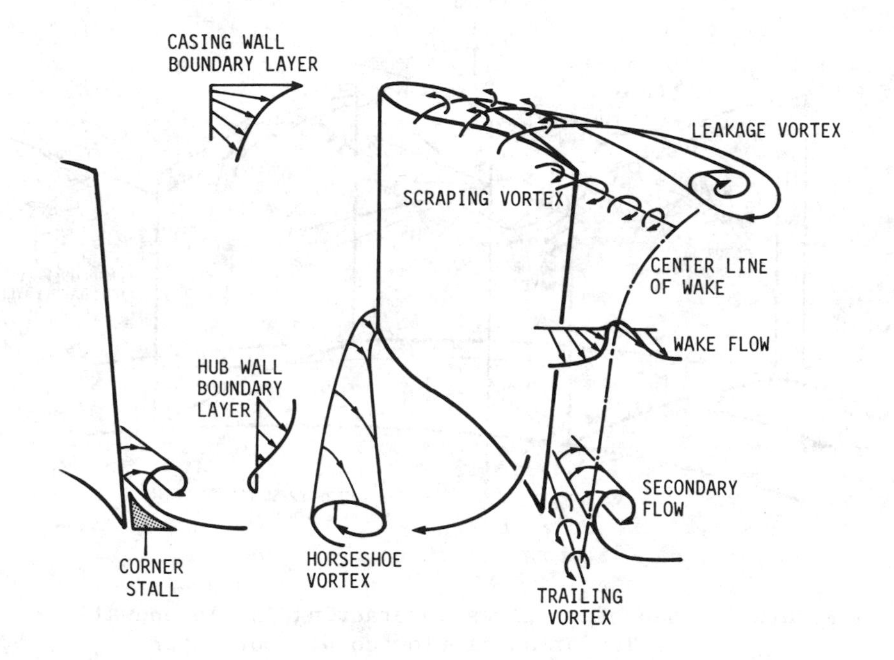

Figure 7. Combined secondary flow patterns generated by axial-flow compressor rotor as described by Inoue and Kuroumarou (1984).

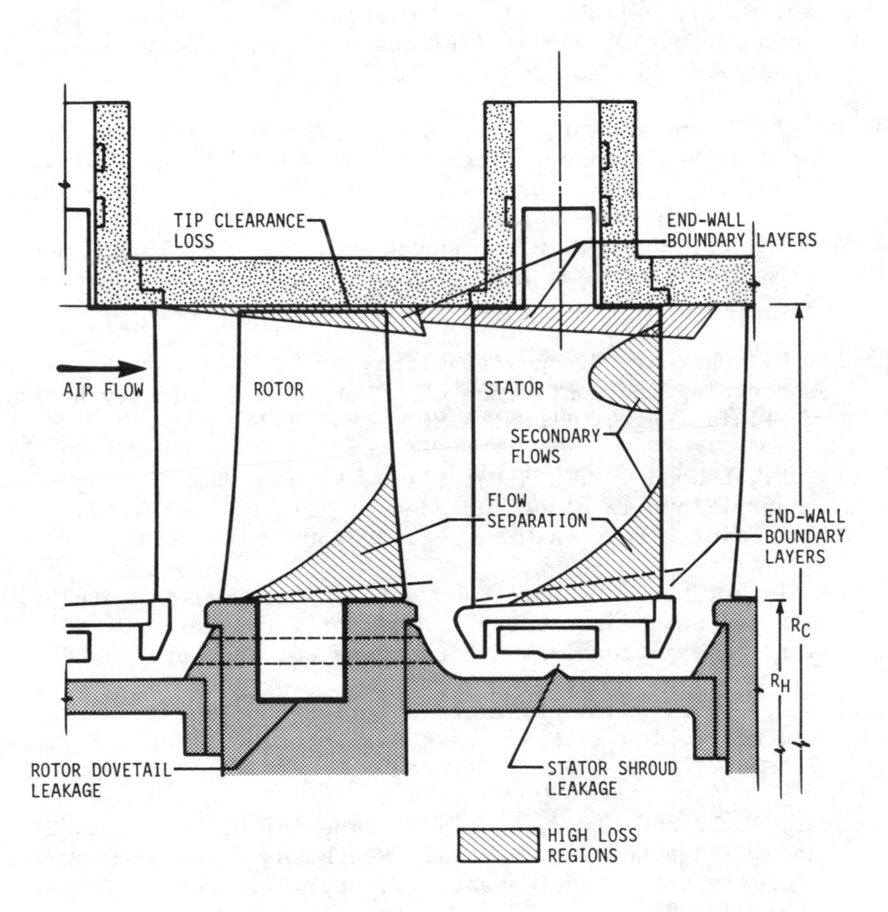

Figure 8. High-loss regions in blading of a compressor stage as discussed by Wisler (1984).

BIBLIOGRAPHY

Adkins, G. G., Jr. and Smith, L. H., Jr. 1982. Spanwise Mixing in Axial-Flow Turbomachines. J. Engr. for Power, Trans. ASME. 104: 97-110.

Andrews, S. J., Jeffs, R. A. and Hartley, E. L. 1956. Tests Concerning Novel Designs of Blades for Axial Compressors. Parts 1 and 2. ARC Rep. and Memo. 2929.

Balsa, T. F. and Mellor, G. L. 1975. The Simulation of Axial Compressor Performance Using an Annulus Wall Boundary Layer Theory. J. Engr. for Power, Trans. ASME. 97: 305-318.

Bardon, M. F., Moffatt, W. C. and Randall, J. L. 1975. Secondary Flow Effects on Gas Exit Angles in Rectilinear Cascades. J. Engr. for Power, Trans. ASME. 97: 93-100.

Bario, F., Charnay, G. and Papailiou, K. D. 1982. A Experiment Concerning the Confluence of a Wake and a Boundary Layer. J. Fluids Engr., Trans. ASME. 104:18-24.

Bario, F., Leboeuf, F. and Papailiou, K. D. 1982. Study of Secondary Flows in Blade Cascades of Turbomachines. J. Engr. for Power, Trans. ASME. 104: 497-509.

Bauermeister, K. J. 1963. Über den Einfluss der Schaufelbelastung auf die Sekundärströmung in geraden Turbinen- und Verdichtergittern. VDI Forschungsheft 496, pp. 19-28.

Belik, L. 1968. An Approximate Solution for Kinetic Energy of Secondary Flow in Blade Cascades. Intl. J. Mech. Sci. 10: 765-782.

Bettner, J. L. and Elrod, C. 1983. The Influence of Tip Clearance, Stage Loading, and Wall Roughness on Compressor Casing Boundary Layer Development. J. Engr. for Power, Trans. ASME. 105: 280-287.

Bois, G., Leboeuf, F., Comte, A., and Papailiou, K. D. 1977. Experimental Study of the Behaviour of Secondary Flows in a Transonic Compressor. In Secondary Flows in Turbomachines. AGARD-CP-214, pp. 3-1 to 3-20.

Breugelmans, F. A. H., Carels, Y. and Demuth, M. 1984. Influence of Dihedral on the Secondary Flow in a Two-Dimensional Compressor Cascade. J. Engr. for Gas Turbines and Power, Trans. ASME. 106: 578-584.

Came, P. M. and Marsh, H. 1974. Secondary Flow in Cascades: Two Simple Derivations for the Components of Vorticity. J. Mech. Engr. Sci. 16: 391-401.

Carter, A. D. S. 1948. Three-Dimensional-Flow Theories for Axial Compressors and Turbines. Proc. IME. 159: 255-268.

Chauvin, J. 1981. Correlation for Secondary Flows and Clearance Effects. In Through Flow Calculations in Axial Turbomachines. AGARD-AR-175. pp. 151-163.

Cheng, P., Prell, M. E., Greitzer, E. M. and Tan, C. S. 1984. Effects of Compressor Hub Treatment on Stator Stall and Pressure Rise. J. Aircraft. 21: 469-475.

Chima, R. A. and Strazisar, A. J. 1983. Comparison of Two- and Three-Dimensional Flow Computations with Laser Anemometer Measurements in a Transonic Compressor Rotor. J. Engr. for Power, Trans. ASME. 105: 596-605.

Comte, A., Ohayon, G. and Papailiou, K. D. 1982. A Method for the Calculation of the Wall Layers Inside the Passage of a Compressor Cascade With and Without Tip Clearance. J. Engr. for Power, Trans. ASME. 104: 657-667.

Debruge, L. L. 1983. The Aerodynamic Significance of Fillet Geometry in Turbocompressor Blade Rows. J. Engr. for Power, Trans. ASME. 102: 984-993.

DeRuyck, J. and Hirsch, Ch. 1981. Investigations of an Axial Compressor End-Wall Boundary Layer Prediction Method. J. Engr. for Power, Trans. ASME. 103: 20-33.

DeRuyck, J. and Hirsch, Ch. 1983. End-Wall Boundary Layer Calculations in Multistage Axial Compressors. In Viscous Effects in Turbomachines. AGARD-CP-351, pp. 19-1 to 19-16.

DeRuyck, J., Hirsch, Ch. and Kool, P. 1977. Hot-Wire Measurements in an Axial Compressor and Confrontation with Theoretical Predictions of Secondary Flows. In Secondary Flows in Turbomachinery. AGARD-CP-214, pp. 7-1 to 7-18.

DeRuyck, J., Hirsch, C. and Kool, P. 1979. An Axial Compressor End-Wall Boundary Layer Calculation Method. J. Engr. for Power, Trans. ASME. 101: 237-249.

Dixon, S. L. 1974. Secondary Vorticity in Axial Compressor Blade Rows. In Fluid Mechanics, Acoustics, and Design of Turbomachinery. NASA SP-304, Part I, pp. 173-191.

Dring, R. P. 1982. Boundary Layer Transition and Separation on a Compressor Rotor Airfoil. J. Engr. for Power, Trans. ASME. 104: 251-253.

Dring, R. P. 1984. Blockage in Axial Compressors. J. Engr. for Gas Turbines and Power, Trans. ASME. 106: 712-713.

Dring, R. P., Joslyn, H. D. and Hardin, L. W. 1979. Experimental Investigation of Compressor Rotor Wakes. AFAPL-TR-79-2107.

Dring, R. P., Joslyn, H. D. and Hardin, L. W. 1982. An Investigation of Axial Compressor Rotor Aerodynamics. J. Engr. for Power, Trans. ASME. 104: 84-96.

Dring, R. P., Joslyn, H. D. and Wagner, J. H. 1983. Compressor Rotor Aerodynamics. In Viscous Effects in Turbomachines. AGARD-CP-351. pp. 24-1 to 24-16.

Ehrich, F. F. and Detra, R. W. 1954. Transport of the Boundary Layer in Secondary Flow. J. Aero. Sci. 21: 136-138.

Elrod, C. W. and Bettner, J. L. 1983. Experimental Verification of an Endwall Boundary Layer Prediction Method. In Viscous Effects in Turbomachines. AGARD-CP-351. pp. 25-1 to 25-21.

Enayet, M. M., Gibson, M. M. and Yianneskis, M. 1982. Measurements of Turbulent Developing Flow in a Moderately Curved Square Duct. Intl. J. Heat and Fluid Flow. 3: 221-224.

Gallus, H. E. and Kümmel, W. 1977. Secondary Flows and Annulus Wall Boundary Layers in Axial-Flow Compressor and Turbine Stages. In Secondary Flows in Turbomachines. AGARD-CP-214, pp. 4-1 to 4-15.

Glynn, D., Spurr, A. and Marsh, H. 1977. Secondary Flow in Cascades. In Secondary Flows in Turbomachines. AGARD-CP-214, pp. 13-1 to 13-12.

Goldstein, S. 1938. Modern Developments in Fluid Dynamics, Vol. II. Oxford. Oxford University Press.

Grahl, K. and Simon, Ch. 1982. Secondary Flow Models for Extending a Two-Dimensional Boundary Layer Theory for Axial Flow Turbomachines. Forsch.-Ing.-Wes. 48: 1-10.

Greitzer, E. M., Nikkanen, J. P., Haddad, D. E., Mazzawy, R. S. and Joslyn, H. D. 1979. A Fundamental Criterion for the Application of Rotor Casing Treatment. J. Fluids Engr., Trans. ASME. 101: 237-243.

Grieb, H., Schill, G. and Gumucio, R. 1975. A Semi-Empirical Method for the Determination of Multistage Axial Compressor Efficiency. ASME Paper 75-GT-11.

Griepentrog, H. 1970. Secondary Flow Losses in Axial Compressors. AGARD LS-39. Paper 5.

Hanley, W. T. 1968. A Correlation of End Wall Losses in Plane Compressor Cascades. J. Engr. for Power, Trans. ASME. 90: 251-257.

Hawthorne, William R. 1951. Secondary Circulation in Fluid Flow. Proc. Roy. Soc. (London). Series A, Vol. 206, No. A 1086: 374-387.

Hawthorne, W. R. 1967. The Applicability of Secondary Flow Analyses to the Solution of Internal Flow Problems. In Fluid Mechanics of Internal Flow. Amsterdam. Elsevier Publishing Company. pp. 238-269.

Herzig, H. Z. and Hansen, A. G. 1955. Visualization Studies of Secondary Flows with Applications to Turbomachines. Trans. ASME. 77: 249-266.

Herzig, H. Z. and Hansen, A. G. 1957. Experimental and Analytical Investigation of Secondary Flows in Ducts. J. Aero. Sci. 24: 217-231.

Herzig, H. Z., Hansen, A. G. and Costello, G. R. 1954. A Visualization Study of Secondary Flows in Cascades. NACA TR 1163.

Hirsch, Ch. 1974. End-Wall Boundary Layers in Axial Compressors. J. Engr. for Power, Trans. ASME. 96: 413-426.

Hirsch, Ch. 1981. End-Wall Boundary Layer Calculation Methods. In Through Flow Calculations in Axial Turbomachines. AGARD-AR-175. pp. 137-150.

Hirsch, Ch. and Kool, P. 1977. Measurement of the Three-Dimensional Flow Field Behind an Axial Compressor Stage. J. Engr. for Power, Trans. ASME. 99: 168-180.

Horlock, J. H. 1963. Annulus Wall Boundary Layers in Axial Compressor Stages. J. Basic Engr., Trans. ASME. 85: 55-65.

Horlock, J. H. 1977. Recent Developments in Secondary Flow. In Secondary Flows in Turbomachines. AGARD-CP-214. pp. 1-1 to 1-18.

Horlock, J. H. and Lakshminarayana, B. 1973. Secondary Flows: Theory, Experiment, and Application in Turbomachinery Aerodynamics. Annual Review of Fluid Mechanics. Vol. 5, pp. 247-280.

Horlock, J. H., Louis, J. F., Percival, P. M. E. and Lakshminarayana, B. 1968. Wall Stall in Compressor Cascades. J. of Basic Engr., Trans. ASME. 90: 637-648.

Horlock, J. H. and Marsh, H. 1982. Fluid Mechanics of Turbomachines: a Review. Intl. J. Heat and Fluid Flow. 3: 3-11.

Horlock, J. H. and Perkins, H. J. 1974. Annulus Wall Boundary Layers in Turbomachines. AGARDograph 185.

Hubert, G. 1963. Untersuchungen über die Sekundärverluste in axialen Turbomaschinen. VDI Forschungsheft 496, pp. 5-18.

Hunter, I. H. and Cumpsty, N. A. 1982. Casing Wall Boundary Layer Development Through an Isolated Compressor Rotor. ASME Paper 82-GT-18.

Inoue, M. and Kuroumaru, M. 1984. Three-Dimensional Structure and Decay of Vortices Behind an Axial Flow Rotating Blade Row. J. Engr. for Gas Turbines and Power, Trans. ASME. 106: 561-569.

Jansen, W. 1967. The Application of End-Wall Boundary-Layer Effects in the Performance Analysis of Axial Compressors. ASME Paper 67-WA/GT-11.

Joslyn, H. D. and Dring, R. P. 1984. Axial Compressor Stator Aerodynamics. ASME Paper 84-GT-90.

Koch, C. C. and Smith, L. H., Jr. 1976. Loss Sources and Magnitudes in Axial-Flow Compressors. J. Engr. for Power, Trans. ASME. 98: 411-424.

Kool, P., DeRuyck, J. and Hirsch, Ch. 1978. The Three-Dimensional Flow and Blade Wake in an Axial Flow Compressor Rotor. ASME Paper 78-GT-66.

Kreskovsky, J. P., Briley, W. R. and McDonald, H. 1981. Prediction of Laminar and Turbulent Primary and Secondary Flow in a Strongly Curved Duct. NASA CR-3388.

Lakshminarayana, B. 1970. Methods of Predicting the Tip Clearance Effects in Axial Turbomachines. J. Basic Engr., Trans. ASME. 92: 467-482.

Lakshminarayana, B. and Horlock, J. H. 1963. Review: Secondary Flows and Losses in Cascades and Axial-Flow Turbomachines. Intl. J. Mech. Sci. 5: 287-307.

Lakshminarayana, B. and Horlock, J. H. 1965. Leakage and Secondary Flows in Compressor Cascades. ARC Rep. and Memo. 3483.

Lakshminarayana, B. and Horlock, J. H. 1967. Effects of Shear Flows on the Outlet Angle in Axial Compressor Cascades. J. Basic Engr., Trans. ASME. 89: 191-200.

Lakshminarayana, B. and Horlock, J. H. 1973. Generalized Expressions for Secondary Vorticity Using Intrinsic Coordinates. J. Fluid Mech. 59(1): 97-115.

Lakshminarayana, B., Murthy, K., Pouagare, M. and Govindan, T. R. 1983. Annulus Wall Boundary Layer Development in a Compressor Stage, Including the Effects of Tip Clearance. In Viscous Effects in Turbomachines. AGARD-CP 351, pp. 21-1 to 21-17.

Lakshminarayana, B. and Ravindranath, A. 1982. Interaction of Compressor Rotor Blade Wake With Wall Boundary Layer/Vortex in the End-Wall Region. J. Engr. for Power, Trans. ASME. 104: 467-478.

Leboeuf, F., Comte, A. and Papailiou, K. D. 1977. Calculations Concerning the Secondary Flows in Compressor Bladings. In Secondary Flows in Turbomachines. AGARD-CP-214, pp. 2-1 to 2-9.

Leboeuf, F., Bario, F., Bois, G. and Papailiou, K. 1982. Experimental Study and Theoretical Prediction of Secondary Flows in a Transonic Axial Flow Compressor. ASME Paper 82-GT-14.

Louis, J. F. 1958. Secondary Flow and Losses in a Compressor Cascade. ARC Rep. and Memo. 3136.

Mager, A., Mahoney, J. J. and Budinger, R. E. 1952. Discussion of Boundary Layer Characteristics Near the Wall of an Axial Flow Compressor. NACA Report 1085.

Marchal, Ph. and Sieverding, C. H. 1977. Secondary Flows Within Turbomachinery Bladings. In Secondary Flows in Turbomachines. AGARD-CP-214, pp. 11-1 to 11-20.

Marsh, H. 1974. Secondary Flow in Cascades: The Effect of Axial Velocity Ratio. J. Mech. Engr. Sci. 16: 402-407.

Marsh, H. and Horlock, J. H. 1972. Wall Boundary Layers in Turbomachines. J. Mech. Engr. Sci. 14: 411-423.

McMillan, O. J. 1982. Mean-Flow Measurements of the Flow Field Diffusing Bend. NASA CR-3634.

Meldahl, A. 1941. The End Losses of Turbine Blades. Brown Boveri Rev. 28: 356-361.

Mellor, G. L. and Strong, R. E. 1967. End-Wall Effects in Axial Compressors. ASME Paper 67-FE-16.

Mellor, G. L. and Wood, G. M. 1971. An Axial Compressor End-Wall Boundary Layer Theory. J. Basic Engr., Trans. ASME. 93: 300-316.

Moore, R. W., Jr. and Richardson, D. L. 1957. Skewed Boundary Layer Flow Near the End Walls of a Compressor Cascade. Trans. ASME. 79: 1789-1800.

Murthy, K. N. S. and Lakshminarayana, B. 1984. Laser Doppler Velocimeter Measurements in the Tip Region of a Compressor Rotor. AIAA Paper 84-1602.

Pandya, A. and Lakshminarayana, B. 1983. Investigation of the Tip Clearance Flow Inside and at the Exit of a Compressor Rotor Passage - Part I: Mean Velocity Field. J. Engr. for Power, Trans. ASME. 105: 1-12.

Pandya, A. and Lakshminarayana, B. 1983. Investigation of the Tip Clearance Flow Inside and at the Exit of a Compressor Rotor Passage - Part II: Turbulence Properties. J. Engr. for Power, Trans. ASME. 105: 13-17.

Papailiou, K. 1975. Correlations Concerning the Process of Flow Deceleration. J. Engr. for Power, Trans. ASME. 97: 295-300.

Papailiou, K. D., Flot, R. and Mathieu, J. 1976. Secondary Flows in Compressor Bladings. ASME Paper 76-GT-57.

Papailiou, K., Flot, R. and Mathieu, J. 1977. Secondary Flows in Compressor Blading. J. Engr. for Power, Trans. ASME. 99: 211-224.

Pouagare, M. and Lakshminarayana, B. 1982. Development of Secondary Flow and Vorticity in Curved Ducts, Cascades, and Rotors, Including Effects of Viscosity and Rotation. J. Basic Engr., Trans. ASME. 104: 505-512.

Pouagare, M., Galmes, J. M. and Lakshminarayana, B. 1984. An Experimental Study of the Compressor Rotor Blade Boundary Layer. PSU/TURBO R84-1.

Pratap, V. S. and Spalding, D. B. 1975. Numerical Computations of the Flow in Curved Ducts. Aero. Quarterly. 26: 221-228.

Railly, J. W. and Howard, J. H. G. 1962. Velocity Profile Development in Axial-Flow Compressors. J. Mech. Engr. Sci. 4: 166-176.

Railly, J. W. and Sharma, P. B. 1977. Treatment of the Annulus Wall Boundary Layer Using a Secondary Flow Hypothesis. J. Engr. for Power, Trans. ASME. 99: 29-36.

Ravindranath, A. and Lakshminarayana, B. 1982. Rotor Wake Mixing Effects Downstream of a Compressor Rotor. J. Engr. for Power, Trans. ASME. 104: 202-210.

Renken, J. H. 1977. Corner Boundary Layer and Secondary Flow Within a Straight Compressor Cascade. In Secondary Flows in Turbomachines. AGARD-CP-214, pp. 21-1 to 21-16.

Salvage, J. W. 1974. A Review of the Current Concept of Cascade Secondary Flow Effects. VKI TN-95.

Salvage, J. W. 1974. Investigation of Secondary Flow Behaviour and End Wall Boundary Layer Development Through Compressor Cascade. VKI TN-107.

Scholz, N. 1954. Secondary Flow Losses in Cascades. J. Aero. Sci. 21: 707-708.

Sieverding, C. H. 1984. Recent Progress in the Understanding of Basic Aspects of Secondary Flows in Turbine Blade Passages. ASME Paper 84-GT-78.

Sitaram, N. and Lakshminarayana, B. 1983. End Wall Flow Characteristics and Overall Performance of an Axial Flow Compressor Stage. NACA CR-3671.

Smith, A. G. 1957. On the Generation of the Streamwise Component of Vorticity for Flow in Rotating Passages. Aero. Quarterly. 8: 369-383.

Smith, G. D. J. and Cumpsty, N. A. 1984. Flow Phenomena in Compressor Casing Treatment. J. Engr. for Gas Turbines and Power, Trans. ASME. 106: 532-541.

Smith, L. H., Jr. 1955. Secondary Flow in Axial-Flow Turbomachinery. Trans. ASME. 77: 1065-1076.

Smith, L. H., Jr. 1970. Casing Boundary Layers in Multistage Axial-Flow Compressors. In Flow Research and Blading. Amsterdam. Elsevier. pp. 275-304.

Smith, L. H., Jr. 1974. Some Aerodynamic Design Considerations for High Bypass Ratio Fans. Second ISABE. Sheffield, England. March 24-29, 1974.

Soderberg, Olof. 1958. Secondary Flow and Losses in a Compressor Cascade. MIT Gas Turbine Lab. Rep. 46.

Squire, H. B. and Winter, K. G. 1951. The Secondary Flow in a Cascade of Airfoils in a Nonuniform Stream. J. Aero. Sci. 18: 271-277.

Stanitz, J. D., Osborn, W. M. and Mizisin, J. 1953. An Experimental Investigation of Secondary Flow in an Accelerating Rectangular Elbow with 90° of Turning. NACA TN 3015.

Stratford, B. S. 1967. The Use of Boundary Layer Techniques to Calculate the Blockage from the Annulus Boundary Layers in a Compressor. ASME Paper 67-WA/GT-7.

Swainston, M. J. C. 1968. Development of Spanwise Velocity Profiles Through Cascades and Axial-Flow Turbomachines. Proc. IME. 183(1): 189-204.

Taylor, A. M. K., Whitelaw, J. H. and Yianneskis, M. 1981. Measurements of Laminar and Turbulent Flow in a Curved Duct with Thin Inlet Boundary Layers. NASA CR 3367.

Taylor, A. M. K. P., Whitelaw, J. H. and Yianneskis, M. 1982. Curved Ducts With Strong Secondary Motion: Velocity Measurements of Developing Laminar and Turbulent Flow. J. Fluids Engr., Trans. ASME. 104: 350-359.

Vavra, M. H. 1960. Aero-thermodynamics and Flows in Turbomachines. New York. John Wiley and Sons.

Wagner, J. H., Dring, R. P. and Joslyn, H. D. 1983. Axial Compressor Middle Stage Secondary Flow Study. NASA CR-3701.

Wagner, J. H., Dring, R. P. and Joslyn, H. D. 1984. Inlet Boundary Layer Effects in an Axial Compressor Rotor: Part I. Blade-to-Blade Effects. ASME Paper 84-GT-84.

Wagner, J. H., Dring, R. P. and Joslyn, H. D. 1984. Inlet Boundary Layer Effects in an Axial Compressor Rotor: Part II. Throughflow Effects. ASME Paper 84-GT-85.

Wennerstrom, A. J. 1984. Experimental Study of a High-Throughflow Transonic Axial Compressor Stage. J. Engr. for Gas Turbines and Power, Trans. ASME. 106: 552-560.

Weske, J. R. 1948. Secondary Flows in Rotating Blade Passages at High Reynolds Numbers. Proc. Seventh Intl. Cong. Appl. Mech. Vol. 2, pt. 1, pp. 155-163.

Wisler, D. C. 1984. Loss Reduction in Axial-Flow Compressors Through Low-Speed Model Testing. ASME Paper 84-GT-184.

Woods, J. R., Jr. 1971. The Analytical Treatment of Secondary Flows and Associated Losses in Axial-Flow Turbomachines. NPS - 57W071121A.

Wu, C.-H. 1951. Survey of Available Information on Internal Flow Losses Through Axial Turbomachines. NACA RM E50J13.

SECONDARY FLOWS IN STRAIGHT AND ANNULAR TURBINE CASCADES

C.H. Sieverding

von Karman Institute for Fluid Dynamics

1. INTRODUCTION

From a review of cascade data on secondary losses in turbines in 1970 Dunham [1] concluded that "... the flow pattern associated with secondary losses is adequately understood, but the magnitude of the losses not". His belief in further improvements on secondary loss correlations was apparently not very strong, since he suggested that the most practical way of advancing the state of the art would be a fresh approach using three dimensional (3D) boundary layer theory, supported by detailed measurements. This recommendation was not made in vain since several experimental research programs on secondary flows were started in the early seventies with the particular aim to provide the information needed to develop appropriate flow models for the theoretical analysis. These investigations led to detailed interpretations of hitherto neglected flow features like horseshoe vortices and 3D separation and reattachment lines. Referring to this work Tall [2] did not hesitate to declare in 1976 "It certainly dispels any remaining notion that secondary flows in cascades are understood".

The present paper is an attempt to summarize the results of the many valuable experimental secondary flow investigations carried out over the last 10 to 12 years in order to give a full picture of our present knowledge and uncertainties of basic secondary flow aspects. Whenever necessary, of course, reference is made to previous work.

Chapters 1 to 4 present with minor changes the content of a paper prepared by the author for the 1984 ASME Gas Turbine Conference in Amsterdam [3].

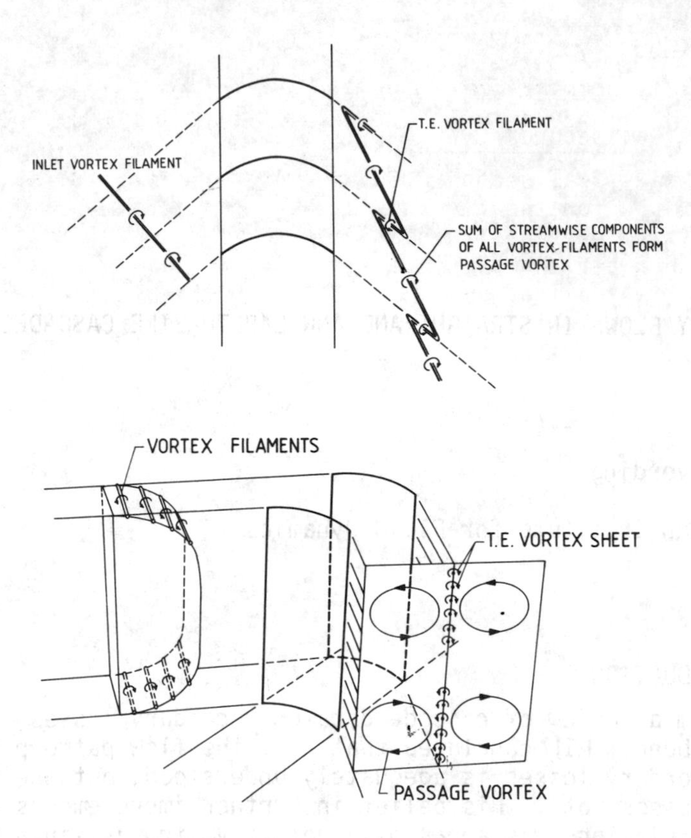

FIG. 1 - CLASSICAL SECONDARY FLOW MODEL OF HAWTHORNE

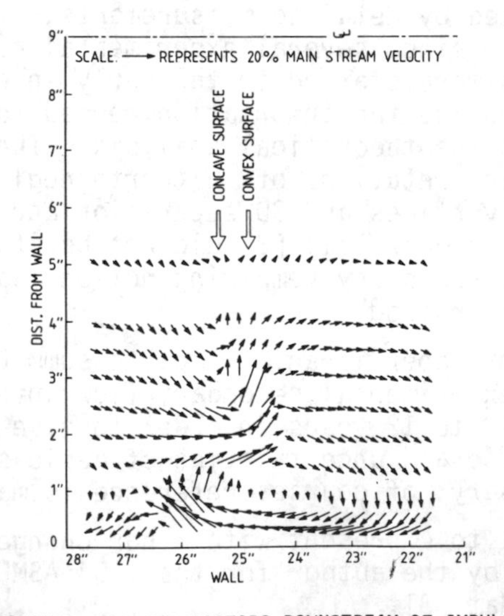

FIG. 2 - SECONDARY VELOCITY VECTORS DOWNSTREAM OF IMPULSE BLADE
(from Armstrong [6])

2. VORTEX STRUCTURE

2.1 Classical secondary flow model

The classical secondary flow vortex system as depicted in Fig.1 was described for the first time by Hawthorne in 1955 [4]. The vortex system presents the components of vorticity in the direction of flow when a flow with inlet vorticity is deflected through a cascade. The so-called passage vortex presents the distribution of secondary circulation which occurs due to the distortion of the vortex filaments of the inlet boundary layer passing with the flow through a curved passage.

The vortex sheet at the trailing edge (Fig.1b) is composed of :
(a) the trailing filament vortices which arise due to the stretching of the inlet vortex filaments when passing through the cascade with different velocities between suction side and pressure side (see Fig.1a);
(b) the trailing shed vorticity which is due to the spanwise change of the blade circulation.
The sense of rotation of the trailing filament and trailing shed vorticities is opposite to that of the passage vortex. Smoke visualizations by Herzig et al. [5] in 1954 demonstrated clearly the existence of the passage vortex while secondary velocity vector plots by Armstrong in 1957 [6] from measurements behind an impulse blade present evidence of both the passage vortex and blade shed circulation (Fig.2). The resemblance with the theoretical model is, however, poor.

2.2 Horseshoe vortex

The rolling up of the endwall boundary layer in front of a cylinder on a flat plate into a vortical motion, called horseshoe vortex due to its particular shape when flowing on both sides past the cylinder, is a well known phenomenon (Fig.3a). Its significance to the flow in turbine bladings has been recognized only very recently.

Oil flow visualizations by Fritzsche in 1955 [7] show evidence of the horseshoe vortex in accelerating cascades, but to the author's knowledge it is Klein in 1966 [8] who first mentions the existence of what he calls a stagnation point vortex. Without any further comments, Klein presents a cascade vortex model with both the passage and horseshoe vortices (Fig.3b). This flow phenomenon attracted little interest from the scientific turbomachinery community until the early seventies apart from exceptions such as Bölcs [9], who demonstrated in 1969 in a water table test the equivalence of the near wall patterns round a cylinder and round the leading edge of a turbine blade.

In the early seventies both aerodynamicists and heat transfer people started to show an increasing interest in the leading edge vortex. In an experimental study of the heat transfer on large scale turbine endwalls by Blair in 1973 [10], increased heat transfer rates were shown to occur near the leading edges of the pressure

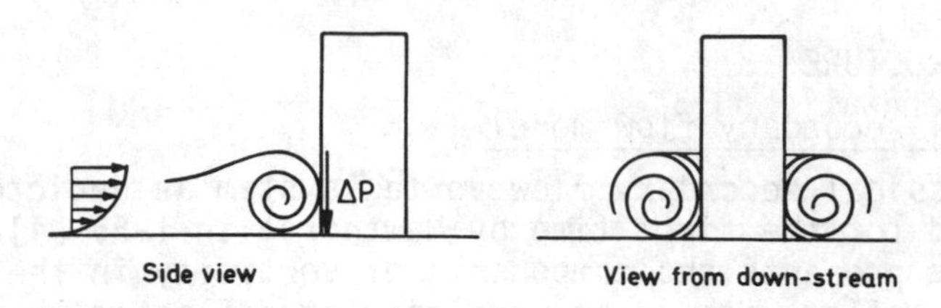

FIG. 3a– ROLLING UP OF ENDWALL BOUNDARY LAYER INTO A VORTEX IN FRONT OF A CYLINDER – FORMATION OF HORSESHOE VORTEX

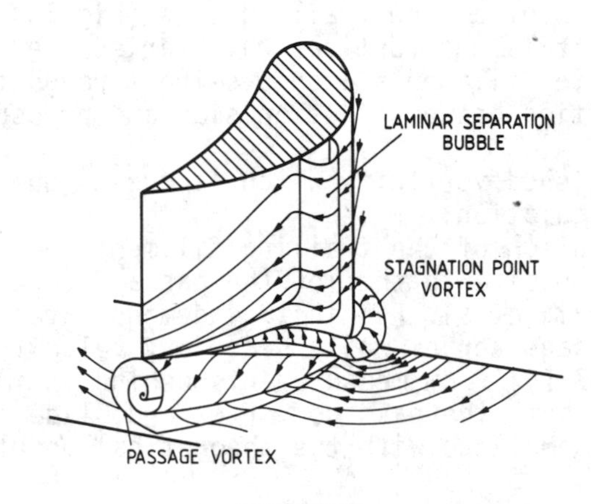

KLEIN'S MODEL

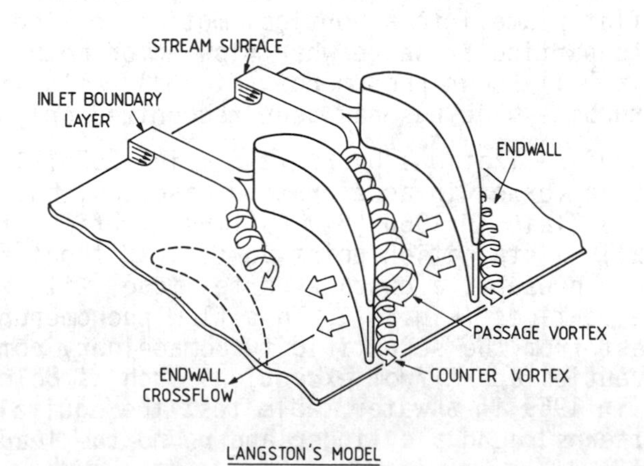

LANGSTON'S MODEL

FIG. 3b- ENDWALL FLOW MODELS BY KLEIN [8] AND LANGSTON [13]

and suction surfaces and the author attributed this to the distortion of the endwall boundary layer by the leading edge vortex. The aerodynamicists discovered an interest in the leading edge vortex as a result of a new approach to secondary flow research. Contrary to the hitherto prevailing control volume approach in which the flow survey is limited to measurement of the up- and downstream flow conditions, the new approach consisted in looking at the secondary problem as an endwall boundary layer problem, which required a detailed study of the evolution of the flow throughout the cascade. The papers of Langston et al [11], presented in 1976, and of Sjolander, 1975 [12] are pioneering works for the detailed analysis of secondary flow patterns in turbine cascades in general and of the role of the leading edge horseshoe vortex in particular. The cascade vortex model (Fig.3b) derived from the measurements in [11] (3D pressure probe measurements and visualizations of limiting streamlines by ink traces) is presented by Langston [13] in a later paper. The main differences between the models of Klein and Langston are twofold :
(a) Langston clearly postulates that the pressure side leg of the leading edge horseshoe vortex, H_p, which has the same sense of rotation as the passage vortex, merges with and becomes part of the passage vortex;
(b) Langston sees the suction side leg of the leading edge horseshoe vortex, H_s (called counter vortex in Fig.3b), which rotates in the opposite sense to the passage vortex, continuing in the suction side endwall corner, while the presentation of Klein suggests that this vortex is gradually dissipated in contact with the passage vortex.

In 1977, Marchal & Sieverding [14] introduced the light sheet technique in secondary flow research, which allowed them to take sectional views of the flow in selected planes of the blade passage. This technique is demonstrated in Fig.4. It shows (a) the generation of the horseshoe vortex at the blade leading edge and (b) the combination of horseshoe and passage vortices near the trailing edge. The authors agree with Langston that the leading edge vortex, H_p, and the passage vortex seem to merge to form a single vortex but contrary to Langston, they find the counter-rotating vortex, H_s, in the trailing edge plane on the midspan side of the passage vortex rather than in the corner. However, a clear description of the merging process between the H_p branch of the leading edge vortex and the passage vortex and of the interaction between the H_s branch of the leading edge vortex and the passage vortex cannot be given yet.

More details on the horseshoe vortex were obtained by Gaugler & Russell in 1980 [15]. By injecting neutrally buoyant helium filled soap bubbles into the upstream boundary layer of a cascade they were able to visualize the corkscrew flow path of streamlines caught in the horseshoe vortex. The streamwise stretching of the rotational motion of these streamlines is put into evidence.

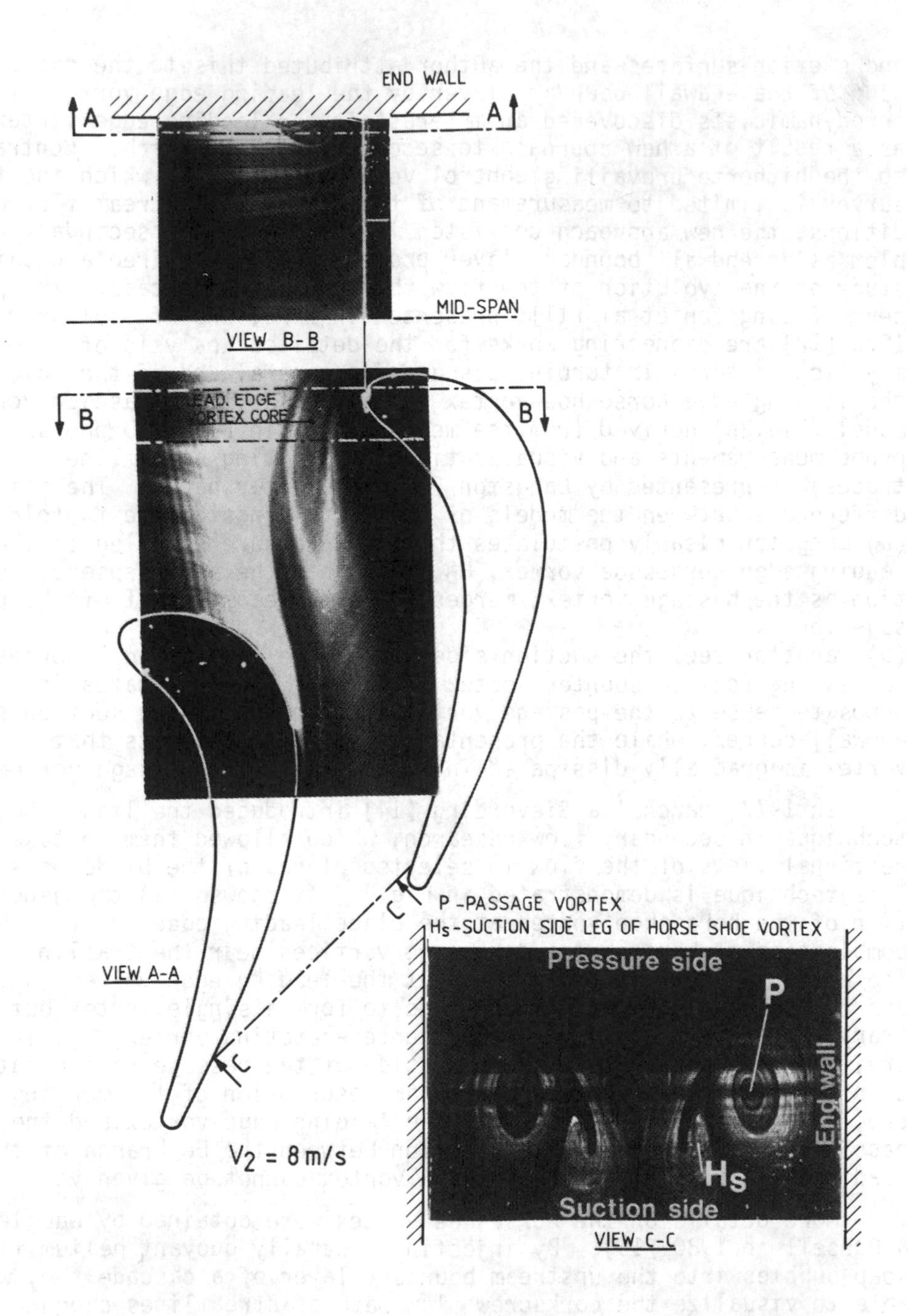

FIG. 4 - VISUALIZATION OF HORSESHOE AND PASSAGE VORTICES USING THE LIGHT SHEET TECHNIQUE

2.3 Description of the synchronous evolution of horseshoe and passage vortex

An essential piece of information about the evolution of horse shoe and passage vortices through cascade passages was contributed by Moore & Smith in 1983 [16] who measured the flow trajectories by ethylene detection in the exit plane of a replica of Langston's cascade. The authors found that ethylene injected at the location of the suction side branch of the horseshoe vortex near the blade leading edge was convected around the passage vortex core, while ethylene injected into the pressure side branch of the horseshoe vortex was found in the center of the passage vortex.

A more detailed explanation of the synchronous evolution of horseshoe and passage vortices is finally given in late 1983 by Sieverding & Van den Bosche [17]. The authors used a coloured smoke wire technique to visualize the evolution of entire stream surfaces through cascades. Based on photographs and direct observations, they presented the flow model in Fig.5. It shows the shape of two stream surfaces SS_1 and SS_2, SS_1 starting upstream inside the endwall boundary layer and SS_2 outside the endwall boundary layer at different axial positions in the blade passage. Approaching the leading edge the lateral extremities of stream surface SS_1 start to roll up into the two counter-rotating branches of the horseshoe vortex, H_p and H_s, the main part of the stream surface being nearly undisturbed. Behind the leading edge plane the whole stream surface starts slowly to rotate. All parts of the stream surface including the vortices H_p and H_s take part in this vortical motion which gradually develops into what is called a passage vortex. The flow visualizations show that the pressure side branch of the horseshoe vortex, H_p, follows basically a smooth curve through the passage without any noticeable vortical motion, which would indeed suggest that its core coincides with that of the passage vortex, while the suction side branch of the horseshoe vortex H_s wraps itself around the passage vortex core which confirms the results of Moore & Smith in their ethylene detection tests. The position of the H_s vortex depends on the rotational speed of the passage vortex, which in turn depends on the cascade geometry and the overall flow conditions. This explains why Marchal & Sieverding locate the H_s vortex on the midspan side of the passage vortex, while Moore & Smith can detect it on the endwall side. The counter vortex of Langston right in the endwall suction side corner is believed to be of different origin as seen in the next section.

Except in the entrance region, nobody has succeeded so far in using pressure probes to trace the flow path of the H_s vortex through the cascade. The reason for this is different in the front and rear parts of the passage. In the accelerating front part, the vortex moves away from the endwall but stays close to the suction surface. Its small size combined with a strong stretching in the streamwise direction makes it difficult to detect it. In the rear decelerating part it grows in size but it has lost its intensity in

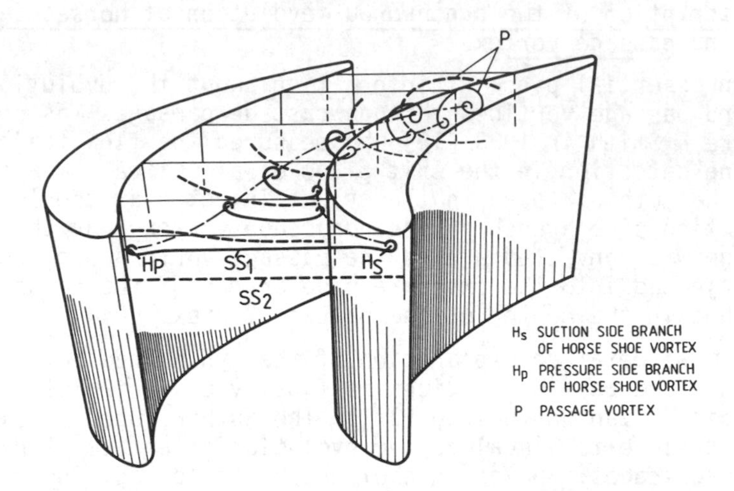

FIG. 5 - SYNCHRONOUS EVOLUTION OF HORSESHOE AND PASSAGE VORTICES AFTER SIEVERDING AND VAN DEN BOSCH [17]

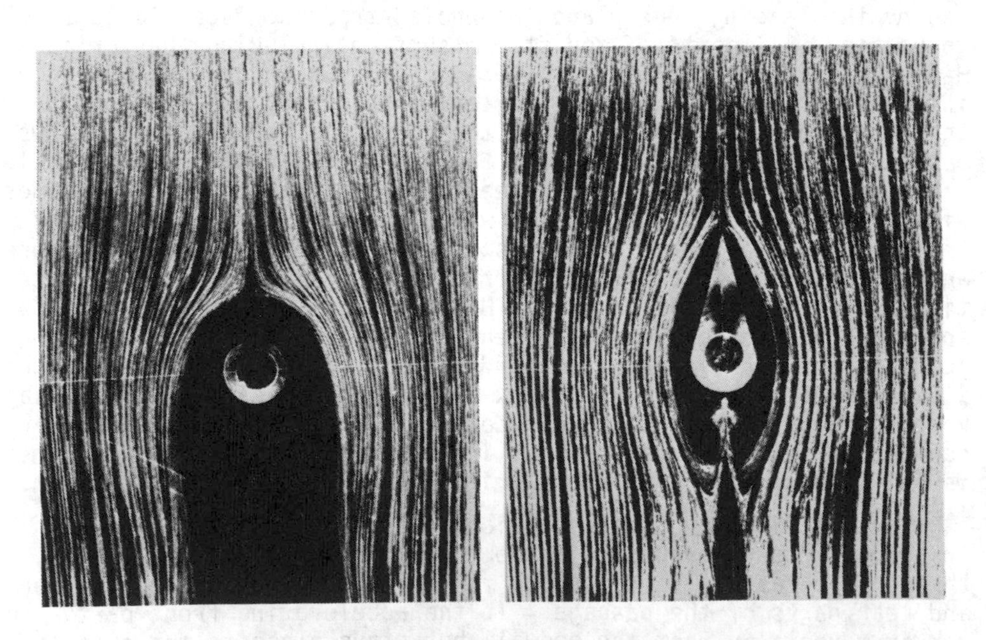

FIG. 6 - EFFECT OF LEADING EDGE GEOMETRY ON ENDWALL LIMITING STREAMLINES (M_1=0.48; Re_d=2.5×10^5; d=20 mm, VKI)

contact with the passage vortex due to the dissipating action of the shear forces.

Moore & Ransmayer [18] investigated the effect of the leading edge geometry on secondary flows by replacing the standard cylindrical leading edge of their cascade by a wedge-shaped leading edge. They found that neither the static pressure nor the overall losses were significantly affected by this modification. This is a somewhat surprising results since a priori it could be thought that the addition of a sharp leading edge should suppress the formation of the horseshoe vortex and thereby affect the further evolution of the flow field. Since this is apparently not the case one may conclude that
(a) the horseshoe vortex is only of local importance in the vicinity of the leading edge and does not affect the overall flow patterns further downstream in the blade passage, i.e., strength and size of the horseshoe vortex are of secondary importance only, or
(b) the addition of the sharp leading edge introduces new flow phenomena which compensate for the reduction of the horseshoe vortex effects. The comparison of the endwall limiting streamlines around a cylinder and a cylinder with a 30° wedge added to it is therefore of particular interest, Fig. 6. The wedge reduces the upstream standoff-distance of the endwall boundary separation line to a small fraction, but the lateral standoff-distance at the center of the cylinder decreases only by 50% due to the action of a corner vortex.

2.4 Corner vortex

Here we want to refer to a vortex which rotates in the opposite sense to the passage vortex and is located invariably right in the endwall/suction side corner. Because of its relatively small size it is difficult to visualize but its existence is often put into evidence in the spanwise angle distribution behind highly turning cascades by a characteristic reduction of the overturning near the endwall, as shown in Fig.7 extracted from a recent paper (1983) by Gregory-Smith & Graves [19]. To trace the corner vortex back to its origin is more difficult, but the oil flow visualization in Fig.8 from Belik in 1975 [20], showing the limiting streamlines in a very high turning impulse type cascade, is enlightening. The overturning at the endwall is such that the limiting streamlines interfere almost at a right angle with the blade suction surface near the position of maximum surface curvature. It is certainly not too farfetched to compare this interaction with that causing the boundary layer ahead of the leading edge to roll up into a vortical motion. Without going further into the analysis of limiting streamlines (see following chapter), let us note that :
(a) the line S_3 in Fig.8 presents a 3D separation line which is characteristic of the existence of a vortex turning in the opposite direction of the passage vortex;
(b) this separation line starts near the point where the cross flow interferes the first time with the suction surface.
Similar, but in general less violent, situations occur in most cas-

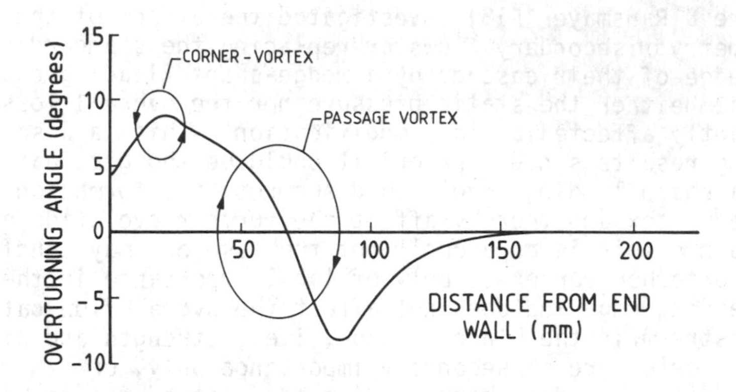

FIG. 7 - REDUCTION OF OVERTURNING NEAR ENDWALL DUE TO PRESENCE OF CORNER VORTEX (Extracted from Gregory-Smith and Graves [19])

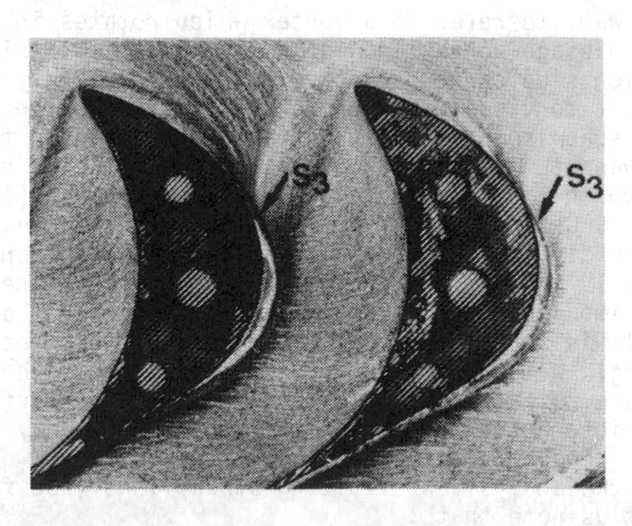

FIG. 8 - ENDWALL LIMITING STREAMLINES INDICATING GENERATION OF CORNER VORTEX IN AN IMPULSE BLADE PASSAGE (from Belik [20])

cades with significant cross flows (see oil flow visualizations by Marchal & Sieverding [14] and by Moore [21]).

3. EFFECT OF VORTEX STRUCTURES ON ENDWALL BOUNDARY LAYERS

3.1 Limiting streamlines

The visualization of limiting streamlines by coating the blade and endwall surfaces with a mixture of oil and T_iO_2-powder before starting the wind tunnel has been widely used. However, the high viscosity of this mixture presents a certain drawback when using it at low speed. Tests at both low and high speed by Marchal & Sieverding [14] showed substantial improvements in the visualization of the limiting streamlines at high speed. Langston et al [11] used ink traces probably for the same reason, but here the spatial resolution is less satisfactory since the ink traces must not be allowed to mix with each other. Recently, Langston & Boyle (1982) developed a new ink dot technique giving more accurate surface streamlines [22].

Figure 9 presents schematically the results of the analysis by Langston et al [11] of the ink traces of the limiting streamlines in the cascade. The flow field is divided into distinct regions through the 3D separation lines S and the reattachment line R (stagnation streamline) with the separation saddle point A at their intersection. The horseshoe and passage vortices form behind the separation line S: the pressure side branch of the horseshoe vortex, H_p, and the passage vortex behind S_p (starting from A) and the suction side branch of the horseshoe vortex, H_s, behind S_s. The passage vortex separates from the blade suction surface along line S_4 (position like in Fig.10).

The oil flow visualization by Belik in Fig.8 and others by Sjolander [12], Marchal & Sieverding [14], Moore [21] and Gotthardt [23] as well as the smoke visualizations in [14] indicate that the endwall limiting streamline patterns might be much more complicated. An attempt is made to present this schematically in Fig.10. There are two major separation lines ahead of the leading edge. The secondary separation line S_2 corresponds to the "liftoff" line of the horseshoe vortex while the separation line S_1 is due to the boundary layer separation ahead of the horseshoe vortex. The secondary separation line is in general easier to detect in cascade endwall flow visualizations than the primary line, which is often rather weak. The observation of both separation lines in cascades is in accordance with flow visualizations around cylinders mounted on a flat plate by Belik [24], Peake et al [26] and East & Hoxey [27]. Smoke visualizations using the light sheet technique [14] confirm the existence of a low energy region between the two separation lines near the leading edge. Based on this experimental evidence, one may postulate the existence of two dividing stream surfaces D_1 and D_2 along which low momentum material is fed into the stagnant separation bubble (see Fig.10c) and two saddle points A_1 and A_2 rather than one as in Fig.9. Sjolander [12] found this separation bubble also in his endwall boundary layer measurement (see next section). More tests are needed

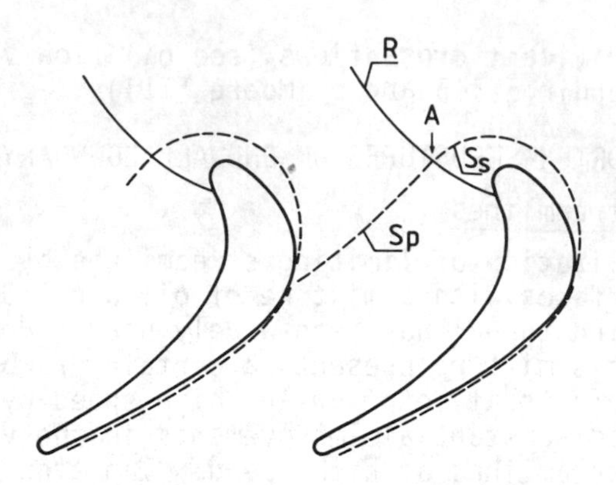

FIG. 9 - ENDWALL THREE DIMENSIONAL SEPARATION AND REATTACHMENT LINES (after Langston [11])

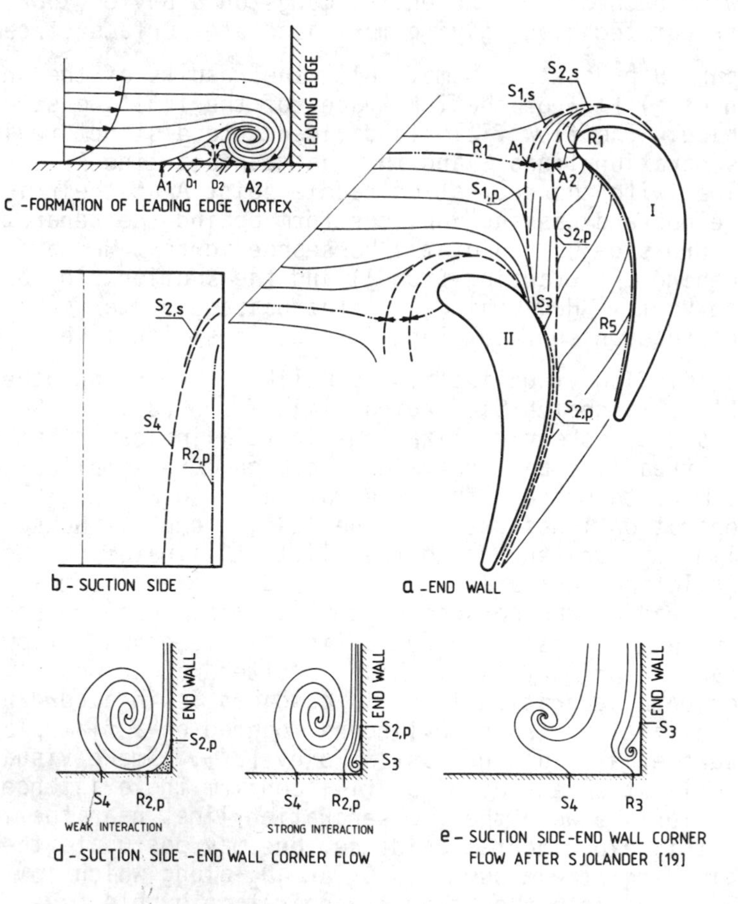

FIG. 10 - GENERAL PRESENTATION OF ENDWALL FLOW CHARACTERISTICS

to investigate how the distances of the two separation streamlines S_1 and S_2 from the leading edge and the lateral extension of the stagnant separation bubble depend on the endwall boundary layer characteristics, the shape of the leading edge and the incidence angle. A change of the incidence angle causes the stagnation streamline to shift around the leading edge with the result that the leading edge appears more or less blunt to the incoming flow (Langston et al [11], Stastny [25]).

At the intersection of the near wall streamlines, contained between separation streamlines, $S_{1,p}$ and $S_{2,p}$, with the suction surface of blade II, one of two things may happen. Depending on the blade loading the interaction may be of the strong or weak type (high or low angle of attack of streamlines with respect to blade surface). In the case of a strong interaction (e.g. Fig.8 or cascade of Moore [21]), a corner vortex is created under similar conditions as at the leading edge, This results in the formation of a separation line S_3 originating in the intersection point of the separation line, $S_{1,p}$ with the suction surface (see Fig.8). The separation line $S_{2,p}$ continues parallel to the line S_3 towards the cascade exit plane (Fig.10a). In the case of a weak interaction (e.g. nozzle blades in [14] and [23]), the oil flow visualizations show only the separation line $S_{2,p}$ running along the suction side endwall corner. Figure 10d shows schematically the suction side corner flow for both types of interaction in a plane normal to the rear suction surface. For comparison, Fig.8e shows also the interpretation of the corner flow patterns by Sjolander [12]. His patterns are overall similar to those of the strong interaction type in Fig.8d. The different shape of the passage vortices is unimportant. What matters is that Sjolander indicates a separation line S_3 which in his flow visualization clearly originates in the region of the intersection of the separation lines $S_{1,p}$ and $S_{2,p}$ with the suction side surface, as shown in Fig.10a. On the contrary, the separation line $S_{2,p}$ was not observed downstream of this region. As a result, Sjolander's interpretation of the position of the separation line S_3 on the rear suction surface differs from that of the present author. Corresponding to the separation line $S_{2,p}$ in Figs.10a and 10d, there is a reattachment line $R_{2,p}$ on the blade surface near the endwall (Fig. 10b). This line was clearly observed in [14] and [23]. Further away from the endwall appears the 3D separation line S_4 of the passage vortex (Fig.10b).

The evolution of the separation lines $S_{1,s}$ and $S_{2,s}$ is dictated by the strong flow acceleration on the front part of the blade suction side and the transverse pressure gradient. The primary separation line $S_{1,s}$ rapidly joins the secondary separation line $S_{2,s}$ which intersects the suction surface after a short distance indicating that the suction side branch of the horseshoe vortex has shifted away from the endwall on to the suction side. The trace of this vortex, i.e., its 3D separation line, can be followed on the suction surface until the vortex eventually moves away from it into the pas-

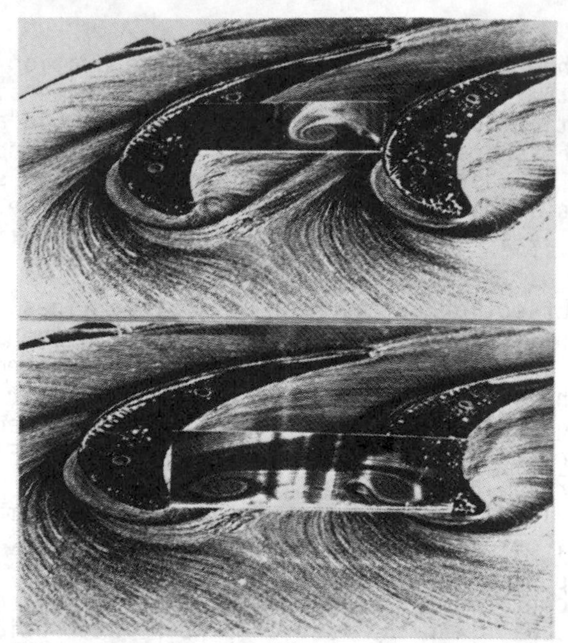

FIG. 11 - PERSPECTIVE VIEW OF ENDWALL LIMITING STREAMLINES AND SMOKE VISUALIZATIONS IN A PLANE NORMAL TO THE ENDWALL (VKI)

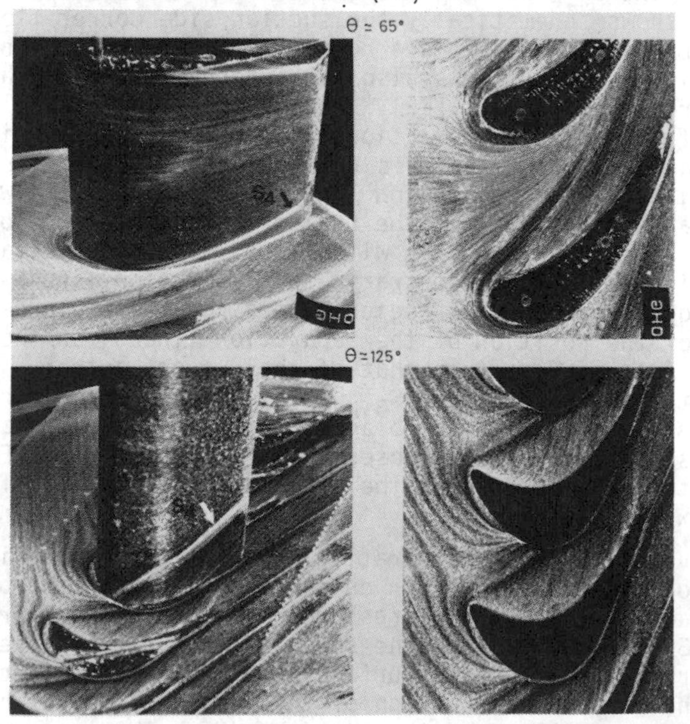

FIG. 12 - EFFECT OF FLOW TURNING ANGLE ON LIMITING STREAMLINES (top - VKI; bottom - ANSALDO)

sage under the influence of rotational motion of the passage vortex to which it is linked as seen in Fig. 5.

Finally, Fig.10a shows a reattachment line R_5 which gives evidence of a pressure side endwall corner vortex, the importance of which probably increases with increasing positive inlet flow angle. Its origin is found in a downwash motion observed on the front part of the pressure side. This downwash seems to be in general small, except in the rotor blade flow visualizations [14] which were made at 15° positive incidence. The pressure side corner vortex rotates in the same sense as the suction side corner vortex and thus contributes to reduce the downstream overturning angle near the endwall.

The relation between vortices and endwall limiting lines is best demonstrated in Fig.11 (cascade :α_1=45°, α_2=-69°, g/C=0.69) which shows two photographic montages of a combination of oil flow and smoke visualization pictures, the latter being placed normal to the first. The good agreement between the smoke and the oil flow visualizations is somewhat surprising since the Reynolds numbers for the tests differed by a factor of 20 (smoke visualization : $Re=0.6\times10^5$, oil flow visualization : $Re=1.2\times10^6$). However, tests by Peake et al [26] for both laminar and turbulent endwall flows around oval cylinders indicate that the changes in the endwall streamlines are quantitative rather than qualitative in nature. In particular, the existence of both the primary and secondary separation lines ahead of the cylinder was not affected. Nevertheless, it should be mentioned that in laminar flow more than one vortex may occur. A similar phenomenon also occurred occasionally in the smoke tests by Marchal & Sieverding.

To conclude, Fig.12 shows the evolution of limiting endwall and suction surface streamlines with increasing flow turning angle.

3.2 Endwall boundary layer characteristics

One of the most important tasks of current secondary flow research is to provide test cases for viscous flow calculations. This implies the provision of information on the nature of endwall boundary layer : the type of boundary layer profile, separation regions, wall shear stresses and turbulence throughout the blade passage. Before the mid 70's, Senoo [28] was apparently the only investigator to carry out boundary layer measurements within a turbine blade passage. He studied the flow in a high turning nozzle cascade. He concluded that the endwall boundary layer in the throat is laminar no matter what the state and thickness of the upstream boundary layer and ascribed this to the steep favourable gradient.

Flow surveys including detailed endwall boundary layer traverses inside the cascade passages have become more frequent since the mid 70's. Examples of such experiments with collateral inlet boundary layers are those of Langston et al [11], Marchal & Sieverding [14], Bario-Leboeuf-Papailiou [29], Bailey [30] and Gregory-Smith & Graves [19] in straight cascades and Sjolander [12], Sieverding-Boletis-Van

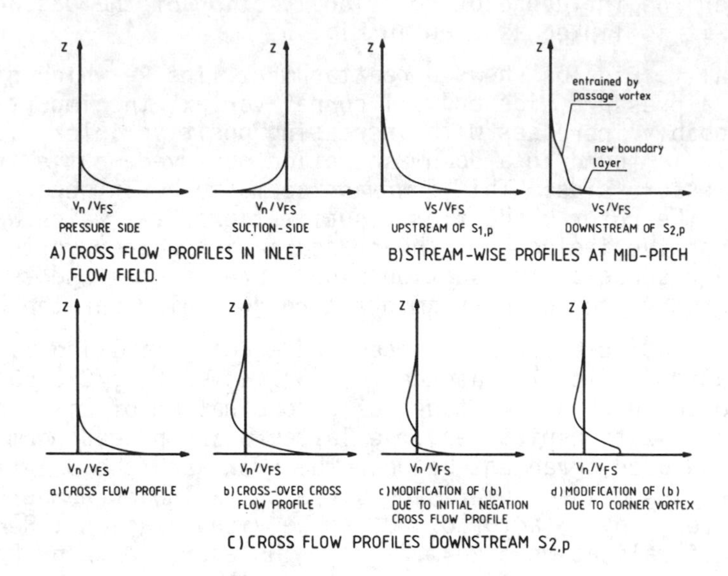

FIG. 13 - VARIOUS TYPES OF ENDWALL BOUNDARY LAYER PROFILES

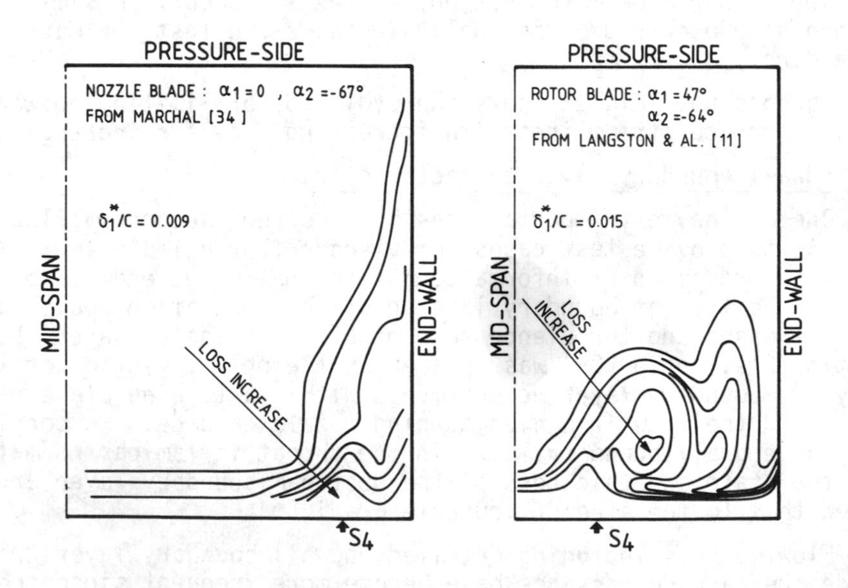

FIG. 14 - INFLUENCE OF CASCADE GEOMETRY ON LOSS CONTOUR PLOTS IN TRAILING EDGE PLANE

Hove [31] in annular cascades. Only a few results have been presented in the form of endwall velocity profiles. In most cases the authors have preferred to present their results as total and static pressure coefficient loss contour plots. All tests are done at low speed with Reynolds numbers in the range $Re=2.5\times10^5$ to 10^6. The state of the inlet boundary layer was in most cases turbulent and in a few cases transitional (range of shape factors from H=1.3 to 1.7). The inlet free stream turbulence level was in general of the order of 0.5% to 1% (in one case 3%).

The boundary layer profiles are in general presented under the form of streamwise and cross flow velocity profiles, except for Sjolander who plots total pressure deficit coefficients and yaw angles. A short description of the most characteristic velocity profiles which may occur in a turbine blade passage is given below. For their positioning in the cascade the reader is referred to Fig. 10a.

- Upstream flow field : the upstream effect of the cascade imposes a positive cross flow component on the incoming pressure side streamlines and a negative component on the incoming suction side streamlines (positive direction : from pressure side to suction side), see Fig.13a [31]. The streamwise component conserves a 2D character.

- Passage entrance field up to the primary separation line : the cross flow components are reinforced. The streamwise component remains essentially 2D in character. The strong convergence in the suction side leading edge region (ahead of the separation lines $S_{1,s}$ and $S_{2,s}$) causes a local thickening of the boundary layer. This effect is rapidly reduced through the action of the strong streamwise acceleration of the flow in this region.

- Region between separation lines $S_{1,p}$ and $S_{2,p}$: only Sjolander [12] refers specifically to boundary layer measurements in this region. One of his traverses is positioned approximately halfway between the separation lines $S_{1,p}$ and $S_{2,p}$, at lateral distance from A_1A_2 (saddle points defined in Fig.10) approximately equal to the distance $A_1 \rightarrow A_2$. He concludes that the stagnation separation bubble still has a radial extension of about one fifth of the boundary layer thickness at this point.

- Downstream of separation line $S_{2,p}$: the vortical motion of the horseshoe and/or passage vortices convects fluid from the outer boundary layer or even from the free stream towards the endwall and reenergizes the near endwall layers. To which degree and over what part of the passage this happens depends on the intensity of the passage vortex. A new boundary layer is growing. This effect is most apparent from the streamwise velocity component as schematicaly shown in Fig.13b. In addition to this, the passage vortex imposes the characteristic over- and underturning on the cross flow profile of the endwall flow. The resulting profile varies significantly from the pressure side to the suction side, changing from a single cross flow profile at the pressure side to a crossover cross flow profile.

Near the suction side the crossover cross flow profile may be further complicated by (a) the influence of an initially significant negative cross flow component at the passage entrance as shown by Sieverding et al [31] and (b) a suction side corner vortex like that apparent in the profiles measured by Langston [13]. The above described profiles are presented in Fig.13c. Langston found that the cross flow behaviour of the profiles (a) and (b) in Fig.13c could be correlated in the form of polar plots by the expression :

$$\frac{V_n}{V_s} = \tan\left[(\varepsilon_w - az)\, e^{-\gamma z}\right]$$

The constants ε_w, a and γ (ε_w represents the yaw angle deviation with respect to free stream conditions at the wall, a is a measure for the wall distance at which ε changes sign and γ is related to the maximum underturning) have to be determined experimentally. Since both the strength and position of the passage vortex depend on the blade loadin loading and the inlet endwall boundary layer thickness (to mention the most important parameters), it will be difficult to find a general correlation for these constants. Cross flow profiles of types (c) and (d) in Fig.13c are clearly excluded at once from any approach based on experimental correlations.

The convection of high energy fluid to the endwall through the action of the horseshoe and passage vortices results in an increase of the skin friction and wall shear stresses as stated by Belik [32] who measured the variation of the wall shear stresses along the centerline of two high turning nozzle cascades (∿70° turning) using thin film gauges. He finds a strong increase of the shear stresses with the flow path length, the maximum stresses occurring near the point of maximum streamwise pressure gradient. Measurements of endwall boundary layer noise confirm that this position corresponds closely to the start of a laminar boundary layer. Except for Belik [32] and Senoo [28] all investigators find the viscous layer behind the separation line $S_{2,p}$ too small for it to be possible to ascertain the state of the boundary layer.

So far only few data have been published on turbulence measurements and unfortunately they are rather contradictory. In an investigation of the flow through a single large scale turbine vane passage with an inlet turbulence level varying from a root mean square value of 0.7% at midspan to 6% near the wall, Bailey [30] states that turbulent stresses are insignificant in large regions of passage vortex, suggesting that a laminar flow calculation procedure may provide a reasonable accurate prediction of the secondary flow structure far from the wall. Gregory-Smith & Graves [19] found, for an upstream free stream turbulence of 3%, peak turbulence intensities of 30% in the vortex core. The authors concluded that some of the total pressure loss through the cascade appears as an increase in turbulent kinetic energy before being dissipated by viscous action. Unfortunately, the authors did not quote the turbulence increase through their rather thick inlet endwall boundary layer. Langston et al [11] present turbulence measurements in the trailing edge plane, but no data within the high loss core in the endwall suction side region. The one traverse located just at the edge of the loss core gives a slight hint that the turbulence intensity might actually increase in the loss core.

4. ORIGIN, GROWTH AND SPATIAL DISTRIBUTION OF LOSSES

4.1 Generation and redistribution of low momentum flow in cascades

Analysis of the vortex structures and their effects on the endwall boundary layer gives a fairly clear idea about the factors contributing to the generation and spatial distribution of low momentum flow through turbine cascades. The most important loss contributions are summarized below :

(a) Natural increase of the inlet endwall boundary layer up to the separation lines.
(b) Stagnant separation bubble in the leading edge region between the two separation lines.
(c) Growth of "new boundary" layer behind the separation line $S_{2,p}$.
(d) Corner losses in both pressure side and suction side endwall corners, the latter being most important.
(e) Shear stress effects along all 3D separation lines.
(f) Losses due to the shear action of the passage vortex on the blade suction side and the mixing process between the cross flow and the blade flow along the 3D separation line S_4.
(g) Dissipation of all vortices and complete mixing of the non uniform outlet flow field downstream of the cascade.

Gregory-Smith calculated secondary losses by taking into account only the losses due to the new boundary layer (with a 2D boundary layer method) and the secondary kinetic energy (as presented by classical secondary flow theory), which was considered to be lost entirely. Adding to this the losses due to the inlet boundary layer, he found fair agreement with experimental data in at least two cascades [33, 19]. Calculations of the kinetic energy of the measured secondary flow motion in the trailing edge plane of a rotor blade (α_1=30°, α_2=-60°, H/C=0.83) tested at VKI showed that the ratio of secondary to total kinetic energy varied from 0.6% for an inlet boundary layer with δ^*/C=0.011 to 1.2% for δ^*/C=0.028 with a distribution of 55% for the spanwise secondary velocity component and 45% for the component normal to the streamwise direction. It is probably correct to say that the assumption that the entire secondary kinetic energy is lost is exaggerated and partially accounts for other losses, such as corner losses and mixing losses along the separation line S_4.

The pitch- and spanwise distribution of secondary losses under the influence of the secondary flow vortex structures is one of the most characteristic aspects of secondary flows. The limiting streamlines in Fig.10a indicate that the inlet low momentum boundary layer material in the immediate vicinity of the endwall is directly de-

flected to the blade suction side along the separation line $S_{1,p}$ except in front of the leading edge. Here part of it is fed into the stagnation separation bubble from where it is directed between the two separation lines $S_{1,p}$ and $S_{1,p}$ also towards the suction surface. The particles in the remaining part of the inlet boundary layer can take very different paths. If they get involved with the horseshoe vortex, either they migrate to the passage vortex center or are convected around the passage vortex core depending on whether they are part of the pressure or suction side part of the horseshoe vortex. Inlet boundary layer material which is not involved in the leading edge vortex and which makes its way across the region between the two separation lines $S_{1,p}$ and $S_{2,p}$ will be either entrained by the passage vortex along the secondary separation line without necessarily migrating to the vortex center or it is turned behind the separation line towards the endwall where it takes part in the formation of the new boundary layer. The motion of low momentum boundary layer material towards the blade suction side continues downstream of the second separation line, including even parts of the pressure surface boundary layer. All endwall boundary layer material which has been pushed throughout the passage onto the blade surface is driven together with parts of the regular blade surface boundary layer to the separation line S_4 as indicated by the limiting streamlines in Fig.12. However, not all of the boundary layer material which has moved across the passage migrates to the suction surface;some of it continues in th suction side endwall corner and may appear downstream as a non-negligible local loss core from which low momentum material is continuously fed into the main stream [19]. Apart from the remaining boundary layers on the endwalls there are three characteristic types of loss concentrations which may be found in the trailing edge plane :
(a) corner loss;
(b) loss core associated with passage vortex (a coincidence of the center of the loss core with the center of the passage vortex is, however, fortuitous -see [19]), and
(c) loss core along passage vortex separation line S_4.
Depending on the inlet boundary layer thickness and blade loading, these loss cores are more or less superimposed. Two examples are given in Fig.14.

The position of the passage vortex in the trailing edge plane is obviously of primary importance for the loss distribution. The following general tendencies have been observed :
(a) Increasing the inlet flow angle α_1 at constant α_2 and $\delta_1^{\star}/C$ results in a shift of the passage vortex center towards the blade suction side. This pitchwise shift is accompanied by a spanwise displacement (see nozzle blades [14, 23] and high turning rotor blade sections [11, 14, 19, 29]).
(b) A variation from a thick to a thin inlet boundar layer causes a decrease of the distance separating the vortex center from the endwall and a shift of the vortex towards the blade suction side ([30], [34] and from pressure measurements of Graziani et al in [41]).

(c) A decrease of the aspect ratio below the critical value (mutual interference of secondary flow regions from both blade ends) moves the passage vortex closer to the endwall (Bailey [30]). This could partially explain the nonlinear variation of secondary losses at low aspect ratios.
(d) A variation of the outlet Mach number (in the subsonic range) does not seem to affect significantly the position of the passage vortex (tests on nozzle blade by Sieverding & Wilputte [42] from M_2=0.1 to 0.8).

4.2 Loss growth through cascade

Various authors have presented the growth of the pitch- and spanwise averaged losses within their cascades. Gregory-Smith & Graves [19] find a fairly steady growth of the losses throughout their cascade. On the contrary, Langston [11] and Marchal & Sieverding [14] conclude that the losses remain fairly constant up to the axial position of the maximum suction side velocity and then grow rapidly from there to the trailing edge. The growth is attributed to the interaction of the endwall cross flow with the suction side boundary layer and to the overall effect of the rear suction side flow deceleration on the entire secondary flow behaviour. It is interesting to note that the calculation of the loss growth through Langston's cascade by Hah [43] with a Navier-Stokes computer code shows almost constant losses up to 30% of the axial chord followed by a rapid increase of the losses up to the trailing edge.

Figure 15 presents the loss growth inside the blade passage for three different cascades measured by Marchal [34]. By approximating the loss increase for each cascade by a straight line he was able to correlate the slope of the curves with a parameter, a, representing the maximum pressure difference across the blade passage, divided by the corresponding distance between pressure and suction sides and non dimensionalized by the mean kinetic energy of the main flow :

$$a = \frac{C_{ax}}{S} \cdot \frac{\Delta P_{max}}{\frac{1}{2}\rho \cdot V_\infty^2}$$

Taking into account the distance of the minimum suction side pressure from the leading edge plane, x_{min}, Marchal further proposed a quality factor :

$$Q = \left(1 - \frac{X_{min}}{C}\right)\left(\frac{C_{ax}}{S} \cdot \frac{\Delta P_{max}}{\frac{1}{2}\rho V_\infty^2}\right)^{1.25}$$

which allows a qualitative evaluation of the effect of blade design on secondary losses. This factor has been used to analyze the performance of blades with the same overall load but different load

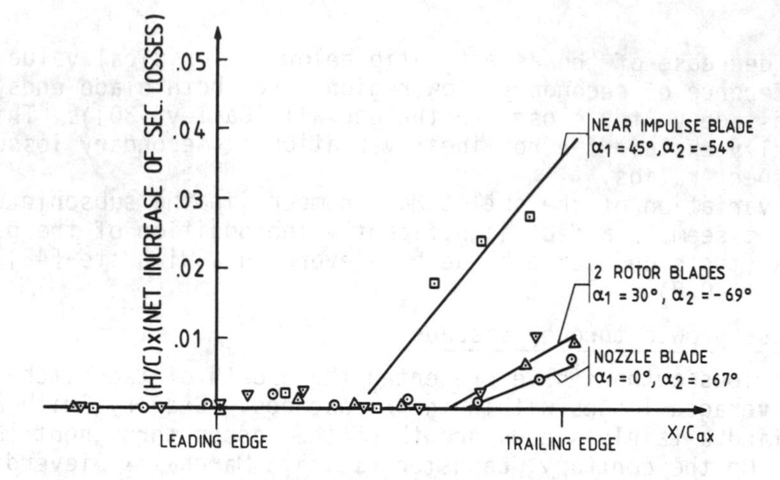

FIG. 15 - LOSS GROWTH THROUGH CASCADES from Marchal [34])

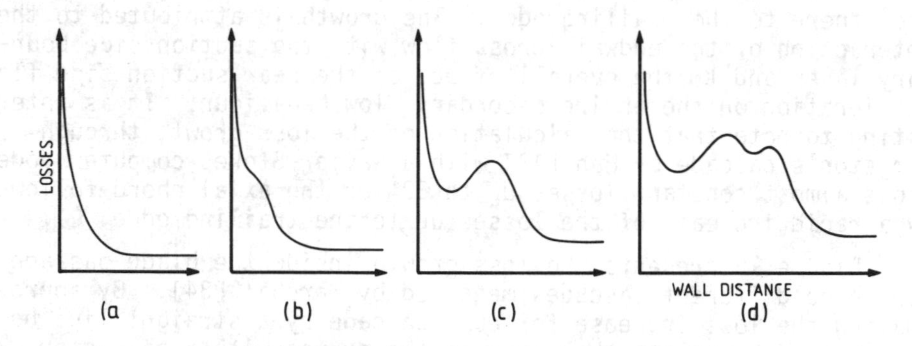

FIG. 16 - VARIOUS TYPES OF SPANWISE LOSS DISTRIBUTIONS

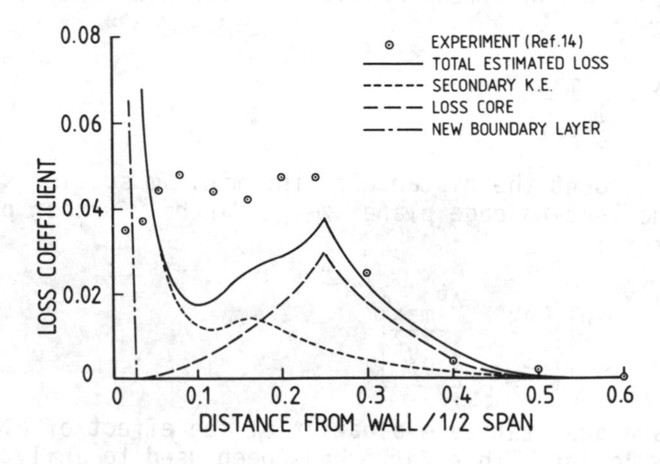

FIG. 17 - PREDICTION OF SPANWISE LOSS DISTRIBUTIONS BY GREGORY-SMITH [33]

distributions. An increase of Q indicates an increase of secondary losses. In the particular case of two bladings tested at VKI, a front loaded blade A and a rear loaded blade B (designed for α_1=30°, α_2=-69°, g/C_{ax}=0.87), the quality factor Q_A=Q_B=1.44 explained why the modification of the blade geometry did not affect the secondary losses.

4.3 Downstream spanwise loss distribution

The downstream spanwise loss distribution is naturally closely related to the position of the passage vortex. However, the actual shape does not only depend on the blade loading (Groschup [35], Marchal [34]) but also on the inlet boundary layer (Wolf [37], Came [36]) and the downstream distance (Armstrong [6], Wolf [37], Gregory-Smith & Graves [19]). The most frequent types of distribution are shown in Fig.16. Everything else being the same, the loss distribution will change from left to right with increasing loading. The bump in the loss distribution in Fig.16b and the loss cores occurring well away from the endwall in Figs 16c and 16d are evident results of the displacement action of the passage vortex on the endwall boundary layer as described before. A change from pattern (b) to (c) may also occur as a result of a strong thinning of the inlet boundary layer [36, 37]. The appearance of a double peak as shown in Fig.16d can have two reasons : the existence of important losses along the suction side separation line S_4 next to the loss core associated with the passage vortex (see [19]) or the splitting of the main loss core by the shearing action of the passage and trailing edge vortices as shown by Armstrong [6].

The best physical approach to spanwise modelling is probably that of Gregory-Smith [33] which is illustrated in Fig.17. The sharp peak of the loss core which represents the kinetic energy deficit is due to the assumption of a triangular distribution of these losses. A distribution function of the type e^{-z} as proposed by Groschup [35] would probably further improve the model. Since the model is based partially on classical secondary flow theory the extension of the secondary flow region is naturally taken proportional to the inlet boundary layer thickness.This is,however,not at all the general case. Figure 16 presents data on the relation : extension of secondary loss zone b with respect to the inlet boundary layer thickness δ_1 by Wolf [38], Came [36], Groschup [35], Marchal [34] and Gregory-Smith [19]. In cases where the begin of the secondary loss region was not clearly defined, it was arbitrarily deciced to put it at a wall distance, where the losses reached 1.1 times the value of the profile losses. In spite of the fact that the downstream measurement plane varied considerably between the various investigators (see Fig.18), several conclusions can be drawn :

(a) except for thick boundary layer and low turning blades, the ratio b/δ_1 is always greater than 1;

(b) the relation $b=f(\delta_1)$ is not linear; the rate at which b/δ_1 changes decreases rapidly to reach asymptotically a constant value for large

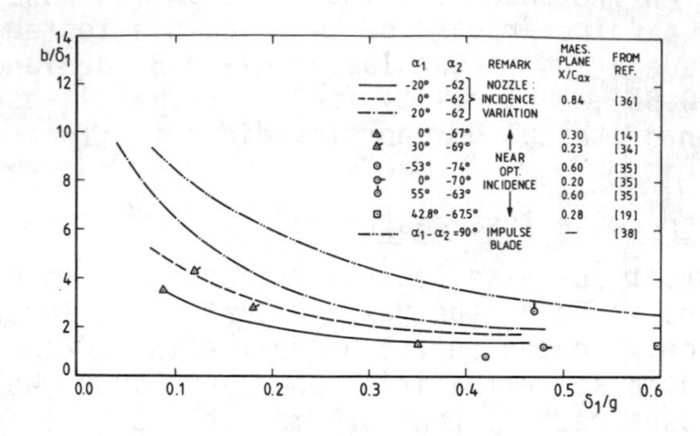

FIG. 18 - DEPENDENCE OF EXTENSION OF SECONDARY FLOW REGION ON INLET BOUNDARY LAYER THICKNESS

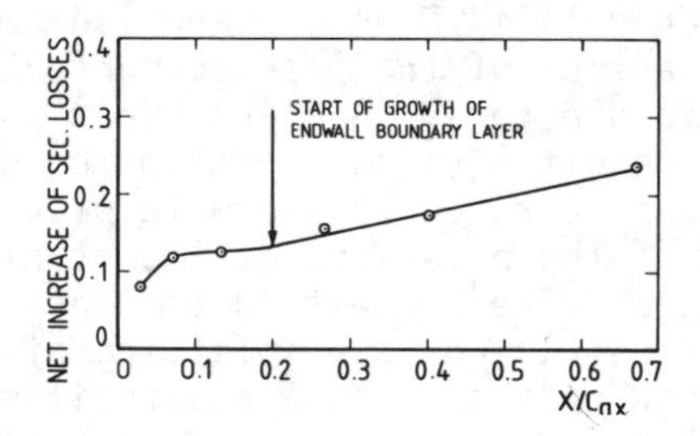

FIG. 19 - EXAMPLE OF GROWTH OF SECONDARY LOSSES DOWNSTREAM OF CASCADE
(from Groschup [35])

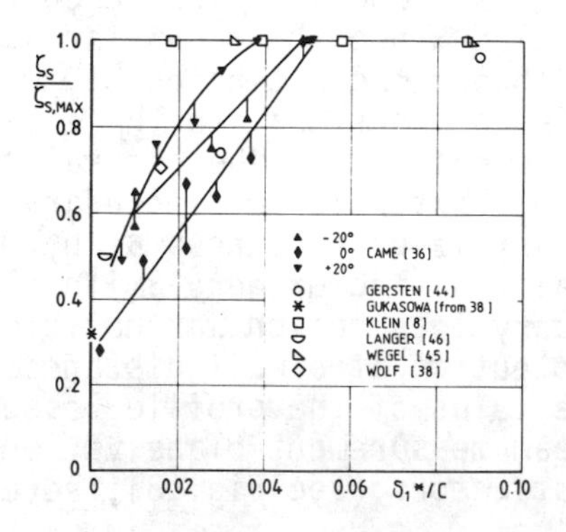

FIG. 20 - VARIATION OF SECONDARY LOSS WITH INLET DISPLACEMENT THICKNESS

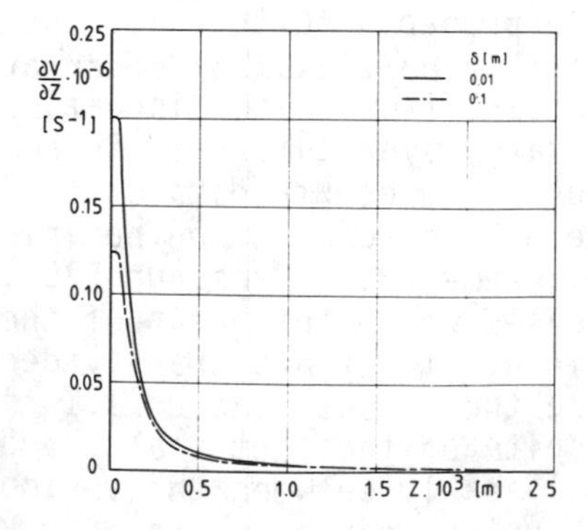

FIG. 21 - VELOCITY GRADIENTS FOR TURBULENT FLAT PLATE BOUNDARY LAYERS [37]

values of δ_1/g;
(c) the ratio b/δ_1 increases with blade loading at constant δ_1/g.

4.4 Downstream growth of losses

The fact that the endwall losses continue to rise with increasing downstream distance due to the action of wall shear stresses, makes an exact definition of secondary losses difficult. The situation is further complicated by the fact that the low momentum material produced by the wall shear stresses does not necessarily appear under the form of a normal boundary layer profile at the endwall, but is often fed into the main flow for some downstream distances [19, 23, 38]. This is reflected in the spanwise loss distribution by the absence of a "normal" wall loss profile which has been assumed to exist in all loss distributions in Fig.16. Under these circumstances the spanwise loss distribution pattern might, for instance, change from a single loss maximum to two in Fig.16c and from a double peak in Fig.16d to a triple. Tests by Wolf [38] on a cascade with α_1=32°, α_2=-58° and g/C=0.8 show that this feeding process continues up to a downstream distance X/C_{ax}=0.8, preventing up to there the development of a full endwall boundary layer. For bigger downstream distances, the secondary losses start to rise rapidly. Tests on a nozzle blade by Gotthardt [23] - Fig.19, indicate the same characteristic change of the secondary loss evolution to occur at X/C_{ax}=0.2 (the figure presents the net increase of all secondary losses occurring downstream of the trailing edge plane). The way in which the losses mix out is entirely unclear. There is not a single test which shows the complete downstream mixing process. Obviously the passage and trailing edge vortices play a significant role but their mutual interference is not clear. The fact that the passage vortex changes its position with respect to the trailing edge vortex as a function of the loading should play an important role for the downstream mixing.

In a real machine the increase of secondary losses due to wall shear stresses is interrupted when the flow enters the following blade row. To ensure the applicability of secondary loss data from straight cascades to a turbine stage, it is therefore advisable to select the downstream traverse plane at a distance corresponding to the distance between the blade rows. Such an approach was taken by Groschup [35]. In the absence of a direct application to a particular turbine, an axial distance equal to the throat seems to be adequate in most cases, since according to Dejc & Trojanosky [39] this distance presents an optimum value for the axial interspace between two blade rows. The tests by Came [36] and Wolf [38] are made at about twice this distance. In any case, secondary loss data should also include mixing losses occurring downstream of the measurement plane. An idea of the order of magnitude may be obtained from the extreme case of Woods [40] who measured the flow conditions behind a low aspect ratio impulse cascade of 135° turning and g/C=0.59 at a downstream distance of only 5% of the axial chord. Applying the momentum theorem to the flow between the measurement station and the uniform flow far downstream, the author calculates an increase for the sum of profile plus secondary losses of about 15% for an aspect ratio H/C=1.0. As pointed out by Woods, the calculation does not take into account the shear stress integral in the measurement plane.

5. EFFECT OF UPSTREAM BOUNDARY LAYER THICKNESS ON SECONDARY LOSSES

The dependence of the extension of the secondary flow region on the inlet boundary layer thickness presented in Fig.18 suggests that also the magnitude of secondary losses will depend on δ_1. Figure 20 summarizes all available data to illustrate the relation $(\zeta/\zeta_{max})_s = f(\delta_1^\star/C)$. The tests by Gersten [44], Klein [8], Came [36], Wegel [45] and Langer [46] are done on nozzle blades with 65° to 70° turning. Came changed the blade loading by varying the incidence angle by ±20°. The data of Wolf include both impulse blades (90° turning) and reaction blades.

The graph indicates that a variation of $(\zeta/\zeta_{max})_s$ occurs only for small values of $\delta_1^\star/C$, but it is not clear below which $(\delta_1^\star/C)_{cr}$ the loss ratio starts to decrease. Came's data indicate a critical value of $(\delta_1^\star/C)_{cr}=0.05$ while Klein does not find any variation down to $\delta_1^\star/C=0.02$. For $(\delta_1^\star/C)\to 0$, the loss ratio decreases to $(\zeta/\zeta_{max})_s = 0.3\to 0.5$.

The observation that the increase of the secondary losses with $\delta_1^\star/C$ reaches a maximum value is intriguiung. An explanation for it is given by Wolf [37]. The secondary losses must be proportional to the strength of the passage vortex and thereby to the magnitude of the vorticity up- and downstream of the cascade. The downstream vorticity is given by the equation :

$$\xi_2 = - \frac{\partial V_1}{\partial z} (\alpha_1 - \alpha_2)$$

where $(\partial V_1)/(\partial z)$ is the velocity gradient of the upstream boundary layer and $(\alpha_1-\alpha_2)$ the turning angle. Figure 21 shows the distribution of the velocity gradients for a flat plate turbulent boundary layer of 0.01 m and 0.1 m thickness calculated by Wolf. The gradient is biggest for z=0 and decreases rapidly. This suggests that the biggest part of the boundary layer is only of secondary importance for the generation of secondary flows, the main contribution comes from a small inner layer. The variation of $\partial V_1/\partial z$ will be biggest for thin boundary layers and reduces gradually with increasing δ. Furthermore, an artificial thickening of the inlet boundary layer will not affect the $\partial V_1/\partial z$ distribution in the same way as a natural growing of the boundary layers. This would explain for example the differences between the results of Klein and Came. Klein obtained a thickening of his boundary layer by means of stacks of plates arranged upstream of the cascade on the side walls of the entry duct, while Came varied δ_1 by removing or adding sections of a "false" side wall.

It is not clear whether the relation $(\zeta/\zeta_{max})_S=f(\delta_1/C)$ is affected by blade loading. The incidence angle variation of ±20° in Came's experiments presents a significant variation in blade loading but his data do not show clear tendencies. This is perhaps due to a superposition of several effects : variation of total blade loading, variation of load distribution and incidence angle effects on the formation of the horseshoe vortex.

6. SECONDARY FLOWS IN ANNULAR CASCADES

The description of fundamental aspects of secondary flows presented in the preceding chapters is based almost exclusively on experiments in straight cascades. In a few cases reference was made to experiments in annular cascades [12, 31] but so far we have not yet referred to the specific character of flows in annular cascades.

Depending on the blade design the flow field in an annular cascade is characterized by a more or less strong radial pressure gradient. This gradient has a non negligible effect on the evolution of secondary flows along the outer and inner shroud. Hence, the spanwise loss and angle distributions will differ from those of straight cascades having the same blade sections. Glynn & Marsh [47] have pointed to another fundamental difference between the flows through straight and annular cascades. "If a uniform flow is turned in an annular cascade which is not of free vortex design, then there is a variation of lift along the span of the blades. A trailing vortex sheet is then present and to be consistent with this vortex sheet there must be small velocity components across the blade passage and along the span of the blades".

Figure 22 shows qualitatively the velocity patterns in the trailing edge plane induced by these small velocity components and compares them with secondary flow patterns caused by turning a shear flow. When both effects are present, they tend to reinforce each other at one endwall and oppose each other at the other endwall.

6.1 Early experiments at NASA

The first extensive research program on secondary flows in annular turbine cascades was carried out in the early 1950's at NASA by Rohlik et al [48]. Their experiments included testing of different types of high aspect ratio stator designs at both subsonic and supersonic downstream Mach numbers. Pressure probes were used to survey the inlet and exit flow conditions. Smoke and oil flow visualizations added some information on the flow inside the blade channel. Their results are summarized below :

- Provided that the suction surface velocity distribution was smooth at all spanwise positions, the radial velocity distribution was qualitatively the same for a free vortex blade and a constant-discharge-angle blade, i.e., low losses at the tip and high losses at the hub (Fig.23a). The loss accumulation at the inner shroud was attributed to a radial inward flow of boundary layer material from the outer

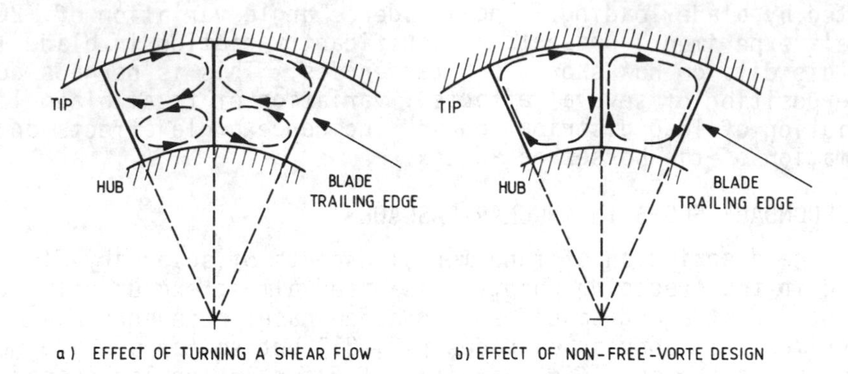

FIG. 22 - SECONDARY FLOW PATTERNS IN ANNULAR CASADES [47]

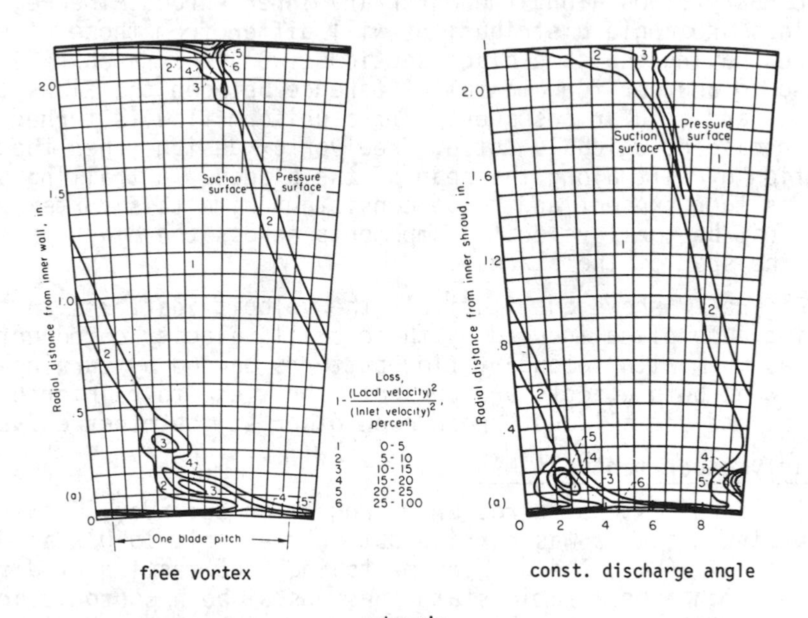

FIG. 23a - KINETIC ENERGY LOSS DISTRIBUTION AT EXIT OF STATOR BLADES WITH SMOOTH SUCTION SIDE VELOCITY DISTRIBUTIONS AT ALL RADII [48]

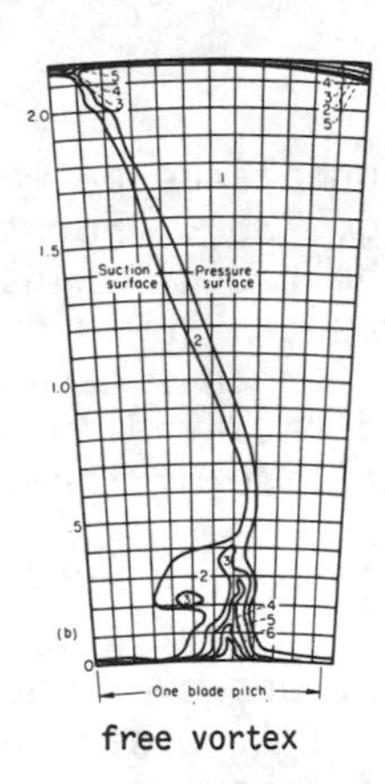

free vortex

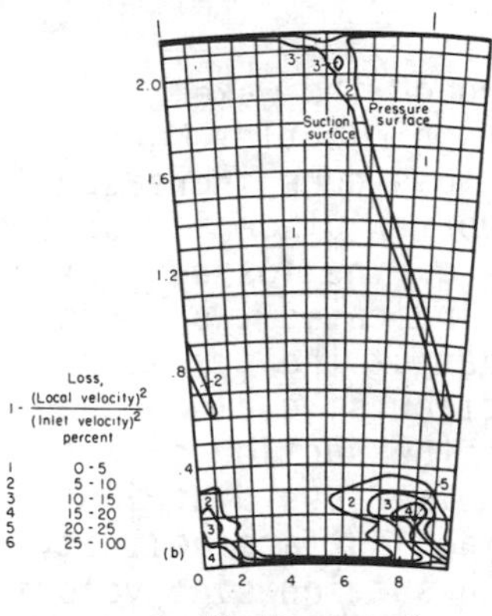

const. discharge angle

FIG. 23b - KINETIC ENERGY LOSS DISTRIBUTIONS FOR BLADES OF FIGURE 23a AT SUPERSONIC OUTLET [48]

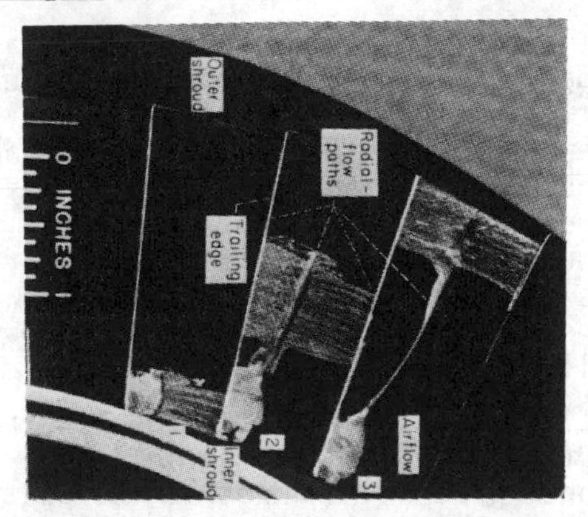

FIG. 23c - OIL FLOW VISUALIZATION IN FREE VORTEX STATOR AT SUPERSONIC DOWNSTREAM MACH NUMBERS [48]

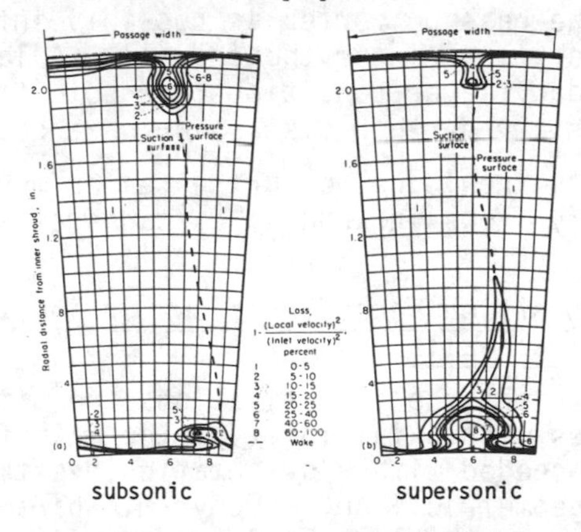

subsonic supersonic

FIG. 23d - EXIT KINETIC ENERGY LOSS DISTRIBUTION FOR CONSTANT DISCHARGE ANGLE BLADES WITH PEAKY SUCTION SIDE VELOCITY DISTRIBUTION IN TIP REGION [48]

shroud through the wake.

- At supersonic outlet Mach numbers, the unbalance between the outer and inner loss core increased further (Fig. 23b). For the case of the free vortex blade it was shown that the radial transport accounted for 65% of the losses at the inner shroud. Flow visualizations (Fig.23c) indicated two radial flow paths for this case : in addition to that at the trailing edge a second radial flow path appeared on the suction side along the interference line of the trailing edge shock with the suction side boundary layer.
- The spanwise loss distribution was significantly affected by the type of blade surface velocity distribution. Contrary to blades with smooth suction side velocity distributions everywhere, it was demonstrated that a peaky suction side velocity distribution at the tip of the constant-discharge-angle cascade resulted, at subsonic outflow, in high tip losses and low hub losses. This tendency changed, however, for supersonic outlet Mach numbers (Fig.23d).

From the last experiment it was concluded that the state of the suction side boundary layer near the endwall played a significant role in the formation of the passage vortex. Due to the authors a thick suction side boundary layer (caused by a peaky velocity distribution) contributes to the formation of the passage vortex. The boundary layer is entrained by the vortex. Hence, the material available for the secondary flow radial transport is reduced. At supersonic Mach numbers the shock-thickened radial flow path on the suction surface drains off some of the tip region boundary layer material before it rolls up into the passage vortex.

This explanation is not very satisfactory since the formation of the passage vortex is related to the inlet endwall boundary layer and the cross channel gradient and not to the suction side boundary layer. A change in the blade pressure distribution affects both the cross channel gradient and the blade suction side boundary layer. The evolution of the passage vortex is possibly influenced by the suction side boundary layer near the endwall. A lack of information on the actual boundary layer conditions for both types of velocity distributions makes any further discussion hazardous.

To the author's knowledge no further experiments have been published on secondary flows in annular cascades at supersonic downstream conditions.

6.2 Further experiments on secondary flows in annular cascades with cylindrical walls

In nearly all cases the investigations are limited to secondary flows in guide vanes with axial inlet of the main flow. Apparently nobody has ever proceeded with a systematic investigation of the effect of important geometrical and aerodynamic parameters. All research programs are more or less isolated experiments which are of little direct help to the turbine designer. Therefore, the main purpose of secondary flow research in annular cascades is to highlight particular secondary flow aspects like radial migration of low momentum boundary layer material, downstream mixing process, effects due to neighbouring blade rows, effect of cooling, endwall contouring, etc. The value of the experiments depends more than ever on a judicious choice of blade and cascade geometry, measurement technique, test conditions and test program.

6.2.1 Annular cascades with collateral inlet boundary layers.

6.2.1.1 Brief review of published test results. On several occasions NASA has published performance measurements of low aspect ratio annular cascades for application in high temperature ratio turbines. A typical example of this type of work is published by Goldman & McLallin in 1975 [49] for the case of a low aspect ratio ($H/C=0.65$; $D_H/D_T=0.85$) untwisted stator vane with a mean outlet flow angle of $\alpha_2=67°$ and a mean pitch to chord ratio of $g/C=0.69$. As usual at NASA the tests are limited to a survey of the inlet and exit flow conditions at one axial position. A lack of information on the inlet endwall boundary layer limits further the value of these experiments for secondary flow research. An interesting issue raised by the authors is that the stator angle maldistribution may produce rotor incidence losses. In the present case the rotor incidence variation is evaluated at ±6 to 8°. The authors report that such a deviation of the incidence angle from its design value has only a negligible effect on the rotor profile loss. However, the effect of the stator secondary flows on the rotor secondary loss is not investigated.

Binder & Romey [50] investigated the evolution of the 3D flow field downstream of a very low aspect ratio stator with twisted blades ($H/C=0.56$, $D_H/D_T=0.76$, $\alpha_{2,m}=68.9°$). The authors determine the position and nature of the secondary vortices on the basis of iso-α-lines and analyze the effect of these vortices on the radial motion of low momentum flow and the mixing out of the wake. Both Binder and Goldman made their experiments at high subsonic outlet Mach numbers.

Measurements of the evolution of the 3D flow throughout stator vane passages are made by Sjolander [12] and Sieverding-Boletis-Van Hove [31]. Sjolander made his experiments in a low speed, low turning cascade with cylindrical blades which are not representative of actual blade design philosophy ($D_H/D_T=0.75$; $H/C=0.66$; $(g/C)_{mean}=0.92$, $\beta_{2,mean}=45°$, T-6 profile). His measurements cover only half of the blade passage and do not show the specific character of flows in annular cascades. Sieverding et al studied the flow field in a stator vane with a blade and cascade geometry very similar to that of Goldman & McLallin [49]. The cascade geometry is described in the next section. Measurements of the entire flow field from far upstream to far downstream allow a detailed description of the secondary flow patterns. The tests are done at low speed and rather low Reynolds number ($Re_1=10^5$). To eliminate to a large extent Reynolds number effects on the blade boundary layer, the suction side boundary layer was tripped at the point of minimum pressure.

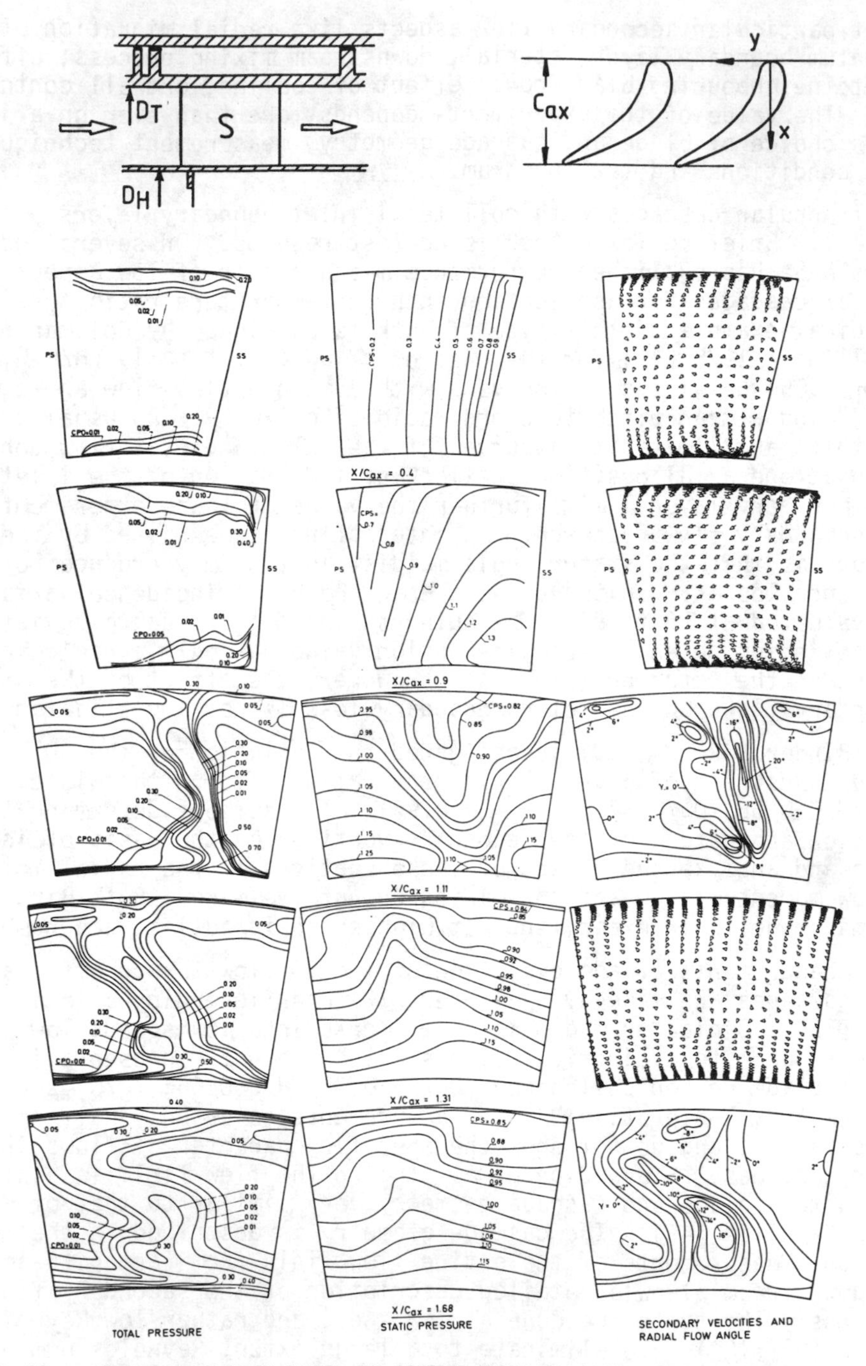

FIG. 24a - SECONDARY VELOCITY CHARTS AND CONTOUR PLOTS OF LOSSES, STATIC PRESSURES AND RADIAL FLOW ANGLE AT DIFFERENT AXIAL PLANES FOR VKI ANNULAR GUIDE VANE WITH COLLATERAL INLET BOUNDARY LAYER [31]

6.2.1.2 Description of annular cascade flow patterns for collateral inlet boundary layers. The following description of secondary flows in annular cascades will focus on the different evolution of the endwall boundary layers along the outer and inner shroud.

Figure 24a presents a combination of secondary flow charts and contour plots of total pressure, static pressure and radial flow angle of the flow in the annular cascade at VKI (from [31]). The secondary velocity vector is obtained by decomposing the actual velocity vector in a streamwise vector and a vector normal to it. The streamwise direction in the whole flow field is calculated with a singularity method applied to the flow in the blade to blade planes at different blade heights. The radial flow angle γ is calculated from the ratio of radial and axial velocity components : $\gamma = \mathrm{arctg}(V_r/V_{ax})$. The most important geometrical and aerodynamic characteristics of the cascade are :

$D_H/D_T = 0.8$	$D_T = 0.71$ m
$H/C = 0.6$	$C = 0.12$ m
$g/C = 0.71; 0.80; 0.89$	(hub, mean, tip)
$\arccos(0/g) = 69°; 67.8°; 66.8°$	(hub, mean, tip)
$\delta_1^{\star}/H = 0.017; 0.025$	(hub, tip)
$Re = 10^5$	(based on inlet velocity & chord)

Spatial distribution

Flow inside the blade passage : the flow conditions are presented in two axial planes, situated at $X/C_{ax}=0.4$ and 0.9. The secondary flow charts indicate clearly in both planes the existence of the passage vortex which is the main driving force for the tangential and, near the suction side, the radial displacement of the endwall boundary layer material. The differences at hub and tip are considerable and have to be attributed to (a) the difference in vortex strength at hub and tip; (b) the radial pressure gradient and (c) the leaning of the blade surfaces. The radial pressure gradient which is still extremely weak at $X/C_{ax}=0.4$ has taken important proportions at $X/C_{ax}=0.9$, in particular near the suction side where the pressure difference reaches 30% of the outlet dynamic head. The big difference in the pitchwise position of the vortex centers at hub and tip is somewhat surprising and it seems unlikely that it can be attributed entirely to the different blade loading. The difference might in fact point to the inability of a 2D method to predict correctly the streamwise direction of the inviscid flow field near the endwalls. The measurements reveal little about the horseshoe vortex and its significance outside the entrance region of the blade passage. There is still some evidence in plane $X/C_{ax}=0.4$ for the existence of the suction side leg of the horseshoe vortex but this is about as far as

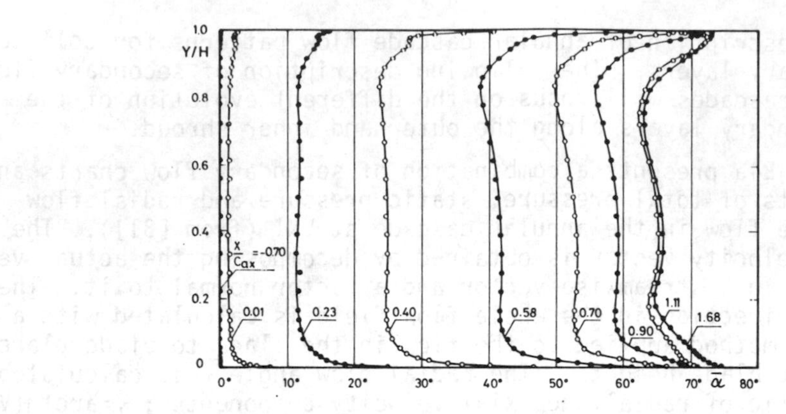

FIG. 24b - AXIAL EVOLUTION OF SPANWISE DISTRIBUTION OF PITCHWISE INTEGRATED FLOW ANGLE $\bar{\alpha}$ FOR VKI ANNULAR GUIDE VANE WITH COLLATERAL INLET BOUNDARY LAYER

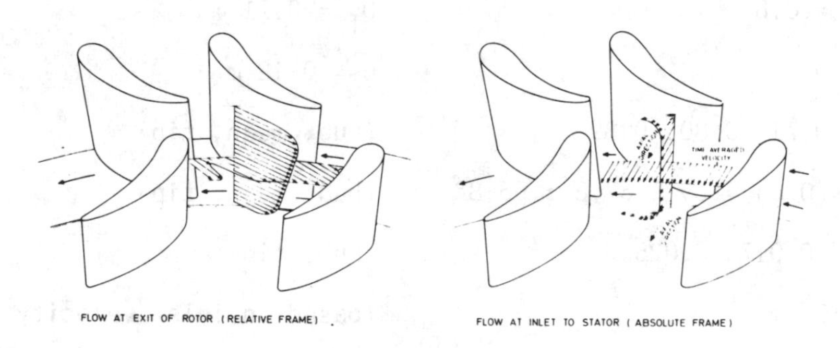

FIG. 25 - TRANSITION OF FLOW FROM ROTOR OUTLET TO STATOR INLET [52]

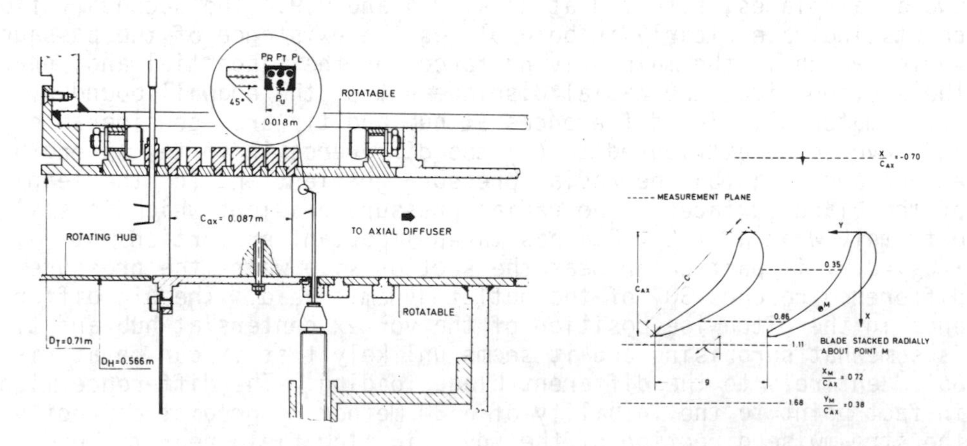

FIG. 26a - VKI ANNULAR CASCADE WITH PROVISION FOR GENERATING SKEWED INLET BOUNDARY LAYER AT HUB

one can go. As for the pressure side leg of the horseshoe vortex, we do not expect it to appear as a distinct vortex except in the immediate vicinity of the leading edge plane, since its sense of rotation coincides with that of the passage vortex.

Flow downstream of blade passage : the influence of the radial pressure sure gradient on the outlet flow field is the most striking aspect in all three downstream planes. The first plane is situated about 10% of the axial chord away from the trailing edges (X/C_{ax}=1.11). Outside the wake region the constant total pressure lines are widely spaced at tip and closely spaced at the hub with considerably higher losses at hub than at tip. Within the wake region one observes a radial transport of low momentum material from the endwall and possibly also from the blade boundary layer. This radial migration from tip to hub is confirmed by the distribution of the radial flow angle γ. Values of up to -20° are recorded for γ in the center of the wake slightly above midspan. Also, the wake region is characterized by significant tangential pressure variations which contribute probably to the dissipation of the losses in pitchwise direction when proceeding further downstream.

The radial convection of losses continues in the measuring plane X/C_{ax}=1.31. The corresponding secondary velocity chart is chosen to underline this feature. A superposition of loss contour plot and secondary velocity chart contributes also to a better understanding of the pitchwise shift of the constant loss lines at hub and tip.

In the last plane (X/C_{ax})=1.68) the endwall boundary layer have grown significantly at both hub and tip and the width of the wake is almost twice that in the preceding plane. The radial pressure gradient is nearly the same as before and keeps the migration of low momentum material through the wake going. The radial flow angles are still very significant although somewhat less than before. Compared to plane X/C_{ax}=1.11, the position of the maximum negative angle has shifted towards the hub

Spanwise distribution of pitch-averaged flow angles : the axial evolution of the spanwise distribution of the pitchwise integrated flow angle α is presented in Fig.24b. The early evolution of α at hub and tip is influenced considerably by the differences of the upstream boundary layer profiles at hub and tip. Further downstream, $X/C_{ax} \geqslant 0.58$, the distribution of $\overline{\alpha}$ is dominated by the presence of the passage vortices. The different radial evolutions of $\overline{\alpha}$ at hub and tip are a clear sign for the different vortex strength, i.e., the different blade loading at hub and tip. Downstream of the blade the angle distribution varies rather little with increasing distance from the trailing edge plane except close to the hub where a further increase of α is observed. The reason for this local angle increase is probably to be looked for in a reduction of the axial velocity component due to the appreciable rise of losses.

6.2.2 Annular cascades with skewed inlet boundaries

6.2.2.1 Brief review of research work. The occurrence of inlet skew at the entrance of a stator is illustrated in Fig.25. The

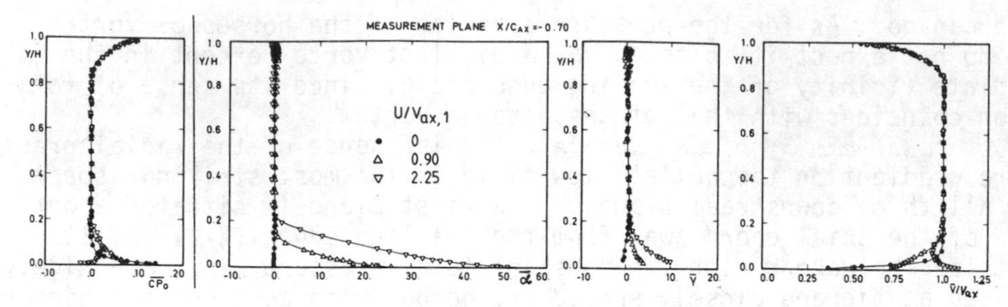

FIG. 26b - UPSTREAM FLOW CONDITIONS FOR ANNULAR CASCADE TESTS WITH SKEWED INLET BOUNDARY LAYER

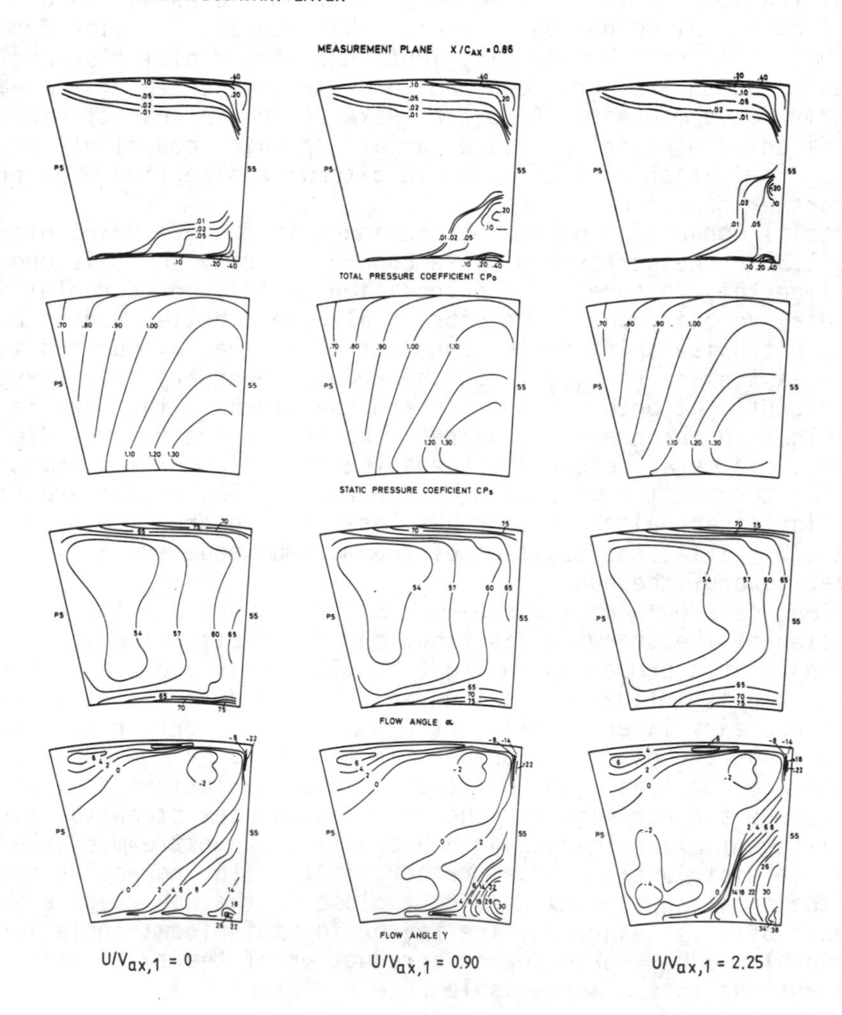

FIG. 26c - EFFECT OF INLET SKEW ON FLOW CONDITIONS IN VKI ANNULAR GUIDE VANE MEASUREMENT PLANE AT x/C_{ax} = 0.86 [52]

velocity flow field at the exit of a rotor blade row is transformed from the rotating to the stationary system fixed with respect to the downstream stator. This carry over of the velocities from the relative to the stationary frame causes significant changes of the incidence angle to the next blade row. The same thing occurs of course when passing from the stationary frame of the stator to the rotating frame of the rotor. The incidence angle variations related to the endwall boundary layer profile are much bigger than those caused by the secondary flow angle variation referred to by Goldman and McLallin [49].
Klein [8], Bindon [51] and Boletis et al [52] investigated the effect of inlet skew on the flow through annular cascades by rotating the hub extending upstream of the cascade. Bindon carried out his experiments on the low turning cascade of Sjolander. Boletis et al used the same high turning gas turbine nozzle blade as in [31]. The blade of Klein is a high pressure steam turbine stator blade (very thick leading edge, 66° turning, thin trailing edge). In all three cases the tests are done at low speed. Apart from the difference in the cascade geometry there is a difference in the positioning of the slot separating the rotating hub from the fixed stator hub. In the experiments of Klein and Bindon the slots are right at the leading edge of the blade while Boletis et al placed it at a distance $X/C_{ax}=0.15$ from the leading edge, which corresponds to twice the distance of the endwall boundary layer saddle point from the leading edge.

In the case of the slot at the leading edge special care must be taken to avoid that due to the tangential pressure gradient flow is drawn near the pressure side and discharged near the suction side. This slot effect was present in the tests of Bindon as indicated by the author. An analysis of his data in function of inlet skew only is therefore rather difficult.

6.2.2.2 Example of secondary flow patterns in an annular cascade with inlet skew. The following illustrations are taken from [52], see §6.2.1.2 for cascade geometry and general flow conditions. The inlet flow conditions are given in Fig.26a,b. The maximum skew near the wall amounts to 25° at $U/V_{ax}=0.9$ and 50° for $U/V_{ax}=2.25$ (U=peripheral speed of the rotating hub, V_{ax}=inlet axial velocity at midspan). For the last case, the total pressure coefficient $CP_0=(P_{01,m}-P_{0,\ell})/(P_{01}-P_{S_2})$ takes negative values at the wall due to the kinetic energy supply by the rotating hub ($P_{01,m}$=inlet total pressure at midspan, $P_{0,\ell}$=local total pressure in boundary layer, $(P_{01}-P_{S_2})$= isentropic outlet dynamic head). The effect of inlet skew on the flow conditions within and downstream of the cascade are shown in Figs. 26c and 26d. Without going into a detailed discussion of the results, one can say that inlet skew amplifies all characteristic features associated with the usual endwall vortex system for a collateral inlet boundary layer. In particular it augments the concentration of low momentum fluid in the suction side endwall corner and generates an important outward radial flow along the rear blade suction surface. Downstream of the cascade, this motion counteracts the effect of the radial pressure gradient on the migration of low momentum boundary

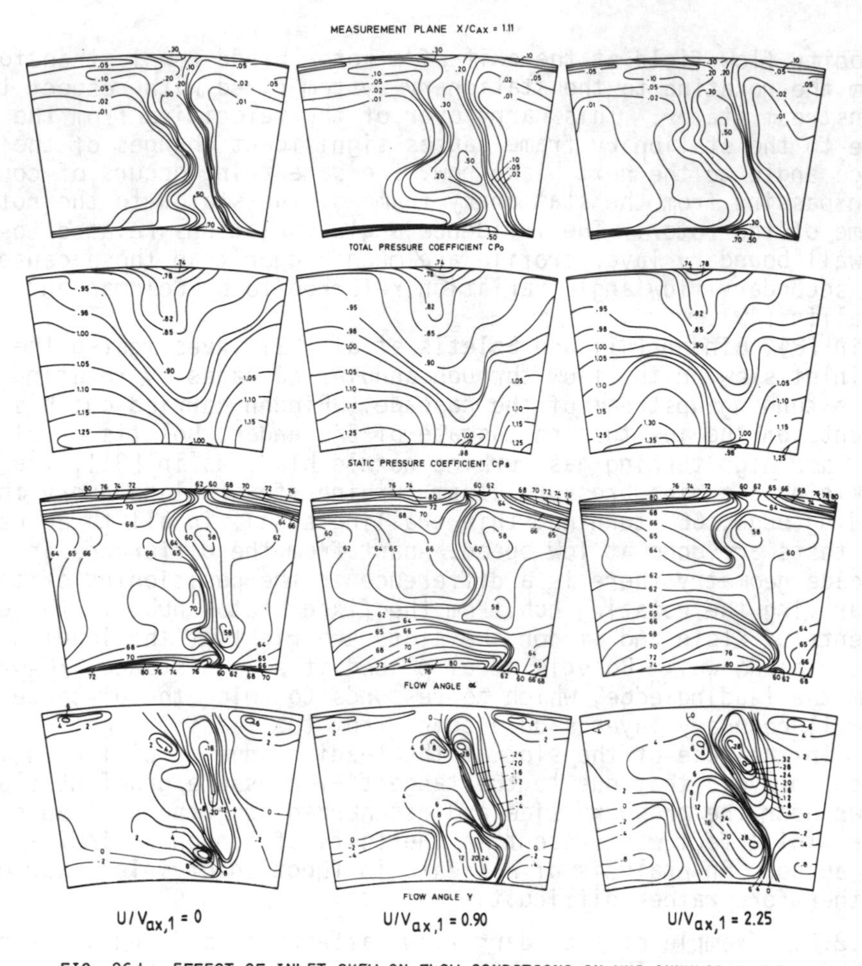

FIG. 26d - EFFECT OF INLET SKEW ON FLOW CONDITIONS IN VKI ANNULAR GUIDE VANE MEASUREMENT PLANE AT x/C_{ax} = 1.11 [52]

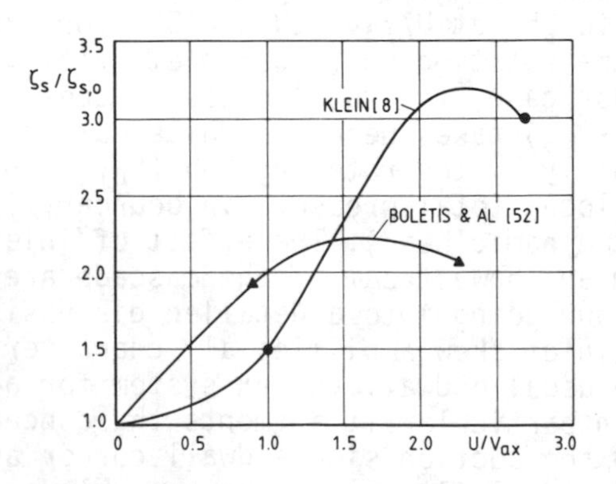

FIG. 27 - VARIATION OF SECONDARY LOSS IN FUNCTION OF ROTATION PARAMETER u/V_{ax}

layer through the wake. The gradual shift of the maximum wake losses from the hub towards midspan with increasing skew underlines this feature. Figure 27 presents, for the experiments of Klein [8] and Boletis et al [52], the variation of the secondary losses in function of the rotation parameter U/V_{ax}. The principle geometrical and aerodynamic parameters of the two cascade are given below.

	D_H/D_T	H/C	$(g/C)_m$	α_2	Re_1	M_2
Klein [8]	0.75	1.43	0.85	66°	1.4×10^5	low speed
Boletis et al [52]	0.8	0.6	0.8	68°	10^5	low speed

	U/V_{ax}	$\delta^\star_{1,x}/C$	$\delta^\star_{1,y}/C$	H
Klein [8]	0	0.012	0	1.63
	1.0	0.019	0.021	1.45
	2.0	0.026	0.060	1.31
	2.7	0.029	0.094	1.30
Boletis et al [52]	0	0.010	0	1.67
	0.9	0.011	0.011	1.27
	2.26	0.015	0.041	1.25

Both curves of Fig.27 indicate the strong increase of secondary losses with inlet skew.

The use of U/V_{ax} as correlating parameter is certainly not very satisfactory since it does not give any information about the real inlet boundary layer conditions. The maximum skew angle at the wall certainly would be more appropriate but the angle variation through the boundary layer must also be important. Wegel [45] has proposed to describe the boundary layer skew by the ratio of the secondary tangential momentum to the axial momentum through the boundary layer.

$$(\Delta I_y/I_x)_\delta = \frac{\int_0^\delta \Delta V_y^2(r)\cdot r\cdot dr}{\int_0^\delta V_x^2(r)\cdot r\cdot dr}$$

where ΔV_y represents the deviation of the velocity component with respect to that of the corresponding flow without skew. In this way it is possible to define in a simple way rather complex situations as for instance stator inlet flow conditions resulting from the combined effect of endwall flows and tip clearance flows of a preceding rotor blade row.

CONCLUSIONS

The combination of new types of flow visualization methods and very detailed 3D probe measurements has contributed to an improved understanding of the secondary flow vortex system and its impact on the flow characteristics in turbine blade passages. However, important areas need further investigations. Some problems, like leading edge geometry, high incidence angles and the influence of the vortex system on the downstream-mixing process may be studied in straight cascades. The research in annular cascades has to be extended to take into account inlet flow characteristics which are typical for a multistage environment, i.e., non-axisymmetric, unsteady inlet flows with significant spanwise gradients of all flow properties, where clearance flows of the upstream rotor play an important role.

REFERENCES

1. Dunham, J.: A review of cascade data on secondary losses in turbines. J.Mech.Engrg Sci., Vol. 12, 1970, pp 48-59.
2. Tall, W.A.: Understanding turbine secondary flow. in "Secondary Flows in Turbomachines", AGARD CP 214, 1977, Paper 14.
3. Sieverding, C.H.: Recent progress in the understanding of basic aspects of secondary flows in turbine blade passages. ASME Paper 84 GT 78; also VKI Preprint 1983-28.
4. Hawthorne, W.R.: Rotational flow through cascades. J. Mech. & Appl. Math., Vol. 3, 1955.
5. Herzig, H.Z.; Hansen, A.G.; Costello, G.R.: Visualization study of secondary flow in cascades. NACA TR 1163, 1954.
6. Armstrong, N.D.: The secondary flow in a cascade of turbine blades. ARC R&M 2979, 1955.
7. Fritsche, A.: Strömungsvorgänge in Schaufelgittern. Techn. Rundschau Sulzer, Nr 3, 1955.
8. Klein, A.: Untersuchungen über die Einfluss der Zuströmgrenzschicht auf die Sekundärströmung in den Beschaufelungen von Axialturbinen. Forsch. Ing., Bd 32, Nr 6, 1966; English translation "Investigation of the entry boundary layer on the secondary flows in the blading of axial turbines". BHRA T 1004, 1966.
9. Bölcs, A.: Flow investigations in a water channel at subsonic and supersonic velocities. Escher Wyss News, Vol. 42, No 1, 1969.
10. Blair, M.F.: An experimental study of heat transfer and film cooling on large scale turbine endwalls. ASME Transact., Series C : J. Heat Transfer, Vol. 96, No 4, Nov 1974, pp 524-529.

11. Langston, L.S.; Nice, M.L.; Hooper, R.M.: Three dimensional flow within a turbine blade passage. ASME Transact., Series A : J. Engrg Power, Vol. 99, No 1, Jan 1977, pp 21-28.
12. Sjolander, S.A.: The endwall boundary layer in an annular cascade of turbine nozzle guide vanes. Carleton U., Canada, TR ME/A 75/4.
13. Langston, L.S.: Crossflows in a turbine cascade passage. ASME Transact., Series A : J.Engrg Power, Vol. 102, No 4, Oct 1980, pp 866-874.
14. Marchal, P. & Sieverding, C.H.: Secondary flows within turbomachinery bladings. in "Secondary Flows in Turbomachines", AGARD CP 214, 1977, Paper 11; also VKI Preprint 77-11.
15. Gaugler, R.E. & Russell, L.M.: Streakline flow visualization study of a horseshoe vortex in a large scale, 2D turbine stator cascade. ASME P 80 GT 4.
16. Moore, J. & Smith, B.L.: Flow in a turbine cascade. Part 2 : Measurement of flow trajectories by ethylene detection. ASME Transact., Series A : J.Engrg for Gas Turbines and Power, Vol. 106, No 2, April 1984, pp 409-413.
17. Sieverding, C.H. & Van den Bosch, P.: The use of colored smoke to visualize secondary flow in a turbine-blade cascade. J. Fluid Mechanics, Vol. 134, Sept 1983, pp 85-89; also VKI Prepr. 82-26.
18. Moore, J. & Ransmayr, A.: Flow in a turbine cascade. Part 1 : Losses and leading edge effects. ASME Transact., Series A : J. Engrg for Gas Turbines and Power, Vol. 106, No 2, April 1984, pp 400-408.
19. Gregory-Smith, D.G. & Graves, C.P.: Secondary flows and losses in a turbine cascade. in "Viscous Effects in Turbomachines", AGARD CP 351, 1983, Paper 17.
20. Belik, L.: Secondary losses in turbine blade cascade with low aspect ratio and large deflection. Proc. 6th Conf. on Steam Turbines of Large Power Output, Plzen, Czechoslovakia, Sept 1975.
21. Moore, J.: Flow trajectories, mixing and entropy fluxes in a turbine cascade. in "Viscous Effects in Turbomachines", AGARD CP 351, 1983, Paper 5.
22. Langston, L.S. & Boyle, M.T.: A new surface streamline flow visualization technique. J. Fluid Mechanics, Vol. 125, Dec 1983, pp 53-57.
23. Gotthardt, H.: Theoretische und experimentelle Untersuchungen an ebenen Turbinengittern mit Pfeilung und V-Stellung, Dissertation TU Braunschweig, Germany, 1983.
24. Belik, L.: The secondary flows around a cylinder mounted on a flat plate. Aero. Qua., Vol. 24, Part 1, Feb 1973, pp 47-54.
25. Stastny, H.: Visualization of some phenomena connected with non-potentiality of the flow in steam turbines. Int. Symp. on Flow Visualization, Bochum, Germany, Sept 1980, pp 261-266.
26. Peake, D.J.; Galway, R.D.; Rainbird, W.J.: The 3D separation of a plane, incompressible, laminar boundary layer produced by a Rankine oval mounted normal to a flat plate. NRC LR 446, 1965.
27. East, L.F. & Hoxey, R.P.: Low speed 3D turbulent boundary layer data. RAE TR 69041, 1969.

28. Senoo, Y.: The boundary layer on the endwall of a turbine nozzle cascade. ASME Transact., Series A : J.Engrg Power, Vol. 80, 1958.
29. Bario, F.; Leboeuf, F.; Papailiou, K.D.: Study of secondary flows in blade cascades of turbomachines. ASME Transact., Series A : J.Engrg Power, Vol. 104, No 2, April 1982, pp 497-509.
30. Bailey, D.A.: Study of mean and turbulent velocity fields in a large scale turbine-vane passage. ASME Transact., Series A : J.Engrg Power, Vol. 102, No 1, Jan 1980, pp 88-97.
31. Sieverding, C.H.; Boletis, E.; Van Hove, W.: Experimental study of the 3D flow field in an annular turbine nozzle guide vane. ASME Transact., Series A : J.Engrg for Gas Turbines and Power, Vol. 106, No 2, April 1984, pp 437-444.
32. Belik, L.: Three dimensional and relaminarization effects in turbine blade cascades - an experimental study. JSME-ASME Joint Gas Turbine Conf., Paper 37, 1977.
33. Gregory-Smith, D.G.: Secondary flows and losses in axial flow turbines. ASME Transact., Series A : J.Engrg Power, Vol. 104, No 4, Oct 1982, pp 819-822.
34. Marchal, P.: Etude des écoulements secondaires en grille d'aubes de détente. Ph.D. Thesis, U. Libre de Bruxelles, 1980.
35. Groschup, G.: Strömungstechnische Untersuchung einer Axialturbinenstufe in Vergleich zum Verhalten der ebenen Gitter ihrer Beschaufelung. Dissertation TU Hannover, Germany, 1977.
36. Came, P.M.: Secondary loss measurements in a cascade of turbine blades. Inst. Mechanical Engrs, C33/73.
37. Wolf, H.: Ein Beitrag zum Problem der Sekundärströmung in Schaufelgittern. Wiss.Z. Techn. Hochschule Dresden, Bd 7, Heft 4, 1958/59.
38. Wolf, H.: Die Randverluste in geraden Schaufelgittern. Wiss.Z. Techn. Hochschule Dresden, Bd 10, Heft 2, 1961.
39. Dejc, M.E. & Trojanovskij, B.M.: Untersuchung und Berechnung axialer Turbinenstufen. Berlin, VEB Verlag Technik, 1973.
40. Woods, J.R.: An investigation of secondary flow phenomena and associated losses in a high deflection turbine cascade. Ph.D. Thesis, Naval Post Graduate School, 1972.
41. Graziani, R.A.; Blair, M.F.; Taylor, J.R.; Mayle, R.E.: An experimental study of endwall and airfoil surface heat transfer in a large scale turbine blade cascade. ASME Transact., Series A : J.Engrg Power, Vol. 102, No 2, Apr. 1980, pp 257-267.
42. Sieverding, C.H. & Wilputte, P.: Influence of Mach number and endwall cooling on secondary flows in a straight nozzle cascade. ASME Transact., Series A : J.Engrg Power, Vol. 103, No 2, April 1981, pp 257-264; also VKI Preprint 1980-02.
43. Hah, C.: A Navier-Stokes analysis of 3D turbulent flows inside turbine blade rows at design and off-design conditions. ASME Transact., Series A : J.Engrg Gas Turbines and Power, Vol. 106, No 2, April 1984, pp 421-429.
44. Gersten, K.: Über den Einfluss der Geschwindigkeitsverteilung in der Zuströmung auf die Sekundärströmung in geraden Schaufelgittern. Fors.Geb.Ing., Bd 23, Heft 3, 19 , pp 95-101.

45. Wegel, S.: Strömungsuntersuchungen an Beschleunigungsgittern im Windkanal und in der Axialturbine. Dissertation TU Darmstadt, 1970.
46. Langer, L.: Über den Einfluss der Profillänge auf die Randverluste in einem Düsengitter. Maschinenbautechn., Bd 15, Heft 6, 1966, pp 303-306.
47. Glynn, D.R. & Marsh, H.: Secondary flows in annular cascades. Int.J.Heat & Fluid Flow, Vol. 2, No 1, 1980, pp 29-33.
48. Rohlik, H.E; Kofskey, M.G.; Allen, H.W.; Herzig, H.Z.: Secondary flows and boundary layer accumulations in turbine nozzles. NACA TR 1168, 1954.
49. Goldman, L.J. & McLallin, K.L.: Cold air annular cascade investigation of aerodynamic performance of core-engine-cooled turbine vanes. NASA TM X 3224, 1975.
50. Binder, A. & Romey, R.: Secondary flow effects and mixing of the wake behind a turbine stator. ASME Transact., Series A : J.Engrg Power, Vol. 105, No 1, Jan 1983, pp 40-46.
51. Bindon, J.P.: Exit plane and suction surface flow in an annular turbine cascade with a skewed inlet boundary layer. Int. J.Heat & Fluid Flow, Vol. 2, No 2, 1980, pp 57-66.
52. Boletis, E.; Sieverding, C.H.; Van Hove, W.: Effects of a skewed inlet endwall boundary layer on the 3D flow field in an annular turbine cascade. in "Viscous Effects in Turbomachines", AGARD CP 351, 1983, Paper 16; also VKI Preprint 1983-12.

LIST OF SYMBOLS

A	area (A_{ax} - annular area)
b	spanwise extension of secondary flow region
c	chord
C_ℓ	lift coefficient
CP_0	local total pressure loss coefficient, $\frac{(P_{01,m}-P_{0,\ell})}{(P_{01,m}-\overline{\overline{P}}_{s,2})}$
CP_S	local static pressure coefficient, $\frac{(P_{01,m}-P_{s,\ell})}{(P_{\theta 1,m}-\overline{\overline{P}}_{s,})}$
D	diameter
g	pitch
H	blade height or boundary layer shape factor ($H=\delta^{*}/\theta$)
I	momentum
M	Mach number
P	pressure
P_0	total pressure
P_S	static pressure
Re	Reynolds number based on chord and inlet (outlet) velocity
r	radius
S	distance across blade passage
U	peripheral speed
V	velocity
V_∞	mean velocity (between inlet and outlet)
x	axial direction

Subscripts

a	annulus
ax	axial
cr	critical
FS	free stream
H	hub
is	isentropic
ℓ	local
m	mean (mean blade height or mean diameter)
n	normal to streamwise direction
p	profile
r	radial
s	static
T	tip
1	upstream
2	downstream

Superscripts

$\overline{}$	pitchwise mass averaged value
$\overline{\overline{}}$	pitch- and spanwise mass averaged value

DESCRIPTION OF SECONDARY FLOWS IN RADIAL FLOW MACHINES

R. Van den Braembussche

von Karman Institute for Fluid Dynamics

1. INTRODUCTION

Most of the analysis methods for inviscid flows are based on two dimensional assumptions and assume that the flow is restricted to two types of surfaces [1]. A first one assumes that the flow changes from hub to shroud, but is constant in the circumferential direction (axial symmetric solution). A second one is a surface of revolution on which the conditions vary from one blade to the other, but are constant normal to the surface (blade to blade solution). The streamlines of the first are used to construct the axisymmetric surface for the blade to blade solution.

It has been shown by Vavra [2] that the flow through stationary or rotating cascades, cannot be axisymmetric if the flow must exert a moment on the blades (except in some very special cases). Real streamsurfaces are not surfaces of revolution. Strictly speaking, this difference between quasi 3D and full 3D calculations should not be considered as secondary flows. However, it is worthwhile to have some discussion on it because secondary flows are superposed to it.

Secondary flows are defined as the difference between the full three dimensional inviscid solution and the real flow occurring in a component of the compressor. This departute from the inviscid solution has been observed since 1900 by Smith [3] and Carrard [4] indicating that the flow has a tendency to separate from the blades and to form a jet and wake flow pattern. Carrard [4] studied the real flow by a quasi two dimensional method, assuming a jet and wake region (Fig. 1) and obtained a remarkably good agreement with measured performances (Fig. 2).

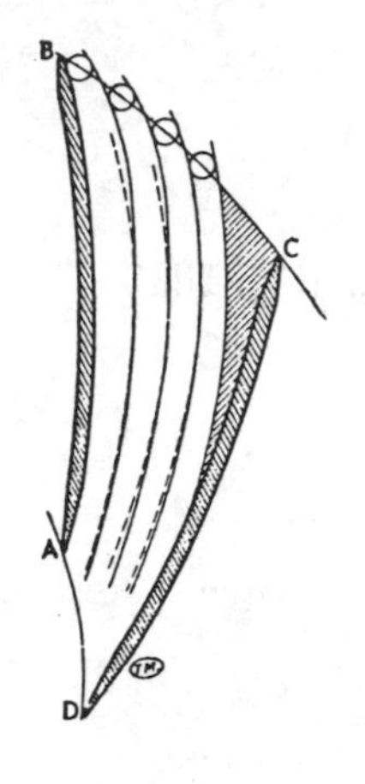

FIG. 1 - FLOW PATTERN [4]

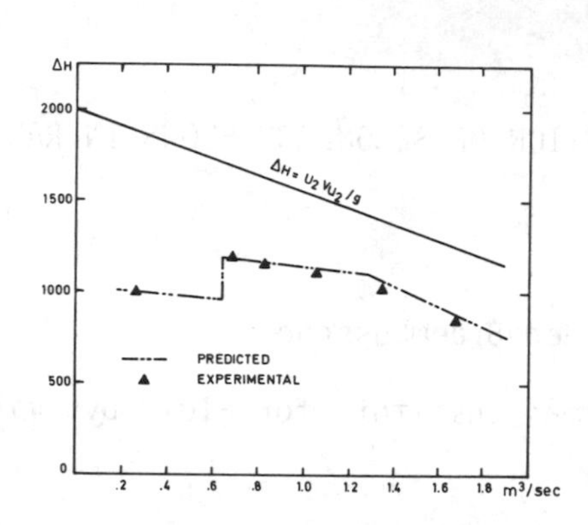

FIG. 2 - PERFORMANCE CURVE [4]

Detailed measurements by Moore [5] revealed that the jet-wake formation is essentially a three dimensional phenomenon governed by streamwise vorticity. The boundary layers are not at all two dimensional but have an important contribution to the transport of low energy fluid to a wake type region.

This three dimensional model is highly perturbed by the leakage flow at the impeller shroud and by the boundary layer on the non rotating shroud. This will be discussed in a separate chapter.

Secondary flows also contribute to the distorted diffuser inlet flow conditions, which in turn are responsible for important downstream perturbations.

2. INVISCID THREE DIMENSIONAL FLOW WITH UNIFORM INLET CONDITIONS

These considerations are made for flows with uniform inlet conditions only, because any vorticity perpendicular to the mean flow direction will result in an additional streamwise vorticity.

The full three dimensional flow in an impeller with radial blades has been calculated by Ellis and Stanitz [6], assuming the fluid is inviscid and incompressible. They have visualized the three dimensional flow by integrating the velocity along the surfaces to obtain the surface streamlines.

Their results in figures 3a,b are for a forced vortex type impeller (pure radial blades) and indicate that the streamlines

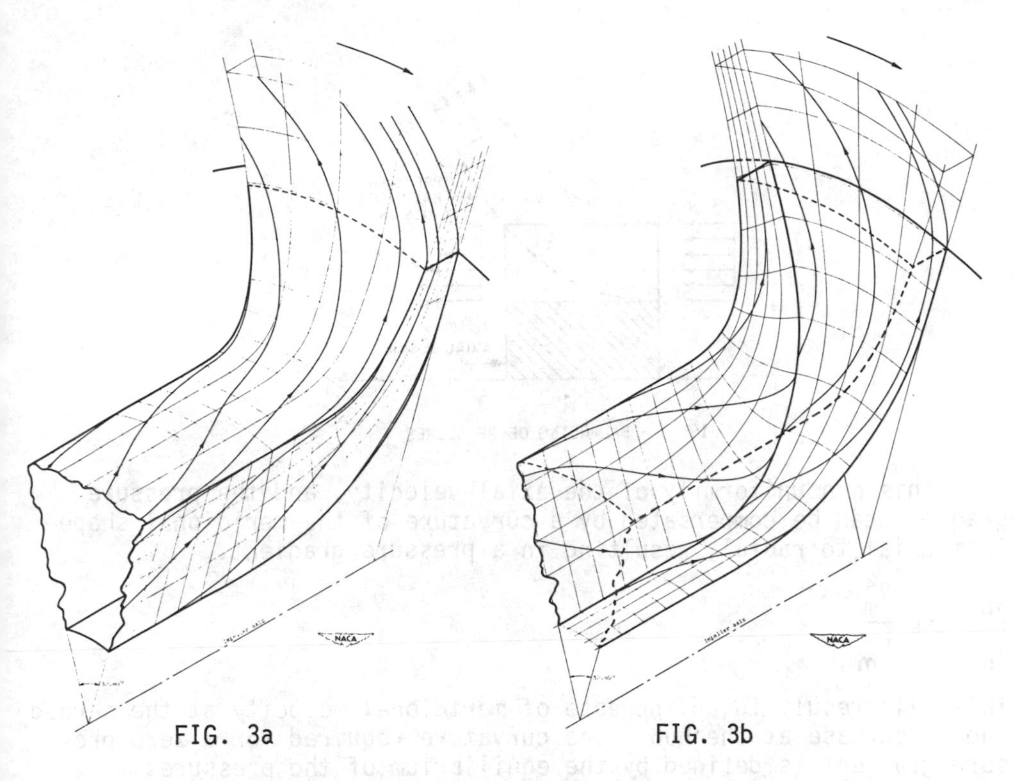

FIG. 3a FIG. 3b

FIG. 3 - THREE DIMENSIONAL INVISCID FLOW PATTERN [5]

shift to the hub on the pressure side, and to the shroud on the suction side. This corresponds to a relative flow rotation, opposite to the impeller rotation. The resulting flow is very different from the quasi 3D solution, but it is generally accepted that the pressure distribution on the blades will not be very different.

The same problem was also studied by Balje [7] using an analytical model in a simplified geometry, to explain how the observed variations of streamlines are function of the existing pressure gradients.

Assume that the flow is turned by the inducer to the axial direction, resulting in a solid body rotation of the fluid. The inducer exit flow is a forced vortex and the hub to shroud distribution of axial velocity can be calculated from the radial equilibrium (Fig. 4). The radial pressure gradient is due to the centrifugal forces and relates to the tangential velocity component

$$\frac{\partial p}{\partial R} = \rho \frac{V_u^2}{R}$$

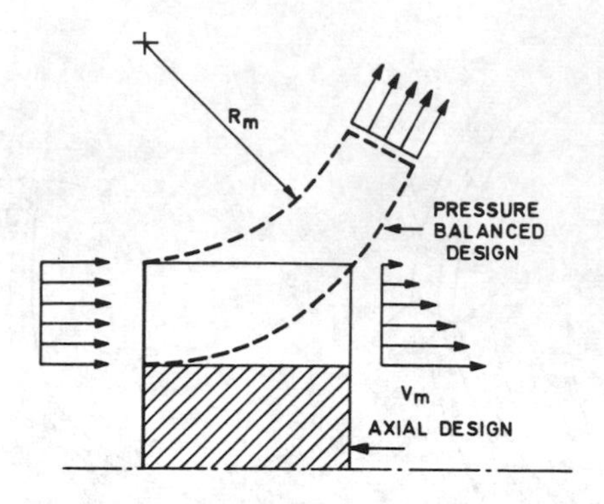

FIG. 4 - BALANCING OF PRESSURES

This non uniformity of the axial velocity, and the pressure gradient can be compensated by a curvature of the meridional shape from axial to radial, resulting in a pressure gradient.

$$\frac{\partial p}{\partial n} = -\rho \frac{V_m^2}{R_m}$$

This will result in an increase of meridional velocity at the shroud and a decrease at the hub. The curvature required for a zero pressure gradient is defined by the equilibrium of the pressures

$$\frac{V_m^2}{R_m} = \frac{V_u^2}{R} \cos\delta$$

This equation can be satisfied for the mean velocity between pressure and suction side. However, V_m is smaller on the pressure

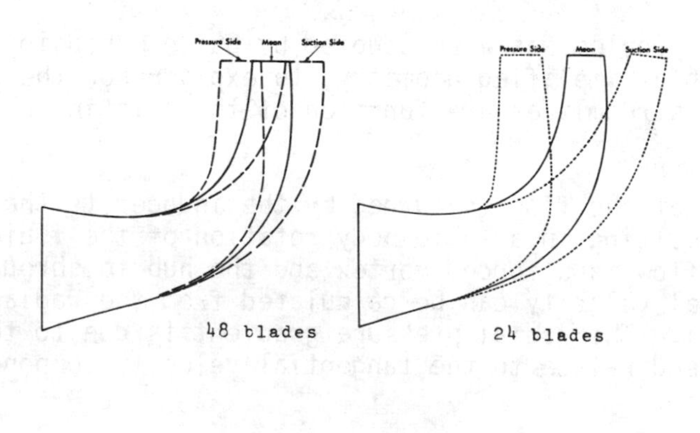

FIG. 5 - ZERO PRESSURE GRADIENT MERIDIONAL FLOW PATH [7]

side than on the suction side, resulting in a different curvature radius. The optimum meridional flow path calculated by Balje [7] is therefore different at pressure and suction side (Fig. 5). The difference depends on the blade number, because this has a direct influence on the suction to pressure side velocity difference.

It is clear that the pressure gradient cannot be zero everywhere and that the three dimensional effects cannot be avoided, but they can be minimized

- by designing the meridional shape in function of the mean velocity variation to obtain a more uniform outlet flow;
- by providing a high blade number in that region where the pressure and suction side streamlines begin to diverge, up to the radial section.

3. SECONDARY FLOW IN A SHROUDED IMPELLER

Insight into the basic phenomena of secondary flow can be gained from experimental results of simplified models, such as the rotating duct of Moore [5]. This model is schematically shown on figure 6 and represents one channel of a centrifugal impeller.

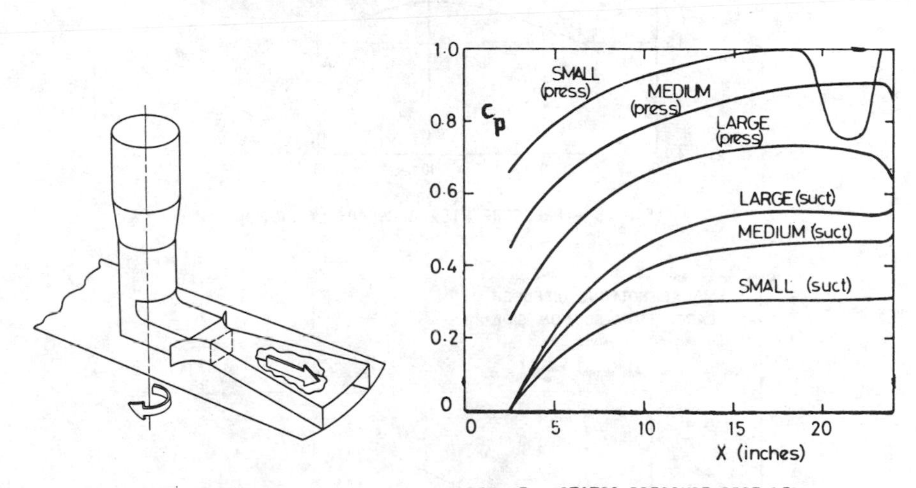

FIG. 6 - ROTATING CHANNEL [5]

FIG. 7 - STATIC PRESSURE RISE [5]

The detailed three dimensional flow at small, medium and large mass flow has been measured by Moore, using hot wire and five hole probes. The main results of this investigation are related to the influence of rotation on the boundary layers, and the influence of the boundary layers on the inviscid flow core.

The total static pressure rise from inlet to outlet is much larger on the pressure side than on the suction side because of the lower velocity induced by the passage vortex (Fig. 7). However, the boundary layer remains attached on the pressure side and separates from the suction side. This is due to the effect of Coriolis forces on the boundary layer turbulence as explained by Johnston [8]. Boundary layers on a suction side or convex surface have a lower turbulence and are more sensitive to flow separation than boundary layers on a pressure side or concave surface who have a larger turbulence intensity. Calculations have shown [9] that the

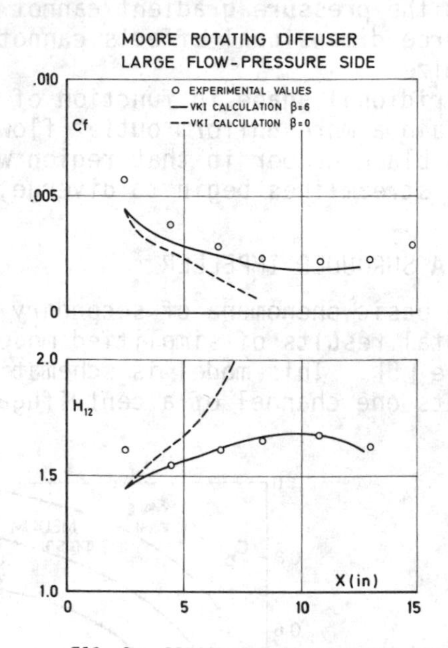

FIG. 8 - PRESSURE SIDE BOUNDARY LAYER [9]

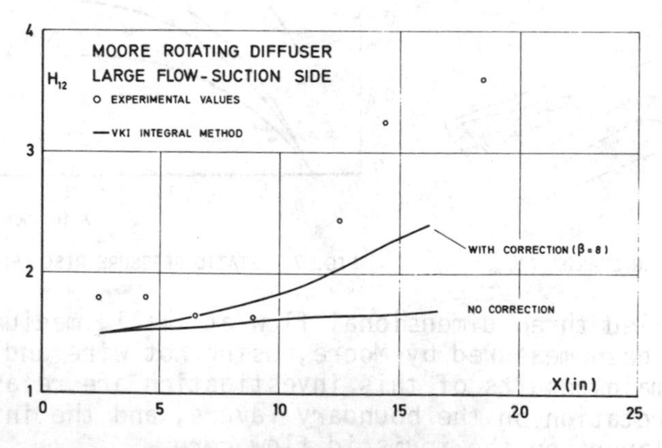

FIG. 9 - SUCTION SIDE BOUNDARY LAYER [9]

pressure side boundary layer can be correctly predicted by a boundary layer method, if one accounts for this change in turbulence. This is possible by a change in dissipation factor or mixing length (Fig. 8). Similar calculations for the suction side do not show the same agreement because the flow is not two dimensional (Fig. 9).

The measured suction side separation is a wake type accumulation of low energy fluid and does not show return flow. Measurements close to the walls reveal a transport of fluid to the suction side along the top and bottom walls, under the influence of the transversal pressure gradient in the channel. This results in a streamwise vorticity with two passage vortices rotating in opposite directions (Fig. 10). The measured vorticity is not symmetric

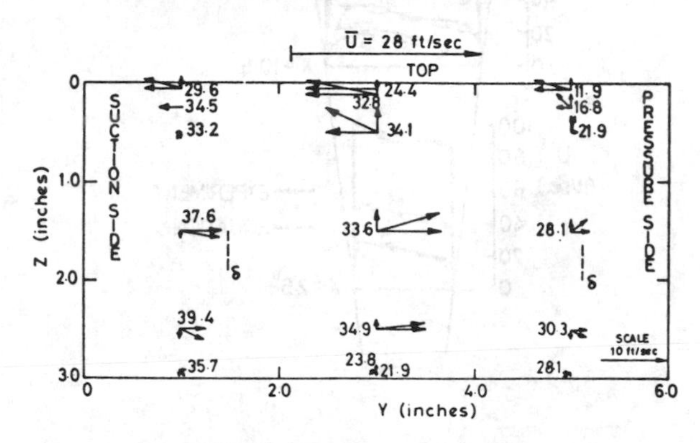

FIG. 10 - SECONDARY FLOW VELOCITY VECTORS [5]

around the mid section because :
- It is superposed to some initial vorticity, already present at the inlet, also when the channel is not rotating.
- A difference in boundary layer thickness on top and bottom wall results in a different vortex strength.
- Also the inviscid flow with uniform inlet conditions will have a velocity component perpendicular to the mean stream direction.
The migration of low energy fluid and the resulting vorticity is the main problem in secondary flows.

The largest influence of the boundary layers on the flow is the displacement and acceleration of the potential core by the suction side wake. The pressure to suction side velocity gradient remains almost unchanged (Fig. 11), but the mean velocity of the potential core near the exit, is 30% higher than those calculated by a potential theory, because of blockage. This explains, why relatively simple models, in which one only accounts for boundary layer and wake blockage [4,10] can give relatively good predictions of the real flow. Secondary flows change the distribution of low energy fluid at the impeller exit through the streamwise vorticity but have a minor influence on the level of the inviscid core velocity and pressure. Losses and performances are integrated values and do not directly depend on the detailed distribution at impeller exit.

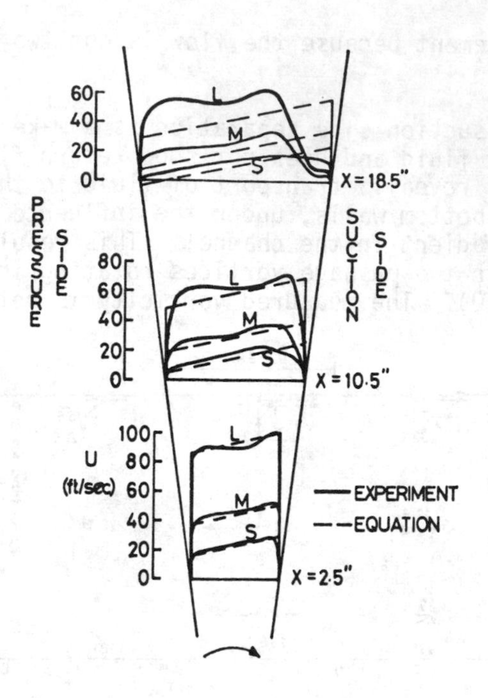

FIG. 11 - MERIDIONAL VELOCITY VARIATION [5]

The equations describing the development of the streamwise vorticity along a relative streamline were derived by Smith [11] but the most practical form is the one of Hawthorne [12]

$$\frac{\partial}{\partial s}\left(\frac{\Omega s}{W}\right) = \frac{2}{\rho W^2}\left(\frac{1}{R_n}\frac{\partial p^\star}{\partial b} + \frac{\omega}{W}\frac{\partial p^\star}{\partial z}\right)$$

where $p^\star$ is the rotary stagnation pressure defined by

$$p^\star = p + \frac{1}{2}\rho\,(W^2-\omega^2R^2)$$

s, n, b are the streamwise, normal and binormal directions defined by figure 12.

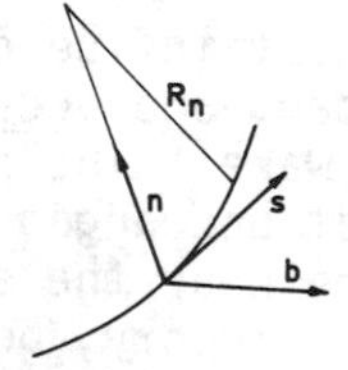

FIG. 12 -

The Hawthorne equation shows two contributions to the streamwise vorticity. The first one expresses the generation of vorticity due to curvature when there is a gradient of rotary stagnation pressure in the binormal direction. The second one expresses the generation of streamwise vorticity due to rotation when there is a gradient of rotary stagnation pressure in the axial direction.

Rotary stagnation pressure is constant along a streamline in isentropic flow and reduces because of losses. Low p^* values and gradients are therefore found in the boundary layers. Although the existence of secondary flows is associated with boundary layers in the impeller, many theoretical studies use an inviscid flow theory. The gradient of p^* is then introduced on the walls at the inlet. Because these gradients are conserved throughout the impeller, it allows to investigate secondary flows with an inviscid model.

Three contributions to the secondary flow can be observed in a centrifugal impeller. In the inducer the relative flow is turned from the inlet direction to almost axial. The low p^* fluid of the shroud and hub boundary layers migrates to the suction surface, because of blade curvature. The low p^* fluid on the pressure side is unstable and migrates to the hub and shroud. The resulting

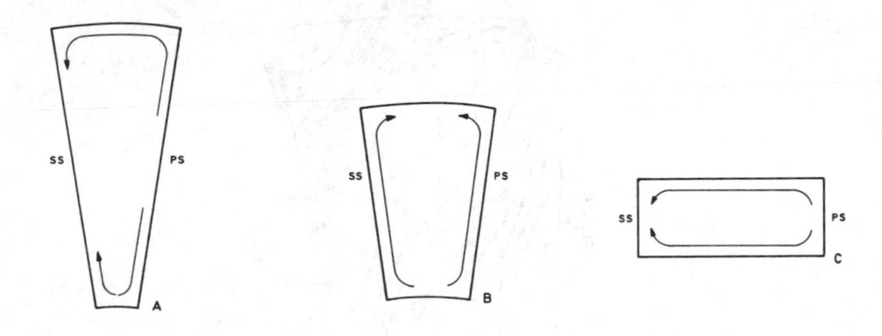

FIGS. 13a,b,c - COMPONENTS OF SECONDARY FLOW

vorticity is schematically shown in figure 13a. In the axial to radial bend, low p^* fluid moves along the pressure and suction side to the shroud. The resulting vorticity is shown in figure 13b. The impeller rotation contributes to the migration of low p^* fluid from hub and shroud to the blade suction side, as shown in figure 13c.

The secondary flow has been measured by Johnson and Moore [13] in a centrifugal impeller (Fig. 14). The relative velocity at section 3 is influenced by the blade curvature in the inducer and the axial to radial bend. The expected secondary flows are a combination of the vortices shown in figures 13a and b. The measured secondary flow (Fig. 15a) shows a strong anti-clockwise vorticity and a weaker clockwise vorticity in the hub-pressure side corner. This clockwise vorticity is smaller because up to section 3 there is almost no diffusion on the hub and pressure side walls, resulting in thinner boundary layers. At section 5, close to the impeller exit, the vorticity due to rotation (Fig. 13c) is predominant as

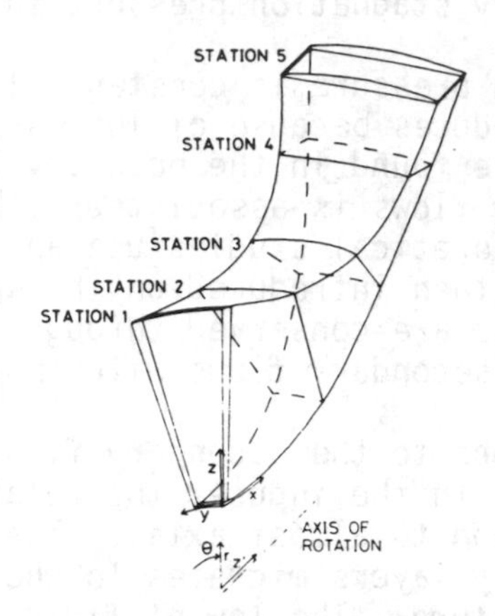

FIG. 14 - IMPELLER GEOMETRY [13]

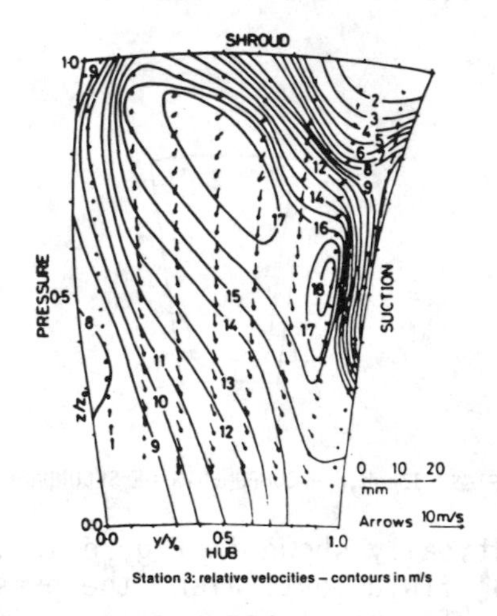

FIG. 15a - SECONDARY FLOW VELOCITY VECTORS [13]

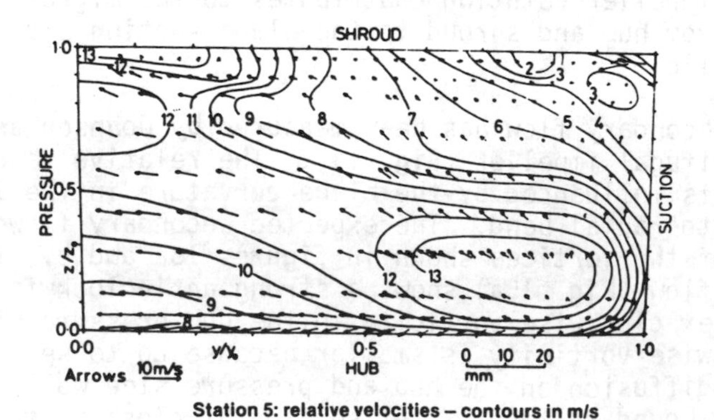

FIG. 15b - SECONDARY FLOW VELOCITY VECTORS [13]

shown by the measurements (Fig. 15b). The counter-clockwise vorticity at the hub is smaller for the same reason as mentioned before. The clockwise vorticity at the shroud has resulted in a region of low energy fluid (2.-3. m/s) in the shroud suction side corner, which corresponds to the wake. Outside the wake, the velocity is almost constant (10-13 m/s).

The location of the wake is defined by the relative importance of the vorticity due to the axial to radial turning (Fig. 13b) and the vorticity due to rotation (Fig. 13c). This is expressed by the Rossby number

$$R_0 = \frac{W}{\omega R_n}$$

which is nothing else that the ratio of the two terms in the Hawthorn equation (assuming that the gradients of p^* are equal).

Low Rossby numbers correspond to important coriolis forces and low meridional curvature. According to figure 13c the wake can be expected on the blade suction side. High Rossby numbers correspond to small coriolis forces and strong meridional curvature. The wake can be expected on the shroud as shown in figure 13b.

An increase in rotational speed (decrease of R_0) moves the wake location towards the suction side. An increase in mass flow (increase of R_0) moves the wake location towards the shroud. This is shown by figures 16a and 16b obtained from [14]. In Carrard's impeller [4], the Rossby number is zero because there is no meridional curvature and the wake region is on the suction side. This explains the good prediction obtained with a two dimensional model.

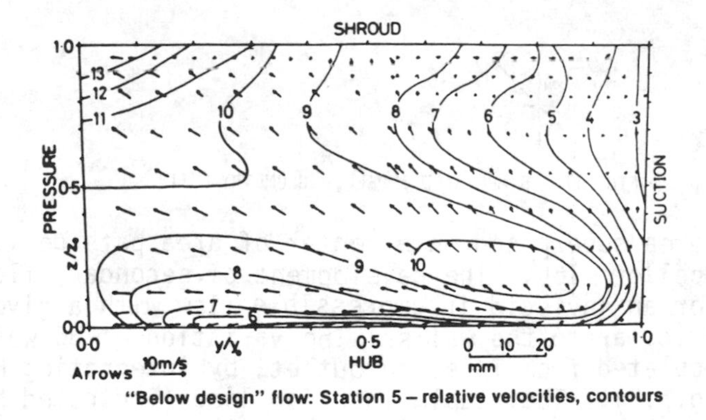

FIG. 16a - SECONDARY FLOW VELOCITY VECTORS AT LOW MASS FLOW [14]

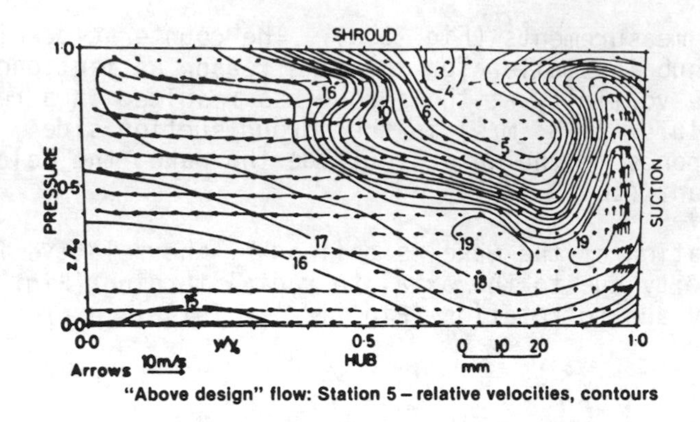

FIG. 16b - SECONDARY FLOW VELOCITY VECTORS AT HIGH MASS FLOW [14]

The Hawthorn equation has been used by Johnson [15] to calculate the streamwise vorticity in the fictitious impeller shown in figure 17. This impeller is constructed with elliptical sections

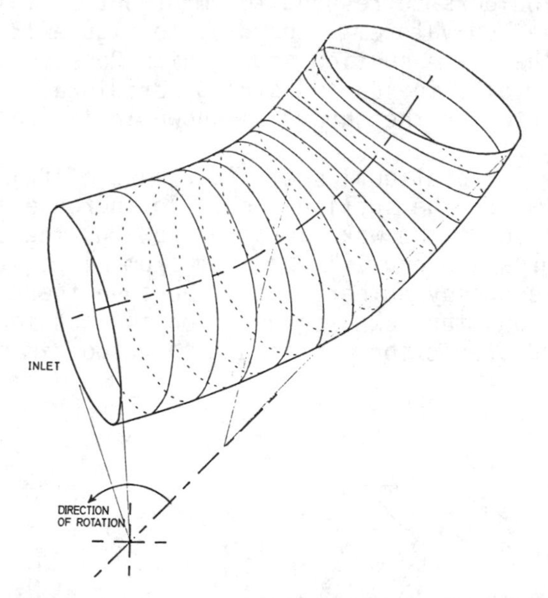

FIG. 17 - SIMPLIFIED IMPELLER GEOMETRY [15]

of constant area superposed on a center of area path derived from Eckardt's impeller [16]. The development of secondary flow is calculated for an inviscid incompressible flow with a given p^* gradient perpendicular to the walls. The variation of 36 wall streamlines is calculated from inlet to outlet, by integrating Hawthorn's equation (Fig. 18). The stable location line (indicated by a dashed line) is the locus of points where the streamwise vorticity generation is zero, and where the low p^* fluid tends to accumulate.

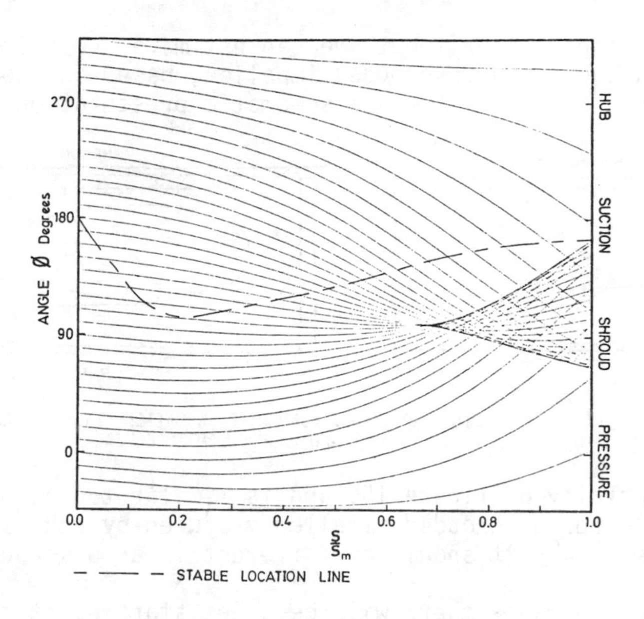

FIG. 18 - WALL STREAMLINES IN ROTATING BEND [15]

This stable location changes from the suction side at the inlet to the shroud at S/S_{max} = .20 and back to the suction side at the exit, where the coriolis forces become predominant. The full lines indicate the migration of low p* fluid, from the initial position in the direction of the stable location. The slope of the curves at the inlet is due to the difference in tangential velocity, between hub and shroud. In the relative frame, this results in an initial pressure side, hub, suction side, shroud rotation.

The suction side streamlines do not converge on the stable location line but cross it and converge closer to the shroud. This is due to the absence of any viscosity in the model so that the flow continues by inertia until opposite secondary flow forces call it back to the stable position. The predicted wake position therefore does not coincide with the stable position.

Downstream S/S_{max} = .7, the secondary flow streamlines cross each other. In reality, there will be a sharp turn when the fluid collides and accumulation of low p* fluid in the wake. The crossing streamlines downstream S/S_{max} = .7 are therefore a good presentation of the wake.

4. SECONDARY FLOW IN AN UNSHROUDED IMPELLER

As already discussed, the vorticity and wake location changes with mass flow, geometry and RPM as illustrated by figures 16a,b. It is therefore not possible to draw a general picture of the vorticity at impeller exit but one can estimate the difference between a shrouded and unshrouded impeller, based on theoretical considerations. Figure 19a is a schematic presentation of the

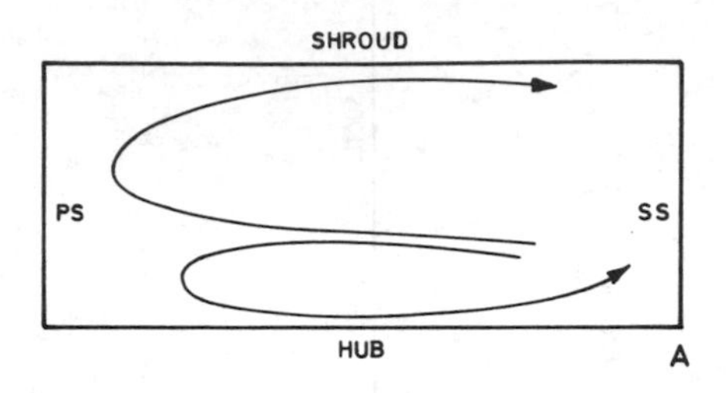

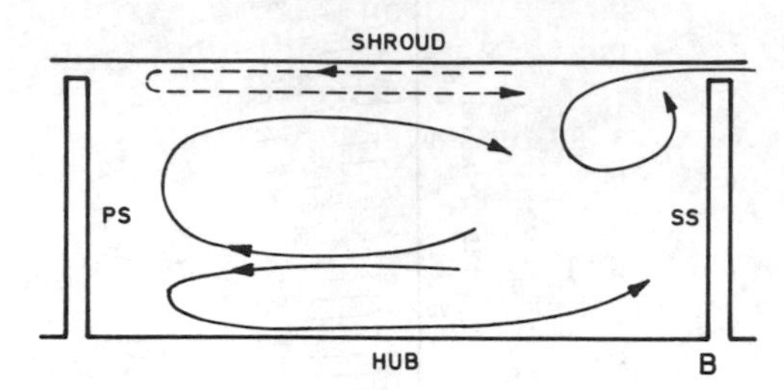

FIGS. 19a,b - SECONDARY FLOW PATTERN IN SHROUDED AND UNSHROUDED IMPELLER

measured vorticity of figure 15b and is similar to the one derived by Ellis [17] for a shrouded impeller measured by Ash [18]. There are two reasons why it should be different in an unshrouded impeller (Fig. 19b) :

- Due to tip clearance there will be a jet starting at the suction-shroud corner which blows the wake into the pressure side direction along the shroud. This jet is counteracted by the shroud passage vortex and therefore rolls-up in the shroud suction side corner.
- Next to the shroud, the fluid in the boundary layer has been slowed down, in the absolute frame, by the shear with the shroud. In the relative frame this results in a movement of the fluid from suction to pressure side, opposite to the shroud passage vortex. Some additional vorticity is therefore expected close to the shroud.

The secondary flow has been measured by Eckardt and Krain [19] in an unshrouded impeller with a 30° backward lean angle. The β distribution at impeller exit (Fig. 20) is not measured close enough to the shroud wall to see this additional vorticity. However, an extrapolation of β suggests that close to the wall the fluid is moving from suction to pressure side as indicated in figure 19b.

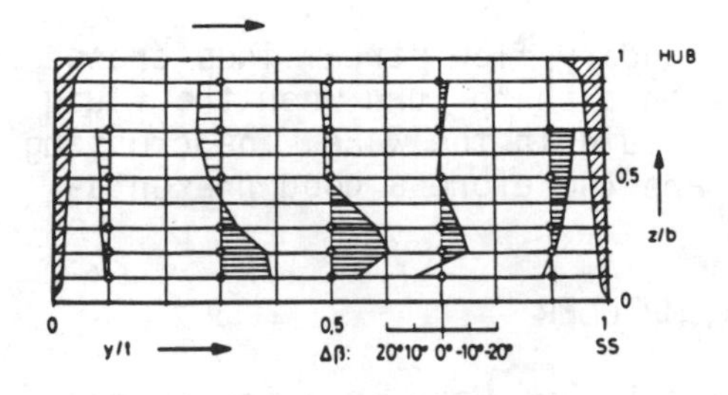

FIG. 20 - RELATIVE FLOW ANGLE DISTRIBUTION [17]

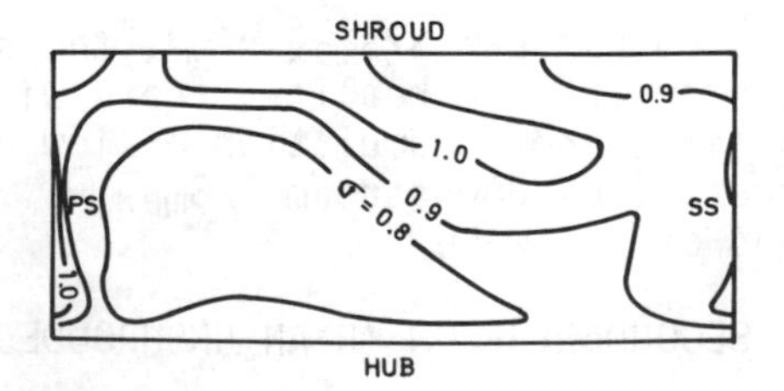

FIG. 21 - SLIP FACTOR DISTRIBUTION [13]

An interesting result of Eckardt and Krain is the variation of the slip factor at the exit of a radial bladed impeller (Fig. 21). Due to the secondary flows, the slip factor locally increases to 1 in the shroud region. This corresponds to a higher work output at the shroud and is reflected by a larger tangential velocity V_u at diffuser inlet. The slip factor predicted by correlations corresponds to a mass average value over the passage and does not predict any variation of V_u at the diffuser inlet.

5. SECONDARY FLOW IN DIFFUSERS

The detailed secondary flow distribution at the impeller exit has an important influence on diffuser performances because they define the inlet conditions. The tangential variations of the impeller outlet flow result in diffuser inlet mixing losses. Axial variations result in three dimensional boundary layers. The available calculation models are basically two dimensional and suppose a uniform flow in the third direction. They assume that the stream surfaces remain perpendicular to the axis of rotation (or close to it).

It has been shown by Ellis [17] and Rebernick [20] that axial variations of the diffuser inlet conditions can cause an important shift of streamlines between the diffuser walls resulting in local flow separation. In this context flow separation does not mean a reversal of the total velocity but a change of sign of the radial velocity component.

Figure 22a shows the axial variation of diffuser inlet flow conditions in function of mass flow as measured by Rebernick in a centrifugal pump. Figure 22b shows the corresponding flow path in a vaneless diffuser, illustrating how small changes in difuser inlet conditions can result in drastic changes of the diffuser flow.

These variations in flow pattern are explained by Rebernick by means of the equation of motion for steady, axisymmetric, incompressible and inviscid flow.

$$\frac{\partial V_R}{\partial z} - \frac{\partial V_z}{\partial R} = \frac{1}{V_R}\frac{\partial P_0/\rho}{\partial z} - \frac{1}{V_R}\cdot\frac{RV_u}{R^2}\cdot\frac{\partial(RV_u)}{\partial z}$$

Assuming that the second term of the LHS is negligible compared to the first one, it may be recognized that a positive pressure gradient will cause a positive gradient of the radial velocity. A total pressure, increasing towards one side, will cause the flow to go to the same side. The second part of the RHS has an opposite sign. An increase of RV_u towards one side will push the flow away from that side.

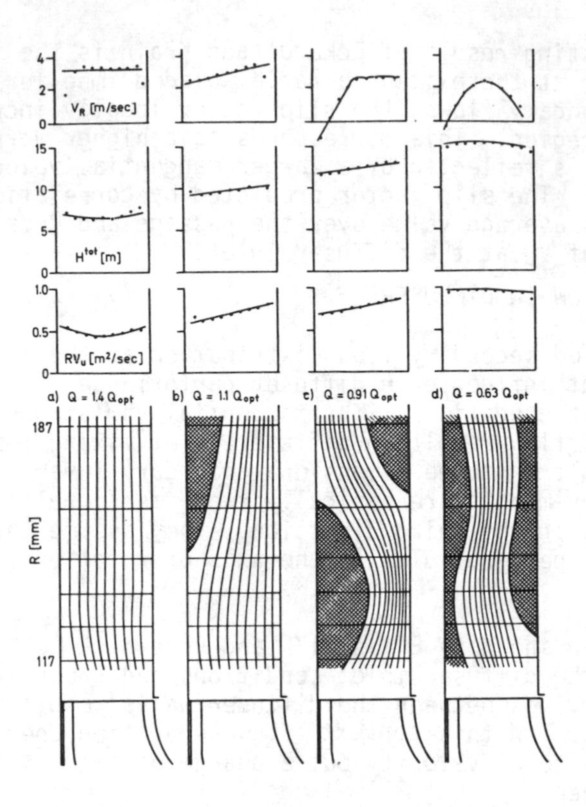

FIGS. 22a,b - DIFFUSER INLET CONDITIONS AND FLOW PATTERN [20]

At Q = 1.4 Q_{opt} the effect of the pressure gradient is stronger than the momentum gradient, and the flow remains attached to both side walls. At Q = 1.1 Q_{opt}, the gradient of angular momentum and pressure rise push the flow in opposite directions. At increasing radius, the pressure gradient term becomes dominant and the streamlines shift to the shroud side, causing flow separation at the hub. At Q =.63 Q_{opt}, the gradients are opposite to the ones at Q = 1.4 Q_{opt} and the streamlines shift to the center, producing separation on both sides. At Q = .91 Q_{opt}, the flow is already separated at the diffuser inlet and this separation is amplified by the total pressure gradient similar to Q = 1.16 Q_{opt}. However, at larger radius ratios the flow reattaches at the hub and separates from the shroud. This can only be explained by a change of momentum and pressure gradient.

On the walls where the flow is attached, the total pressure and momentum decreases rapidly because of friction. In the separated flow region total pressure and momentum remains almost constant. This results in a complete change in pressure and momentum gradient downstream in the diffuser as shown in figure 23. The downstream pressure gradient pushes the flow from the shroud to the hub.

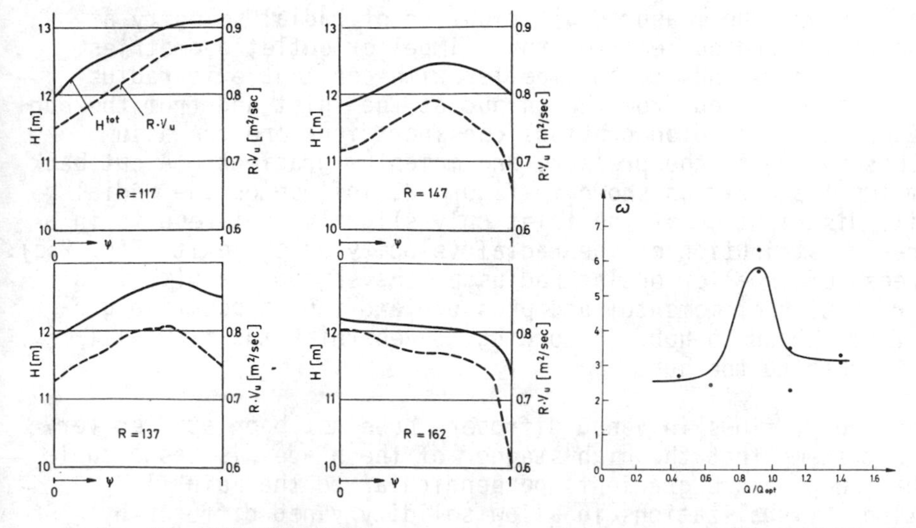

FIG. 23 - VARIATION OF FLOW CONDITIONS [20]

FIG. 24 - VARIATION OF LOSSES [20]

Rebernick was able to recalculate the measured flow pattern by means of his inviscid model. However, at $Q = .91\ Q_{opt}$ agreement could only be obtained if a radial variation of momentum and pressure gradient was introduced.

The different types of flow in the diffuser as shown in figure 22, result in a non linear change of diffuser total pressure losses (Fig. 24). A similar variation was also observed by Ellis [17] who claims that the influence of separation on diffuser losses is strongly dependent on the location where separation occurs. Local discontinuities in diffuser losses can result in a local positive slope of the compressor performance curve and limit the flow range because of surge.

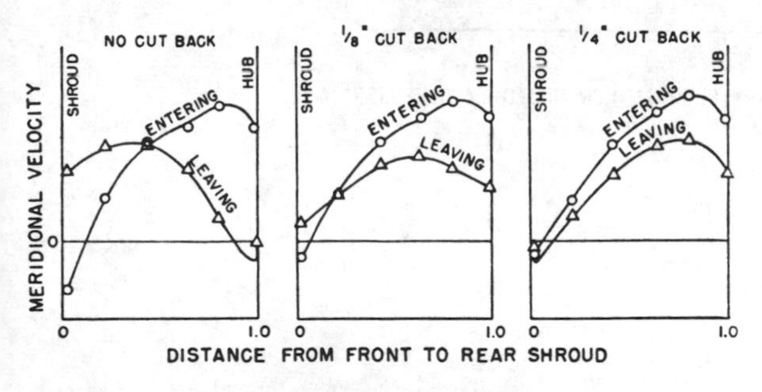

FIG. 25 - INFLUENCE OF OUTLET GEOMETRY

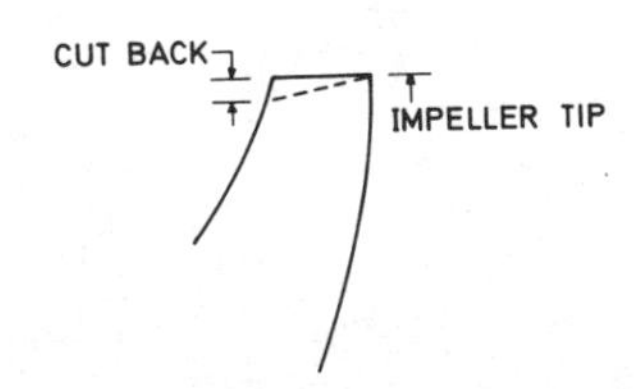

FIG. 26 - DEFINITION OF CUT-BACK

Secondary flows in vaneless diffusers can be influenced by modifying the impeller outlet geometry as demonstrated by Ellis [17].

Figure 25 shows the measured distribution of radial velocity at diffuser inlet and outlet, for three impeller outlet geometries. Figure 25a corresponds to an impeller with constant exit radius. The flow is separated from the shroud at the inlet and from the hub at the outlet. Based on previous considerations one can assume that this is due to the pressure and momentum gradient. A cut back of the impeller exit as shown in figure 26 influences the radial velocity distribution at the inlet only slightly, but results in a different distribution of the radial velocity at the exit (Fig. 25c). A decrease of impeller outlet radius at the shroud, results in a local reduction of momentum and pressure and thus a positive gradient from shroud to hub. According to Rebernick the flow will stay attached to the hub.

Secondary flows in vaned diffusers have not been studied very much. It seems that the high stagger of the blades is responsable for the low pressure gradients perpendicular to the main flow direction. Visualizations in a low solidity vaned diffuser by Senoo [21] shows that this can result in a negative secondary flow in which the low energy fluid flows from suction side to pressure side (Fig. 27). The suction side boundary layer becomes very thin and flow separation is delayed. It is possible that the low energy fluid, accumulated on the vane pressure side, is responable for diffuser stall.

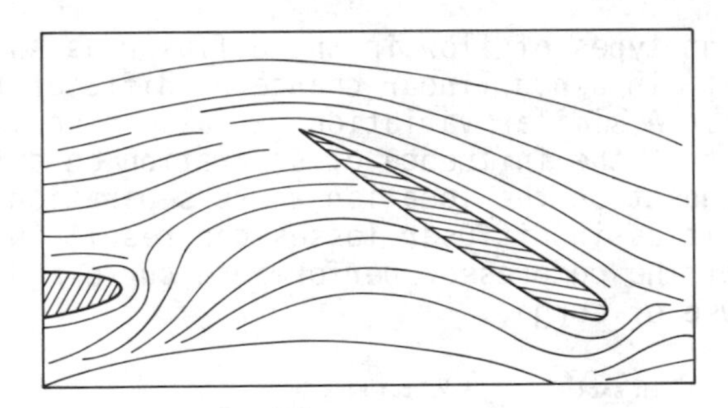

FIG. 27 - SECONDARY FLOW IN LOW SOLIDITY DIFFUSER [21]

REFERENCES

1. ADLER, D.: Status of centrifugal impeller internal aerodynamics. Part I - Inviscid flow prediction methods. ASME Trans., Series A - J. Engineering Power, Vol. 102, No. 3, July 1980, pp 728-737.
2. VAVRA, M.: Aero-thermodynamics and flow in turbomachines. Krieger Publ. Co. Inc., 1974.
3. SMITH, J.H.: Notes on some experimental researches on internal flow in centrifugal pumps and allied machines. Engineering, Vol. LXXIV, 1902, p 763.
4. CARRARD, A.: Sur le calcul des roues centrifuges. La Technique Moderne, Tome XV, N° 3, 1923, pp 65-71; Tome XV, No. 4, 1923, pp 100-104. Translated by J. Moore, Cambridge University Dept. of Engineering, Report No. CUED/A-TURBO/TR 73, 1975
5. MOORE, J.: A wake and an eddy in a rotating, radial flow passage. Part 1 - Experimental observations. Trans. ASME, Series A - J. Engineering Power, Vol. 95, No. 3, July 1973, pp 205-212.
6. ELLIS, G.O. & STANITZ, J.D.: Comparison of two and three dimensional potential-flow solutions in a rotating impeller passage. NACA TN 2806, 1952.
7. BALJE, O.E.: Loss and flow path studies on centrifugal compressors - Part II. ASME Paper 70-GT-12b, 1970.
8. JOHNSTON, J.P.: The effects of rotation on boundary layers in turbomachine rotors. NASA SP 304, Part 1, pp 207-250.
9. VAN DEN BRAEMBUSSCHE, R. & ZUNINO, P.: Corrections for streamline curvature and coriolis force in a boundary layer integral method. ASME Paper 81-GT-97, also VKI Preprint 1980-22.
10. CIPOLLONE, R.: Viscous flow calculations in a centrifugal impeller. VKR PR 1984-24.
11. SMITH, A.G.: On the generation of a streamwise component of vorticity in a rotating passage. Aero. Quart., Vol. 8, 1957, pp 369-383.
12. HAWTHORNE, W.R.: Secondary vorticity in stratified compressible fluids in rotating systems. Cambridge University, Dept. of Engineering, Report N° CUED/A Turbo/TR 63, 1974.
13. JOHNSON, M.W. & MOORE, J.: Secondary flow mixing losses in a centrifugal impeller. ASME Trans., Series A - J. Engineering Power, Vol. 105, No. 1, Jan. 1983, pp 24-32.
14. JOHNSON, M. W. & MOORE, J.: The influence of flow rate on the wake in a centrifugal impeller. ASME Trans., Series A - J. Engineering Power, Vol. 105, No. 1, Jan. 1982, pp 33-39.
15. JOHNSON, M.W.: Secondary flow in rotating bends. Cambridge U., Dept. of Engineering, Report N° CUED/A-Turbo/TR 92, 1978.
16. MOORE, J.: Eckardt's impeller - a ghost from ages past. Cambridge U., Dept. of Engineering, Report N° CUED/A-Turbo/TR 83, 1976.

17. ELLIS, G.O.: A study of induced vorticity in centrifugal compressors. ASME Trans., Series A - J. Engineering Power, Vol. 86, No. 1, Jan. 1964, pp 63-76.
18. ASH, J.E.: Measurements of relative flow distributions in mixed flow impellers. ASME Paper 56-1205, 1956.
19. ECKARDT, D. & KRAIN, H.: Secondary flow studies in high-speed centrifugal compressor impellers. In: "Secondary Flows in Turbomachines", AGARD CP 214, 1977, paper 18.
20. REBERNICK, B.: Investigation on induced vorticity in vaneless diffusers of radial flow pumps. Proc. 4th Conf. on Fluid Machinery, Budapest 11-14 Setp. 1972.
21. SENOO, Y.; HAYAMI, H.; UEKI, H.: Low solidity tandem cascade diffusers for wide flow range centrifugal blowers. ASME Paper 83-GT-3.

LIST OF SYMBOLS

b	binormal direction coordinate channel width
C_p	static pressure rise coefficient
H	total manometric head
n	normal direction coordinate
p	static pressure
P_0	total pressure
p^*	rotary stagnation pressure
Q	volume flow rate
R	radius
R_0	Rossby number
s	streamwise direction coordinate
V	absolute velocity
W	relative velocity
z	axial direction coordinate
δ	angle between meridional direction and axis of rotation
σ	slip factor
ω	rotational speed
Ω_s	streamwise vorticity
$\overline{\omega}$	diffuser losses

Subscripts

m	meridional component
max	maximum value
n	normal component
opt	optimum value
R	radial component
u	peripheral component
z	axial component

END-WALL BOUNDARY LAYER CALCULATIONS IN MULTISTAGE AXIAL COMPRESSORS

J. De Ruyck, Ch. Hirsch

Vrije Universiteit Brussel, Department of Fluid Mechanics, Belgium

INTRODUCTION

Three main components can be distinguished in the flow passing through a blade row. A main component 'far' from blades and end-walls is in general considered as an inviscid part of the flow. The flows 'close' to the blade surfaces and to the end-walls are subject to viscous effects since the presence of the blades and the end-walls both give rise to boundary layers which moreover present a high three-dimensional character. A direct consequence of the presence of boundary layers is a blockage effect phich can rise to 10 or even 20 % of the passage area. The three-dimensional character of the end-wall boundary layers (EWBL) is moreover shown to have an influence on the energy exchange and efficiency of the machine [1].

A current approach to solve for these three distinct parts of the flow consists in the splitting of the complete flow into blade-to-blade flow and meridional through flow. The meridional through flow is obtained by considering an averaged blade-to-blade flow in the meridional plane. In this plane radial equilibrium is to be expressed and the EWBL are to be taken into account. Because of the averaging of the blade-to-blade flow, the through flow approach consists in an <u>axisymmetric approach</u> of the turbomachine flow.

The present chapter illustrates a method for the determination of the EWBL effects [13]. This is done through a three-dimensional integral boundary layer approach, where a special attention is given to the blade-to-blade averaging procedure. Blade force variations inside the EWBL are shown to have important effects and are taken into account in a semi-empirical way. Since the present method is based on an integral boundary layer approach, it delivers integral boundary layer thicknesses only. Spanwise velocity profiles are however reconstructed from these integral parameters through the introduction of velocity profile models, allowing a complete through flow prediction, including the EWBL regions.

1. SUMMARY OF THE EWBL THEORY

The present method originates in Mellor and Wood's approach [2] where rigorous integral boundary layer equations were written for the end-wall boundary layers, including force defect thicknesses. In this work, simplified assumptions such as constant shape factor or skin friction were made and the equations were integrated from inlet to outlet of a blade row. This approach was extended by Hirsch [3] to include simple velocity profile models and further developed by the introduction of more sophesticated velocity profile models and improved correlations for the defect forces by De Ruyck and Hirsch [4]. Although the results presented in these papers were predicting correctly most of the phenomena associated with secondary flow and EWBL effects, detailed comparisons with experimental velocity profiles indicated some discrepancies, in particular in presence of tip clearances. An improved correlation for the force defect thicknesses has been introduced and discussed in the Ph.D. thesis of De Ruyck [5], where a detailed description of the whole theory is presented.

In the present section a summary is given of the equations which are used for all the applications. These equations are written in the meridional coordinate system m,n,u (fig 1). 'c' denotes an absolute velocity and 'w' a relative velocity component.

The boundary layer momentum and entrainment equations are :

$$\frac{d}{dm}\,\rho r c_m^2\,\theta_{mm} + \rho r c_m\,\delta_m^*\,\frac{dc_m}{dm} - \rho c_m^2\,\frac{dr}{dm}\,(\delta_u^*\,\mathrm{tg}\alpha + \theta_{uu}) = r\tau_m + rF_m \tag{1}$$

$$\frac{d}{dm}\,\rho r c_m^2\,\theta_{um} + \rho r c_m\,\delta_m^*\,\frac{dc_u}{dm} + \rho c_m^2\,\frac{dr}{dm}\,(\delta_m^*\,\mathrm{tg}\alpha + \theta_{um}) = r\tau_u + rF_u \tag{2}$$

$$\frac{1}{\rho r c_m}\,\frac{d}{dm}\,\rho r c_m(\delta - \delta_m^*) = \frac{E(H_k^*)}{\cos\alpha} \tag{3}$$

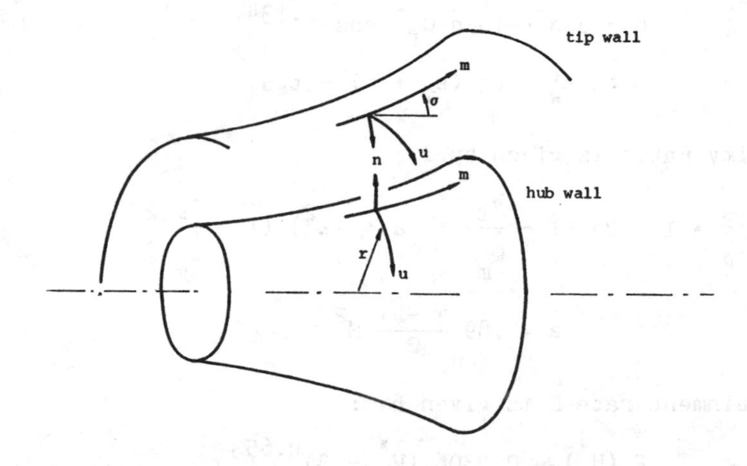

Figure 1 : Meridional coordinate system m,n,u

Equations 1 to 3 are written in the absolute reference system. All the variables are pitch-averaged. The boundary layer thicknesses in these equations are defined as :

$$\theta_{ij} = \int_0^{\delta} \left(\frac{\hat{w}_i}{\hat{w}_m} - \frac{w_i}{\hat{w}_m} \right) \frac{\rho w_j}{\hat{\rho}\hat{w}_m} \, dn \tag{4}$$

$$\delta_i^* = \int_0^{\delta} \left(\frac{\hat{\rho}\hat{w}_i}{\hat{\rho}\hat{w}_m} - \frac{\rho w_i}{\hat{\rho}\hat{w}_m} \right) dn \tag{5}$$

These thicknesses are defined in the wall reference system. The velocities with an overhead carat (^) denote the 'outer' or 'inviscid' reference velocities. Corrections for variations of $\hat{w}$ with r are not considered in eqs 1 and 2. The velocity ratios are modelled through :

$$\frac{w_m}{\hat{w}_m} = 1 - b \left(1 - \frac{y}{\delta}\right)^n \tag{6}$$

$$\frac{w_u - w_m \mathrm{tg}\alpha}{\hat{w}_m} = (1 - b) \, \mathrm{tg} \, \varepsilon_w' \left(1 - \frac{y}{\delta}\right)^n \tag{7}$$

where

$$b = \exp(-10\, n\, C_f^{.5} \cos\alpha^{-.134}) \tag{8}$$

$$tg\, \varepsilon'_w = tg\, (\varepsilon_w + \alpha) - tg\alpha \tag{9}$$

The density ratio is given by :

$$\frac{\rho}{\hat{\rho}} = 1 - 2a\, (1 - \frac{w_m}{\hat{w}_m}) + (a + 4a^2)\, (1 - \frac{w_m}{\hat{w}_m})^2 \tag{10}$$

$$a = .89\, \frac{\gamma - 1}{2}\, M^2 \tag{11}$$

The entrainment rate E is given by :

$$E\, (H_k^*) = 0.0306\, (H_k^* - 3)^{0.653} \tag{12}$$

$$H_k^* = \frac{\delta - \delta_{mk}^*}{\theta_{mmk}} \tag{13}$$

The mainstream and cross flow shear stresses are obtained from :

$$\tau_s = C_f\, \frac{\rho \hat{w}_s^2}{2} \tag{14}$$

$$\tau_t = \tau_s\, tg\varepsilon_w \tag{15}$$

where :

$$C_f = 0.246\, Re_{\theta_{mm}}^{-0.268}\, (\frac{\hat{T}}{T^*})\, \exp\, (-\, 1.56\, H_k) \tag{16}$$

$$H_k = \frac{\delta_{mk}^*}{\theta_{mmk}}$$

$$\frac{T^*}{\hat{T}} = 1 + 0.72\, a \tag{17}$$

The 'kinematic' thicknesses δ_{mk}^* and θ_{mmk} are defined as

$$\delta_{mk}^* = \int_0^{\delta} (1 - \frac{w_m}{\hat{w}_m})\, dn \tag{18}$$

$$\theta_{mmk} = \int_o^{\delta} \left(1 - \frac{w_m}{\hat{w}_m}\right) \frac{w_m}{\hat{w}_m} \, dn \tag{19}$$

Equations (4) to (19) are written in the wall reference system. The thicknesses θ_{mm}, θ_{um}, δ_m^* and δ_u^* and the parameters b and n have the same values in both absolute and relative systems whereas the wall skewing angles and θ_{uu} are related by the jump relations

$$(tg\varepsilon'_w)^{abs} = (tg\varepsilon'_w)^{rel} + \frac{1}{1-b}\left(tg\alpha^{rel} - tg\alpha^{abs}\right)$$

$$\theta_{uu}^{abs} = \theta_{uu}^{rel} + U\,\delta_u^* / \hat{c}_m \tag{20}$$

The blade mainstream defect force is found from :

$$\frac{F_s}{\rho} = L \frac{\hat{w}_s^2}{2\cos\gamma} \sigma C_L^2 \tag{21}$$

$$L \approx .01 \tag{22}$$

The transverse lift defect is found from :

$$\frac{F_t}{\rho} = -\frac{k}{s} w_s^2 \theta_{tm} \tag{23}$$

in absence of a tip clearance and from :

$$\frac{f_y}{\rho} = \frac{t_c}{\cos\alpha} w_m \frac{dw_u}{dm} - \frac{k'}{s} w_s^2 \theta_{tm} \tag{24}$$

in presence of a tip clearance. The constants k and k' are given by :

$$k \approx 3. \qquad k' \approx .5 \tag{25}$$

The cross flow thickness is defined through :

$$\theta_{tm} = -\cos\alpha \int_o^{\delta} \frac{w_t}{\hat{w}_s} \frac{w_m}{\hat{w}_s} \, dn$$

$$\theta_{tm} = (\theta_{um} - \theta_{mm}\,tg\alpha)\cos^2\alpha \tag{26}$$

Equations (21) to (26) are written in the <u>blade reference</u> system.

The basic equations (1), (2) and (3) are integrated in the meridional direction using a fourth order Runge-Kutta method (Merson) with as complementary relations equations (4) to (26). The values for k, k' and L are obtained as the best overall values from a fit to the large number of applications discussed in the following.

2. EWBL EFFECTS ON THE OVERALL PERFORMANCE

The important impact of the end-wall boundary layers on the overall machine efficiency has been demonstrated in the early papers of Smith [1] and Mellor and Wood [2]. According to De Ruyck [5], the stage efficiency can be approached by

$$\eta = \hat{\eta}\, \eta_{bl} \tag{27}$$

where :

$$\hat{\eta} = \frac{Q\, \overline{\Delta\hat{p}_t}}{\hat{p}}$$

$$\eta_{bl} = \frac{1 - \sum\limits_{h,t} \delta^{**}/h}{1 - \sum\limits_{h,t} (\nu - \delta_m^*) / h} \tag{28}$$

where :

$$Q = 2\,\pi\, r\hat{c}_m h$$

$$\hat{p} = 2\,\pi\omega r^2 h \hat{f}_u \Delta m$$

The thicknesses δ^*, δ^{**} and ν are respectively the meridional displacement thickness, the absolute energy loss thickness and the angular defect force and they are defined as

$$r\hat{c}_m\, \delta_m^* = \int_0^\delta (\hat{c}_m - c_m)\, r dr \tag{29}$$

$$\Delta\hat{p}_t r\hat{c}_m \delta^{**} = \int_0^\delta (\hat{p}_t - p_t)\, c_m\, r dr \tag{30}$$

$$\hat{f}_u \nu = F_u = \int_0^{\delta} (\hat{f}_u - f_u) dr \tag{31}$$

$\Delta \hat{p}_t$ and $\hat{P}$ denote respectively the overall pressure increase and the input shaft power which would be found in absence of the EWBL. The energy thickness was in general not considered in efficiency corrections in the past, but in the authors opinion, the variation $\Delta\delta^{**}$ may not be neglected since an important reenergizing of the EWBL can occur in a rotor and the absolute value of δ^{**} may vary in an important way. It was found that $\Delta\delta^{**}$ may be of the same order of magnitude as $\nu - \delta_m^*$.

It is important to observe that ν and δ_m^* have the same value in both absolute and relative reference systems, but the energy thickness δ^{**} not.

In an incompressible approach the absolute and relative energy thicknesses are related as follows :

$$Ec_m \delta^{**}\Big|_{rel} = Ec_m \delta^{**}\Big|_{abs} - U\rho c_m^2 \theta_{um} \tag{32}$$

Hence, if the energy exchange occurs without extra loss in the EWBL, giving no increase in the energy loss thickness in the steady relative reference system, a variation in absolute energy thickness is found through equation (32). It was found that the three contributions δ^{**}, δ^* and ν are not independent and the efficiency decrease due to the EWBL directly depends on the energy loss only. The defect force thickness however has an important indirect effect on the efficiency, since non-zero defect forces can increase or even decrease the energy loss thickness, through the boundary layer momentum equations. It can be shown that in the absolute system :

$$\Delta\delta^{**\text{underturned}} > \Delta\delta^{**\text{collateral}} > \Delta\delta^{**\text{overturned}} \tag{33}$$

It should finally be observed that the efficiency equation (27) is an approximation. A correct way to estimate the total efficiency is discussed in [5].

3. APPLICATIONS

The theory has been tested in its different aspects. The base boundary layer models were first tested through simple 2D boundary layers along flat plates and conical diffusers where strong adverse pressure gradients are present. Three dimensional boundary layers are next considered in a radial angular diffuser. The force defect correlations are tested in cascades and single stage compressors where the tip clearance effects are emphasized. The details of all these tests can be found in [5] and the most interesting results are reported in the present paper. The theory is finally applied to a NASA two stage compressor [12].

The following test cases are considered :

2D boundary layers with given pressure gradient

The boundary layer development in 2D flows is predicted using experimental pressure gradients as input. The theory is tested in presence of positive and negative pressure gradients. Two conical diffuser tests are considered where the flow is close to separation, which can be considered as a severe test for the used model equations.

3D boundary layers in a radial vaneless diffuser

A radial diffuser with a 3D boundary layer at inlet is considered. A simple viscous-inviscid interaction algoritm is introduced in order to predict the diffuser blockage level as well as the meridional pressure gradient. The overall angular velocity is determined through radial equilibrium. 3D profiles, boundary layer thicknesses and external velocities are compared with experimental data.

Cascades of blade rows

The behaviour of the EWBL in a single compressor blade row is analysed. These tests allow the analysis of the defect force correlations in absence of a tip clearance. External pressure gradients are based on experimental data at inlet and outlet of the considered cascades.

Single stage compressors with variable tip clearance

Two test cases are considered in order to analyse the defect force assumption i in presence of a tip clearance. These tests are performed in experimental turbomachines, since such cascade data are not available. Even in the case of turbomachine flows, detailed data about the tip clearance flows are not profuse.

NASA two stage compressor

The method is finally applied to a low aspect ratio two-stage fan where detailed experiments are available.

3.1. Stanford turbulent boundary layer test cases [6]

The first elements which are tested are the 2D momentum conservation equation (1), complementary equations (3), (12), (16) and the profile model equations (6) and (8). All the test flows are incompressible. Test flows with negative, constant and positive external pressure gradient are considered. For the experiments where the momentum balance is not satisfied in an experimental way, no good agreement can be expected between calculation and experiment. The input pressure gradients have therefore been determined from the experimental momentum thickness, shape factor and skin friction in such a way that the momentum balance is satisfied.

On figures 2 and 3 some results obtained with the Stanford Turbulent Boundary layer data are shown. Most of the Stanford data were analysed and the results obtained at the most downstream position are summarized on figure 2. The different markers in figure 2 discern the zero-pressure gradient, the negative and positive gradient and the diffuser tests. According to figure 2 the momentum thicknesses are correctly predicted for Reynolds numbers based on the momentum thickness from 1000 to 100.000. The shape factor presents some scatter. The friction coefficient is correctly predicted except for test case 3600 where an abrupt positive gradient is present. On figure 3 velocity profiles are compared at exit of two conical diffuser test cases. The experimental profiles are quite well predicted, although these profiles are close to separation.

3.2. Radial vaneless diffuser [7]

In the present application three-dimensional effects are introduced. The second momentum equation equation (2)nd the cross flow velocity profile model equation (7) are verified.

Since the present method is based on an axisymmetric approach of a turbomachine flow simple 3D channel flows are not suited as test cases. Axisymmetric flows such as pipe flows or diffuser flows are to be considered. In the present section, the flow through a radial vaneless diffuser has been selected (Gardow [7]). The test diffuser is drafted on figure 4. A rotating screen was used at inlet of the diffuser in order to obtain a flow angle of about 45 degrees. Experimental data are available for all the boundary layer thicknesses, skewing angle, friction coefficient, external velocity and detailed 3D velocity profiles at several radii.

The experimental external velocities are not used as input for the boundary layer calculation but are predicted through radial equilibrium and continuity. The radial velocity is given as function of blockage through :

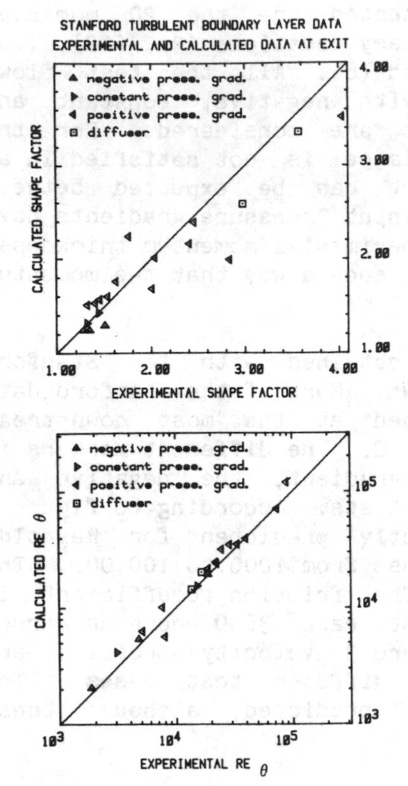

Figure 2 :
Stanford Turbulent Bound. Layer Data
Calculated and experimental data at the most downstream position

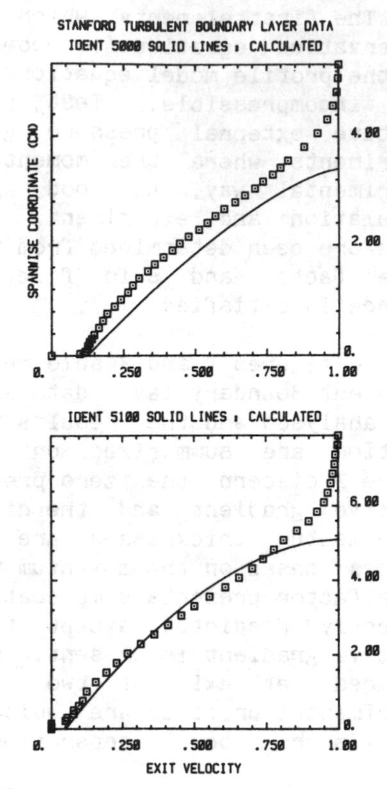

Figure 3 :
Stanford Turbulent Bound. Layer Data
Velocity profiles at downstream position
Conical diffuser test cases 5000, 5100

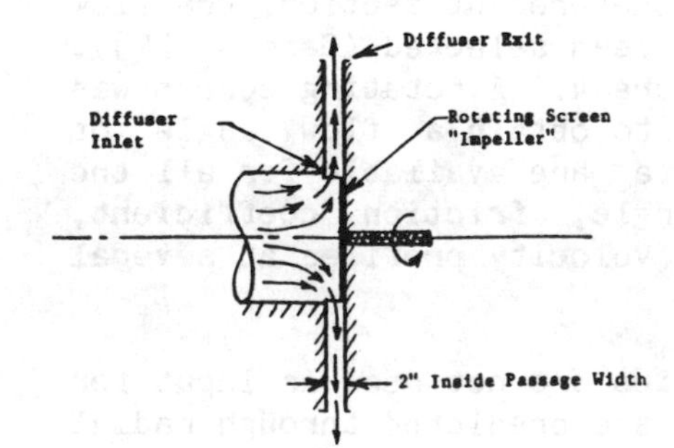

Figure 4 : Gardow vaneless diffuser (ref 7)

$$\hat{c}_m = \frac{Q}{S\ (1 - B)} \tag{34}$$

where B denotes the blockage. The pitchwise velocity is given by radial equilibrium through :

$$r\hat{c}_u = cte \tag{35}$$

An iterative procedure is to be used since equation (34) depends on the blockage which is still unknown. The boundary layers are assumed to be identical at both walls of the radial diffuser.

The results are reported on figure 5. On figure 5 four momentum and two displacement thicknesses, the skin friction coefficient and the wall skewing angle are compared with the experimental data. A fair agreement is in general found. This figure illustrates the possibility to predict <u>all</u> the integral boundary layer thicknesses through the use of an integral boundary layer method with profile models. The reconstruction of all these thicknesses would not be possible without the use of velocity profile model equations.

Less agreement is found between the prediction and the experimental streamwise and displacement thicknesses for the most downstream data points. This is probably due to the confluence of the boundary layers, as reported by Gardow. The mainstream displacement thickness (and hence the shape factor) is somewhat underestimated. This is in contradiction with the good result for the radial velocity since both values contain the same information through equation (35). The experimental data are therefore not fully compatible which may be due to an asymmetry in both end-wall boundary layers (only one has been measured).

The present test gives a fair result which is moreover obtained <u>without any experimental input</u>, since blockage level, cross flow intensity and external velocities are obtained through calculation. The calculation approaches the procedure which is used in real turbomachines where the radial equilibrium is performed in a less straightforward way.

3.3. Salvage cascade data [8]

In the previous applications the basic equations and velocity profile models have been tested in flows where no blade interactions and defect forces are present. These interaction terms are introduced in the present application where single cascades of compressor blades are considered.

A large amount of experimental cascade data have been compiled by Salvage and presented in a VKI report [8]. In these test cases, several solidities, blade camber, stagger and Reynolds numbers were

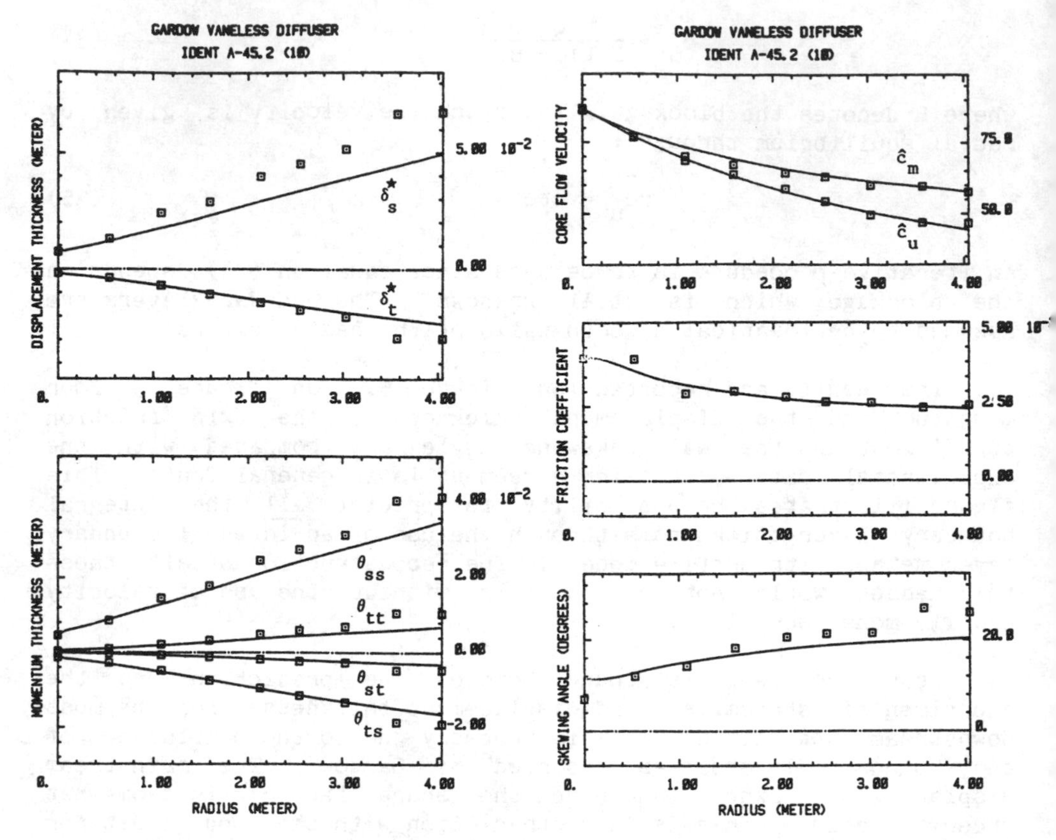

Figure 5 : Gardow vaneless diffuser (ref 7)
Solid lines : calculated, symbols : experiment

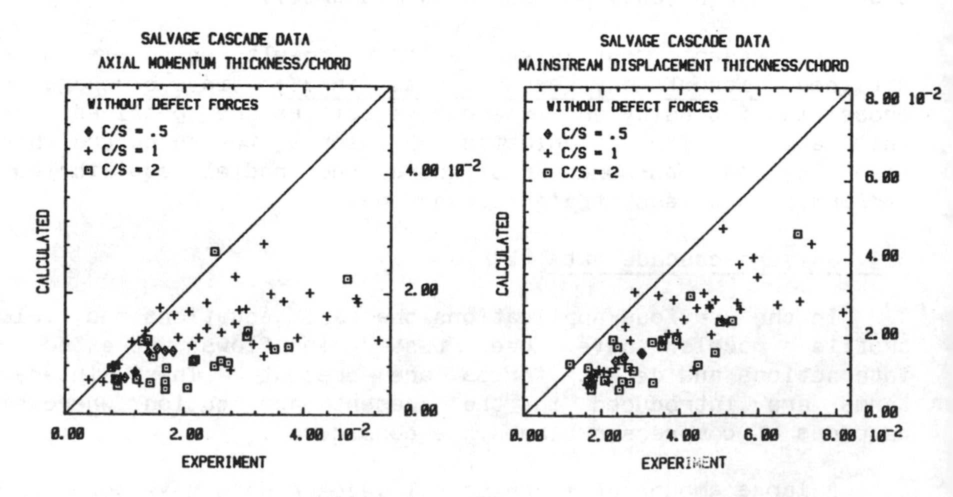

Figure 7 : Salvage cascade data (ref 8)
Calculated and experimental data at exit of the cascades
Results obtained with a zero defect force assumption

considered. All inlet boundary layers can be considered as quite collateral. Experimental data are available at inlet and outlet of the cascades along the center streamline. It is to be accepted that these local data are representative for the pitch averaged data.

Results are summarised on figures 7 to 9. Much scatter is present on these results which is to be explained by measurement incertainty and errors due to the poor knowledge of the experimental pressure gradient. Quite similar test cases sometimes give strongly different experimental results.

On figure 7 results are shown where zero defect forces were assumed. Although much scatter is present, the calculated data show a clear tendency to an underestimation of the exit boundary layer thickness. This underestimation is the most pronounced at the high thickness over chord ratios. The defect force equations (21) and (23) were used for figure 8 where the agreement between experiments and calculated data is improved when compared with figure 7. The increase in the predicted thicknesses is mainly due to the secondary loss, through the drag defect force equation (21).

On figure 9 calculated flow angle profiles are shown for some of the considered test cases. The outlet cross flows are overestimated in all cases when no defect forces are considered. A fair agreement is found for all the outlet flow profiles when equations (21) and (23) are used for drag and lift defect forces. From figure 9 it appears how the cross flow intensity is retricted by the lift defect force through equation (23).

Although much scatter is present on the results, the importance of the defect force and the validity of the used assumptions is well illustrated in the present test cases.

<u>3.4. Cascade with 3D inlet boundary layer</u> [9]

The inlet boundary layers in the Salvage cascade data can in general be considered as collateral. In the present application, a cascade is considered where a three-dimensional upstream boundary layer is present (Moore & Richardson [9]). Experimental thicknesses and velocity profiles are available at several chordwise positions along the centerline of the cascade flow. These data are not pitch averaged data and it is to be accepted that they are representative for the pitch averaged flow. On figure 10 experimental and calculated thicknesses and velocity profiles are compared. The blade leading and trailing edges are indicated by a mixed line.

Results with and without defect forces are shown. The cross flow profiles are well approached when using the defect force assumptions and the behaviour of the other thicknesses is improved

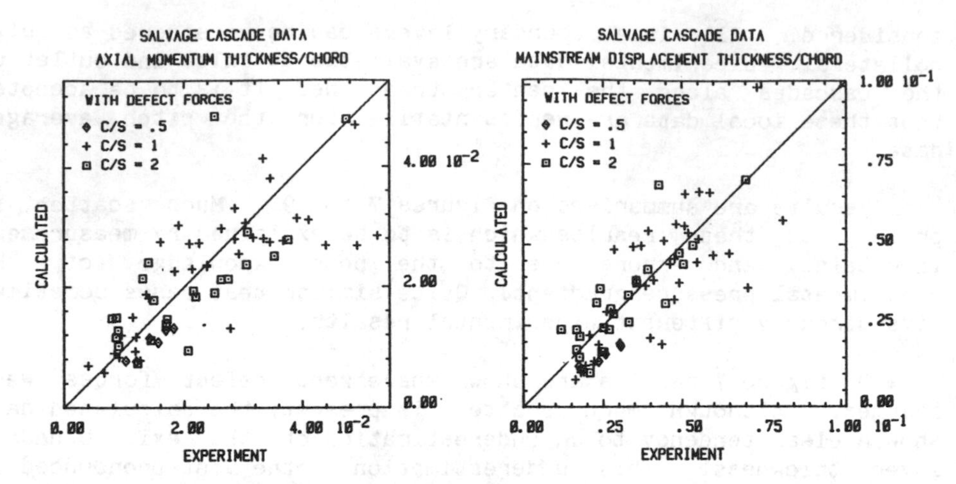

Figure 8 : Salvage cascade data (ref 8)
Calculated and experimental data at exit of the cascades
Results obtained when using the defect force eqs 21 and 23

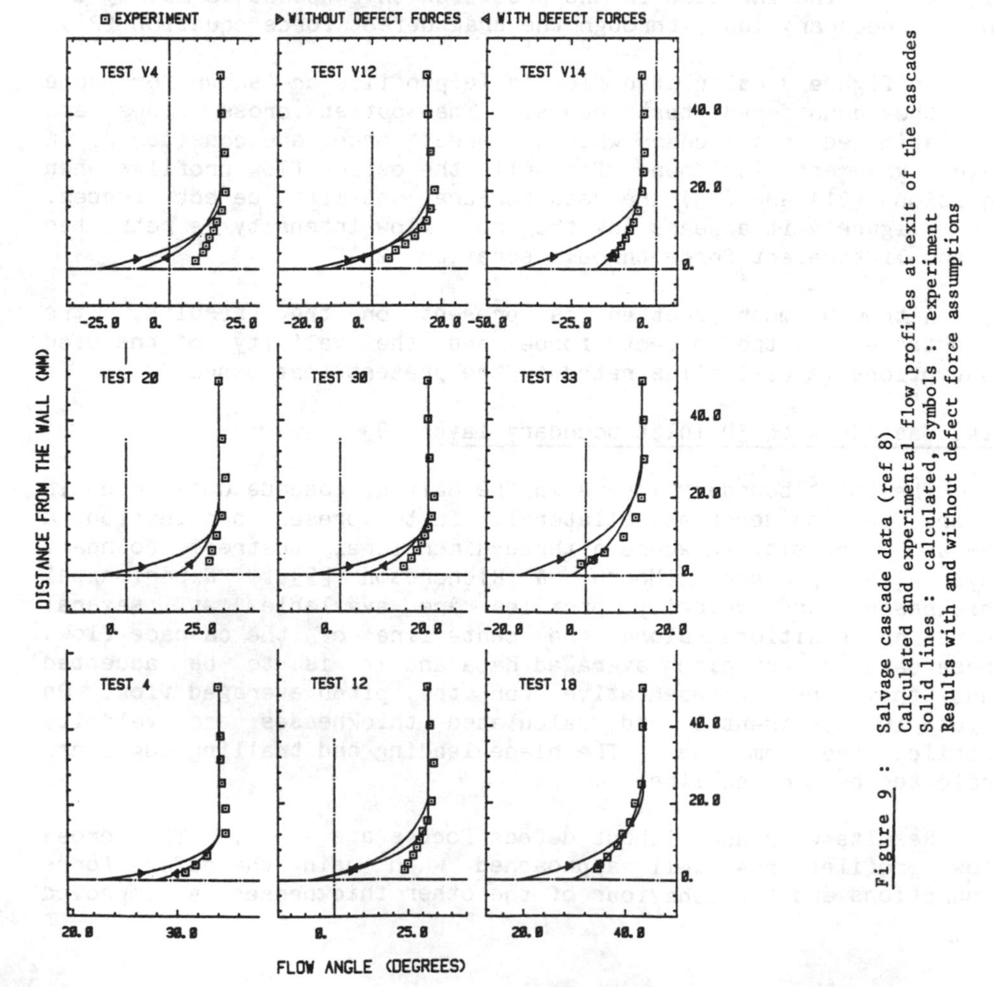

Figure 9 : Salvage cascade data (ref 8)
Calculated and experimental flow profiles at exit of the cascades
Solid lines : calculated, symbols : experiment
Results with and without defect force assumptions

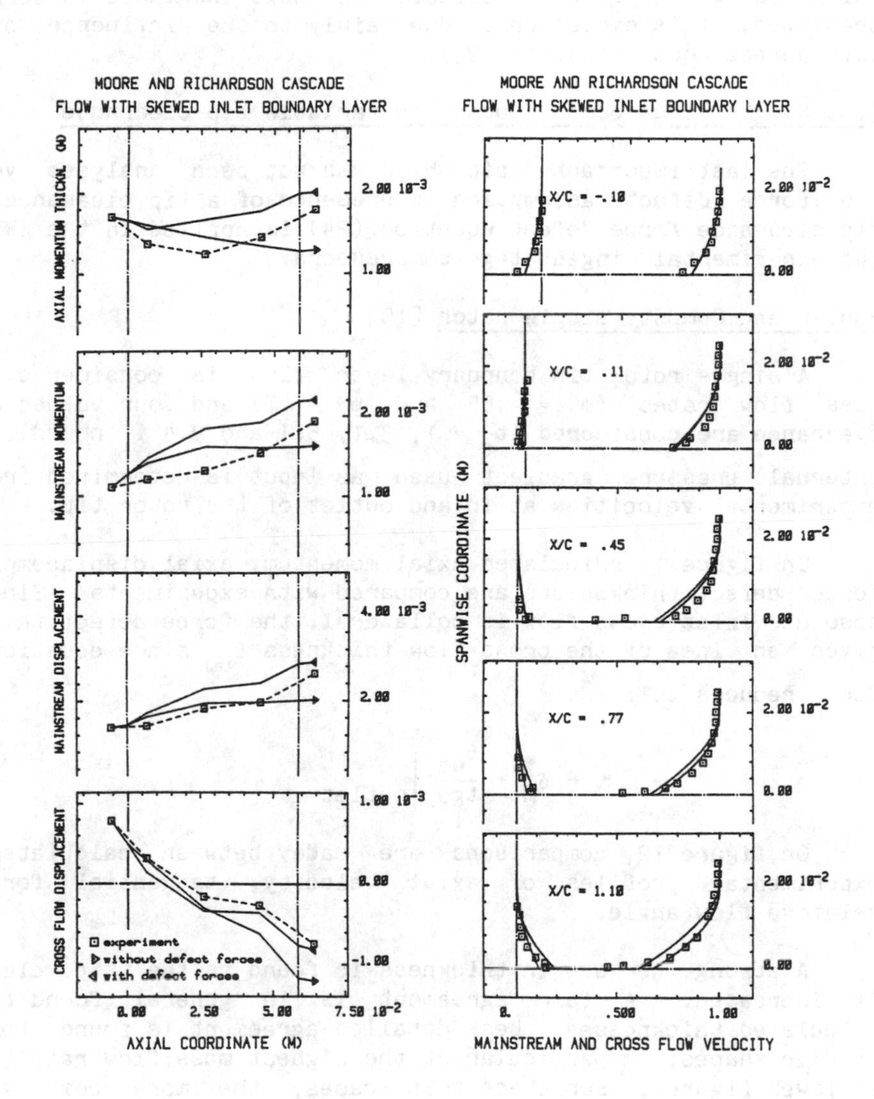

Figure 10 Moore and Richardson cascade (ref 9)
Solid lines : calculated, symbols : experiment
Results with and without defect force assumptions

when compared with the zero defect force assumption. In particular, the qualitative behaviour of the axial momentum thickness which first decreases and next increases is correctly predicted. This evolution is due mainly to the influence of the lift defect force equation (23).

3.5. Single stage compressors with variable tip clearance

The last important topic which has not been analysed yet is the force defect assumption in presence of a tip clearance. The tip clearance force defect equation (24) is applied in the EWBL of two experimental single stage compressors.

Hunter and Cumpsty single rotor [10]

A single rotor tip boundary layer flow is considered. Two mass flow rates (ψ = .55 and ψ = .7) and four values of tip clearance are considered (t_c = 1, 3.1, 5.1 and 9.4 % chord). The external pressure gradient used as input is determined from the experimental velocities at in and outlet of the rotor tip.

On figure 11 calculated axial momentum, axial displacement and force defect thicknesses are compared with experiments. Since the absolute inlet cross flow is collateral, the force defect thickness gives an idea of the cross flow thickness θ_{um} since equation (31) for ν reduces to :

$$\nu = \delta_m^* + \frac{\theta_{um}}{tg\alpha}\bigg|_{outlet} \tag{36}$$

On figure 12, comparisons are made between calculated and experimental profiles of axial velocity, tangential force and relative flow angle.

A strong increase in thickness is found as the tip clearance is increased. A fair agreement is in general found for the calculated thicknesses. Less detailed agreement is found for the profile shapes, in particular at the highest mass flow rate (figure 12 lower figure). For these test cases, the more complex flow patterns can no more be approached by simple profile model equations such as those used in the present method.

The present test shows the important blade force variations in presence of a tip clearance. This effect is shown to be approached by the defect force equation (24). Less detailed agreement is however found for the detailed velocity profiles inside the EWBL. The present test has been used as a calibration for the value $k' \sim .5$.

Bettner and Elrod single stage compressor [11]

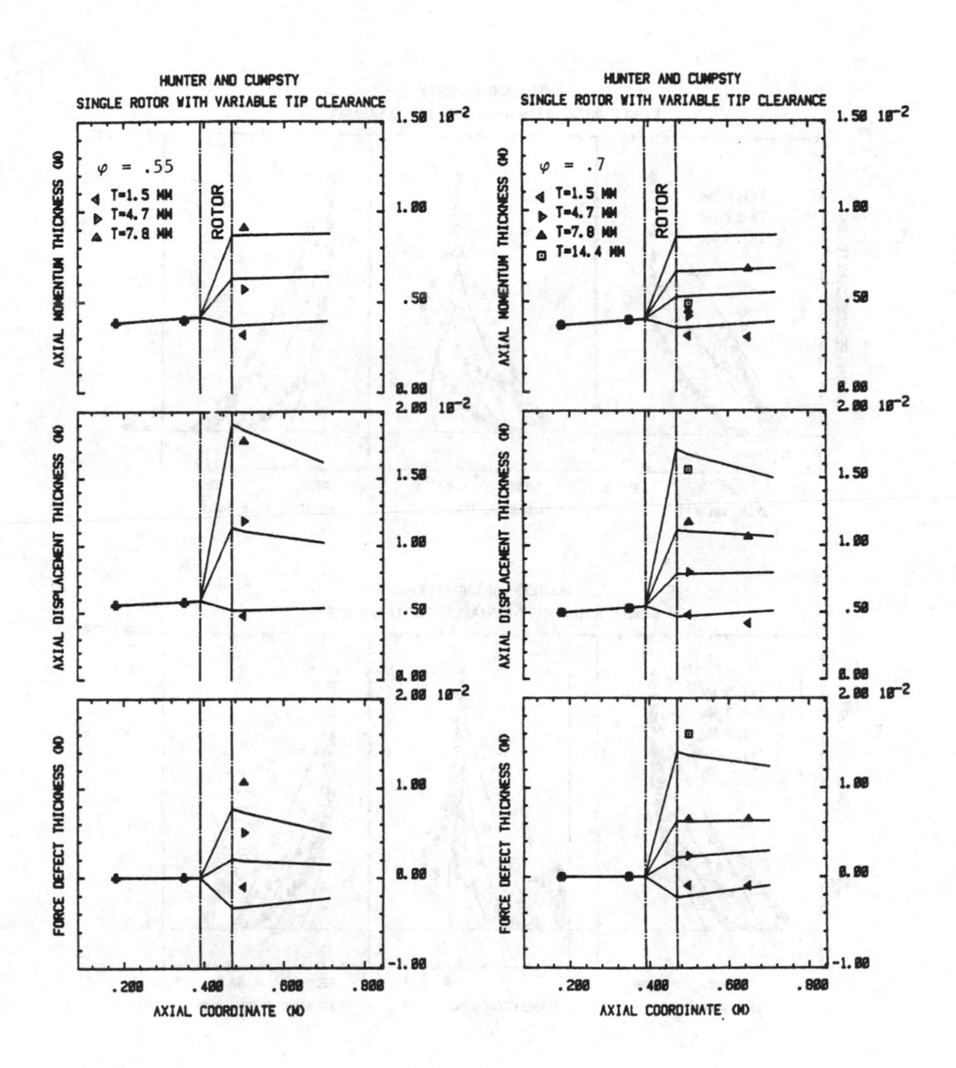

Figure 11 : Hunter & Cumpsty single rotor with variable tip clearance (ref 10) Solid lines : calculated, symbols : experiment

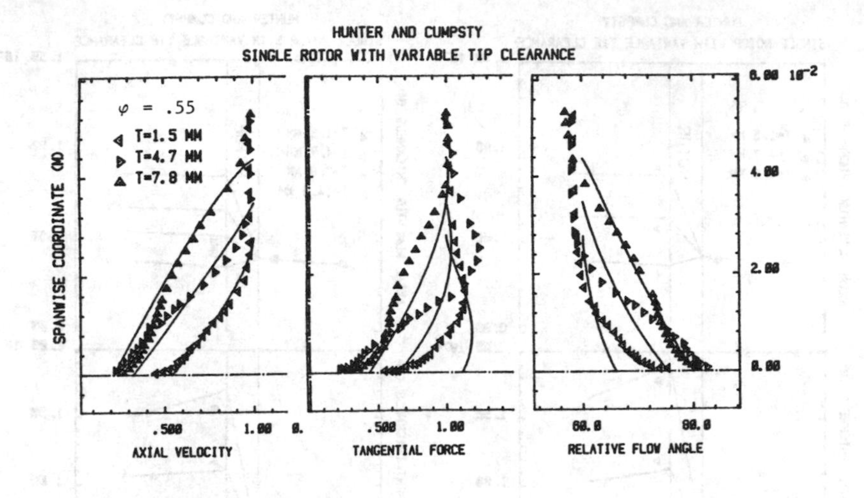

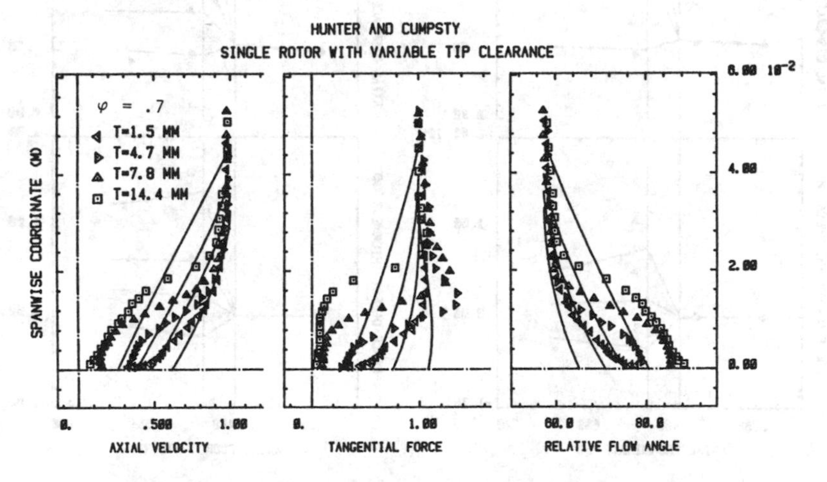

Figure 12 : Hunter & Cumpsty single rotor with variable tip clearance (ref 10)
Solid lines : calculated
Symbols : experiment

In the present test the EWBL through a single stage compressor has been measured at rotor tip. The stage consists of a rotor where the upstream boundary layer is two dimensional in the absolute reference system and a stator which restores the rotor outlet flow into the meridional direction. The results are reported on figures 13 and 14.

Two values of the tip clearance are considered at the design mass flow rate. The tip clearances are as large as the observed displacement thicknesses and the present test is therefore to be considered as severe.

On figure 13 calculated and experimental values of δ_m^* , θ_{ss} , δ_s^* and δ_t^* are compared. All the thicknesses are strongly underestimated at the exit of the rotor when no defect forces are assumed. When the lift and drag defect forces are added, a better agreement is found. At the high tip clearance value an overestimation in boundary layer thickness is found, indicating less reliability at such high clearance values, which are unprobable in practice however.

On figure 14 a comparison is made between calculated and experimental velocity profiles. No agreement is found for the angular velocity at exit of the rotor, but the agreement is improved as soon as the stator is entered. The agreement at stator exit is fair for this test configuration where the tip clearance has a small value (1.13 % span).

Although the qualitative behaviour of the EWBL is improved through the defect force assumptions, less agreement is found in the present test case. A better agreement can be found when the values of the coefficients k, k', and L are modified (equations 22 and 25). These values were however obtained as the best overall values and are kept constant for all the present applications.

3.6. Two stage compressor [12]

Detailed experimental flow and pressure gradient distributions in all the stages of a multistage compressor are practically inexistant at the present time. Past applications of the EWBL theory in multi-stage compressors were therefore performed in connection with the prediction of the complete meridional compressor flow. Results of such applications were presented in the past [4]. The quality of the results obtained in these applications depends not only on the accuracy of the EWBL method, but also on the accuracy of the profile loss and the wall velocity field estimations.

Detailed experimental data in a two stage compressor were recently presented [12] and the present theory can be applied in this multistage machine without interference of other calculations.

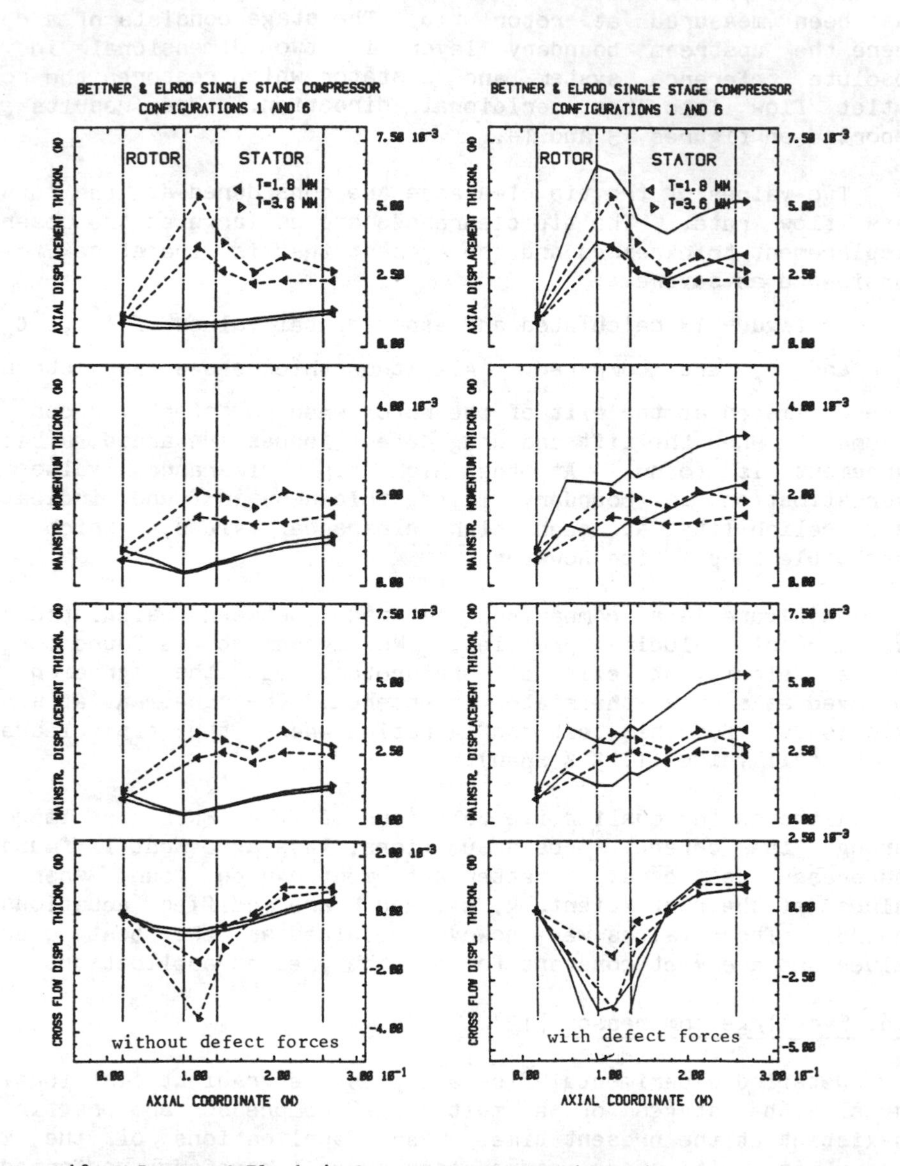

Figure 13 : Bettner & Elrod single stage compressor (ref 11)
Left side figures : results obtained with a zero defect force ass.
Right side figures : results obtained with defect force eqs 21, 24

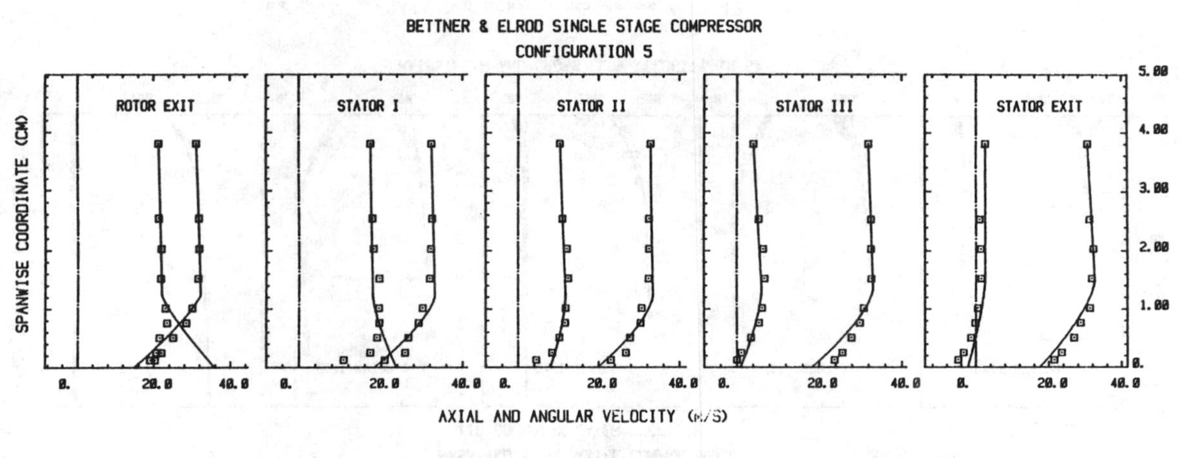

Figure 14 : Bettner & Elrod single stage compressor (ref 11)
Solid lines : calculated, symbols : experiment

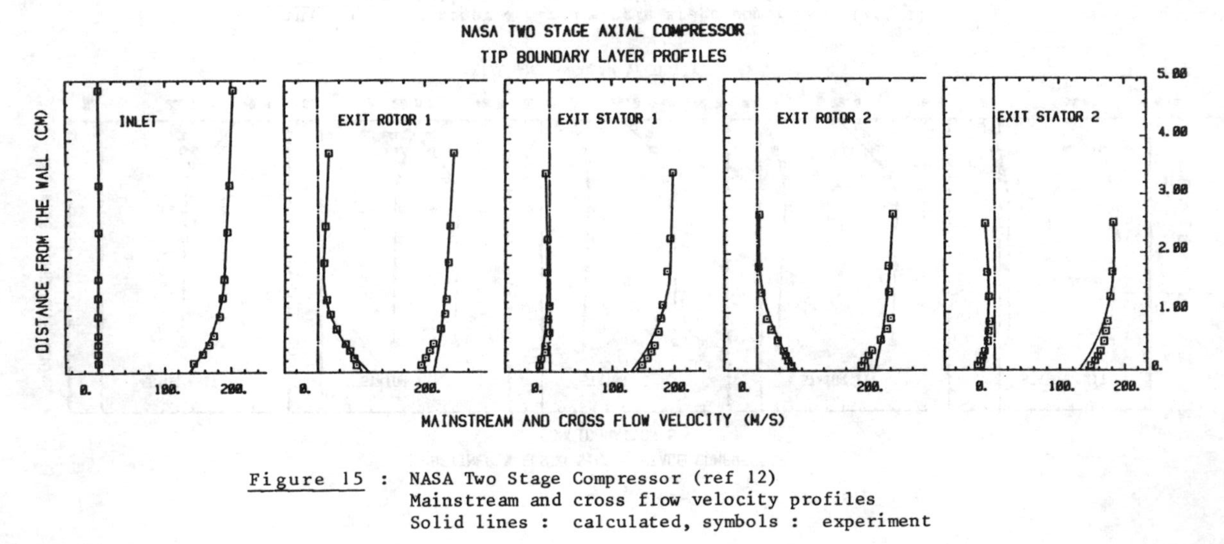

Figure 15 : NASA Two Stage Compressor (ref 12)
Mainstream and cross flow velocity profiles
Solid lines : calculated, symbols : experiment

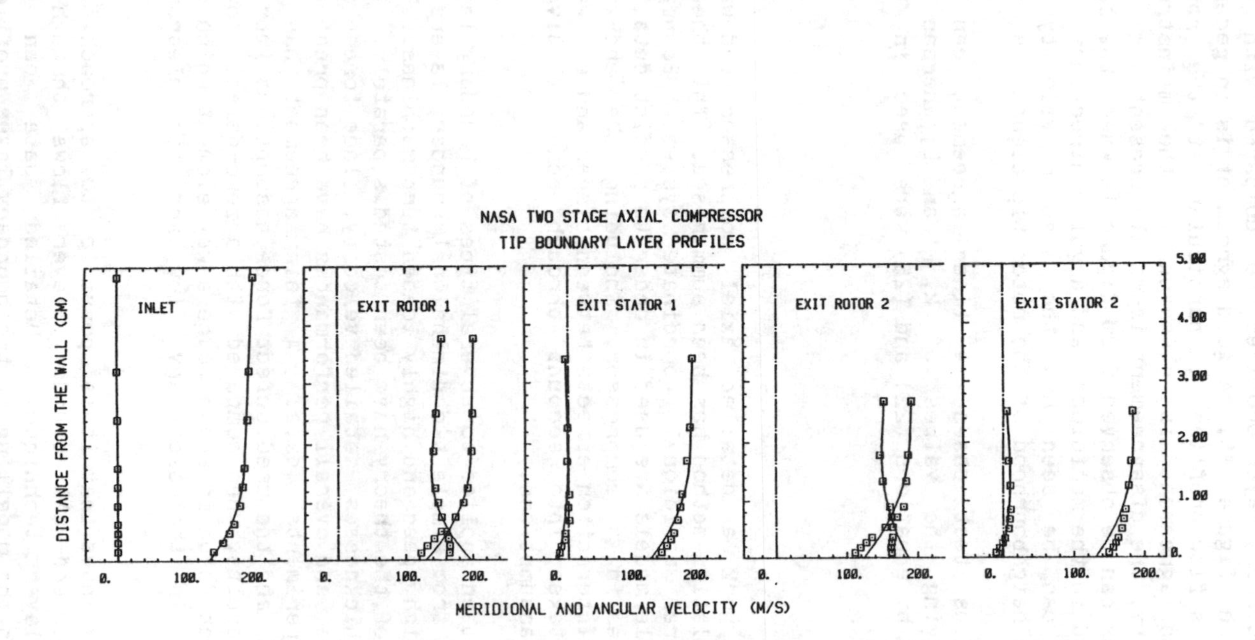

Figure 16 : NASA two stage compressor (ref 12)
Meridional and angular velocity profiles
Solid lines : calculated, symbols : experiment

This NASA compressor is a highly loaded machine with a low aspect ratio in the first rotor.

The predicted boundary layer profiles are compared with the experimental data on figure 15. A good agreement is in general observed for the cross flow profiles, in particular at the rotor exits, whereas less agreement is observed in the mainstream direction. In reality, the disagreement is mainly present in the angular direction, as can be observed from figure 16 where the same results are projected in the meridional and angular directions. On this figure it can be seen how the angular velocity is overestimated in the neighbourhood of the rotor tip clearances.

As in the previous test cases, a finer agreement can be obtained when changing the values of k, k' and L, whereas the overall values given by equations (22) and (25) were used in the present applications.

4. CONCLUSIONS

In the present work a detailed axial compressor end-wall boundary layer calculation method has been summarised. This theory is defined in a general meridional coordinate system. Boundary layer velocity profile models are used in order to predict detailed spanwise flow profiles in a compressor, including the end-wall regions. All the interaction effects between blades and end-wall boundary layers are taken into account through some relatively simple defect force assumptions.

The method has been applied to several types of boundary layer flows which range from simple two dimensional boundary layers to boundary layers in high speed and highly loaded turbomachines. The different aspects of the theory have been tested separately. All the boundary layer thicknesses, detailed velocity, blade force and flow angle profiles, and overall performances have been predicted and compared with experimental data. A fair agreement has in general been found and the used defect force assumptions improve the quality of the results, when compared with a zero-defect force approach. The defect force is shown to be a key element which can play a dominant role in the secondary flow and tip clearance effects.

The present work shows that it is possible to approach the complex turbomachine end-wall boundary layer flows through an integral boundary layer technique. Detailed data can be reconstructed through the modelling of the boundary layer profiles. Although all the details of the flow cannot be predicted accurately, the quality of the results can be sufficient for engineering purposes and the computer requirements for such an approach are small.

NOMENCLATURE

b	boundary layer parameter
c	absolute velocity component, chord
E	entrainment rate
F	force defect thickness
f	blade force
H	shape factor
H^*	Head's shape factor
h	annulus height
m	meridional coordinate
n	coordinate normal to the wall
P	power
p	pressure
Q	volume flow
r	radius
s	streamwise coordinate, pitch
T	temperature
T^*	Eckert reference temperature
t	transverse coordinate
t	tip clearance
U	wheel speed
u	pitchwise coordinate
W	velocity in the blade reference system
w	relative velocity component
y	coordinate normal to the wall
α	flow angle
γ	stagger angle
δ	physical boundary layer thickness
δ^*	displacement thickness
ε_w	wall skewing angle
η	efficiency
η_{bl}	boundary layer efficiency
θ	momentum thickness
ν	defect force thickness
ρ	density
σ	solidity
τ	shear stress
ψ	mass flow coefficient c_m/U
ω	angular speed

Subscripts

k	kinematic
w	at the wall

m,n,u in meridional coordinates
s,y,t in streamline coordinates

REFERENCES

1 SMITH L.H., 1969, "Casing Boundary Layers in Multistage Axial Flow Compressors" Brown Boveri Symposium, Flow Research on Blading, Elsevier, 1969

2 MELLOR G.M., WOOD G.L., 1971, "An Axial Compressor End-Wall Boundary Layer Theory" ASME Journal of Basic Engineering, Series d, Vol 93, 1971, pp 300-316

3 HIRSCH C., 1974, "Flow Prediction in Axial Flow Compressors Including End- Wall Boundary Layers" ASME Journal of Engin. for Power, Vol 96, 1974, pp 413-426

4 DE RUYCK J., HIRSCH CH., 1981, "Investigations of an Axial Compressor End-Wall Boundary Layer Prediction Method" ASME Journal of Eng. for Power, Vol 103, no 1, pp 20-33, 1981

5 DE RUYCK J., 1982, "Computation of End-Wall Boundary Layers in Axial Compressors" Ph.D. Thesis, Vrije Universiteit Brussel, Dept of Fluid Mechanics, 1982

6 COLES D.E., HIRST E.A., 1968, "Computation of Turbulent Boundary Layers" Afosr - Ifp - Stanford Conference Proceedings, Volume 2, Compiled Data, Dept of Mech. Eng., Stanford Univ., California, 1968

7 GARDOW E., 1958, "The Three Dimensional Turbulent Boundary Layer in a Free Vortex Diffuser" MIT Gas Turbine Lab. Report 42, 1958

8 SALVAGE J.W., 1974, "Investigation of Secondary Flow Behaviour and End-Wall Boundary Layer Development Through Compressor Cascades" Von Karman Institute, Technical Note n° 107, 1974

9 MOORE R.W., RICHARDSON D.L., 1956, "Skewed Boundary Layer Flow Near the End- Walls of a Compressor Cascadë ASME Paper no 56-a-131, 1956

10 HUNTER I.H., CUMPSTY N.A., 1982, "Casing Wall Boundary Layer Development ThRotorugh an Isolated Compressor Rotor" ASME Paper Number 82-gt-18, 1982

11 BETTNER J.L., ELROD C., 1982, "The Influence of Tip Clearance, Stage Loading, and Wall Roughness on Compressor Casing Boundary Layer Development" ASME Paper Number 82-gt-153, 1982

12 GORRELL T., 1983, "Detailed Flow Measurements in the Casing Boundary Layer of a 427 Meter per Second Tip Speed Two Stage Fan" AVRADCOM, Lewis Research Center, report to be published.

13 DE RUYCK J., HIRSCH Ch., 1983, "End-wall boundary layer calculations in multistage axial compressors", in : "Viscous effects in turbomachines", AGARD Conference proceedings n° 351.

DESIGN METHODS AND PEFORMANCE PREDICTION

AXIAL FLOW COMPRESSOR PERFORMANCE

Jacques CHAUVIN

Institut de Mécanique des Fluides de Marseille
Laboratoire associé 03 du CNRS

1. INTRODUCTION

In the previous lectures, some of the theoretical, numerical and experimental building blocks, necessary for the understanding of the turbomachinery fluid mechanics have been introduced.

In later lectures, the principle of parameter selection and optimization, and of the computer assisted design and analysis systems will be described. The unstable regime of operation of compressors and their transition to other , often dangerous stable regimes will be dealt with later, in detail.

The present text is devoted to the characterization of axial flow compressor performance, its specification for particular uses. A general description of the flow in axial compressors, including the flow mechanism governing its safe operation will be given. The classical simplified modelling based on Wu's approach and the 3 zones model will be recalled. The architecture of the contemporary design and analysis scheme will be presented, as well as a brief review of one of those methods, recently developed at IMFM.

2. DEFINITION OF THE AXIAL COMPRESSOR PERFORMANCE MAPS

The basic parameters chosen to characterize axial compressor performance [1],[2] are the outlet pressure (total or static), an η (polytropic or isentropic, total to total or static to static) and the enthalpy rise (or the temperature increase, when the gas can be considered to have constant specific heat at constant pressure. This will be assumed in the present paper).

We will use the stagnation conditions, for this presentation and assume that the fluid behaves as a perfect gas.

One usually assumes that those parameters are function of 9 basic parameters

$$p_{o_2}\ ,\ \eta\ ,\ \Delta T_o = f\ (p_{o_1},\ T_{o_1},\ N,\ L,\ \dot{m},\ \rho_{o_1}, \gamma\ ,\ \mu,\ a_{o_1})$$

i.e. depends on the nature of the fluid γ, μ, a_{o_1}
the inlet conditions p_{o_1}, T_{o_1}, a_{o_1}, ρ_{o_1}
operating conditions N, $\dot{m}$
one characteristic linear dimension L (usually a typical diameter)

For a given gas, and a fixed T_{o_1} , a_{o_1} is not an independent variable, and one can write

$$p_{o_2},\ \mu\ ,\ \Delta T_o = f\ (p_{o_1}\ ,\ T_{o_1}\ ,\ N,\ L,\ \dot{m},\ R,\ \gamma\ ,\ \mu)$$

going through a dimensional analysis, this reduces to

$$\frac{p_{o_2}}{p_{o_1}}\ ,\ \eta\ ,\ \frac{\Delta T_o}{T_{o_1}} = f\ (\frac{NL}{R\sqrt{T_{o_1}}}\ ,\ \frac{\dot{m}\sqrt{RT_{o_1}}}{p_{o_1}\ L^2}\ ,\ \frac{p_{o_1}\ L}{\mu\ \sqrt{RT_{o_1}}}\ ,\ \gamma)$$

If the nature of the gas, and hence the value of R and γ (supposed constant through the process) are known, then

$$\frac{p_{o_2}}{p_{o_1}}\ ,\ \eta,\ \frac{\Delta T_o}{T_{o_1}} = f\ (\frac{NL}{\sqrt{T_{o_1}}}\ ,\ \frac{\dot{m}\sqrt{T_{o_1}}}{p_{o_1}\ L^2}\ ,\ \frac{p_{o_1}\ L}{\mu\ \sqrt{T_{o_1}}})$$

which is equivalent to

$$\frac{p_{o_2}}{p_{o_1}}\ ,\ \eta\ ,\ \frac{\Delta T_o}{T_{o_1}} = f\ (\frac{NL}{\sqrt{T_{o_1}}}\ ,\ \frac{\dot{m}\sqrt{T_{o_1}}}{p_{o_1}\ L^2}\ ,\ \frac{\dot{m}}{\mu\ D})$$

The last parameter is a Reynolds number of the flow. The parameter $NL/\sqrt{T_{o_1}}$ is proportional to the ratio of a typical blade speed (for instance, a velocity of rotation if L = D is a typical diameter) to the speed of sound of the entry fluid at stagnation conditions, and hence, a form of blade Mach number. The parameter $\frac{\dot{m}\sqrt{T_{o_1}}}{p_{o_1}L^2}$ characterizes the mass flow in a form involving the flow Mach number at entry.

It is generally accepted that, if the Reynolds number is not too small, nor too high (i.e. turbulent flow regime or "smooth surface" as seen by the flow, respectively), it has little or no influence, and can be dropped. Care has to be taken that this hypothesis is satisfied for the whole range of operations (it would not be true for small helicopter flying at high altitude or for very large pressure ratio (>35) machines).

This form of presentation could be able to summarize in a general way the performance of a large number of compressors if the whole geometry could be adequately characterized by a unique length parameter. This is of could absolutely impossible. The flow description made in other lectures shows that blade shapes and aspect ratio, clearances, blade row spacing, hub and tip contours, and many other form parameters are susceptible to influence the performance. Therefore, the performance map presented as

$$\frac{po_2}{po_1}\ ,\ \eta\ ,\ \frac{\Delta T}{T_o} = f\left(\frac{NL}{\sqrt{T_{o_1}}}\ ,\ \frac{\dot{m}\sqrt{T_{o_1}}}{po_1\ L^2}\right)$$

would be valid only for compressors operating with the same gas, in a domain where γ can be considered as constant, geometrically similar, to the extent that the variations of the form parameters (as clearance, blade twist, etc.) with the operating point be also similar, notwithstanding different sizes. Additionally, the Reynolds number should not affect the performance.

Finally, it is wiser to considerer a performance map for a given compressor, which is then given in the form

$$\frac{po_2}{po_1}\ ,\ \eta\ ,\ \frac{\Delta T}{T_o} = f\left(\frac{N}{\sqrt{T_{o_1}}}\ ,\ \frac{\dot{m}\sqrt{T_{o_1}}}{po_1}\right)$$

as L is a constant.

Additionally $\dfrac{\Delta T}{T_o} = f\left(\dfrac{po_2}{po_1}\ ,\ \eta_p\right)$

as $$\frac{p_{o2}}{po_1} = \left(\frac{T_{o_2}}{T_{o_1}}\right)^{\frac{-\gamma\eta}{\gamma-1}} = \left(1 + \eta_{is}\ \frac{\Delta T}{T_{o_1}}\right)^{\frac{\gamma}{\gamma-1}}$$

FIGURE 1 CLASSICAL COMPRESSOR PERFORMANCE MAP

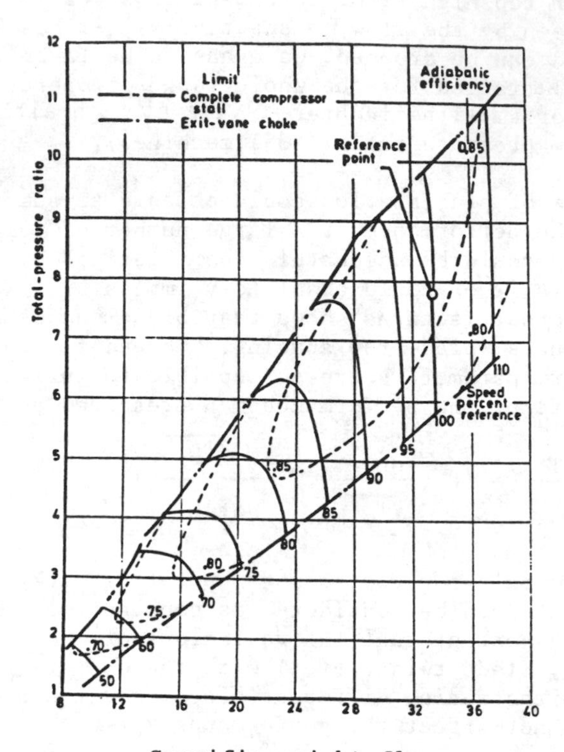

Specific weight flow

The classical performance map, for a given compressor is given in figure 1 (pressure ratio and efficiency in function of reduced mass flow and reduced speed of rotation). Its only generality is to present the performance independently of the inlet flow conditions, as long as the hypothesis of the gas characteristics, the Reynolds number and the similarity of the secondary geometrical characteristics are respected.

In some fast calculation methods, such as some stage stacking methods, a carry over presentation , developed for small pressure ratio machine is still used.

Assuming basically no Mach number effect or of entry flow conditions one can replace

$$\frac{\Delta T_o}{T_o} \quad \text{by} \quad \psi = \frac{C_p \Delta T_o}{U_1^2}$$

and the mass flow by $\varphi = \frac{V_{A_1}}{U_1}$

(where U_1 is a typical rotational velocity at inlet). The rpm disappear as the characteristic curves

$\frac{p_{o_2}}{p_{o_1}}, \psi = f(\varphi)$ collapse on each other for the various speed of rotation (Fig. 2) (with these hypothesis).

FIGURE 2 TYPICAL φ, ψ DIAGRAMS

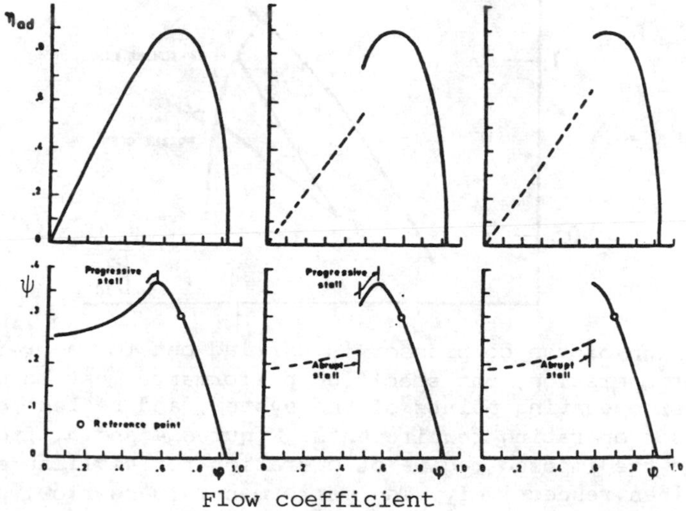

Flow coefficient

(a) Stages 1 to 4 (b) Stages 5 to 9 (c) Stages 9 to 12

3. SYSTEM OPERATION OF AXIAL FLOW COMPRESSOR

The performance map of the compressor, as presented in figure 1, is usually obtained either by calculation or on a test rig, where the compressor stands by itself, between inlet and discharge volumes. The use of a throttle valve and of a variable speed drive allows then to cover the whole map, constant speed line by constant speed line, and throttling the flow progressively on each line. The flow at inlet is axisymmetric.

The operation of the compressor in its real working environment is quite different. It is coupled with other fluid mechanic or mechanical devices, such as piping, volume, throttle, etc., constituting a system which imposes compatibility conditions to the compressor and impose operation on privileged points of the map.

For instance, in a jet engine case, when the first turbine nozzle and the exhaust nozzle are chocked, or in a chocked free gas turbine, the operation is restricted to a single operating curve (Fig.3) running across the constant speed lines, unless geometrical changes are brought to the system (variable stagger to the compressor, bleeds, variable turbine nozzle or variable area exhaust nozzle). The effect of changing the exhaust nozzle area is also indicated in figure 3.

FIGURE 3 AXIAL FLOW COMPRESSOR OPERATING MAP

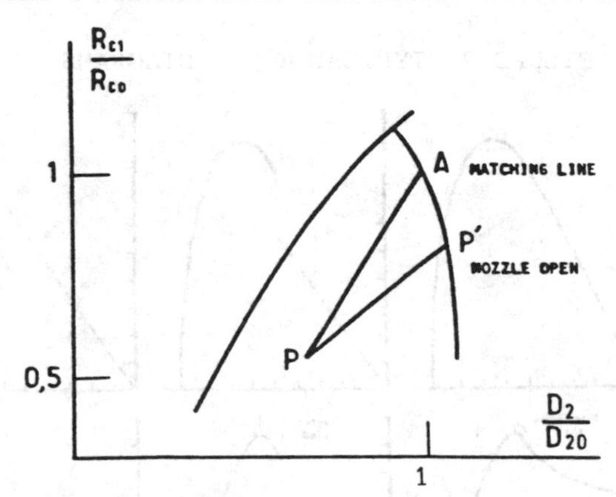

Design of the compressor is carried out for a preferential point of operation, but specified performance must be guaranteed for other operating points of the system, and reflect on the axial compressor operating requirements. Figures 4 and 5, from [3], indicate the imposed regime of operation for civilian and military jet engines respectively. The performan in mass flow, pressure ratio and efficiency are specified within a few per cent, and must be achieved in presence of compressor inlet flow often strongly departing from the ideal uniform compressor inlet flow (inlet distorsion).

FIGURE 4 DEFINITION OF SIGNIFICANT FLIGHT CASES; SELECTION OF DESIGN POINT. Example : COMMERCIAL ENGINES POSITION OF FLIGHT CASES IN HP COMPRESSOR CHARACTERISTICS MAP.

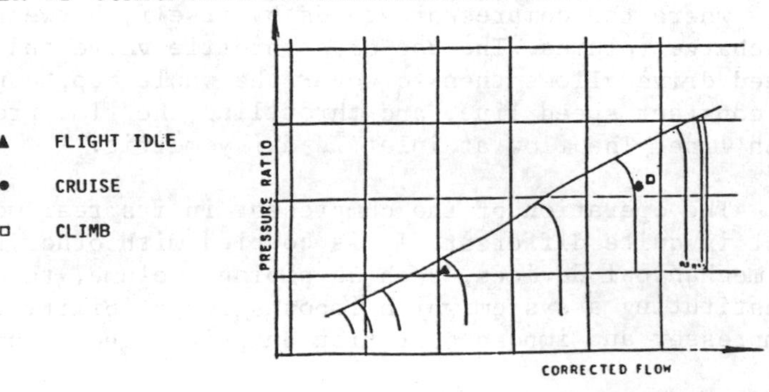

FIGURE 5 DEFINITION OF SIGNIFICANT FLIGHT CASES ;
SELECTION OF DESIGN POINT
EXAMPLE : MILITARY ENGINES

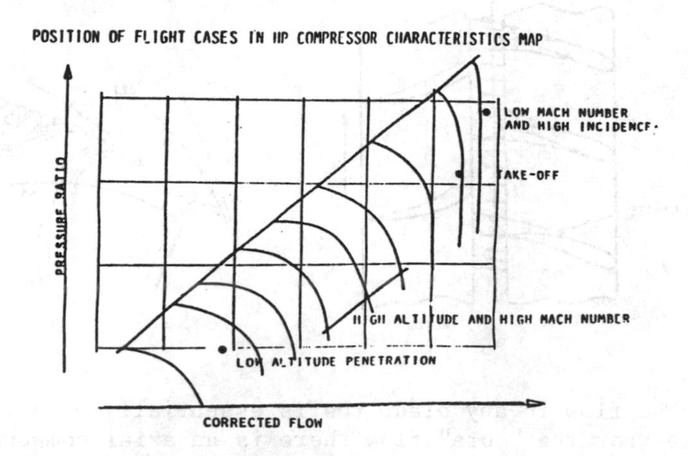

Such exacting requirements impose a high accuracy in the compressor performance prediction scheme, including the determination of the flow regimes which can be dangerous for the compressor integrity (rotating stall, flutter, surge) and which will be discussed later.

Compressor performance prediction modelling must be based on a thorough understanding of the physics of the flow. Our present grasp of it is briefly described in the next section.

4. THE PHYSICS OF FLOWS IN AXIAL FLOW COMPRESSORS

4.1. Operation in the "Sound" Region

By this, we mean the steady operation (constant mass flow, pressure ratio and rpm) in region of the map devoid of dangerous phenomena.

Consider, figure 6, the flow in two consecutive blade rows on a stream surface, far enough from hub and casing to neglect their influence. For a force to exist on the blades, there must be a velocity difference along the pitch between succion and pressure sides. The fluid being viscous, boundary layers develop on pressure end succion side of the blade, and merge in the wake, where the tangential velocity distribution is highly non uniform. The next blade which is in relative motion with respect to the first one, is thus submitted to an unsteady periodic viscous flow (Fig.6 also describes the situation for a radial compressor).

FIGURE 6 PERIODIC UNSTEADY FLOW (SOUND REGIME)

A. Axial compressor

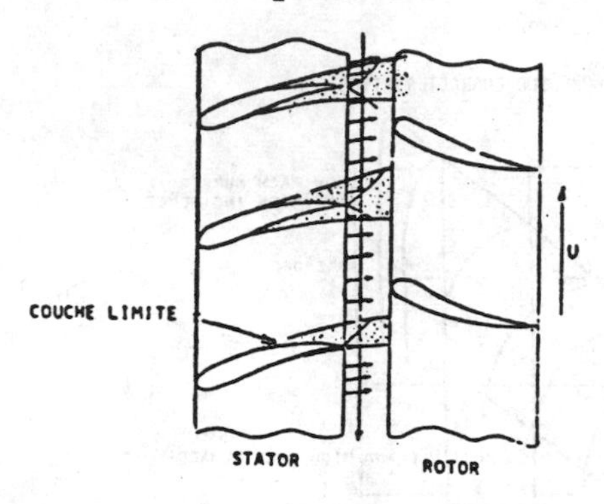

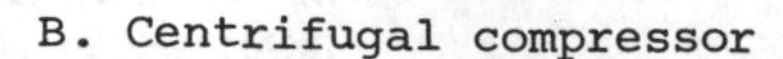
B. Centrifugal compressor

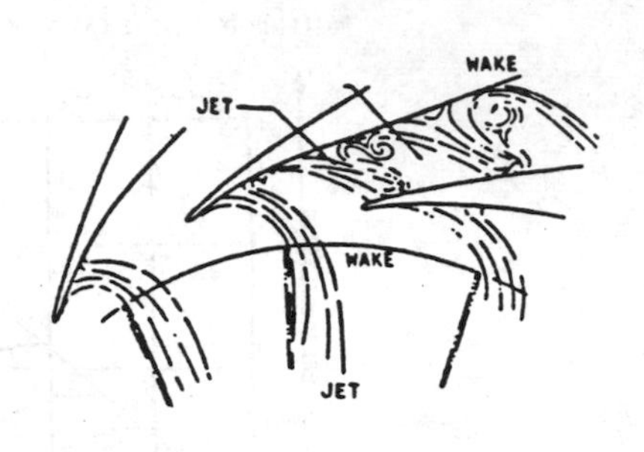

The flow in any blade row is essentially of a three dimensional nature. For the "core" flow there is an axial component, corresponding to the mass flow, a tangential component inherent to the energy transfer and conversion, and radial motion due to the flow path convergence and radial pressure gradient effects. Additionaly (Fig.7), near the hub and casing, a pseudo-boundary layer flow with reduced velocity exists entailing a slip flow between pressure and succion sides, (reduced streamline radius of curvature to balance the succion-pressure side pressure gradient in presence of a reduced velocity), leading in turn to a "corner" vortex. If there is any clearance connecting pressure and succion side of a blade there is a parasitic flow, rolling up in a vortex near the succion side and creating a second vortex. The relative motion between hub

FIGURE 7 THREE DIMENSIONAL FLOW

A. Real

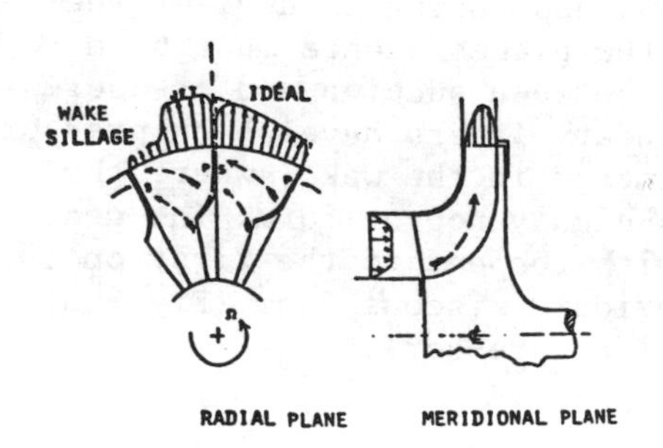

B. Secondary flows

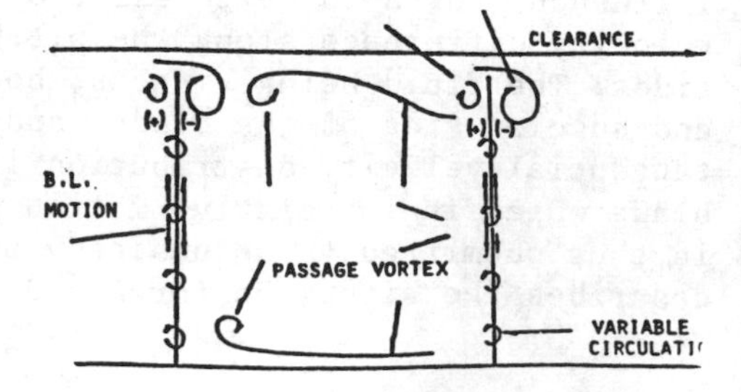

or casing and blade tip creates, on the pressure side a "scraping" vortex. Variable circulation along the blade span creates trailing edge vortices and blade boundary layers move radially due to centrifugal forces and radial pressure gradient. Figures 8 and 9 give a more detailed picture of some of the secondary flows (without clearance or scraping effects). The first one sketches the effects of vorticity transport (governing the outlet angle from the blade) extending much further radially in the flow than the "viscous" regions near hub and casing ; the second one shows the 3D boundary layer and its complex local separation phenomenon. This aspect dominates the loss mechanism. The flow is thus highly tridimensional in any blade row and reinforces the unsteady character of the flow for the next one.

FIGURE 8 NON VISCOUS CASCADE SECONDARY FLOW

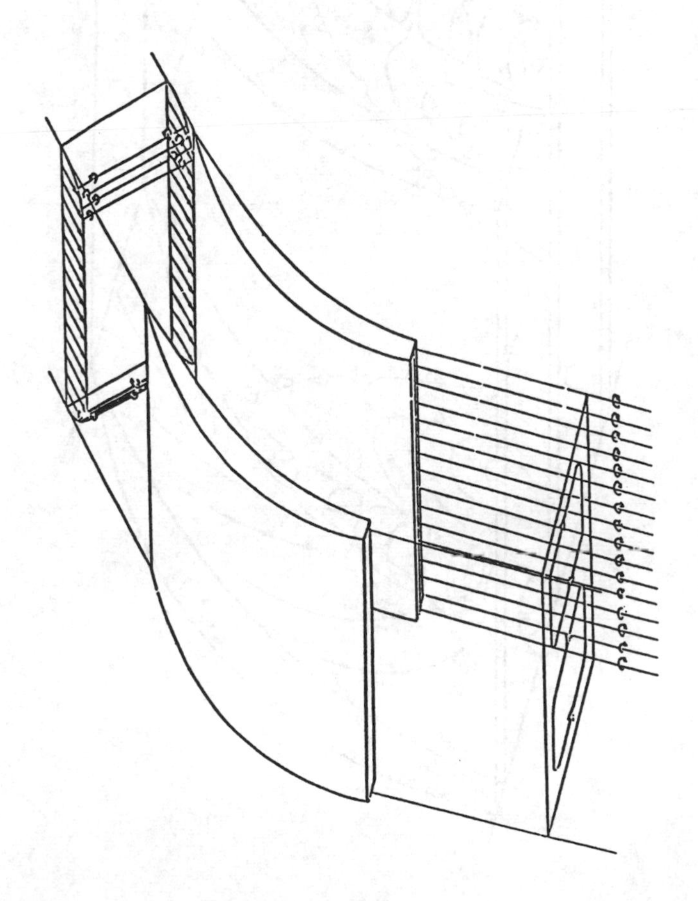

FIGURE 9 VISCOUS CASCADE SECONDARY FLOW

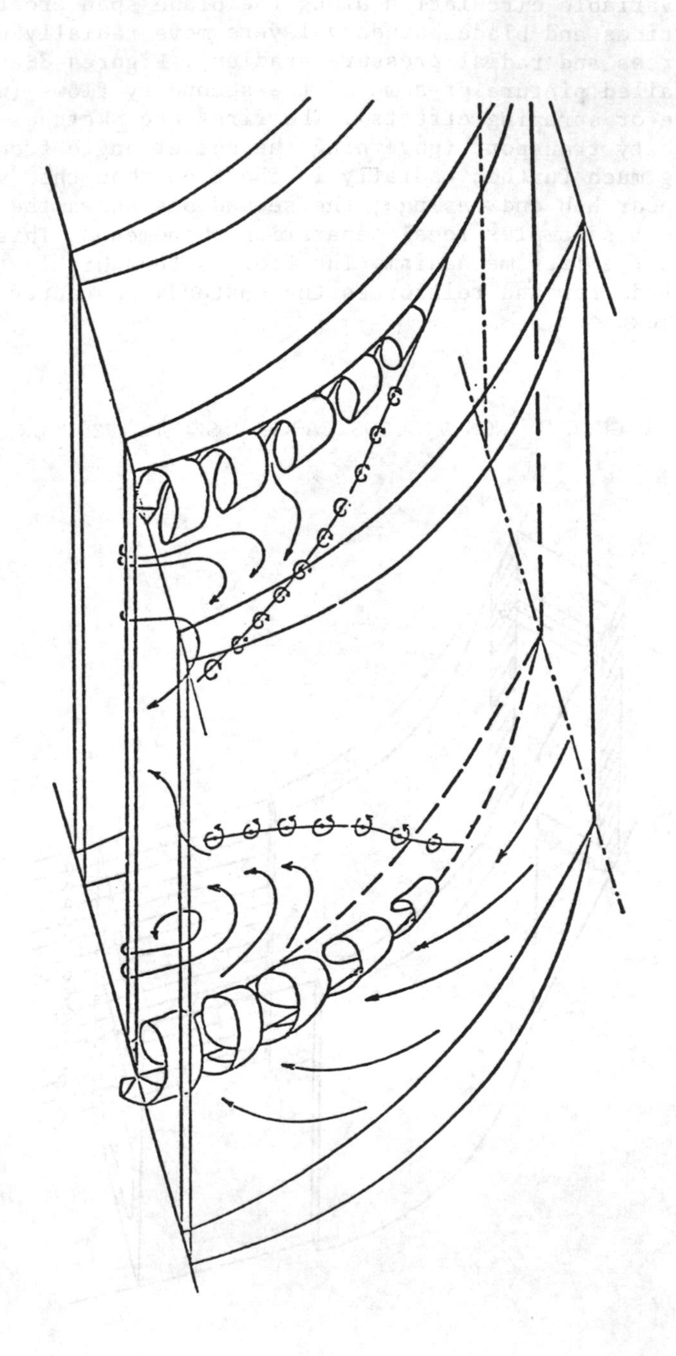

Supposing now that the circumferential averaging techniques discussed in Hirsch's presentation are used near hub and casing ; they lead to the definition of tangentially averaged skewed end wall boundary layers. The corresponding flow is submitted to the violent and almost instantaneous change of referential system depicted in figure 10, tending to generate a less skewed state for the end wall boundary layer.

FIGURE 10 END WALL BOUNDARY LAYERS : RELATIVE MOTION

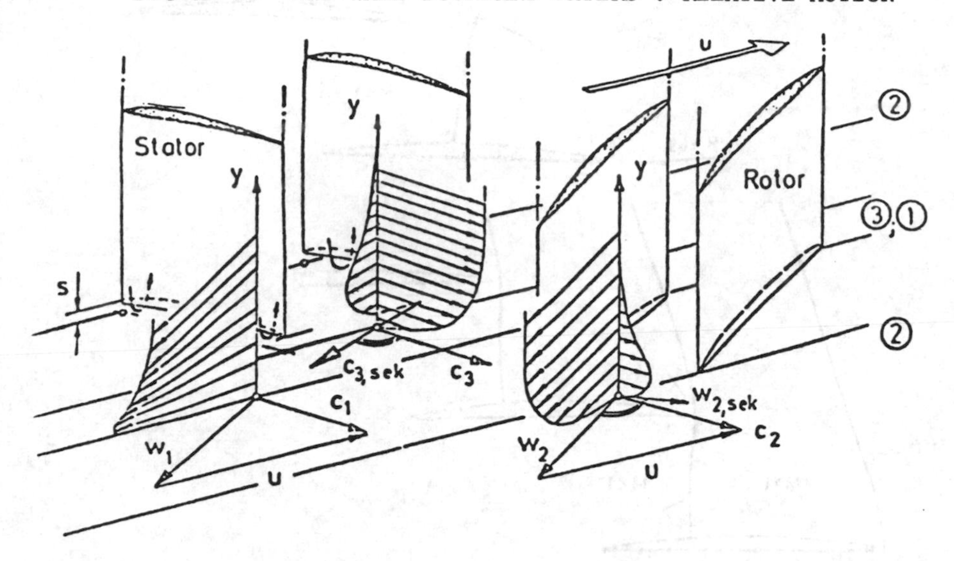

Like the conventional boundary layers, those end wall regions introduce a displacement effect (blockage) and additional losses. They also modify the local blade forces and hence the mechanism of energy transfer. The problems associated with those regions will be delt with in detail in session 6B.

Finally, in most cases, compressibility effects cannot be neglected. Typical blade to blade transonic flows (involving shock and shock boundary layer interaction) are depicted in figure 11.

FIGURE 11 COMPRESSIBILITY EFFECTS : SHOCK WAVES

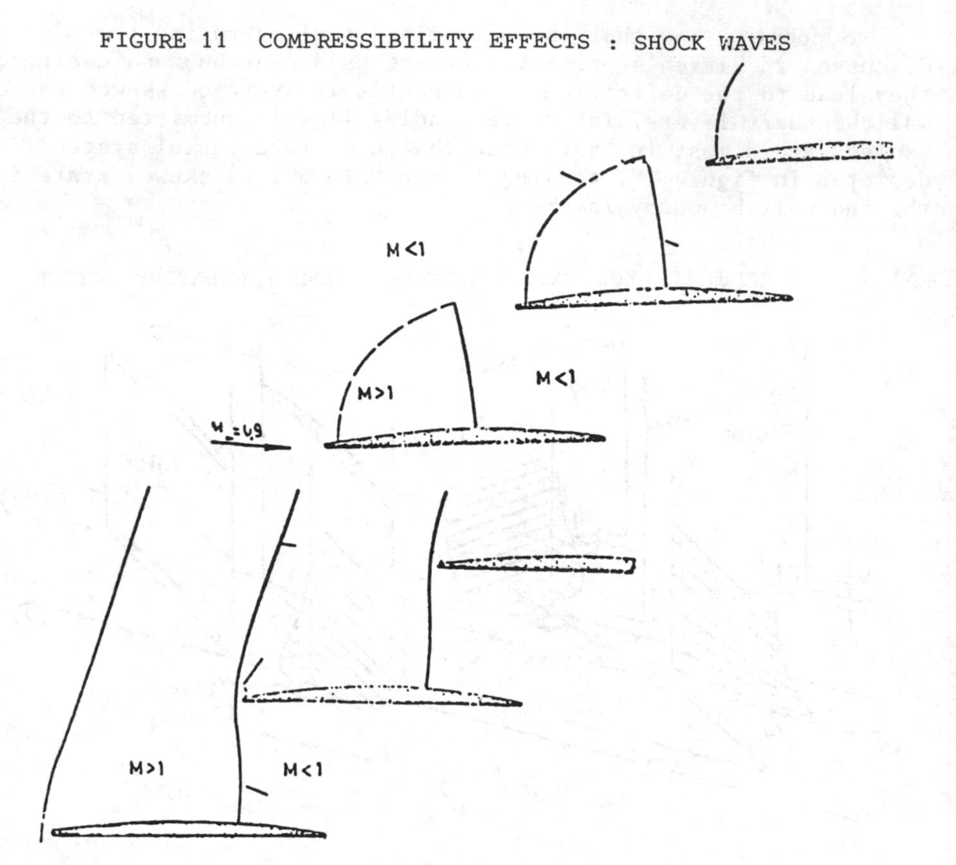

4.2. Flow Modelling for the "Sound Region" of Operation

The real 3-D, unsteady viscous turbulent compressible confined flow in axial flow compressor is intractable in its integrity, in the present state of the art of physical flow understanding and of computer development, notwithstanding the fast developments in progress.

Simplified flow modelling are still required. Those in current use

- postulate a steady flow relative to each blade row (which implies that a representative realistic circumferentially averaged flow must be defined at blade inlet). It is under this postulate that Euler's turbomachinery equation can be written for any circumferential stream surface

$$\Delta H = U_1 \, V_{\theta_1} - U_2 \, V_{\theta_2}$$

for the energy transfer.

Although 3-D methods are progressively coming into operation at least for the case of inviscid, but eventually shocked flows, Wu's basic approach [4] of dividing the three-dimensional flow into two two-dimensional flows (figure 12A) on stream surfaces of the family S_1 and S_2 remains the backbone for flow modelling. Those surfaces correspond respectively to blade to blade and to hub to tip flow.

FIGURE 12

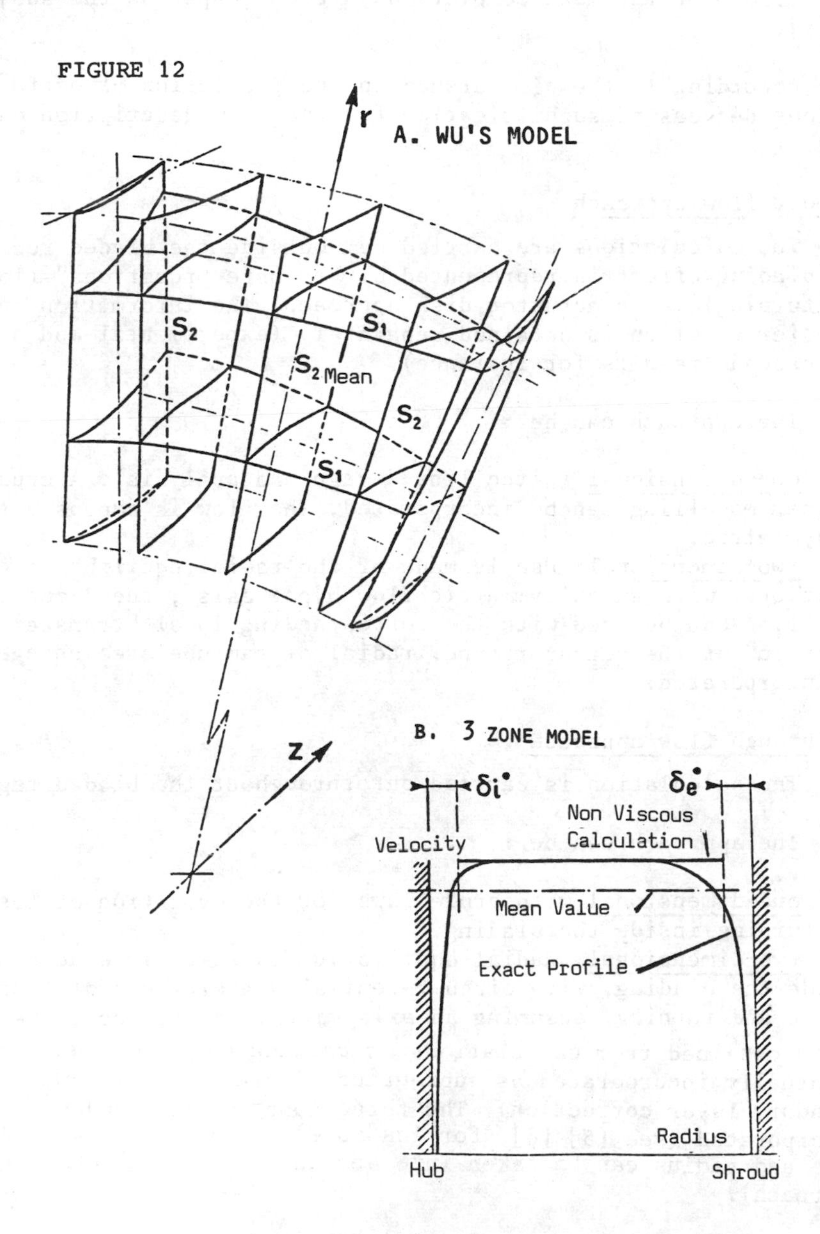

This approach is not sufficient to describe correctly the regions near hub and tip (end wall pseudo boundary layer and vorticity transport). This led to the definition of the "3 zones" approach (figure 12 B) where a "sound" core flow, dealt with according to Wu's approach, is modelled in interaction with a hub and a casing region . Further lectures will describe this model in detail. One of the most complete and recent paper on the subject is [5].

According to the aim pursued in the prediction of performance, various degrees of sophistication for the flow description can be used.

A. Duct flow approach

The calculations are carried out outside the bladed region. The blading effect is represented by "transfer functions" (losses and turning) in an actuator disk approach. The information on the transfer function is obtained separately (experimental and numerical cascade results for instance).

The approach can be :

one-dimensional (pitch line design and analysis). A crude 3 zones modelling can be incorporated. The flow is supposed to be axisymmetric.

two-dimensional. Use is made of the radial equilibrium equations, with an axisymmetric flow hypothesis ; the 3 zones modelling can be used with the corresponding local "transfer function" of the actuator type, radial streamtube area change can be incorporated.

B. Through flow approach

The calculation is carried out throughout the bladed region.

The approach can be :

one-dimensional with crude laws for the evolution of losses and turning inside the blading.

two-dimensional. Radial equilibrium is taken into account inside the blading, with circumferentially averaged evolution for losses and turning, assuming an axisymmetric S1 surface. The average is obtained from correlations or cascade calculations, eventually incorporated as subroutines (potential flow plus boundary layer correction). The three zones model can be incorporated (see [5] [6], for instance), variation of streamline area and radius can be taken into account (so-called quasi 3-D approach).

The definition of the circumferentially averaged streamline can be iterative, including on the "3 zones" aspects (weak S-2/S-1 interaction). A typical flow chart is given in figure 13.

FIGURE 13 FLOW CHART FOR AXIAL COMPRESSOR CALCULATION

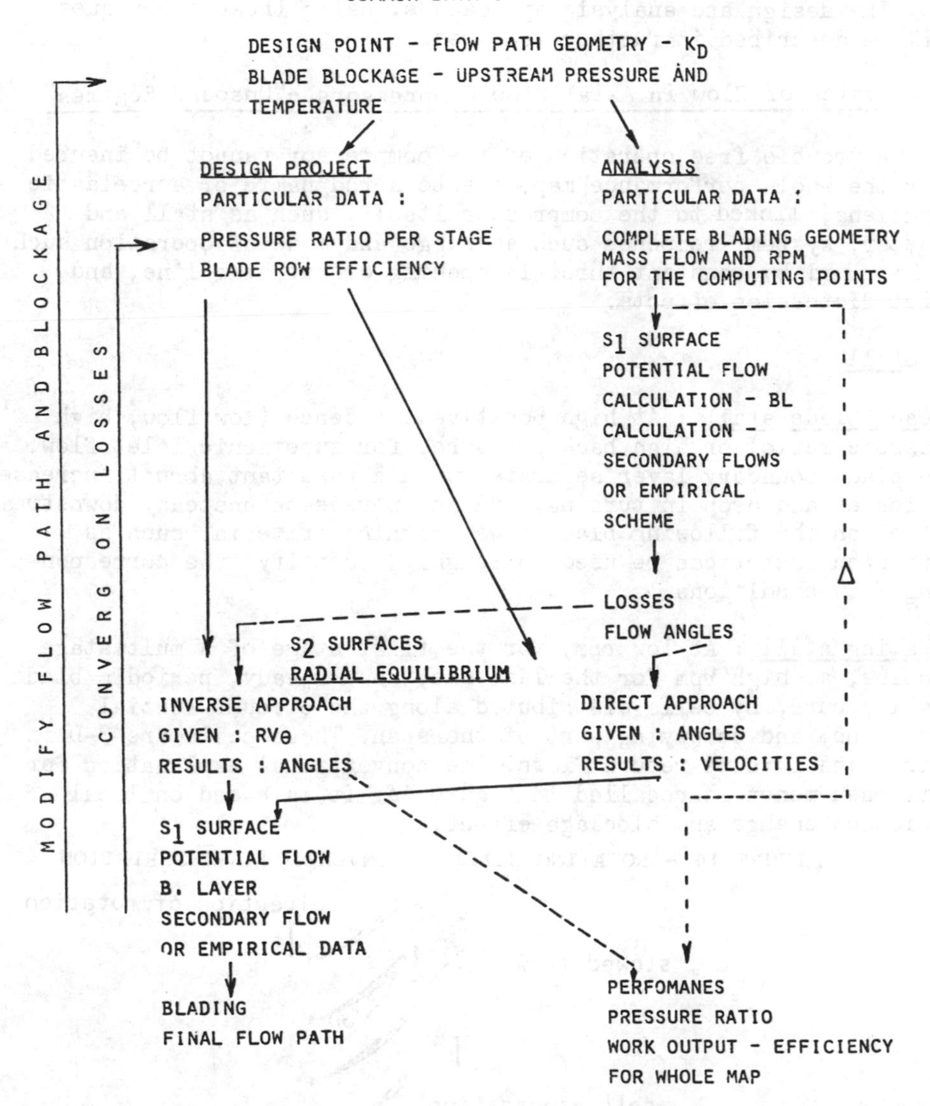

- Full S-2/S-1 iterative calculations - Full 3-D

For the blade to blade treatment, the case of compressible subsonic, transonic and supersonic flow, with shock, boundary layer and shock boundary layer interaction have to be dealt with, in both the inverse (blade design) and direct (known geometry) cases. Such methods, although imperfect, are available, especially for the direct approach but certainly cannot tackle satisfactorily the case of separated flows.

The design and analysis approaches, using those techniques will be described in further lectures.

4.3. Pysics of Flow in Axial Flow Compressors -"Unsound Regimes"

A trouble free operation of the compressor cannot be insured over the whole performance map, due to aerodynamic or aeroelastic phenomena, linked to the compressor itself, such as stall and flutter, system response such as surge and mode of operation such as imposed by constant throttle opening or matching line, and inlet distorsion effects.

A. Stall

Steady blade stall : At high positive incidence (low flow, high pressure ratio) or high back pressure, for supersonic inlet flow, the blade boundary layer separates with a resultant abrupt increase in losses and drop in turning, and an increasing unsteady downstream effect on the following blade rows. Warning criteria, such as diffusion factor can be used to roughly identify the correspon-ding flow conditions.

Rotating stall : At low rpm, for the first stage of a multistage machine, at high rpm for the last one, an unsteady, periodic blade stall occurs, by cells distributed along the circumferential direction, and occupying part of the span. Those cells are 3-D in nature and involve return flows. The conventional explanation for this phenomenon is recalled in figure 14. It is based on local incidence change and blockage effect.

FIGURE 14 - ROTATING STALL - CONVENTIONAL DESCRIPTION

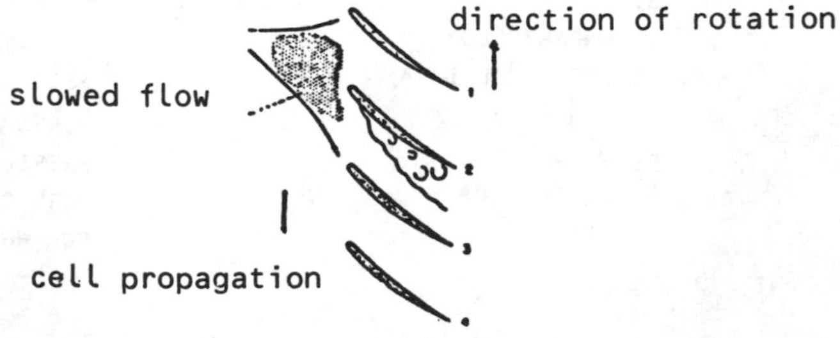

direction of rotation

slowed flow

cell propagation

The real picture is more complex, as shown in figure 15 (from Greitzer [7]) for the axial propagation aspect. The present understanding is that the rotating stall appears due to the inability of the flow to fill the area offered while respecting the radial equilibrium, as suggested by Fabri, among others. Number of cells and speed of rotation can be roughly evaluated by flow stability studies, predicting the transition from a stable (axisymmetric steady flow) to other stable, unsteady periodic flow (see Ferrand [8] for instance). The velocity of rotation is between 0.3 and 0.8 of the rotor velocity. The mechanical effect can be harmful. This type of flow will be studied in the last part of the course.

Depending on the extent of the cells, the overall pressure ratio of a multistage compressor can be affected or is not affected by rotating stall.

FIGURE 15 ROTATING STALL - ACTIVE CELL CONCEPT

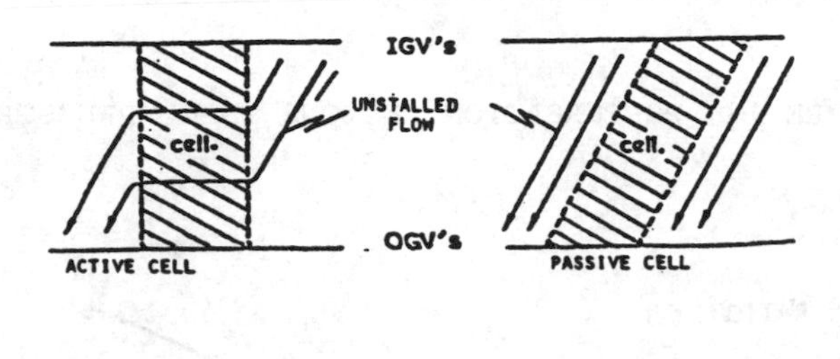

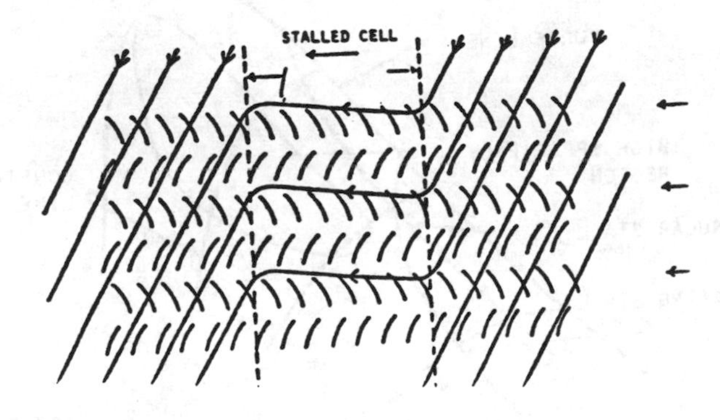

Annular or wall stall : This corresponds to the appearance of an axisymmetric low flow, high loss region near hub or more often, casing, enhanced by clearance effects,appearing in the pseudo-end wall boundary layer. It is, in fact, a self throttling process, like rotating stall, which, at reduced volume flow, allows the equilibrium laws of the main flow to be fulfilled. It can again be predicted by stability studies (zero frequency of rotation of disturbances) [8]. It affects pressure ratio and efficiency.
The type of stall appearing depends on compressor design : loading, basic radial equilibrium law, flow conditions, mode of operation. Figure 16 shows the area where the main stall phenomena can appear.

B. Flutter is an aeroelastic phenomenon whereby the total (aerodynamic plus mechanical damping) of the blade tends to zero or negative value. A self sustained or amplified vibration of the blade in torsion, flexion or both appear. It can be generated by subsonic or supersonic blade stall (a high pressure on a constant speed line) by choking (low pressure end of constant speed line) due to subsequent shock unsteady motion, etc.

Figure 16 also shows the region affected in a compressor map.

FIGURE 16 PERTURBATION REGIONS IN COMPRESSOR MAP

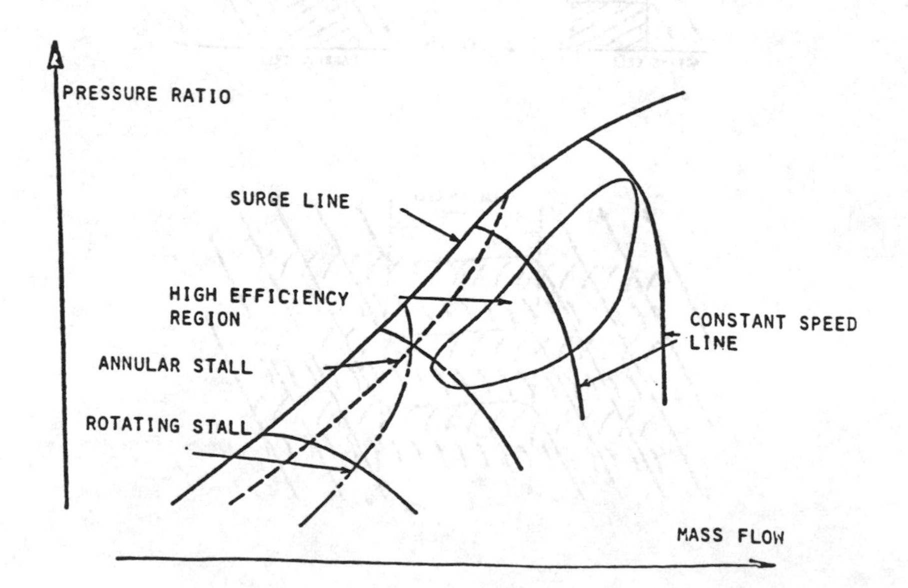

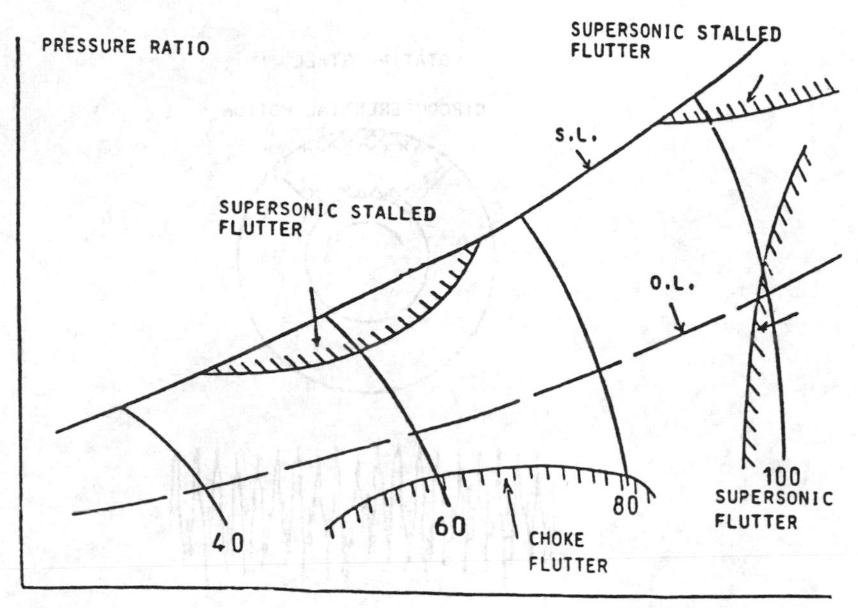

C. Surge

This is a "system" phenomenon (compressor, inlet region, discharge volume, throttling effect) usually generated by one of the stall phenomena described above. It can be approached by stability studies. The result is a global axial flow pulsation, of rather low frequency, which can go up to complete flow reversal. The corresponding blade stresses can be fatal to the mechanical integrity. Figure 17 compares the relative frequencies of rotating stall and surge.

Rotating stall and surge will be dealt with in detail in Greitzer's lecture.

The prediction of stall, flutter and surge is one of of the most important topics for research in the field of axial flow compressor. Progress is fast, but a full understanding is still to be gained.

FIGURE 17 SURGE - ROTATING STALL

ROTATING STALL

CIRCONFERENTIAL MOTION

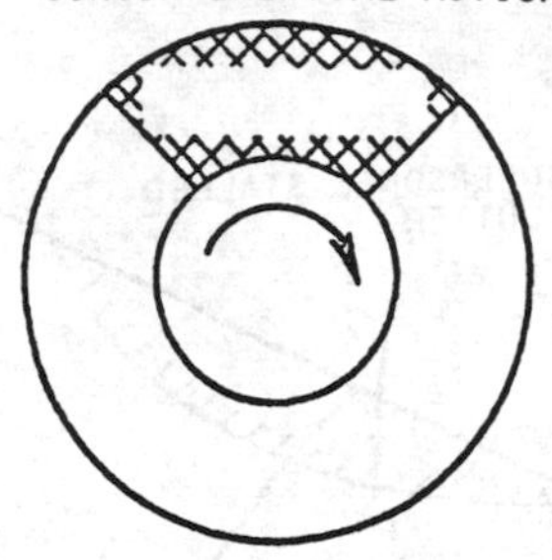

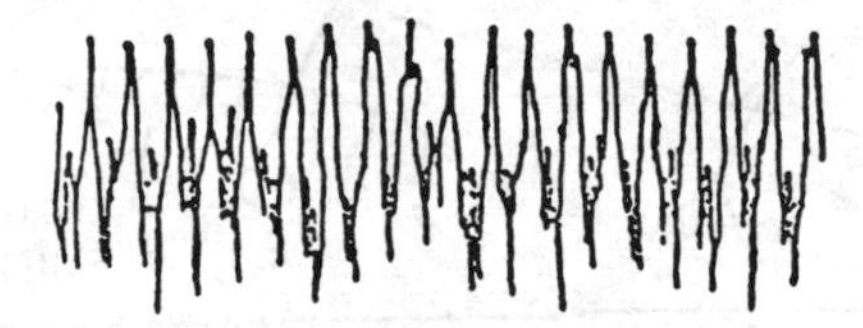

FREQUENCY ~50 - 100 Hz
COMPRESSOR RESPONSE

SURGE
PERIODIC AXIAL MOTION

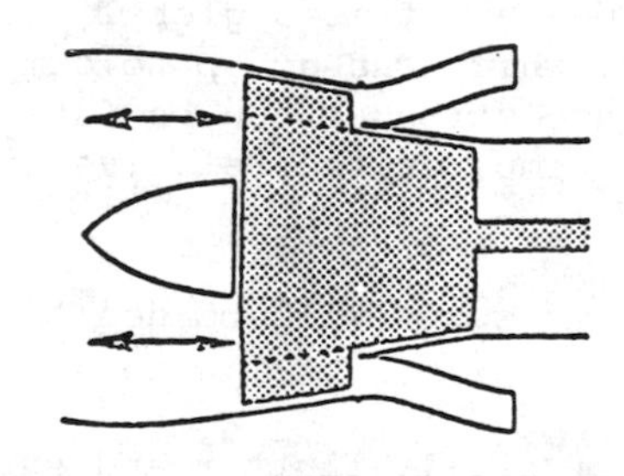

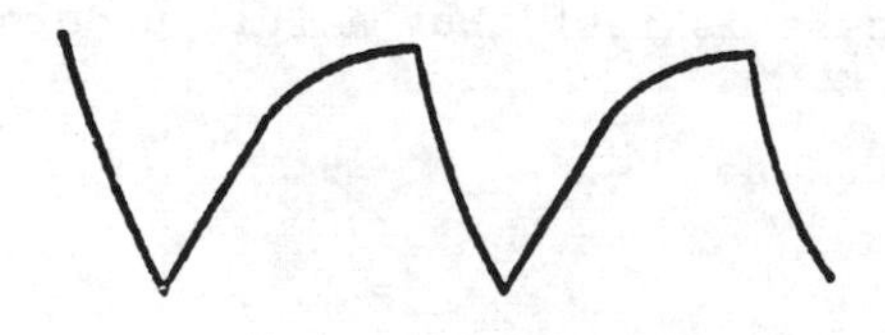

FREQUENCY ~ 3 - 10 Hz

SYSTEM RESPONSE

D. So-called "surge line" (Figures 1, 3, 4,5, 16)

The compressor map is limited, on the high pressure side, by a limit pressure ratio, function of mass flow. At higher value of the pressure, the physical integrity of the compressor cannot be guaranteed. The limiting phenomenon is not necessarily surge but can also be flutter, etc.

C. Influence of operating conditions

Inlet distorsion : In actual operation, the inlet flow to the compressor is not axisymmetric, and sometimes, unsteady (effect of inlet elbows, air intake , where non uniform flows are generated, often associated with local boundary layer separation). Figure 18 explains why a lower local inlet pressure can lower the surge line position (simple parallel compressor approach).

FIGURE 18 INLET DISTORSION

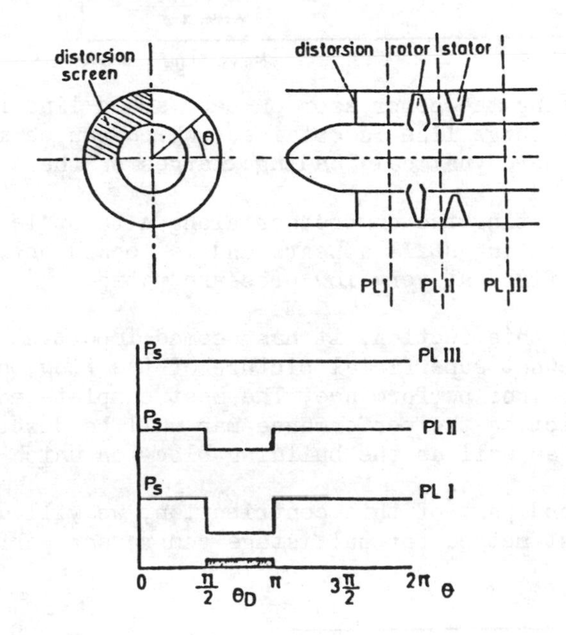

Effect of the path used to vary pressure ratio and mass flow

The transition from a basic stable axisymmetric steady flow to another stable unsteady flow - through a kind of bifurcation process - depends on the mode which is used to change mass flow and pressure ratio.

Figure 19 shows an extreme case of "double" characteristic for a high pressure ratio, five stage compressor (from [6]).

FIGURE 19 CHARACTERISTICS OF A FIVE STAGE COMPRESSOR

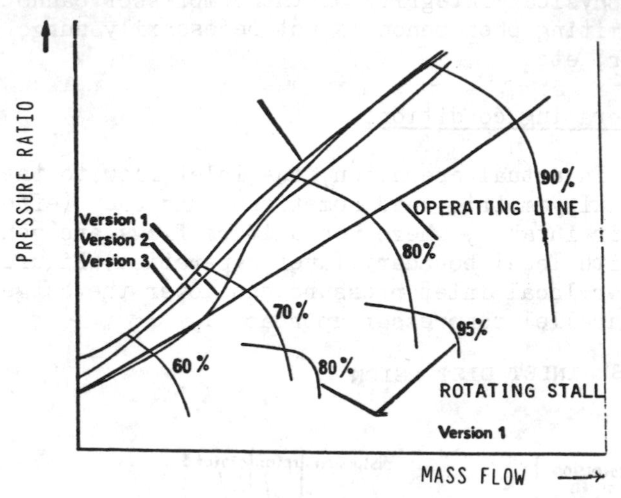

When operating the compressor on each speed line from high to low mass flow, a surge line is obtained, which can be modified by geometrical and aerodynamical loading changes on the compressor.

When accelerating the compressor along a throttle line, a stable, heavy rotating stall appears and is locked on, resulting in a surge line fixed at very low pressure ratio.

To conclude this section, it has seemed important to paint a broad and somewhat superficial picture of the flow phenomena fixing the compressor performance. The most complete modelling system for predicting the performance map will be described in other lectures, as well as the building blocks on which it is based.

In the second part[1] of this contribution, we will describe some approximate, fast method for multistage compressor performance prediction.

---

[1]The approximate, fast method is explained in the paper by H.Minton, Y.Doumandji and J.Chauvin "Simplified method for flow analysis and performance calculation through an axial compressor". ASME paper No. 84-GT-250. The paper will also be published in the Journal of Engineering for Power.

REFERENCES

1. Horlock, J.H. Axial flow compressors. Butterworth 1958.
2. Breugelmans, F. Flow in turbomachines and turbomachinery component characteristics. VKI Course Note 83. Nov. 1971.
3. Karadimas, G. Advanced design and analysis methods for compressor and turbine aerodynamics. Lectures at Athens Polytechnic University and Institut de Mécanique des Fluides de Marseille (1981).
4. Wu, Chung Hua. A general theory of three-dimensional flow in subsonic and supersonic turbomachines of axial, radial and mixed flow types. NACA TN 2604, 1952.
5. Leboeuf F. Contribution theorique et experimentale à l'étude des écoulements secondaires dans un compresseur axial transsonique. Thèse d'Etat, Université Claude Bernard, Lyon. 25/06/1984.
6. Thiaville, J.M. Modèles de calcul de l'écoulement dans les turbomachines axiales. "Through flow calculations in axial turbomachinery". AGARD CP 195, Oct. 1976.
7. Greitzer, E.M. Review. Axial flow compressor stall. MIT publication.
8. Ferrand, P. and Chauvin, J. Theoretical study of flow instabilities and inlet distorsions in axial compressor. ASME paper GT 04 - 1981.

LIST OF SYMBOLS

a Velocity of sound

C_p Specific heat at constant pressure

H_o Stagnation enthalpy

L Characteristic dimension

$\dot{m}$ Mass flow rate

N rpm

p Static pressure

R Gas contants

T Static temperature

T_o Total temperature

U Peripheral velocity

V Absolute velocity

Δ Difference

η_{is} Isentropic efficiency

η_p Polytropic efficiency

ϕ Flow coefficient

ψ Loading coefficient

Subscripts 1 Compressor or wheel inlet

2 Compressor or wheel outlet

o Stagnation.

LIST OF FIGURES

AXIAL TURBINE PERFORMANCE PREDICTION METHODS

C.H. Sieverding

von Karman Institute for Fluid Dynamics

1. INTRODUCTION

The turbine performance prediction methods can basically be split into two major categories : The first category groups together the so-called overall stage methods. The use of these methods does in general not require any details of the bladings and the complex flow patterns in the turbine are deliberately ignored. The turbines are treated more or less as black boxes. These methods are in general derived from overall performance measurements of a certain number of turbines with similar characteristics. The use of these methods is justified in the initial design phase for the selection of the turbine design parameters. The optimization of a turbine for a given duty requires, however, a deep understanding of the flow and only very refined performance prediction methods, which take into account details of the bladings and of the meridional flow channel can be of real help. This is done, to a certain extent by the second category of performance estimation methods which evaluate the total losses through the sum of a great number of individual loss components each of which is influenced by a more or less large number of geometric and aerodynamic parameters. The problem consists in defining clearly the important influence parameters and in making certain that the influence of each parameter can be studied separately. Such systematic variations of the various influence parameters can in general only be done on simplified models. There exist, however, limitations to the degree of simplification beyond which model tests become meaningless. A constant effort has therefore to be made to simulate the real flow conditions as closely as possible. The degree of the remaining differences between the real turbine flow and the model flow determines whether the model test results can be applied to the turbine without any corrections or whether the model test results indicate only the correct tendencies while the absolute loss level needs more or less important correction factors.

2. OVERALL STAGE PERFORMANCE METHODS

2.1 Design point performance

Overall stage methods are in general developed for families of similar turbines. Hence their use is limited to turbines being designed with those characteristics, e.g., same range of degree of reaction, same aspect ratio, same family of blade profiles. The most widely known turbine efficiency correlation is the correlation of Smith [1], (Fig. 1). It is based on data of some 70 HP and LP cold flow model turbine stages. All stages have the following features in common :

- axial velocity ratio of unity
- design point incidence equal to zero
- degree of reaction : for the hp-stages down to 20%, for LP stages up to 60%. The degree of reaction is chosen always sufficiently high to avoid losses associated with recompression at the blade roots.

The turbine efficiencies are presented in function of the stage loading ψ and the flow coefficient ϕ taken at the mean radius. The measured efficiencies were corrected to zero clearance. The Re number effect on the turbine efficiency was found to be negligible in the range $10^5 < RE < 3 \times 10^5$. Smith claims for his correlation an accuracy of $\pm$ 2%.

The presentation of performance data under the form used by Smith, $\eta = f(\phi,\psi)$, has found wide application since such performance maps are particularly easy to use in the early design process to predict stage efficiency changes resulting from the variation of certain basic aerodynamic parameters. Kacker and Okapuu [2] found Smith's correlation so convincing that "it must be accepted as one of the tests which any mean line loss system must satisfy". The authors used this correlation to calibrate the Ainley/Mathieson/Dunham/Came (AMDC) performance prediction method [3]. After a critical analysis of the AMDC method they introduced certain modifications, concerning primarily the Mach number influence and adjusted their impact on the blade performance by a trial and error procedure until a simulation of Smith's curves was obtained. In a second step, Kacker and Okapuu accounted for the progress in blade profile design and demonstrated the prediction capability of their method by comparison with the performance data of some 30 turbines of more recent design.

Also Craig & Cox [4] used Smith's correlation to verify the prediction capability of their own method, which was based primarily on cascade data, and found good agreement.

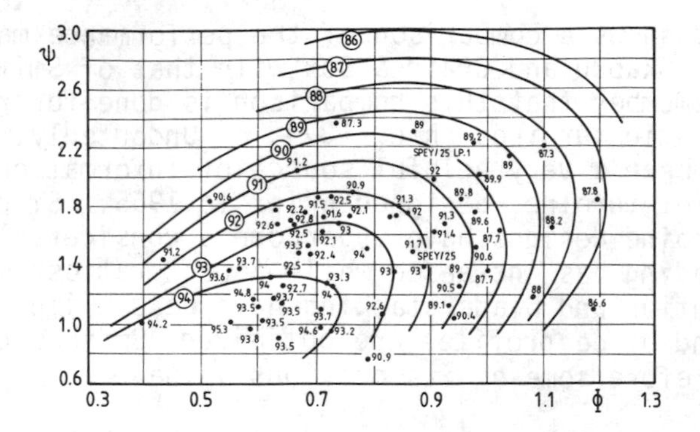

FIG. 1 - TURBINE STAGE EFFICIENCY AT ZERO TIP LEAKAGE, AFTER SMITH [1]

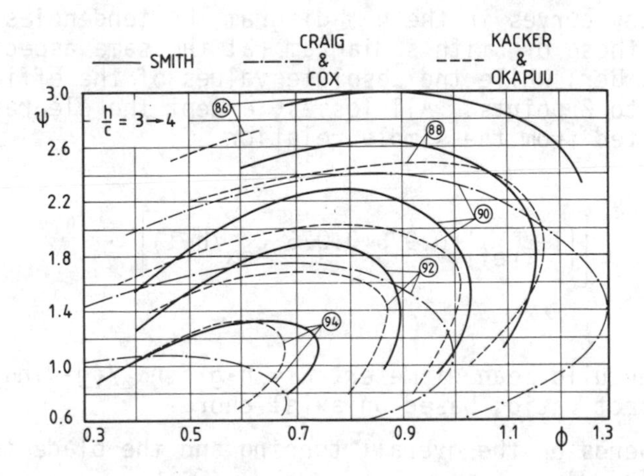

FIG. 2 - COMPARISON OF PREDICTED TURBINE EFFICIENCIES BY CRAIG & COX [4] and KACKER & OKAPUU [2] AT ZERO TIP LEAKAGE WITH SMITH'S CORRELATION

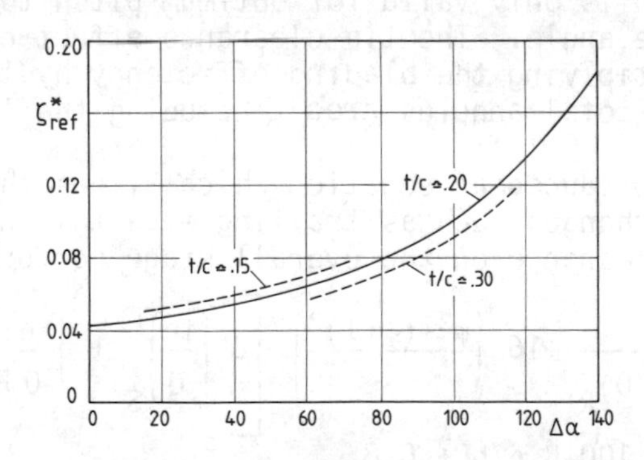

FIG. 3 - SODERBERG'S LOSS COEFFICIENT ζ^{*}_{ref} [5]

Figure 2 shows a comparison of the performance maps calculated by Kacker & Okapuu and Craig & Cox with that of Smith. It is important to remember that this comparison is done for relatively high aspect ratio turbines h/c = 3→4. Undoutedly, Smith's correlation has been a very helpful source of information on turbine performance for turbines designed prior to 1965. Since then, aircraft gas turbine designs have undergone a considerable evolution. The stage loading has increased while aspect ratios decreased. Work distribution and blade stacking have become important design parameters and blade profiles are optimized. Smith's correlation has lost therefore some of its previous value.

Outdated, but in certain cases still used because of it's extreme simplicity is Soderberg's correlation [5] which was derived from high speed (subsonic) turbine tests. Plotted under the form of efficiency curves in the ψ-ϕ diagram the tendencies are very similar to those of Smith's diagram (at the same aspect ratio and Reynolds number) while the absolute values of the efficiencies are low by 0.5 to 2 points. All losses, except the clearance losses are calcualted from the simple relation

$$\zeta^{\star} = \left(\frac{10^5}{R_h}\right)^{1/4} \left[\left(1+\zeta^{\star}_{ref}\right)(0.975+0.075\ C_{ax}/h-1)\right] \qquad (2.1)$$

where

R_h = hydraulic mean diameter, $R = h \cdot g \cdot \sin\alpha_2/(g - \sin\alpha_2 + h)$

C_{ax}/h = aspect ratio, based on axial chord

$\zeta^{\star}_{ref}$ = depends on the overall turning and the blade thickness (Fig. 3)

The relation $\zeta^{\star}$ = f (flow turning angle) is the weakest point of the method, since it does not allow to distinguish between reaction and impulse bladings having the same turning. It should be noted that the above relation is only valid for optimum pitch to chord ratio and zero incidence angle. The tip clearance effect can be accounted for by simply multiplying the blading efficiency by the ratio of the blade area to the total annulus area (including the tip clearance).

Latimer [6] produced an equation which allows the evaluation of the effect of changes such as trailing edge thickness, aspect ratio and blade clearance on the overall stage performance.

$$100-\eta = \frac{42}{(h/0)_S+(h/0)_R} + 16\left(\frac{\psi^2+(\phi+1)^2}{\phi}\right)\left[2\left(\frac{te}{0}\right)_S + \left(\frac{te}{0}\right)_R\right] - 8.5\ \lambda + 100\ B\ \delta/h + 6.8 \qquad (2.2)$$

where :

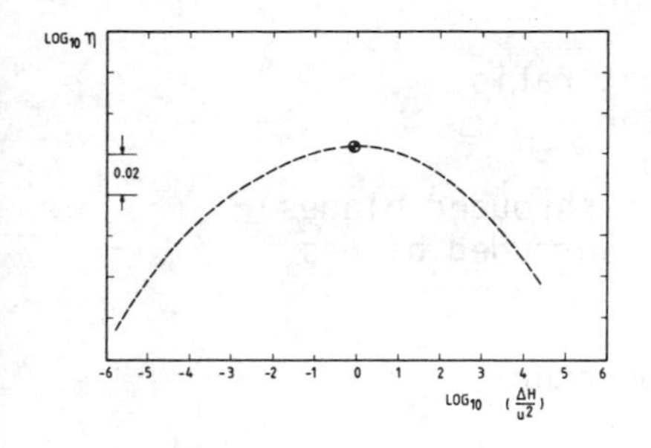

FIG. 4a - LATIMER'S OFF DESIGN PERFORMANCE CORRELATION [6]

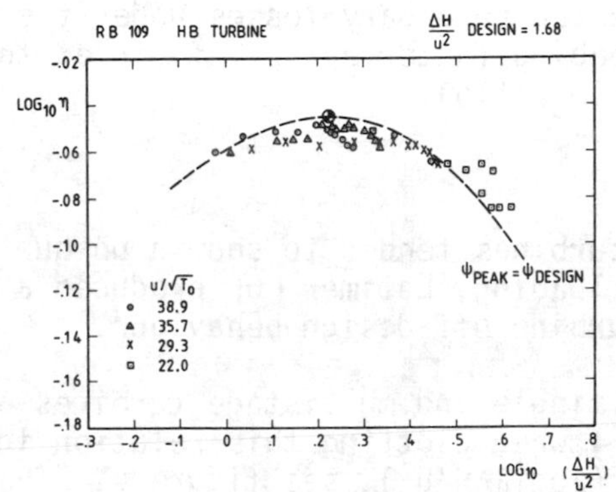

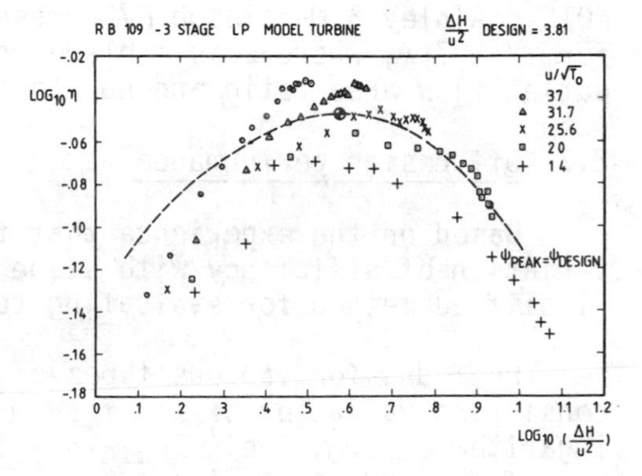

FIG. 4b - COMPARISON OF LATIMER'S CORRELATION WITH TURBINE DATA - PEAK EFFICIENCY AT DESIGN LOADING

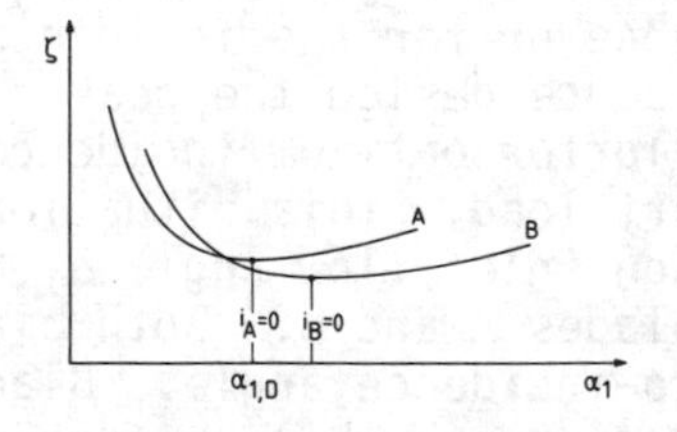

FIG. 5 - PERFORMANCE CURVES FOR TWO BLADE PROFILES WITH DIFFERENT BLADE INLET ANGLES

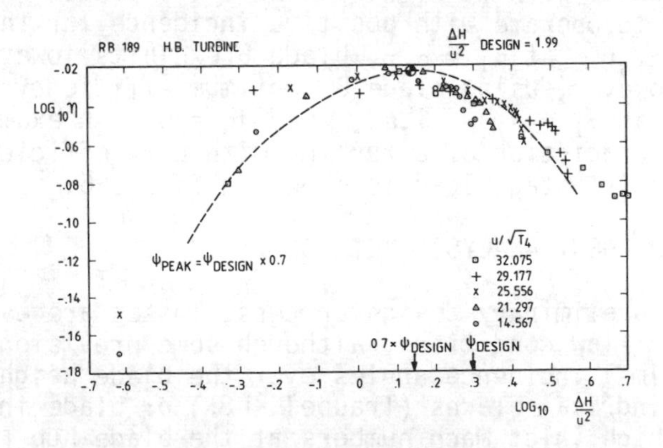

FIG. 6 - COMPARISON OF LATIMER'S CORRELATION WITH TURBINE DATA - PEAK EFFICIENCY AT OFF DESIGN CONDITIONS

h/O = blade height to throat ratio
te/O = trailing edge thickness to throat ratio
λ = Ainley's secondary loss parameter
δ/h = clearance to blade height ratio
B = clearance constant B = 0.5 for unshrouded blades
B = 0.25 for shrouded blades
ϕ = flow coefficient V_x/U
ψ stage loading $\psi = \Delta H/U^2$
subscripts S and R indicate Stator and Rotor

NOTE : Ainley & Mathieson [7] presented secondary losses under the form $Y = Z \cdot \lambda$, where Z is a blade loading factor and λ = f (inlet to outlet flow area ratio and hub to tip ratio).

2.2 Off design performance

Based on the experience that turbines tended to show a unique variation of efficiency with stage loading, Latimer [6] produced a simplified method for evaluating turbine off-design behaviour.

He found for various types of single and multistage turbines a consistent variation of $\eta = f(\Delta H/u^2)$ when plotting this relation in logarithmic form, i.e., $\log_{10} \eta = f(\log_{10}\Delta H/u^2)$, see figure 4. The use of this curve implies a knowledge of the maximum efficiency and the stage loading at which η_{max} occurs. As pointed out by the author the maximum efficiency does not necessarily occur at design point. The turbine performance curves depend strongly on whether the turbines are designed for zero incidence or for the incidence resulting in lowest losses. At zero incidence design the peak efficiency will occur at nominal load while for optimum incidence angle designs the turbine will peak at part load. This situation is illustrated in figure 5. For the design inlet flow angle $\alpha_{1,D}$. the designer has the choice between two blades A and B. Both blades are supposed to have lowest losses at zero-incidence angle. Blade A is designed to work at $\alpha_{1,D}$ with zero incidence, while blade B is designed to operate with positive incidence for the same inlet flow angle $\alpha_{1,D}$. At $\alpha_1 = \alpha_{1,D}$, blade B exhibits lower losses than blade A. However, using blade B, optimum efficiency will not occur at $\alpha_{1,D}$ but at $\alpha_1 > \alpha_{1,D}$, i.e., when $i_B = 0$. An example for the performance prediction of a turbine with peak efficiency at 70% of the design point stage load is shown in figure 6.

3. LOSS COMPONENT ANALYSIS METHODS

In the preliminary design process, losses are evaluated usually for mid-span flow conditions, although some provision may be made for non-optimal incidence angles over the blade height for the case of cylindrical blades (Traupel, [8]) or blade inlet shock losses due to high inlet Mach numbers at the blade hub (Kacker and

Okapuu [2]). For more detailed prediction of the flow conditions along the blade height it is advisable to attempt to predict the loss variation along the blade height using the local velocity triangles and taking into account incidence angle effects when needed.

It is not intended to provide a complete compilation of all loss correlations, which have been used in the past but rather to discuss those few which are widely used at present and point to those aspects which have not yet been adequately resolved.

3.1 Profile losses

Applying the conservation equations between the trailing edge plane 2' and a far downstream plane 2 and neglecting terms of second order importance, it can be shown that the total profile losses are composed of three parts (Traupel [8]),

$$\zeta_p = 2\cdot\bar{\theta} + \left(\frac{\overline{te}}{1-\overline{te}}\right)^2\cdot\sin^2\alpha_2 + CP_b\cdot\overline{te} \tag{3.1}$$

where

$$\begin{aligned} \bar{\theta} &= (\theta_{SS}+\theta_{PS})/(g\cdot\sin\alpha_2') \\ \overline{te} &= te/(g\ \sin\alpha_2') \\ CP_b &= (P_2'-P_b)/(\rho/2\cdot V_2'^2) \end{aligned} \tag{3.2}$$

The first term on the right hand side represents the boundary layer and mixing losses at zero trailing edge thickness, the second term the losses due to the sudden area expansion at the trailing edge in case of $te \neq 0$ and the third term downstream mixing losses due to the difference between the pressure at the trailing edge, P_b, and the average pressure across the channel exit, P_2' ($CP_b \hat{=}$ base pressure coefficient). The boundary layer momentum thickness depends :
(a) on the blade velocity distribution, i.e., on the blade shape, the pitch-to-chord ratio, the incidence angle and the outlet Mach number and,
(b) on Reynolds number, turbulence level and surface roughness.

3.1.1 Basic profile losses : To evaluate the consistency between different loss correlations it is useful to compare them for the most simple case of a cascade at

- optimum pitch-to-chord ratio
- zero trailing edge thickness
- low outlet Mach number (to exclude compressibility effects, $M_2 \leqslant 0.5$)
- standard surface finish, e.g., $k_s/C = 10^{-4}$ (equivalent sand grain roughness)
- Reynolds number above the critical value, taken usually as $Re_{cr} = 2 \times 10^5$.

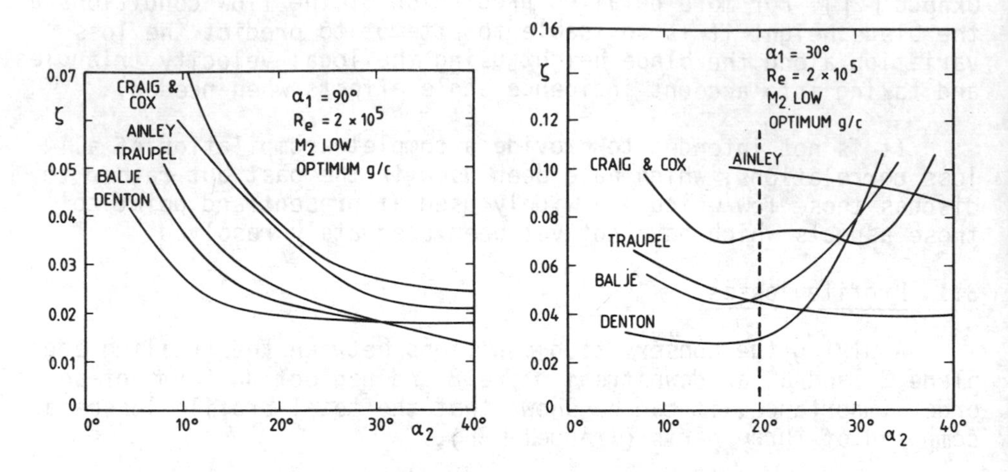

FIG. 7 - COMPARISON OF PROFILE LOSS CORRELATION METHODS FROM DENTON [9]

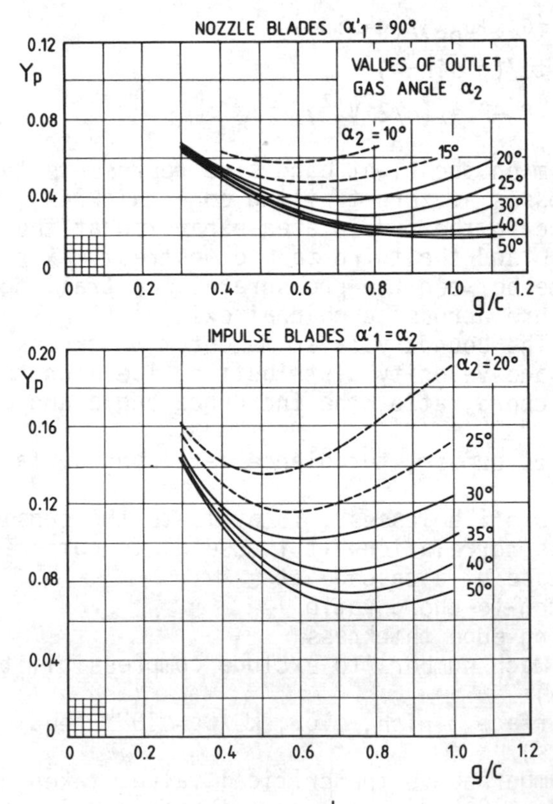

FIG. 8 - PROFILE LOSS COEFFICIENT FOR $\alpha_1'=0$ AND $\alpha_1'=\alpha_2$, $t_{max}/C=0,2$, AFTER AINLEY & MATHIESON [7]

The losses of blades with these chareacteristics are often termed "basic profile losses".

Such a comparison was made by Denton [9] for a total of seven correlations including his own method. The curves obtained with the correlations by Ainley & Mathieson [7], Craig & Cox [4], Balje & Binsley [10] and Traupel [8] which are the best known as well as that of Denton [9] are shown in figure 7. The differences are surprisingly large and there must be some good reasons for these discrepancies. Let us therefore describe briefly the foundations of each method as given by the respective authors.

- Ainley & Mathieson [7] rely on cascade tests as well as on losses derived from overall tests on a variety of turbine stages. The data refer to blade profiles designed prior to 1950, i.e., either blade profiles with contours composed of circular arcs and straight lines or blades which make use of the British blade profile series with circular or parabolic camber lines (e.g. T-6 section). The maximum blade thickness varies between t/c = 0.15 to 0.25. The correlation is based on a series of graphs of total pressure losses versus pitch-to-chord ratio for nozzle and impulse blades (Fig. 8). For any other combination of angles, the losses are calculated as

$$Y_p = \left[Y_{(\alpha_1'=90°)} + \left(\frac{90°-\alpha_1'}{90°-\alpha_2}\right)^2 \left(Y_{(\alpha_1'=\alpha_2)} - Y_{(\alpha_1'=90°)}\right)\right] \left(\frac{t_{max}/c}{0.2}\right)^{\alpha_1'/\alpha_2} \qquad (3.3)$$

Kacker and Okapuu [2] proposed to replace the term $\left(\frac{90°-\alpha_1}{90°-\alpha_2}\right)^2$ by $\left[\frac{90-\alpha_1'}{90-\alpha_2}\right] \times \left(\frac{90-\alpha_1'}{90-\alpha_2}\right)$ to allow for angles $\alpha_1' > 90°$.

The curves in figure 8 are established for a trailing edge thickness of te/g = 0.02 and Re = 2×10^5 (based on chord and outlet velocity). To obtain profile losses at zero trailing edge thickness, Y_p is to be multiplied by 0.91.

- Craig & Cox [4] present a correlation based on experimental data from straight cascades. The basic loss at zero trailing edge thickness depends on a lift coefficient $F_L = f(\alpha_1', \alpha_2)$, the ratio of pitch-to-backbone length g/b and the contraction ratio of the blade channel (Figs. 9). The use of g/b has the advantage of being closely related to the boundary layer developemnt, but it is more difficult to handle than the pitch-to-chord ratio g/c. The channel contraction ratio together with the ratio g/b allow to take into account to a certain degree differences in blade shape for steam and gas turbines and between h_p and ℓ_p turbine stages. Considering that the correlation is based on data prior to 1970 and that the steam

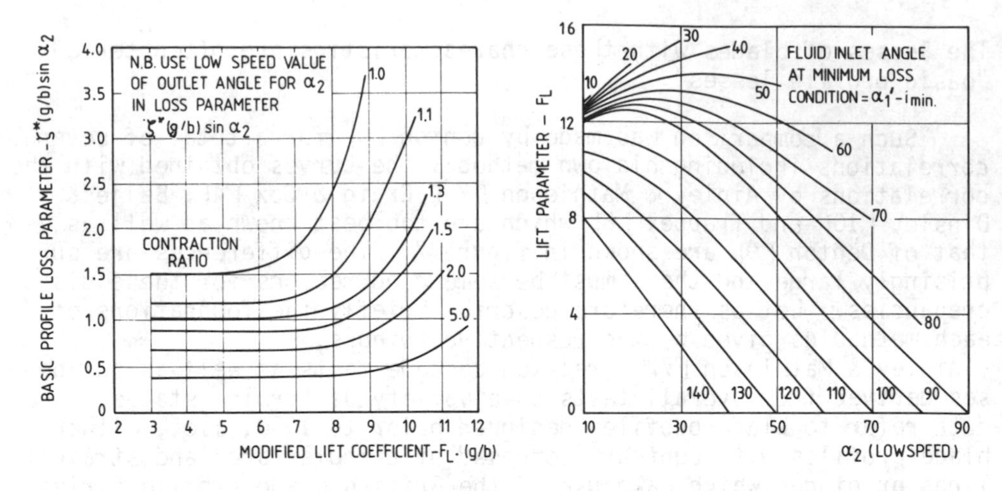

FIG. 9 - BASIC PROFILE LOSS PARAMETER (LEFT) AND LIFT PARAMETER (RIGHT), AFTER CRAIG & COX [4]

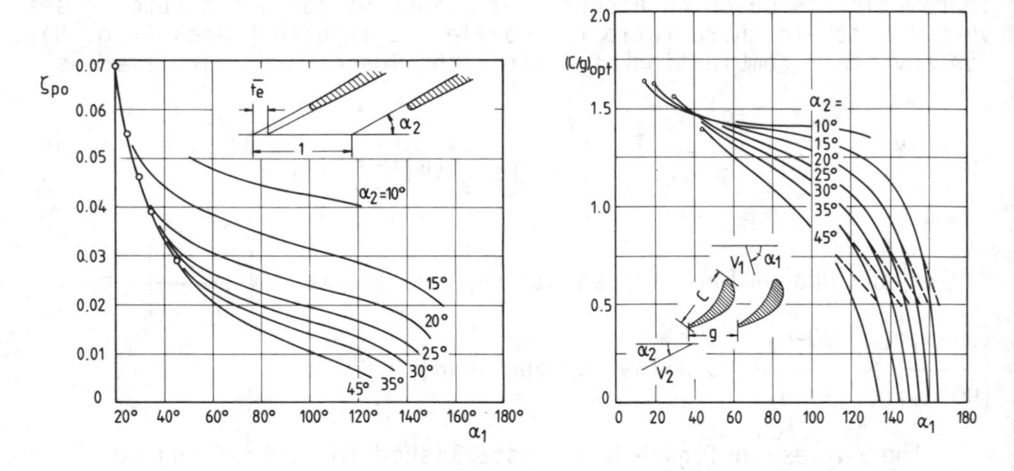

FIG. 10 - BASIC PROFILE LOSS COEFFICIENT (LEFT) AND OPTIMUM PITCH-TO-CHORD RATIO (RIGHT), AFTER TRAUPEL [8b]

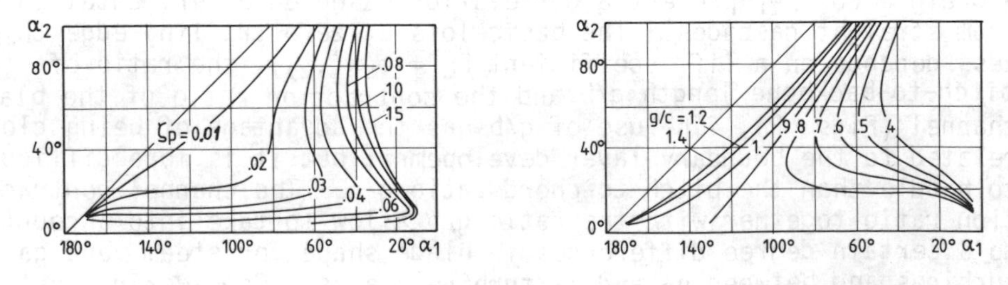

FIG. 11 - PROFILE LOSS COEFFICIENT (LEFT) AND OPTIMUM PITCH-TO-CHORD RATIO (RIGHT), AFTER BALJE & BINSLEY [10]

turbine industry has been rather slow in introducing new blade design concepts, it is likely that this correlation relies to a great extent on conventional blade designs with profiles made up of circles and straight lines.

- Traupel [8] established his loss correlation by calculating the dissipated energy D from an estimated mean velocity for pressure and suction side together with a constant dissipation coefficient c_d.

$$D = C_d \cdot b \cdot \frac{\rho}{2} \cdot \overline{V}^3 \left[(1+U)^3 + (1-U)^3\right] \tag{3.4}$$

where $\overline{V}(1+U) \mathrel{\hat{=}}$ mean suction side velocity
$\overline{V}(1-U) \mathrel{\hat{=}}$ mean pressure side velocity
$b \mathrel{\hat{=}}$ camber line length

Eliminating U by means of the equation for the circulation

$$\Gamma = g \cdot \Delta V_y = 2 \cdot b \cdot \overline{V} \cdot U \tag{3.5}$$

and relating the enthalpy drop $\Delta h_d = D/m$ (where $m = \rho_2 \cdot V_2 \cdot g \cdot \sin\alpha_2$) to the loss coefficient ζ_p, this results in :

$$\frac{\zeta_p}{1-\zeta_p} = \frac{\Delta h_d}{V_2^2/2} = 2\, C_d \left(\frac{b}{g}\right)\left(\frac{\overline{V}}{V_2}\right)^3 \frac{1 + \frac{3}{4}\left(\frac{\Delta V_Y}{\overline{V}} \cdot \frac{g}{b}\right)^2}{\sin\alpha_2} \tag{3.6}$$

Figure 10 was drawn with $C_d = 0.003$, typical values of b/g (i.e., near optimum values) and $\overline{V}$ between V_2 and $(V_1+V_2)/2$. According to the author the losses are typical for well designed blades, i.e., blades with continuous curvature everywhere and low diffusion rates. Particularly favourable blades may have even up to 20% lower losses.

- Balje and Binsley [10] used Truckenbrodt's turbulent boundary layer equation for arbitrary pressure gradients,

$$\theta \left(\frac{V \cdot \theta}{\nu}\right)^{1/n} = V^{-(3+2/n)} \left[C_1 + A \int_{x=X_t}^{X=\ell} V^{(3+2/n)}\, dx\right] \tag{3.7}$$

where C_1 is a constant to be determined from the laminar boundary layer at the point of transition $X = X_t$.

The main assumption made by the authors is that the total boundary layer growth can be reasonably well predicted using a linear variation of the velocity along the mean camberline of the type :

$$V = V_2 \left[\frac{\sin\alpha_2}{\sin\alpha_1} + \frac{x}{b}\left(1 - \frac{\sin\alpha_2}{\sin\alpha_1}\right)\right] \tag{3.8}$$

This assumption was justified on the basis of comparative calculations showing that the total boundary layer growth calculated in this way was essentially the same as the one using a more realistic trapezoidal blade velocity distribution. Integration of equation (3.7) with V from equation (3.8), n = 4, A = constant, $\frac{V_2 \cdot b}{\nu}$ = constant and $X_t = 0$ leads to the simple relation

$$\frac{\theta}{b} = K \left[\frac{1 - \left(\frac{\sin\alpha_2}{\sin\alpha_1}\right)^{4.5}}{1 - \left(\frac{\sin\alpha_2}{\sin\alpha_1}\right)}\right]^{0.8} \tag{3.9}$$

Based on experimental evidence K is taken to be 0.0021.

The profile losses are derived from the momentum thickness using Stewarts [11] relation.

$$\zeta_p = 1 - \left[\frac{\frac{\cos^2\alpha_2(1-\overline{\delta}^\star-\overline{\theta}-\overline{te})^2}{(1-\overline{\delta}^\star-\overline{te})^2} + \sin^2\alpha_2(1-\overline{\delta}^\star-\overline{te})^2}{1+2\sin^2\alpha_2[(1-\overline{\delta}^\star-\overline{te})^2-(1-\overline{\delta}^\star-\overline{\theta}-\overline{te})]}\right] \tag{3.10}$$

where $\overline{\delta}^\star$, $\overline{\theta}$ and $\overline{te}$ are equal $\delta^\star$, θ and te divided by $(g \cdot \sin\alpha_2)$. The profile loss diagram and the corresponding diagram for optimum pitch-to-chord ratios are given in figure 11.

- Denton's method [9] is based on theoretical performance calculations on a very large number of actual cascades. The results are presented as a series of ζ-contours versus α_1 and α_2 at constant g/c. The optimum pitch-to-chord ratio was found to be that at which separation started to occur on the blade surfaces. No further details have been published on his method.

Let us return at present to figure 7. All methods predict for the blades with $\alpha_1 = 90°$ an increase of losses with decreasing outlet angle, i.e., increasing turning, but there is a marked difference between the curves of Ainley & Mathieson and Craig & Cox on one side and Traupel, Balje and Denton on the other side. It is the writer's experience that the level of the profile losses claimed by the second group is characteristic for blades designed with very small suction side and pressure side diffusion rates and furthermore that the losses of such blades are systematically overpredicted by Craig & Cox by a factor 1.5 to 2. This implies that the correlation of Craig & Cox is indeed based on experimental evidence related to outdated blade design methods as indicated earlier. Kácker and Opakuu [2] suggest that the losses predicted by Ainley & Mathieson should be multiplied by 2/3 to account for the progress in blade design. With this correction Ainley's curve would be right in between the curves of Traupel and Balje & Binsley. Comparing further Traupel's and Balje's correlations it appears that the differences between their curves can largely be attributed to higher pitch chord ratios assumed by Balje.

Figure 7b for blades with inlet angles of 30° is divided into two parts by a dotted line at α_2 = 20°. To the left are blades with a positive degree of reaction, to the right blades with near zero and negative degree of reaction. The tendencies on the left side are similar. The big difference in the loss levels is practically due to the progress in blade design mentioned already earlier and to overly optimistic values predicted by Denton's curve. Indeed it seems to be unrealistic that losses for a blade with α_1 = 30° and α_2 = 10°, i.e., 140° turning are nearly the same as for α_1 = 90° and α_2 = 10°, i.e., 80° turning.

To the right of the dotted line in figure 7b, both Ainley and Traupel miss the point that the losses should increase with negative degree of reaction. However, in both cases the authors obviously did not envisage such design conditions. On the contrary, the simple equation $\theta/b = f(\alpha_1,\alpha_2)$ by Balje & Binsley is particularly well suited to handle such situations.

Very low cascade losses in general are indication that the boundary layers are essentially laminar over the entire blade surface. However, in turbines the boundary layers are exposed to strong nonperiodic flows and transition is likely to occur earlier. This is illustrated through experiments by Hodson [12] who measured the wall shear stresses on the suction surface of a rotor blade both in a linear cascade and a turbine (Fig. 12). The author concluded from his tests that the suction side boundary on the linear cascade was laminar, with a separation point close to the trailing edge. On the turbine rotor blade he observed a time dependent transitional boundary layer over most of the suction surface. The first weak velocity deceleration and subsequent constant pressure plateau between 45 and 70 percent of the surface length (Fig. 12a) contributed to accelerate the completion of the transition process as indicated by the sudden increase in shear stresses between 62 and 72 percent surface distance (Fig. 12c). Compared to the linear cascade the losses of the turbine rotor blade increased by 50%.

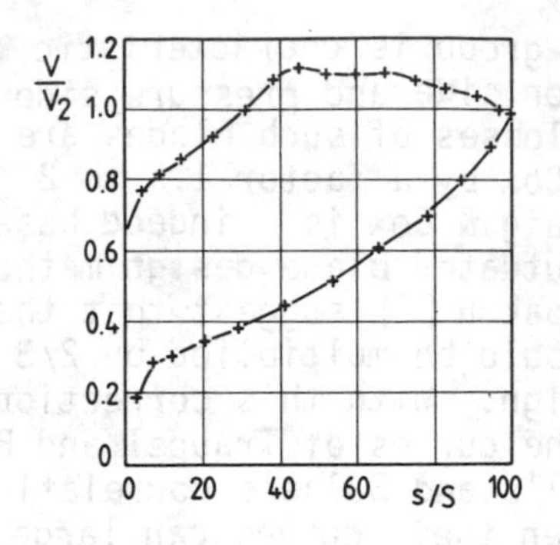

a) Linear cascade velocity distribution

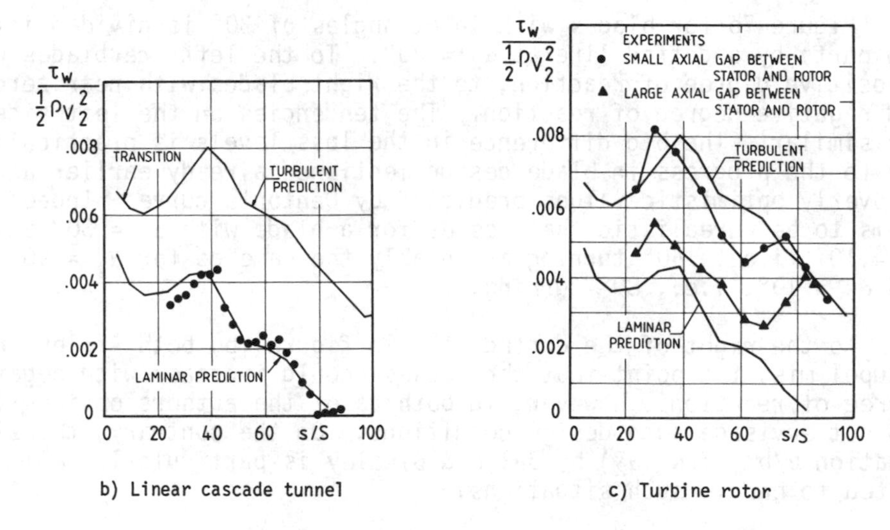

b) Linear cascade tunnel

c) Turbine rotor

FIG. 12 - WALL SHEAR STRESS MEASUREMENTS ON THE SUCTION SURFACE OF A ROTOR BLADE IN A LINEAR CASCADE TUNNEL AND IN A TURBINE STAGE, AFTER HODSON [12]

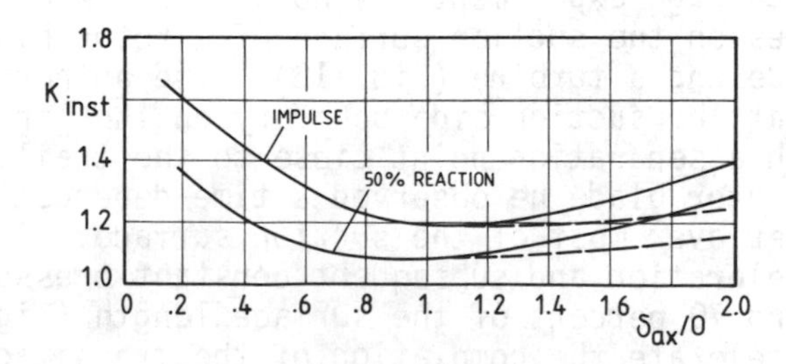

FIG. 13 - INSTATIONARY FLOW EFFECTS ON BLADE PROFILE LOSSES IN FUNCTION OF THE AXIAL DISTANCE BETWEEN TWO BLADE ROWS, AFTER DEJC & TROJANOVSKIJ [13]

Dejc & Trojanovskij [13] propose to apply a correction factor k_{inst} for the use of cascade data in turbine stages (Fig. 13). This correction factor depends on the axial distance between blade rows and on the degree of reaction. Low reaction blades are more sensitive to non-periodic inlet flow conditions. The axial distance is given as a percentage of the throat of the preceeding blade row.

3.1.2 Trailing edge losses : Figure 14 from Bryner [14] illustrates perfectly the situation faced by a designer when evaluating the importance of trailing edge losses for a given machine. The figure contains both experimental data (from cascade and stage measurements) and calculation results. All calculation methods make use of the conservation equations for continuity, momentum and energy. Equation (3.1) which was derived using such an approach indicates that the trailing edge loss is composed of (a) a loss term $\left(\frac{\overline{te}}{1-\overline{te}}\right)^2 \cdot \sin^2\alpha_2$, which is caused by the sudden flow area change at the trailing edge and (b) a loss term "$CP_b \cdot \overline{te}$", which is due to the difference between the pressure at the trailing edge base, P_b, and the average pressure across the channel exit, $\overline{P}_2$. A large uncertainty exists as to the magnitude of the coefficient CP_b, which depends on the trailing edge geometry, e.g., rounded or squared, and on the flow conditions before separation at the trailing edge, i.e., on the state of the boundary layer, the ratio between momentum and trailing edge thicknesses and the Mach numbers on pressure and suction side. The best data base existing so far is that by Sieverding [15]. His data for blades with α_2 = 20° to 30° and te/c = 0.06 to 0.15 confirmed the strong dependence of the base pressure on (a) the downstream Mach number, in particular in the transonic range, and (b) on the Mach number difference across the trailing edge (Fig. 15). The author showed that this Mach number difference was closely related to the rear suction surface turning angle ε and the trailing edge wedge angle δ (in most cases $\delta \simeq \varepsilon$) and correlated his base pressure data for convergent bladings in terms of the parameter $(\varepsilon+\delta)/2$. For convergent-divergent bladings, the internal area increase A/A^* becomes the most influential parameter.

Theoretically, the influence of the momentum thickness on the base pressure P_b is strongest for small ratios of θ/te and decreases gradually with increasing values of θ/te. This might explain why Sieverding did not find a distinct influence of this parameter. On the contrary, Kacker & Okapuu [2] attribute the difference in trailing edge losses for nozzle and impulse bladings (Fig. 16) to different boundary layer momentum thicknesses : "impulse bladings with their thick boundary layers have lower (less negative) base pressure coefficients". For blades other than the two basic types, the trailing edge losses are interpolated in a manner similar to Ainley's profile loss equation.

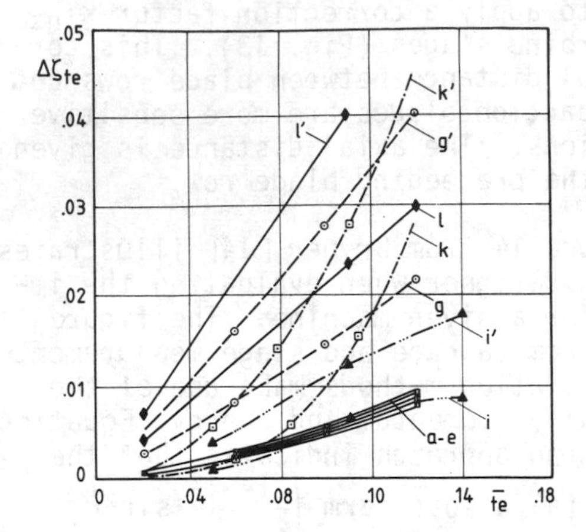

a-e Conservation equations with Cp_b=0 (Dibelius/Traupel,Pfeil,Stewart,BBC)
g-g' Conservation equations with experimental values of CP_b=0.13 and 0.26, resp. (Prust/Helon)
ℓ-ℓ' Turbine stage (Johnston)
k-k' Turbine stage (BBC)

FIG. 14 - COMPARISON OF TRAILING EDGE LOSSES ACCORDING TO VARIOUS AUTHORS, FROM BRYNER [14]

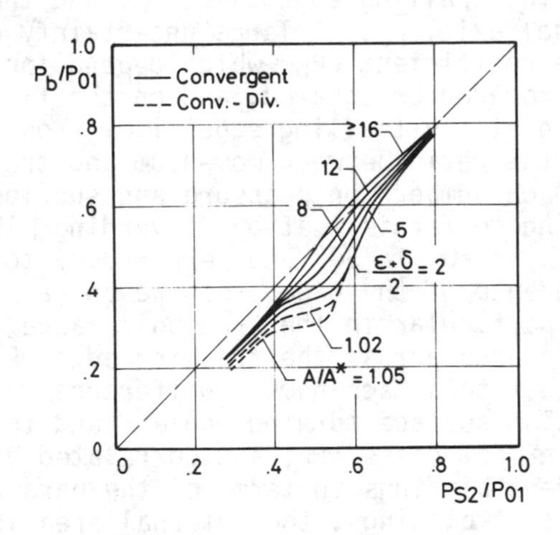

FIG. 15 - BASE PRESSURE CORRELATION, AFTER SIEVERDING et al. [15]

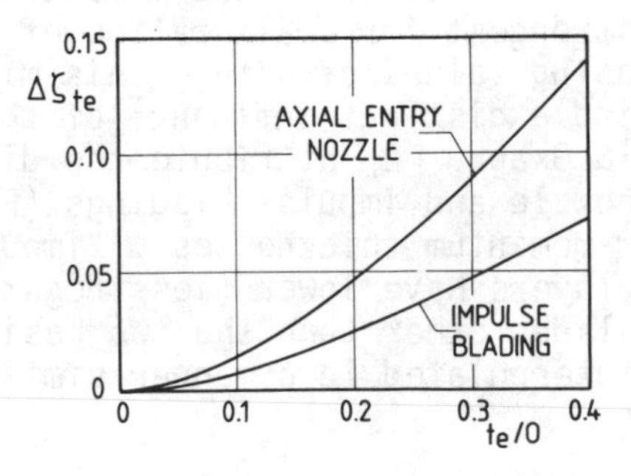

FIG. 16 - TRAILING EDGE LOSSES ACCORDING TO KACKER & OKAPUU [2]

$$\Delta\zeta_{TE} = \Delta\zeta_{TE(\alpha_1'=0)} + \sqrt{\left[\frac{\alpha_1}{\alpha_2}\right]}\left(\frac{\alpha_1}{\alpha_2}\right)\left[\Delta\zeta_{TE(\alpha_1'=\alpha_2)} - \Delta\zeta_{TE(\alpha_1'=0)}\right] \tag{3.11}$$

Some authors investigated the influence of the trailing edge geometry on the TE losses, e.g. Prust & Helon [16] and David [17], (Fig. 17). For a given value of te, the losses for a squared trailing edge are always higher than for a rounded trailing edge. The reasons are : the flow expands around the rounded trailing edge, hence the isobaric base region diminishes in size, and (b) the base pressure coefficient for a squared trailing is higher (more negative).

The ejection of coolant flow from the trailing edge modifies significantly the base pressure as shown by Sieverding [18] for a nozzle blade with small rear suction side curvature, figure 18. The curves present the difference of the base pressure with and without coolant ejection $P_b - P_b^*$ versus the coolant flow ejection pressure ratio P_{0c}/P_b^*. The pressure difference is as high as 15% of the outlet dynamic head.

A final remark concerns the unsteady character of the wake flow. Measurements of the shedding frequency of the von Karman vortices behind turbine cascades show a strong dependence of the Strouhal number on the state of the boundary layer at the trailing edge [19]. Therefore, trailing edge losses should be Re number dependent. This is confirmed by Dejc & Trojanovskij [13], but explicit experimental data are still lacking.

3.1.3 Effect of Mach number : It is useless to discuss the effect of Mach number on profile losses without specifying the blade geometry, in particular the rear suction surface curvature, since the optimum blade shape is strongly Mach number dependent. Already Ainley & Mathieson [7] distinguished between blades with curved rear suction side for operation at subsonic outlet Mach numbers and straight-backed blades (i.e., blades with a straight suction-surface downstream of the throat) for use in the transonic range. The curvature of the rear suction side affects both the position of the maximum velocity V_{max} and the ratio of V_{max}/V_2. As a general rule, in subsonic flow this ratio is higher for straight backed blades than for blades with a curved tail resulting in higher losses for the first group of blades. Within the second group we distinguish between rear loaded and front loaded blades. For rear loaded blades with maximum blade velocity downstream of the throat, the suction side curvature remains strongly convex up to the trailing edge. Compared to the rear loaded blade, the surface curvature of the front loaded blade is strongly reduced downstream of the throat and a non negligible part of the overhang length may be straight. Typical velocity distributions for all three cases are shown in figure 19.

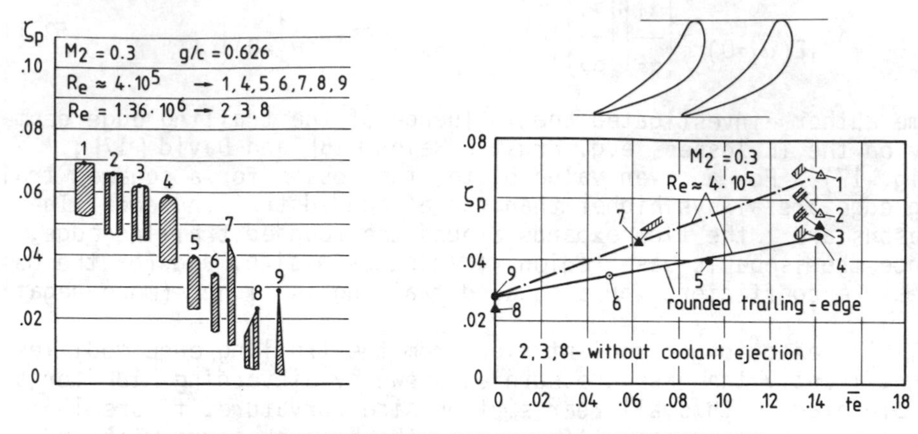

FIG. 17 - EFFECT OF TRAILING EDGE GEOMETRY, AFTER DAVID [17]

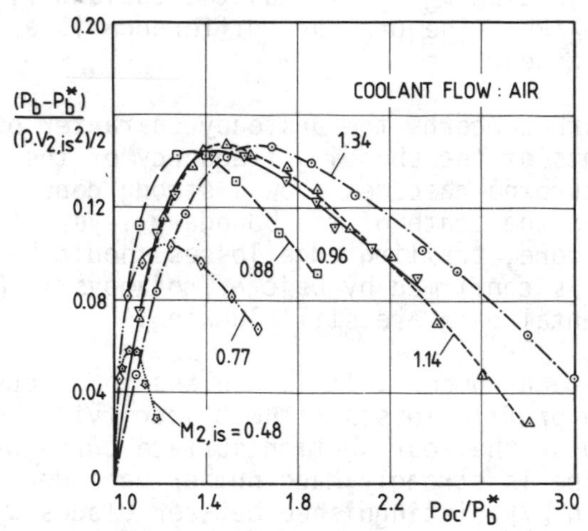

FIG. 18 - EFFECT OF COOLANT EJECTION THROUGH TRAILING EDGE ON BASE PRESSURE, AFTER SIEVERDING [18]

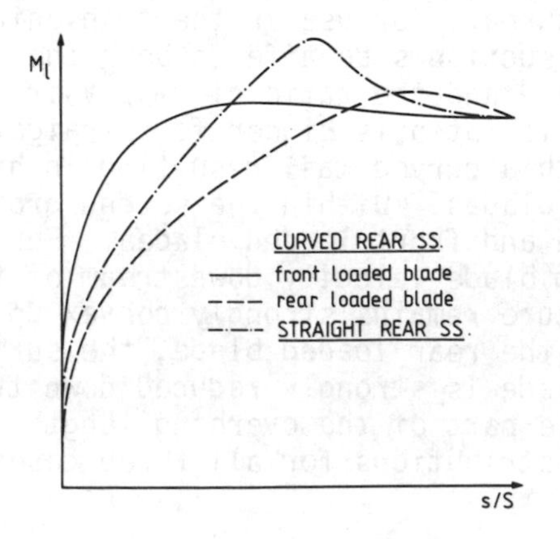

FIG. 19 - EFFECT ON SUCTION SIDE CURVATURE ON BLADE VELOCITY DISTRIBUTION

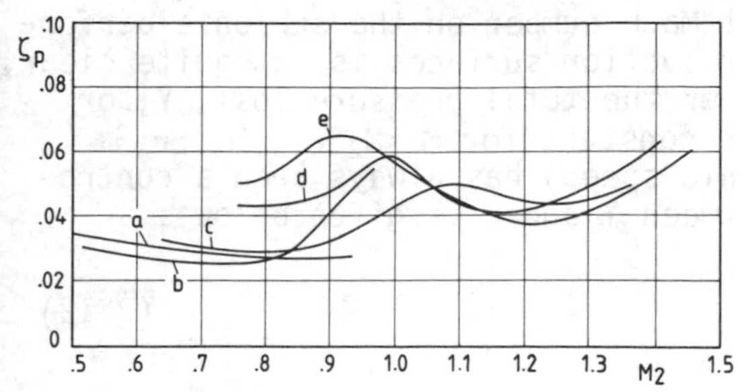

Blade	α_1	α_2^*	ε	te/O	g/C
a-[21]	95	17	10	0.08	0.77
b-VKI	90	22	17	0.11	0.71
c-VKI	60	18	5	0.06	0.70
d-[20]	30	22	8	0.10	0.75
e-VKI	60	22	0	0.075	0.75

FIG. 20 - PERFORMANCE CURVES OF CONVERGENT TRANSONIC TURBINE BLADES

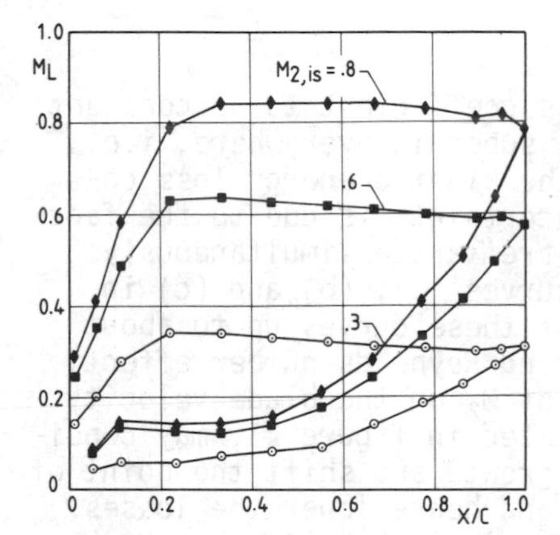

FIG. 21 - EFFECT OF OUTLET MACH NUMBER ON BLADE VELOCITY DISTRIBUTION FOR HIGH TURNING ROTOR BLADE

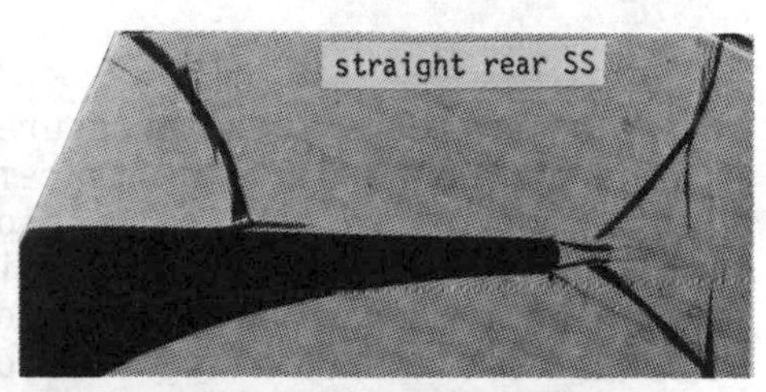

FIG. 22 - TRANSONIC FLOW PATTERNS ON REAR SUCTION SIDE OF TURBINE BLADES

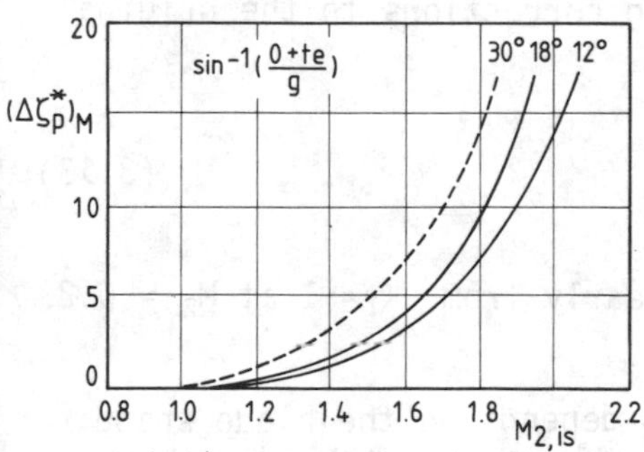

FIG. 23a - MACH NUMBER LOSS FOR CONVERGENT BLADING, AFTER CRAIG & COX [4]

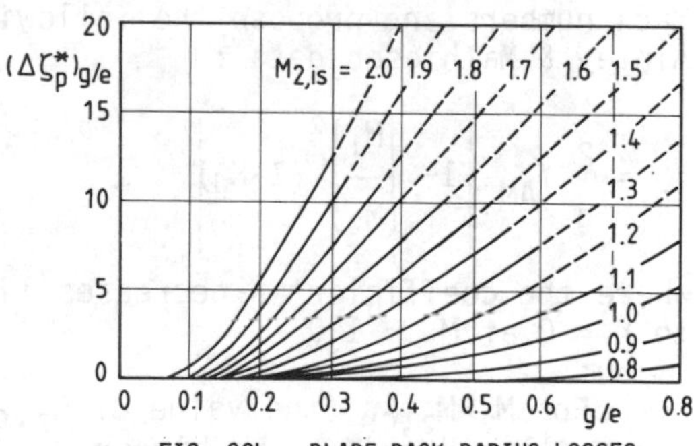

FIG. 23b - BLADE BACK RADIUS LOSSES, AFTER CRAIG & COX [4]

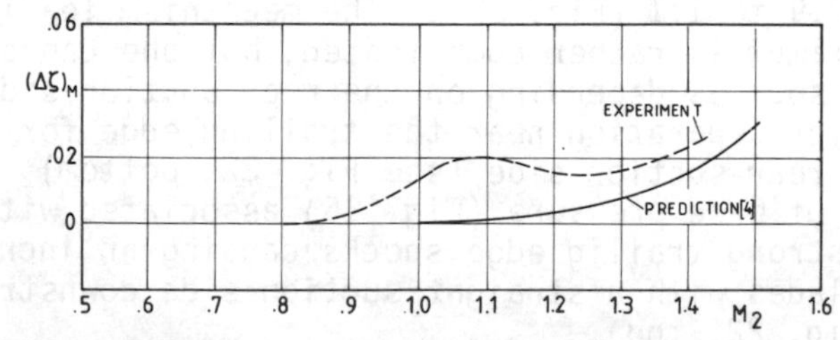

FIG. 24 - PREDICTED MACH NUMBER LOSS USING FIG.23a VERSUS MEASURED MACH NUMBER LOSS FOR BLADE (c) IN FIG. 20

The influence of the outlet Mach number on the subsonic performance of blades with curved rear suction surfaces is not quite clear, in particular the question whether the total pressure loss, Y, or the kinetic energy loss ζ remain constant for $M_2 < M_{2,cr}$ ($M_{2,cr} \triangleq$ local blade velocity exceeds sonic speed) has always been a controversial issue. The relation between Y and ζ is given below :

$$\zeta^\star = \frac{Y}{1 + \frac{\kappa \cdot M_2^2}{2}} \tag{3.12}$$

Ainley assumes the total pressure loss coefficient to be constant as long as the blade velocity remains subsonic everywhere, i.e., $M_2 < M_{2,cr}$ while Craig & Cox consider the kinetic energy loss coefficient $\zeta^\star$ to be constant. This uncertainty is due to the fact that in most cascade tests M_2 and Re are varied simultaneously. This was also the case for the loss curves (a), (b) and (c) in figure 20. The decreasing tendency of these curves up to about $M_2 = 0.8$ could be both a Mach number and Reynolds number effect. The favourable effect of an increase of M_2 on the blade velocity distribution of cascade (a), demonstrated in figure 21, may considerably influence the boundary layer growth and shift the point of transition towards the trailing edge and hence lower the losses. Of course, an increase of Re would have a similar effect. Kacker and Okapuu [2] suggest that the Mach number effect is most pronounced where inlet Mach numbers are only slightly lower than exit Mach numbers and propose the following corrections to the original Ainley & Mathieson data :

$$Y_p = \frac{2}{3} Y_{AM} \left| 1 - \left(\frac{M_1}{M_2}\right)^2 (1 - K_1) \right| \tag{3.13}$$

where the coefficient K_1 decreases linearly from $K_1 = 1$ at $M_2 = 0.2$ to $K_1 = 0$ at $M_2 = 1.0$

For $M_2 > M_{2,cr}$ (the value of $M_{2,cr}$ depends on the blade spape; the smaller V_{max}/V_2, the higher $M_{2,cr}$) all convergent blades show a considerable increase of losses with a relative loss maximum between $M_2 \approx 0.9$ to 1.1 (Fig. 20). The mechanism leading to this local loss maximum is rather complicated, but one can distinguish two main loss sources depending on the rear suction side curvature :
- boundary layer separation near the trailing edge for blades with strong convex rear suction side (see Fig. 22, bottom)
- abrupt drop of base pressure (Fig. 15) associated with rapid development of strong trailig edge shocks causing an increase of V_{max}/V_2 for blades with a straight suction side downstream of the throat (see Fig. 22, top).

Nearly all convergent bladings show a relative loss minimum at outlet Mach numbers between M_2 = 1.10 and 1.25. For higher Mach numbers, shock losses become predominant causing ζ_p to rise continuously.

Neglecting all viscous effects it is possible to calculate the shock losses for straight-backed convergent blades with zero trailing edge thickness (e.g. [8]) using the conservation equations for continuity, momentum and energy between the throat and a far-downstream plane. Figure 23a from Craig & Cox [4] presents these losses. Compared to straight-backed blades, blades with a convex rear suction side tend to overaccelerate the suction side velocity downstream of the throat which results in comparatively higher Mach number differences across the trailing edge and therefore increased trailing edge shock strength and losses. This is expressed by the second graph of Craig & Cox showing the influence of the parameter g/e, i.e., the ratio of pitch-to-back-surface radius.To illustrate the differences between measured and predicted losses at high Mach numbers, figure 24 compares the losses of blade (c) in figure 20 after subtraction of the losses at M_2 = 0.8 with the loss increment $\Delta\zeta = f(M_2)$ calculated after Craig & Cox. Clearly, conventional loss correlations are not suitable for blade performance prediction at transonic outlet Mach numbers.

3.1.4 Effect of Reynolds number, surface roughness and turbulence : The effect of Reynolds number on profile losses as predicted by various authors is summmarized in figure 25. We distinguish roughly three regions characterized by different behaviour of the blade boundary layer :

(a) Blades with predominantly laminar boundary layer for $Re < 10^5$ (based on chord and outlet flow conditions) : losses are strongly Reynolds number dependent;

(b) Blades with transitional boundary layers for $10^5 \leqslant Re \leqslant 10^6$: losses are independent of Reynolds number;

(c) Blades with fully turbulent boundary layers for $Re > 10^6$: losses decrease with increasing Reynolds number as long as the blade surface remains hydraulically smooth.

Operating conditions in the third range are not as frequent in gas turbines as in steam turbines where Reynolds numbers in excess of 10^7 occur in the hp-cylinder of large steam turbines. Theoretically, the skin friction of turbine blades varies as $Re^{-0.2}$ for fully turbulent boundary layers and hydraulically smooth blades. This requires, of course, very small surface roughnesses. The maximum allowable surface roughness is Re-dependent and should not exceed

$$\frac{k_s}{\ell} = \frac{100}{Re} \qquad (3.14)$$

($k_s/\ell \simeq$ equivalent sand grain-roughness) for flat plates

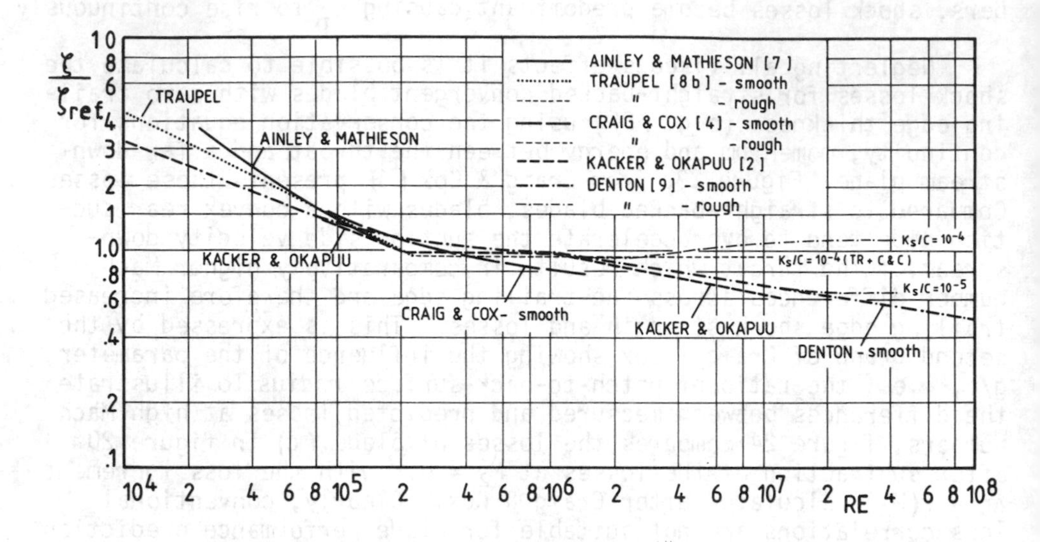

FIG. 25 - COMPARISON OF REYNOLDS NUMBER EFFECTS FOR PROFILE LOSSES BY VARIOUS AUTHORS

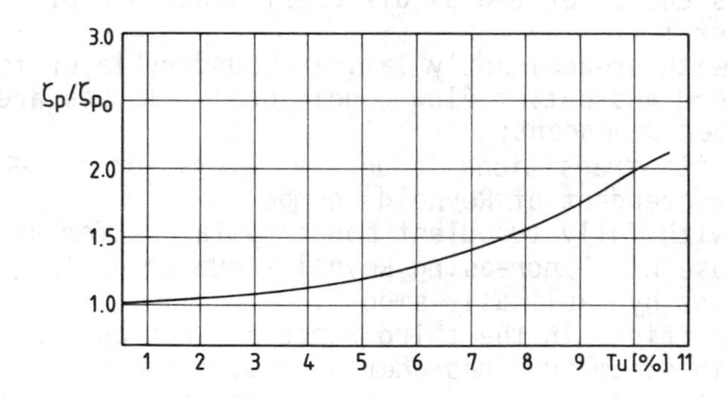

FIG. 26 - EFFECT OF FREE STREAM TURBULENCE ON PROFILE LOSSES FOR REACTION BLADES, AFTER DEJC & TROJANOVSKIJ [13]

according to Schlichting [22].

Denton's curve is based on actual boundary layer calculations on six turbine blades [9]. At low Reynolds number the losses vary as $Re^{-0.4}$, a value which was also adopted by Kacker & Okapuu [2]. It is somewhat low compared to the loss variations suggested by Traupel [8] and in particular Ainley [7] on the basis of experimental evidence. The obvious causes for the increase of the losses at reduced Reynolds number are a general thickening of the laminar boundary layers and a gradual increase of separated laminar flow regions.

The evolution of the losses in the transitional regime between $Re = 10^5$ and 10^6 was found by Denton to be very blade-shape dependent, but altogether the losses continue to decrease slightly in this range. Surface roughness of blades with standard finish of $k_s/C = 10^{-4}$ becomes important only in the initial phase of the full turbulent regime starting at $Re = 10^6$. On the contrary Traupel considers that this roughness is sufficient to accelerate considerably the completion of the transitional phase and proposes that losses already become Reynolds number independent for $Re > 2.2 \times 10^5$. Craig & Cox [4] suggest that loss data do better correlate when defining the Reynolds number with respect to the throat rather than to the chord. Figure 25 shows their curve for a typical value of $O/C = 0.25$. Blade surface roughness of $k_s/C = 10^{-4}$ makes their losses Reynolds number independent for $Re > 4.5 \times 10^5$.

Little has been published on the effect of turbulence on turbine blade losses. Cascade tests are usually conducted at turbulence levels of $Tu \simeq 1\%$ while the turbulence level in machines is estimated to vary from 6% to 20%, the lower values corresponding to the first stage, the high values to intermediate stages. In the absence of any other systematic information, figure 26 from Dejc [13] for reaction bladings may serve as a rough guideline.

3.2 Secondary loss correlations

The number of loss correlations available bears no relation to the number of comprehensive data sets published in the literature. There seem to be only two organizations which have provided any systematic data in this domain, i.e., members of the group of Schlichting and members of the University of Dresden (in particular Wolf). Apart from these there exists a number of published data of very limited value due to (a) incomplete information, (b) restriction to one particular aspect and (c) an insufficient number of cascade configurations making an eventual extrapolation dangerous.

In 1970 Dunham [23] reviewed different loss correlations which are listed hereafter :

Author/year	$Y_{secondary}$
1. Soderberg, 1949	$0.075 \cdot \frac{c}{h}$
2. Ainley, 1951	$\lambda \cdot Z$ with : $Z = \left(\frac{C_\ell}{g/c}\right)^2 \cdot \frac{\cos^2\alpha_2}{\cos^3\alpha_\infty}$; $\lambda = f\left(\frac{A_2^2/A_1^2}{1+D_H/D_T}\right)$
3. Ehrich, 1954	$\frac{0.1178(\alpha_1-\alpha_2)^2}{(1-0.2\ g/h)^3} \frac{g}{h} \frac{\cos^2\alpha_2}{\cos^2\alpha_1}$
4. Scholz, 1954	$a \cdot \frac{c}{h}$ where $a = 0.07$
5. Hawthorne, 1955	$2(\alpha_1-\alpha_2) \frac{g}{h} \frac{\cos^4\alpha_2}{\cos^3\alpha_\infty} f\left(\frac{\delta}{g \cdot \cos\alpha_2}\right)$
6. McDonald, 1956	$\lambda \left(\frac{\ell}{g/c}\right)^2 \frac{\cos^2\alpha_2}{\cos^3\alpha_\infty}$, $\lambda = f\left(\frac{\cos\alpha_2}{\cos\alpha_1'}\right)$
7. Markov, 1958	$\frac{a \times \text{throat}}{h}$, $0.07 < a < 0.18$
8. Vavra, 1960	$0.04 \frac{g}{h} \left(\frac{C_\ell}{g/c}\right)^2 \frac{\cos^2\alpha_2}{\cos^3\alpha_\infty}$
9. Stewart, 1960 Heutreux, 1963	$Y_{profile} \times \frac{\text{blade surface area + wall area}}{\text{blade surface area}}$
10. Boulter, 1962	$0.035 \frac{c}{h} (\tan\alpha_1-\tan\alpha_2)^2 \cdot \cos^2\alpha_2$
11. Scoltack, 1963	$\lambda \left(\frac{C_\ell}{g/c}\right) \frac{\cos^2\alpha_2}{\cos^3\alpha_\infty}$; $\lambda = f\left(\frac{h}{\text{throat}}, \frac{A_2^2/A_1^2}{1/D_H/D_T}\right)$
12. Bauermeister, 1963	$0.07 \frac{c}{h} (\tan^2\alpha_2-\tan^2\alpha_1)$
13. Balje, 1968	$\frac{c}{h} \times f_1\left(\frac{\delta_1}{h}\right) \times f_2(\alpha_1,\alpha_2)$

NOTE : in this table angles are referred exceptionally to axial direction

The formulae are valid for incompressible flow. The losses are expressed by the total pressure loss coefficient :

$$Y_s = \frac{\text{stagnation pressure loss}}{\text{outlet dynamic head}}$$

All formulae are composed of a loading term, representing the effect of cascade loading or deflection and a length ratio term. Dunham compared all formulae with published secondary loss data, especially those of Wolf [24] which by far appear to be the best documented and most comprehensive set of data. Dunham concluded that all available data are best correlated by the equation

$$Y_s = \frac{C}{h} \cdot \frac{\cos(90°-\alpha_2)}{\cos(90°-\alpha_1')} \cdot Z \cdot f\left(\frac{\delta_1^\star}{C}\right) \qquad (3.15)$$

where Z is Ainley's loading parameter :

$$Z = \left(\frac{C_\ell}{g/C}\right)^2 \cdot \frac{\cos^2(90°-\alpha_2)}{\cos^2(90°-\alpha_\infty)}$$

$$C_\ell = 2(g/c)\cdot[\tan(90°-\alpha_1)+\tan(90°-\alpha_1)]\cdot\cos(90°-\alpha_\infty)$$

$$\tan(90°-\alpha_\infty) = \frac{1}{2}[\tan(90°-\alpha_1)-\tan(90°-\alpha_2)]$$

and $f(\delta_1^\star/C)$ presents the inlet endwall boundary layer effect.

Since then a number of additional loss correlations have been published :

14. Traupel in 1968 [8a], revised in 1977 [8b].
15. Dunham & Came in 1970 [25], modification to Ainley-Mathieson method [7].
16. Craig & Cox in 1971 [4].
17. Dejc and Trojanovskij in 1973 [13].
18. Muktarov & Krichiakine, see [26].
19. Hultsch & Sauer in 1978 [27].
20. Kacker & Okapuu in 1982 [2], modification to Ainley-Mathieson-Dunham-Came method

These authors seem to have proceeded in the same way : the basic structure of their loss correlations is derived from straight cascade data while the absolute magnitude of the losses is assessed by comparison with turbine performance measurements. This is important, since a correlation, based on straight cascade data only, is of little use. It would almost certainly underestimate the losses occurring in turbine stages.

3.2.1 Description of most important correlations : The methods most widely used in Europe and the USA are those of :
- Ainley-Mathieson with the modifications by Dunham & Came and more recently by Kacker & Ocapuu (AM-DC-KO);
- Traupel;
- Craig & Cox.

We shall briefly discuss these methods as well as the one of Hultsch and Sauer. The latter is very little known but has some particularly interesting features.

- AM-DC-KO : The basic equation for secondary losses on nozzle blades was mainly derived from test data on bent metal sheet blades. The secondary losses in rotor blades were derived from performance measurements on conventional blading using the correlation already established for profile losses in stator and rotor, for secondary losses in the nozzle and clearance loss in the rotor.

As already mentioned above, Dunham modified Ainley's basic equation $Y_s = \lambda \cdot Z$ to

$$Y_s = \frac{c}{h} \cdot Z \cdot \frac{\cos(90°-\alpha_2)}{\cos(90°-\alpha_1')} \cdot f\left(\frac{\delta_1^\star}{c}\right) \tag{3.15}$$

It is worth noting that Ainley's loading factor Z depends on α_1 and α_2 only and not on the pitch-to-chord ratio. The function $Z = f(\alpha_1,\alpha_2)$ is shown in figure 27.

From his review of secondary loss data Dunham proposed a tentative formula for the boundary layer influence

$$f(\delta_1^\star/C) = 0.0055 + 0.078\sqrt{\frac{\delta_1^\star}{c}} \tag{3.16a}$$

but from a comparison with overall efficiency data of 25 turbines Dunham & Came [25], found that a fixed value of

$$f(\delta_1/c) = 0.034 \tag{3.16b}$$

is more appropriate. From this it can be concluded that
(a) the $\delta_1^\star/c$ in the turbines was above the critical value below which the losses vary with $\delta_1^\star$;
(b) a correlation factor is needed to adapt cascade secondary loss data to the loss level occurring in actual turbine stages.

The secondary loss equation by Dunham & Came suggests that Y_s varies as the reciprocal of the blade aspect ratio over the complete aspect ratio range. This leads to an overestimation of losses at small aspect ratios, as noted by Kacker & Okapuu [2]. These authors propose instead a dependence of Y_s on the aspect ratio for $h/c \leq 2$ of the form

$$f(AR) = \frac{1-0.25\sqrt{2-h/c}}{h/c} \qquad \text{(AR - aspect ratio)} \qquad (3.17)$$

Similar as for profile losses Kacker & Okapuu apply a subsonic Mach number correction factor

$$f(M) = 1-K_3 \left[\left(\frac{M_1}{M_2}\right)^2 (1-K_1)\right]$$

K_1 decreases linearly from $K_1 = 1$ at $M_2 = 0.2$ to $K_1 = 0$ at $M_2 = 0$ and $K_3 = (C_{ax}/h)^2$.

Contrary to Dunham & Came who suggest that a Reynolds number correction of the form

$$\left(\frac{Re}{2\times10^5}\right)^{-0.2}$$

should be applied to both profile and secondary losses, Kacker and Okapuu limit this Reynolds number correction to the profile losses arguing that there is little evidence in the literature that other loss terms are also affected by it.

- Traupel : The structure of Traupel's secondary loss correlation is derived from Wolf's presentation of secondary loss data obtained on straight cascades (see Fig. 30a). However, the quantitative values are obtained by comparison with turbine performance measurements at ETH Zürich. The secondary loss data derived from turbine measurements are about twice as high as Wolf's data. As long as no interference between the secondary loss regions from both endwalls occurs, the losses are expressed as :

$$\zeta_S = \frac{\zeta_p}{\zeta_{p_0}} \cdot F \cdot \frac{g}{h} + \zeta_a \qquad (3.19)$$

with $F = f\left(\Delta\alpha, \frac{V_1}{V_2}\right)$, see figure 28a. The use of the velocity ratio $\frac{V_1}{V_2}$ implies that the correlation does account for compressibility effects.

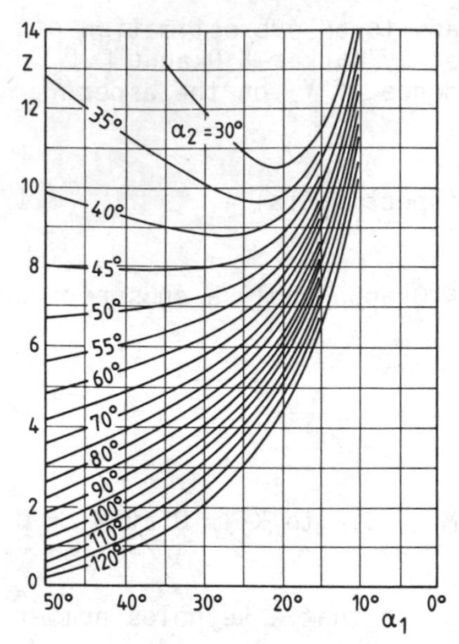

FIG. 27 - AINLEY'S LOADING FACTOR Z FROM [6]

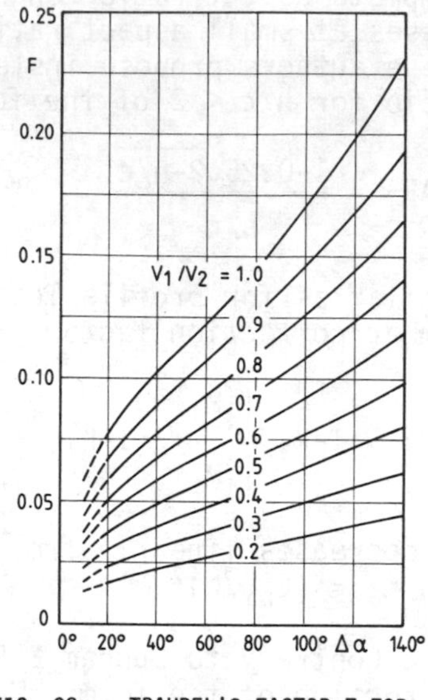

FIG. 28a - TRAUPEL'S FACTOR F FOR CALCULATING SECONDARY LOSSES [8b]

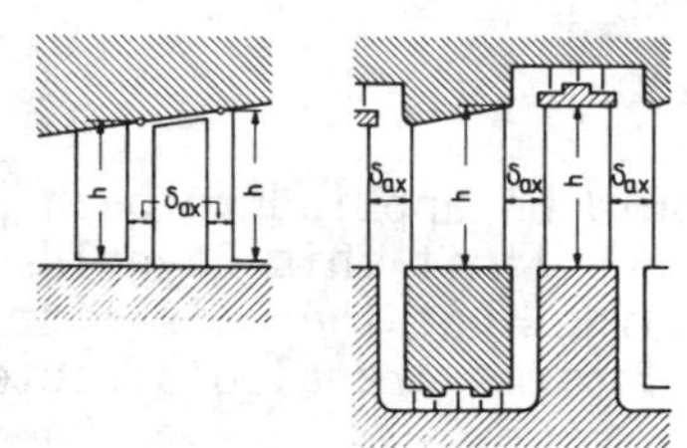

FIG. 28b - UNSHROUDED AND SHROUDED TURBINE BLADE ROWS

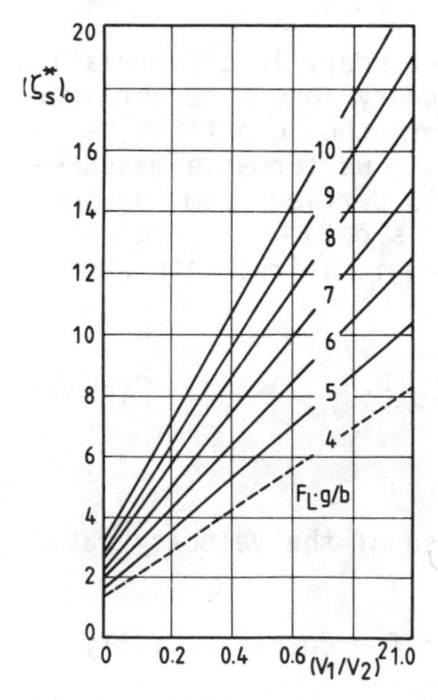

FIG. 29a- BASIC SECONDARY LOSSES, AFTER CRAIG & COX [4]

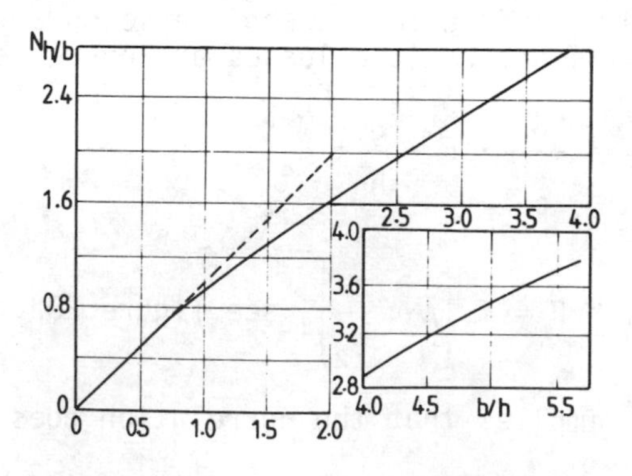

FIG. 29-b - ASPECT RATIO FACTOR [4]

The factor ζ_p/ζ_{p0} (ratio of losses at given Re, M_2 and trailing edge thickness to losses at reference conditions indicates that the secondary losses are affected by the same parameters as the profile losses.

ζ_a is a correction factor which allows for differences in the axial spacing between two blade rows. It accounts for losses due to endwall shear stresses in the case of unshrouded blade rows and losses due to the existence of turbulent free shear layers at hub and tip in the case of shrouded blade rows. Depending on the type of construction (see Fig. 28b) ζ_a is calculated as

$$\zeta_a = \frac{c_f}{\sin\alpha_2}\left(1 \pm \frac{h}{D_M}\right) \cdot \frac{\delta_{ax}}{h} \qquad \text{for unshrouded blades (+ for stator blades; - for rotor blades)} \tag{3.20a}$$

$$\zeta_a = \frac{0.04}{\sin\alpha_2}\,\frac{\delta_{ax}}{h} \qquad \text{for shrouded blades} \tag{3.20b}$$

The equation (3.19) for ζ_s is valid as long as the secondary flow regions from both endwalls do not influence each other. Contrary to Kacker and Okapuu who assumed that this would occur for a fixed value of the aspect ratio $(h/c = 2)$, Traupel relates this critical blade height to the profile losses,

$$(h/g)_{cr} = 7 \rightarrow 10\,\sqrt{\zeta_p} \tag{3.21}$$

This seems to be a physically sound approach since the losses are directly related to the blade loading which in turn determines the spanwise extensiton of the secondary flow regions.

For h/g ratios below the critical value it is impossible to separate secondary losses from profile losses. For such cases it is preferable to evaluate the total losses $\overline{\zeta}_p$ by

$$\overline{\zeta} = \zeta_p + \frac{\zeta_p}{\zeta_{p_0}} \cdot \frac{F}{(h/g)_{cr}} + \zeta_a + A \cdot \frac{c}{g}\left(\frac{1}{h/g} - \frac{1}{(h/g)_{cr}}\right) \tag{3.22}$$

For $h/g < (h/g)_{cr}$ the last term is negative. The constant A takes a value of 0.02 for stator blades and 0.035 for impulse blades.

- Craig & Cox : Their correlation is based on experimental data from straight cascades. From a comparison of calculated and measured turbine performance data the authors concluded that their cascade secondary loss data did not need to be corrected for use in turbine stage calculations.

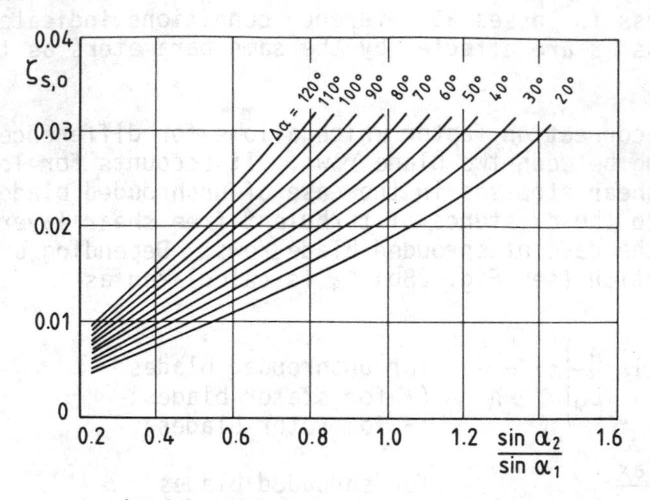

a) Wolf's original secondary loss correlation

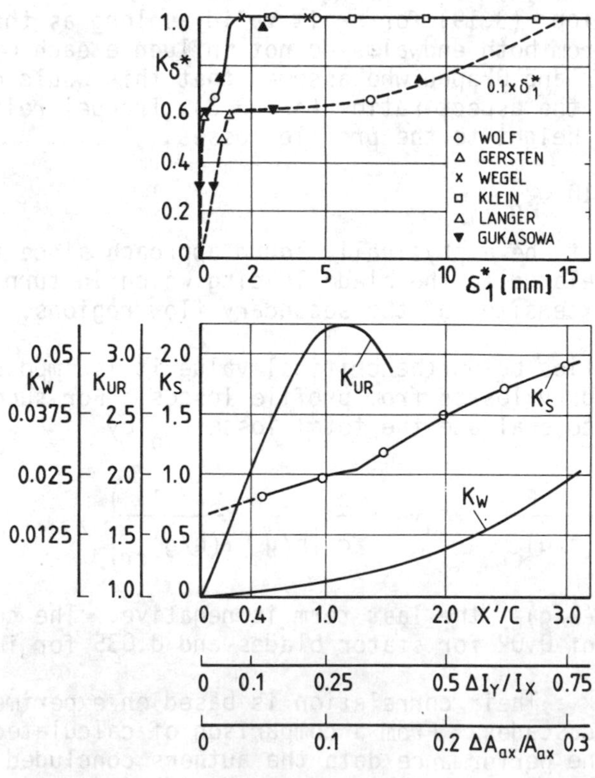

b) Effect of inlet boundary layer thickness (K_{δ^*}), inlet skew (K_{UR}), downstream distance (K_S) and increase of annulus area (K_W)

FIG. 30 - SECONDARY LOSSES AFTER HULTSCH & SAUER [27]

Secondary losses are evaluated on the bases of the equation :

$$\zeta^{\star} = \zeta^{\star}_{S,0} \times N_R \times N_{h/b} \qquad (3.23)$$

$\zeta_{S,0}$ is a basic secondary loss factor depending on the blade loading parameter $F_{\ell} = f(\alpha_1, \alpha_2)$, (Fig. 9b), the ratio of pitch to backbone length g/b and the square of inlet and outlet velocity $(V_1/V_2)^2$ (Fig. 29a).

$(V_1/V_2)^2$ The authors maintain that the correlation holds over a range of incidence angles provided that the correct values are used in evaluating the basic loss coefficient. Compressibility effects are directly included.

$N_{h/b}$ For high values of h/b the losses vary linearly with h/b. For small aspect ratios the rate of increase is less than would be expected from a simple inverse law (Fig. 29b).

N_R The Reynolds number effect is supposed to be the same as for profile losses (Fig. 25)

The above method is for shrouded blades only.

- Hultsch & Sauer : The basic equation used by the authors is :

$$\zeta_S = 1.5 \cdot \zeta_{S0} \quad \frac{h_0}{h} \left(K_{\delta^{\star}} \cdot K_{uR} \cdot K_S + K_W \right) \qquad (3.24)$$

The loss coefficient ζ_{S0} presents Wolf's original data [24] presented in figure 30a for a blade height of h_0 = 100 mm, a chord-length of 45 mm, an inlet boundary layer displacement thickness $\delta_1^{\star}$ = 0.7 mm and a blade roughness $k/h_0 \approx 5 \times 10^{-6}$.

$K_{\delta_1^{\star}}$ This factor presents the influence of the inlet boundary layer thickness

$$K_{\delta_1^{\star}} = \zeta_S / \zeta_{S,max} = f(\delta_1^{\star})$$

This function is shown in figure 30b. Note that $K^{\star}_{\delta 1}$ is plotted against the absolute value of the boundary layer thickness. In spite of variations in the chord length between C = 30 mm for Gersten [28] and 160 mm for Klein [29], all data are fairly well respresented by one single curve. However, the graph does not include the data of Came [30], which deviate considerably from this curve. Came's data show a variation of ζ/ζ_{max} for $\delta_1^{\star} \leqslant 5$ mm (C = 100 mm) compared to 2 mm in figure 30b.

K_{uR} This factor takes into account the effect of inlet skew on secondary losses. Following the suggestion of Wegel [31] the authors use the data of Klein [29] presented under the form

$$\zeta_s/\zeta_{s,0} = f\left(\frac{\int_0^\delta \Delta V_y^2(r)\cdot r\cdot dr}{\int_0^\delta V_x^2(r)\cdot r\cdot dr}\right) = f\left(\frac{I_y}{I_x}\right) \qquad (3.25)$$

where

ζ_S - secondary losses with inlet skew

$\zeta_{S,0}$ - secondary losses with zero skew

ΔV_y represents the deviation of the velocity component with respect to that of the corresponding flow without inlet skew.

The relation $K_{uR} = f(I_y/I_x)$ is also shown in figure 30b.

K_s This factor takes into account the increase of secondary losses with increasing downstream distance. K_S is presented in figure 30b in function of the distance X_2'/C. X_2' is to be taken in streamwise direction. The relation $K_S = f(X_2'/c)$ is that of Wolf [24].

K_W This factor presents the influence of an increase of the annulus area across the blade row (Fig. 30b)

$$K_W = f\ (\Delta A_{ax}/A_{ax})$$

This relation is similar to that proposed by Dejc and Trojanovskij [13] for evaluating the effect of a change of the meridional flow channel on profile losses.

The authors note that an influence of Re number is only noticeable for extremely thin inlet endwall boundary layers. A Mach number effect is acknowledged, but no precise information is given.

3.2.2 Comparison of correlation methods : Before proceeding with some sample calculations, let us make some general statements :

- Blade shape : Since the intensity of the secondary flow depends on the blade loading it is reasonable to expect that the local distribution imposed by the blade shape affects the magnitude of the secondary losses. Traupel considers this possibility by suggesting that a change of the blade shape affects the secondary losses to the same extent as profile losses.
- Pitch-to-chord ratio g/c : Neither the AM-DC-KO method nor the method of Hultsch & Sauer account for a variation of g/c. Wolf's data forming the basis for the latter method did not show any influence of the pitch-to-chord ratio between g/c = 0.3 to 0/7. A possible explanation is that the effect of a change of the total blade loading is off-set by a change of the load distribution. It is premature to draw final conclusions regarding this issue because of an obvious lack of data in the literature.

- Outlet Mach number : All methods, except that of Hultsch & Sauer, consider that compressibility effects influence favourable secondary losses. The parameter V_1/V_2 (velocity ratio) in the methods of Craig & Cox and Traupel would allow to handle blades with supersonic outlet flow, but the use of the methods is intended to be restricted to subsonic outlet Mach numbers. There is no data base in the literature to evaluate secondary losses for transonic turbine blades. One single experiment for a transonic blade with an internal area increase of $A/A^* = 1.01$ is quoted by Dejc & Trojanovskii [13] , see figure 31. It shows a slight drop of ζ_S from $M_2 = 0.8$ to 1.35, followed by a rapid increase of ζ_S for higher Mach numbers. It is interesting to note that the local loss mimimum for ζ_S does not coincide with that of the profile losses.
- Inlet boundary layer thickness : Most authors consider that the thickness of the endwall boundary layer in turbines is well above the critical value below which ζ_S depends on δ^*. The variation of $\zeta_S/\zeta_{S,max} = f(\delta_1^*/C)$ measured by various authors is shown in figure 32 from [32]. The boundary layer thickness at the inlet of the first guide vane may be well below the critical value of δ^*/c.
- <u>Sample calculations</u> : The prediction of the flow correlation methods reviewed are compared in figure 33 where they have been applied to blades with 90° and 45° inlet angles and 10, 20 and 30° outlet flow angles

	h/c	g/c	δ_1^*/h	h[mm]	δ_{ax}[mm]	Re	M_2
α_1=90°	1.0	0.75	0.03	60	equal to throat	2×10^5	low
α_2=45°	1.0	0.65	0.03	60		2×10^5	low

NOTE :
(1) $\delta_{ax} \simeq$ distance between two successive blade rows in a turbine; needed for correlations of Traupel and Hultsch & Sauer
(2) Blade height : constant from inlet to outlet
(3) The ratio of pitch to backbone length needed for the Craig & Cox correlation is calculated from $g/b = (g/c)/(b/c)$ with $b/c \simeq 0.72+0.005\cdot\Delta\alpha$ for $\Delta\alpha > 56°$ [10] .
(4) The blade height to mean diameter ratio needed for Traupel's correlation is taken as $h/D_m = 0.1$. The friction coefficient C_f is taken to 0.04.
In all but one case the losses increase with increasing outlet flow angle. The methods of Craig & Cox, Traupel and Hultsch & Sauer show reasonable agreement at low α_2 but the differences increase with increasing α_2. Compared to these methods the AM-DC-KO method predicts in all but one case much higher losses.

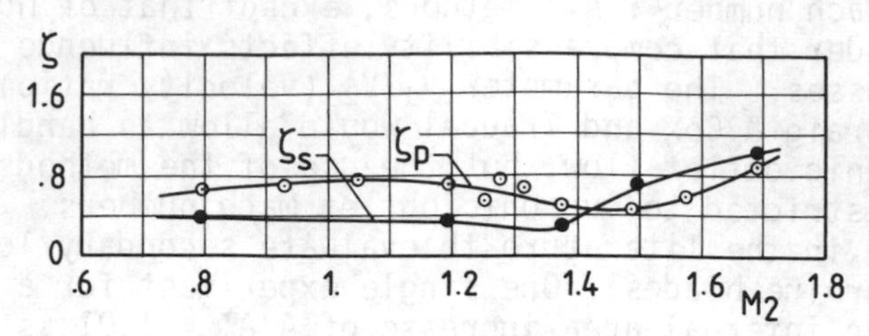

FIG. 31 - EFFECT OF MACH NUMBER ON PROFILE AND SECONDARY LOSSES FOR A TRANSONIC TURBINE BLADE, AFTER DEJC & TROJANOVSKIJ [13]

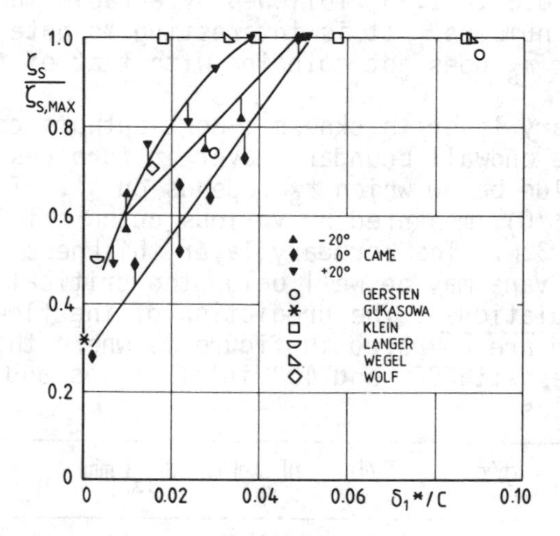

FIG. 32 - VARIATION OD SECONDARY LOSS WITH INLET DISPLACEMENT THICKNESS

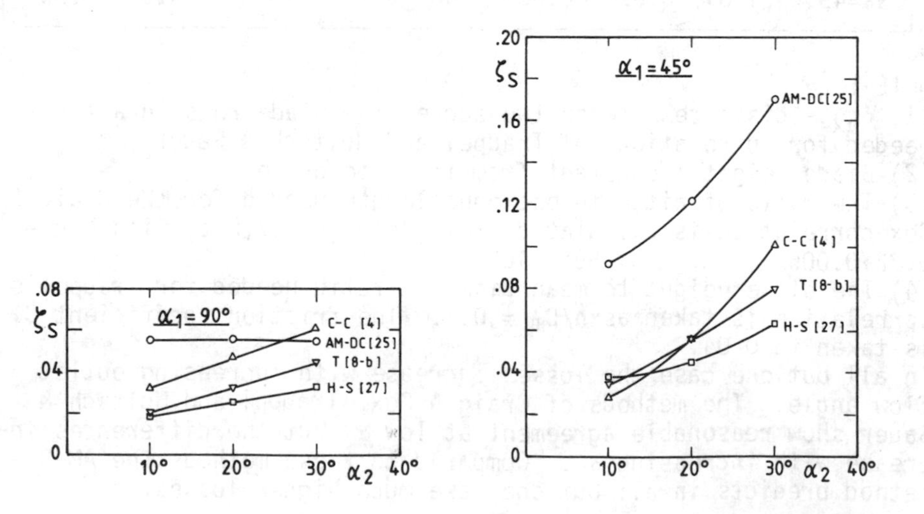

FIG. 33 - COMPARISON OF SECONDARY LOSSES AS PREDICTED BY VARIOUS AUTHORS

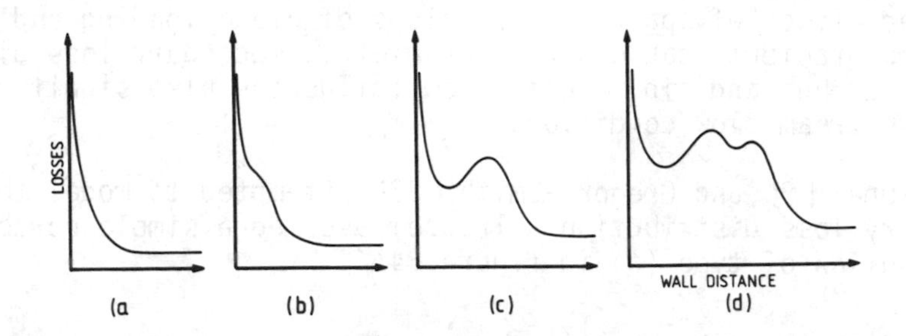

FIG. 34 - VARIOUS TYPES OF SPANWISE LOSS DISTRIBUTIONS, AFTER SIEVERDING [32]

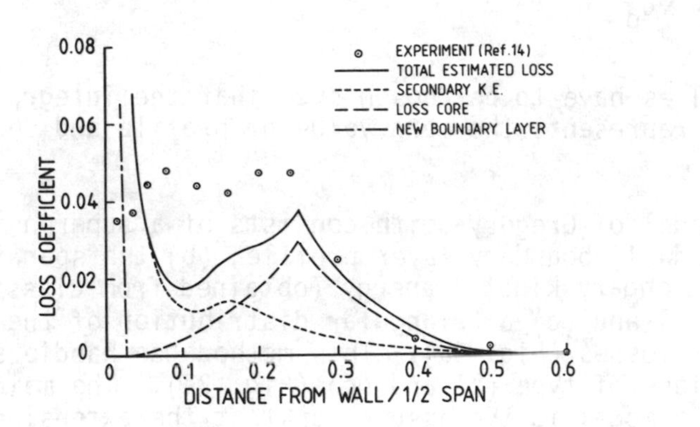

FIG. 35 - PREDICTION OF SPANWISE LOSS DISTRIBUTION BY GREGORY-SMITH [33]

3.2.3 Spanwise loss distributions : It is standard practice to assume secondary losses constantly be distributed over the blade height. This is, of course, an extreme simplification, as demonstrated in figure 34 from [32] which shows various types of loss distributions observed at the exit of straight cascades. In annular cascades the situation is further complicated since the combined effect of spanwise variations of blade loading and radial pressure gradients cause strong asymmetric secondary loss distributions at hub and tip. Inlet skew influences also significantly the downstream flow conditions.

Traupel [8] and Gregory-Smith [33] attempted to model the secondary loss distribution. Traupel assumed a simple parabolic distribution of type (a) in figure 34.

$$\eta(r) = \underbrace{1 - \zeta_p}_{\text{outside of } \underline{b}} - \Delta\eta \left(\frac{z}{b}\right)^2 \qquad (3.25)$$

where η includes only profile and secondary losses. z/b presents the spanwise distance non dimensionalized by the extension of the secondary flow region with

$$\frac{b}{g} = 7 \rightarrow 10 \sqrt{\zeta_p}$$

The $\Delta\eta$ values have to be chosen such that the integration over r correctly represents the mean value of profile and secondary losses.

The model of Gregory-Smith consists of a superposition of (a) an endwall boundary layer profile; (b) the spanwise distribution of secondary kinetic energy (obtained from classical secondary flow theory) and (c) a triangular distribution of the inlet boundary layer losses (Fig. 35). This method can handle secondary loss distributions of type (b) and (c) (Fig. 34). The main disadvantage of his model is the assumption that the extension of the secondary flow region can be calculated using classical secondary flow theory. A compilation of all available data by Sieverding en [32] showed that this assumption is not valid.

3.3 Clearance flow effects

Blade tip clearances affect both the stage efficiency and the mass flow of a turbine and prediction methods must provide information in both areas. The mechanism generating clearance losses is entirely different for shrouded and unshrouded blades and loss correlations must be able to take this into account too. In the case of shrouded rotor blades, the clearance flow is well separated from the flow through the blading. Losses arise then from energy dissipation across the shroud and from mixing of the clearance flow with the main flow downstream of the blade row. For unshrouded rotor blades, a clear separation between secondary and clearance losses is not possible. Therefore clearance losses are defined as the difference between total losses with and without clearance. At zero clearance the flow conditions are only slightly different from those in a fixed blade row except for the scraping effect of the blade suction side on the endwall boundary layers and the skewed inlet flow conditions caused by the transition of the upstream endwall boundary layer from a fixed frame into a rotating frame. With increasing tip clearance we can distinguish various factors contributing to the clearance flow mechanisms [34].

(a) The pressure difference between blade suction and pressure side - primarily an inviscid effect.
(b) The endwall boundary layer - a viscous effect.
(c) The relative movement between blade and endwall boundary layer - primarily a viscous effect. It is opposite to the effect produced by (a).
(d) The height of the radial clearance : depending on the height the factors (a) to (c) are more or less predominant, e.g. for small gaps the pressure forces are balanced by the shear forces while for large gaps the pressure difference is small due to blade unloading [38].

With increasing clearance the ordinary secondary flow patterns described in [32] will be gradually modified. This is illustrated in figure 36 from [35] which shows the spanwise distribution of the relative outlet flow angle α_2 at the exit of a one stage low speed turbine having a high hub-to-tip diameter ratio, high reaction, untwisted stator and rotor blades and 1.5% tip clearance.

Most turbines show a linear drop of stage efficiency with blade tip clearance height as demonstrated by the results of several turbines compiled by Haas & Kofskey [36] in figure 37. The graph includes turbine stages with shrouded rotors, indicated as configuration "SH" and unshrouded rotors with recessed casing and reduced blade height configurations indicated as "RC" and "RBH". In addition to the linear variation of the stage efficiency with δ, the authors concluded from figure 37 that : (1) for a given level of tip reaction the optimum tip clearance configurations from a standpoint of having the smallest tip clearance penalty were the shrouded turbines, followed by the recessed casing configurations, and then the reduced blade height configurations; (2) except for turbine H, the tip clearance loss increased with an increase of the tip reaction for a given tip clearance configuration.

There are different ways of incorporating clearance losses in a performance prediction method : (a) the aerodynamic blade efficiency η_a which accounts already for profile and secondary

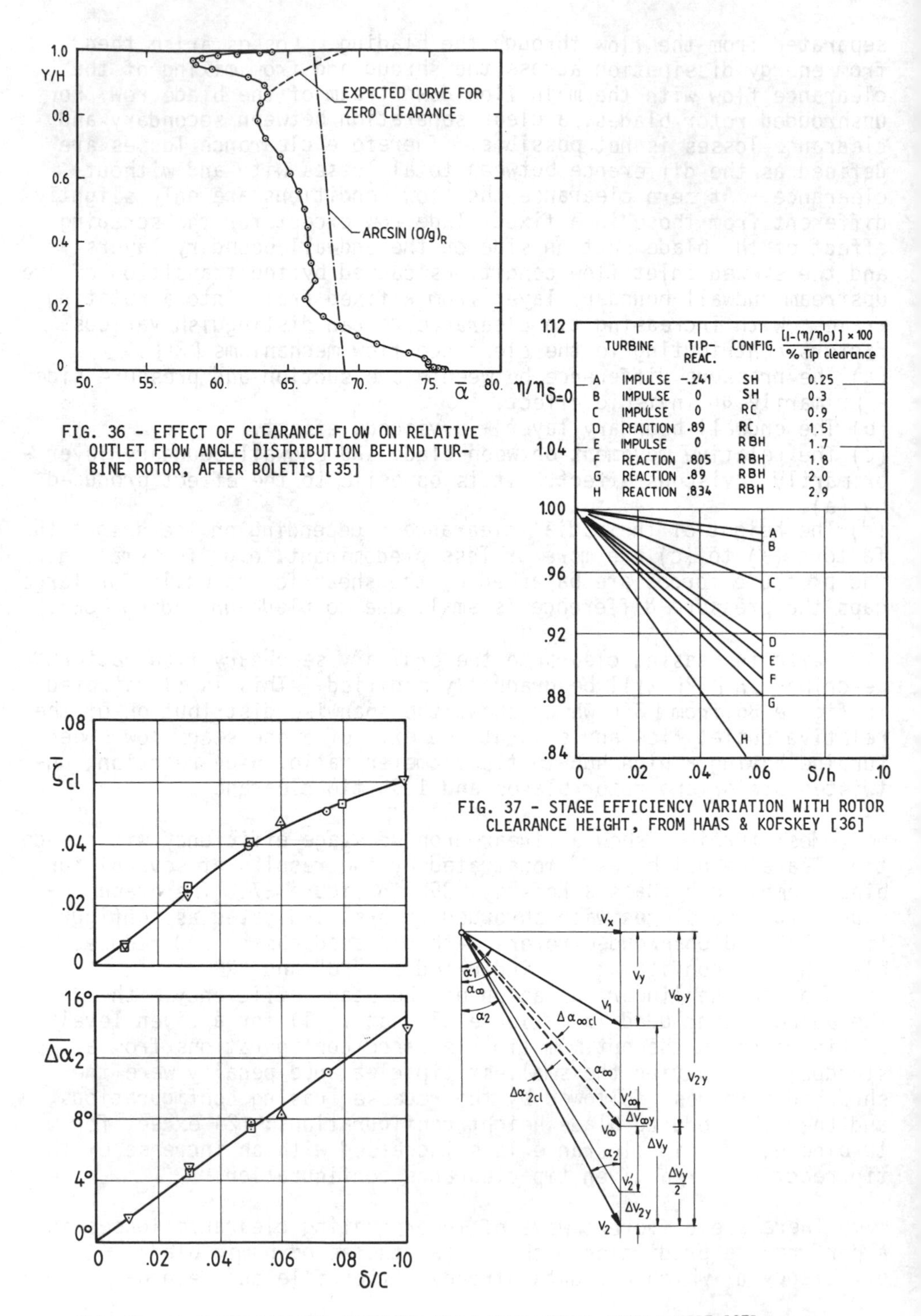

FIG. 36 - EFFECT OF CLEARANCE FLOW ON RELATIVE OUTLET FLOW ANGLE DISTRIBUTION BEHIND TURBINE ROTOR, AFTER BOLETIS [35]

FIG. 37 - STAGE EFFICIENCY VARIATION WITH ROTOR CLEARANCE HEIGHT, FROM HAAS & KOFSKEY [36]

FIG. 38 - CLEARANCE LOSS AND DEVIATION ANGLE CORRELATION, AFTER HUBERT [37]

losses is diminished by an amount $\Delta\eta$ representing the clearance loss $\zeta_{c\ell}$. In this way the clearance loss affects directly the expansion process through the blade rows. It is customary to proceed in this way for unshrouded blades. The stage efficiency η is calculated ultimately by subtracting from η_a a group of miscellaneous losses which do not arise directly from the expansion process: e.g. disc windage losses, partial admission losses, balance hole losses, etc.; (b) The clearance losses are treated in the same way as the above mentioned group of miscellaneous losses. This procedure is in particular justified for shrouded turbines.

3.3.1 Prediction methods for unshrouded blade rows : The most complete experimental studies on clearance flow effects in straight and annular cascades are still those reported by Hubert in 1963 [37]. Clearance losses and outlet flow angle deviations for four different cascades and various clearance heights correlated extremely well using the following relations :

$$\overline{\zeta_{c\ell}} = \frac{\zeta_{c\ell}}{\delta_u \cdot \sin\alpha_\infty} = f(\delta/c) \tag{3.27}$$

with

$$\delta_u = \frac{\Delta V_y}{V_x} = \mathrm{ctg}\alpha_2 - \mathrm{ctg}\alpha_1 \tag{3.28}$$

and the angle notation in figure 38.

$$\overline{\Delta\alpha_{2,c\ell}} = \Delta\alpha_{2,c\ell} \cdot \frac{c}{g} \cdot \frac{\sin\alpha_\infty}{\delta_u \cdot \sin^2\alpha_\infty} = K \cdot \frac{\delta}{c} \tag{3.29}$$

The equations (3.27) and (3.29) are presented in figure 38.

Balje and Binsley [10] rely entirely on Hubert's data for their performance prediction method.

Traupel [8b] relies to a certain extent on the structure of Hubert's correlation but the absolute values of the clearance losses are based entirely on turbine test results. Traupel proposes two sets of equations depending on whether the clearance losses are or are not to be included in the aerodynamic blade efficiency, η_a. In case the losses are included, η_a has to be diminished by :

$$\Delta\eta = K_\delta \frac{(\delta - 0.002\ C) \cdot D_T}{h \cdot D_M} \tag{3.30}$$

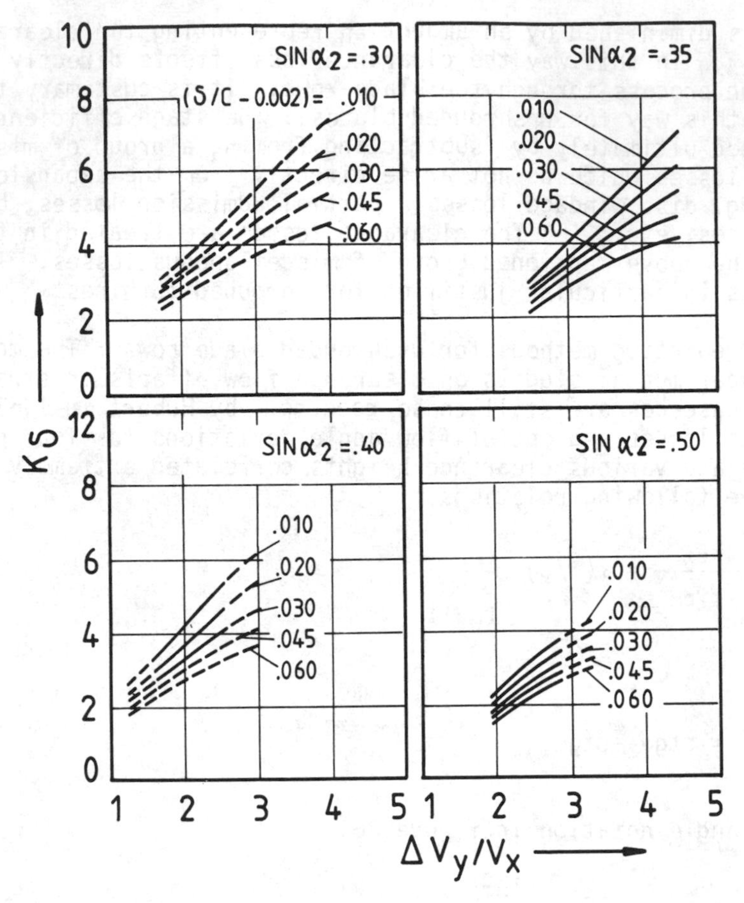

FIG. 39 - EFFICIENCY LOSS COEFFICIENT FOR UNSHROUDED BLADES, AFTER TRAUPEL [8b]

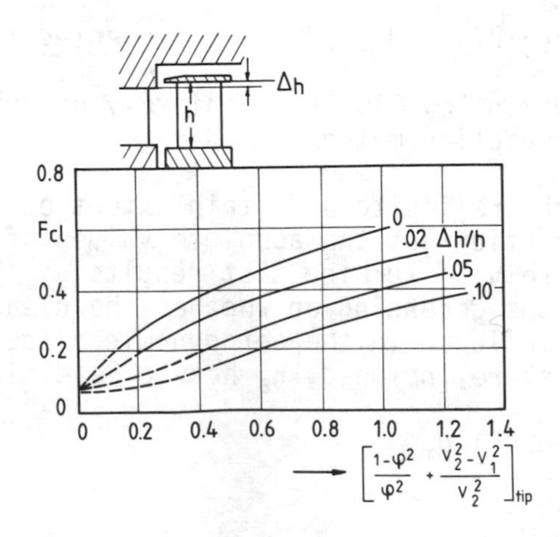

FIG. 40 - SHROUDED EFFICIENCY LOSS, AFTER CRAIG & COX [4]

with $K_\delta = f\left\{\alpha_2, \frac{\Delta V_y}{V_x}, \frac{\delta}{c}\right\}$ in figure 39. The equation shows a linear dependence of the efficiency debit $\Delta\eta$ on the clearance height.

The effect of the clearance flow on the mass flow is accounted for by an outlet angle correction

$$\Delta\alpha_{,c\ell} = 70.5 \cdot \frac{\delta}{h} \cdot \frac{D_T}{D_M} - 0.714 \ \frac{g}{h} - 0.5° \tag{3.31}$$

Traupel specifies that his correlation is based on data taken at operating conditions close to the nominal point.

The structure of the equation for clearance losses by Ainley & Mathieson

$$Y_{c\ell} = k \cdot \frac{\delta}{h} \cdot Z \tag{3.32}$$

$$k = 0.5 \quad \text{(for unshrouded blades)} \tag{3.33}$$

is very similar to that for secondary losses, see equation (3.15).

Craig & Cox [4] use the above equation for their own prediction method, provided that the axial velocity ratio remains approximately constant and that the relative velocities are well below sonic value.

Based on figure 38 from Hubert [37], Dunham and Came [25] changed the linear dependance of loss on tip clearance to

$$Y_{c\ell} = k \cdot \frac{c}{h} \left(\frac{\delta}{c}\right)^{0.78} \cdot Z \tag{3.34}$$

In case of multiple seals

$$\delta = (\delta_{geometric}) \times (\text{number of seals N})^{-0.42}$$

According to Kacker & Okapuu [2] (Eqn. 3.34) overpredicts the leakage losses for "recent turbines" , i.e., turbine stages with transonic relative outlet Mach numbers and strong increase of the axial velocity across the blade row (see comments of Craig & Cox concerning the limitation of the correlation by Ainley & Mathieson). The authors propose instead to calculate the change in turbine efficiency due to changes in tip clearance by

$$\frac{\Delta\eta}{\eta_0} = 0.93 \ \frac{\delta}{h \cdot \cos(90° - \alpha_2)} \cdot \frac{D_T}{D_M}$$

3.3.2 Prediction methods for shrouded blade rows : The mass flow through blade tip gaps in the case of shrouded blade rows can be approximated by the relation

$$\dot{m}_{c\ell} = \alpha \cdot A_{c\ell} \cdot \sqrt{\frac{1}{N}\left(\frac{p_1^2 - p_2^2}{p_1/\rho_1}\right)} \qquad (3.36)$$

where α - 0.7 to 0.8 contraction coefficient
$A_{c\ell}$ - tip clearance area
N - number of seals
p_1, p_2 - upstream/downstream pressures at blade tip

The total mass flow through the blade row is given by

$$\dot{m} = k \cdot \rho \cdot V_x \cdot A \qquad (A \text{ - annulus blade area}) \qquad (3.37)$$

The losses are closely related to the mass flow ratio

$$\mu = \frac{\dot{m}_{c\ell}}{\dot{m} - \dot{m}_{c\ell}} \qquad (3.38)$$

According to Traupel [8b], rotor tip clearance losses are given by the equation

$$\zeta_{c\ell} = \mu \left[1 - \left(\frac{\nu - \nu_D}{\nu_D}\right)^2\right] \qquad (3.39)$$

where :

$\nu = \dfrac{U}{\sqrt{2\Delta h_s}}$ is the so-called "velocity ratio"

U circumferential velocity at mean blade height
Δh_s static enthalpy drop across stage
subscript "D" ≃ design conditions

The effect of the clearance flow on the mass flow is accounted for by the k-factor in continuity equation

$$k = 0.99 + \mu - 0.0125\,\frac{g}{c}\,\mathrm{ctg}\,\alpha_2 \qquad (3.40)$$

As indicated before, the clearance losses in shrouded blades are not included in the aerodynamic blade efficiency η_a.

Craig & Cox [4] express the clearance losses as an efficiency debit

$$\Delta\eta_{c\ell} = F_{c\ell} \cdot \frac{A_{c\ell}}{A} \cdot \eta_a \qquad (3.40)$$

with F_{cl} given in figure 40 as function of the velocity change across the blade row and the overlap $\Delta h/h$. A positive overlap has a double effect : (a) it causes a loss due to the sudden area enlargement and (b) it reduces the static pressure ahead of the blade row which in turn reduces the tip leakage. Craig & Cox suggest that the overall effect is positive and increases constantly with increasing $\Delta h/h$. This is contrary to test results by Roeder [39] (reported by Traupel) showing an optimum overlapping of $\Delta h/h \approx 0.008$. However, the shroud geometry is different; a shroud with three seals for Roeder and without seals for Craig & Cox.

Dunham & Came [3] use equation (3.34) also for shrouded blades, but with k = 0.25 instead of 0.5 as for unshrouded blades. Kacker and Okapuu [2] adopt the same procedure. This is unsatisfactory, since it does not account for the different flow mechanism for shrouded and unshrouded blades.

3.3.3 General remarks : It is relatively easy to obtain for a given turbine configuration good data on clearance losses by systematic variations of the clearance height, but the data can only be used for turbines of similar configurations. The clearance flow mechanism and its effect on the total flow field are still far from being understood. Aerodynamic parameters such as blade load distribution, Mach number and coolant ejection from the blade tip, and geometric configurations such as partial shrouds and recessed casings for unshrouded blades will need much more attention in the future.

LIST OF SYMBOLS

A annulus area
b backbone length or extension of secondary flow region
c chord
c_d dissipation coefficient
c_f friction coefficient
c_ℓ lift coefficient
c_p pressure coefficient
D diameter
g pitch
h static enthalpy or blade height
H total enthalpy
ℓ length
M Mach number
$\dot{m}$ mass flow
O throat
P pressure
R radius
S surface length
t blade thickness
te trailing edge thickness
U peripheral velocity if not otherwise stated
V velocity (for stator-absolute velocity, for rotor-relative velocity)
Y total pressure loss coefficient
α' blade angle (with respect to tangential direction)
α flow angle (for stator-absolute flow angle, for rotor-relative flow angle)
α_2^* gauging angle, $\alpha_2^* = \arcsin o/g$
δ clearance or trailing edge wedge angle
δ^* displacement thickness
ε rear suction side turning angle
ζ kinetic energy loss $(V_{2,is}^2 - V_2^2)/V_{2,is}^2$
ζ^* kinetic energy loss $(V_{2,is}^2 - V_2^2)/V_2^2$
η efficiency
η_a aerodynamic blading efficiency
θ momentum thickness
κ specific heat ratio
ν kinematic viscosity or velocity ratio defined in (3.39)
ρ density
Φ mass flow coefficient
ψ stage loading factor
φ velocity coefficient V/V_{is}

Subscripts

ax axial
b basic or trailing edge base

c coolant
cr critical
cℓ clearance
D design
H hub
is isentropic
M mean blade height
p profile
PS suction side
R rotor
S stator
s secondary
SS suction side
T tip
TE trailing edge
x axial
y circumferential
z spanwise
0 total or basic
1 inlet
2 outlet
∞ infinite

Abbreviations

AM Ainley-Mathieson
DC Dunham-Came
KO Kacker-Okapuu

REFERENCES

1. SMITH, S.F.: A simple correlation of turbine efficiency. J. Royal Aeron. Soc., Vol. 69, 1965, p 467.
2. KACKER, S.C. & OKAPUU, U.: A mean-line prediction method for axial flow turbine efficiency. ASME Trans., Series A - J. Engineering for Power, Vol. 104, No. 1, Jan. 1982, pp 111-119.
3. DUNHAM, J. & CAME, P.M.: Improvements to the Ainley-Mathieson method of turbine performance prediction method. ASME Trans., Series A - J. Engineering for Power, Vol. 92, No. 3, July 1970, pp 252-256.
4. CRAIG, H.R.M. & COX, H.J.A.: Performance estimation of axial flow turbines. Proc. Inst. Mech. Engrs., Vol. 185, 1970-1971.
5. HORLOCK, J.H.: Axial flow turbines. London, Butterworths, 1966.
6. LATIMER, R.J.: Axial turbine performance prediction. In: "Off-Design Performance of Gas Turbines", VKI LS 1978-2, Jan. 1978.
7. AINLEY, D.G. & MATHIESON, G.C.R.: A method of performance estimation of axial-flow turbines. ARC R&M 2974, 1951.
8. TRAUPEL, W.: Thermische Turbomaschinen, Bd I, Springer Verlag. 8a - Ausgabe 1966, 8b - Ausgabe 1977.
9. DENTON, J.D.: A survey of comparison of methods for predicting the profile loss of turbine blades. In: Proc. Inst. of Mech. Engrs. on "Heat and Fluid Flow in Steam and Gasturbine Plant", 1973.
10. BALJE, D.E. & BINSLEY, R.L.: Axial turbine performance evaluation. ASME Trans., Series A - J. Engineering for Power, Vol. 90, No. 4, Oct. 1968, pp 341-360.
11. STEWART, W.L.; WHITNEY, W.I.; WONG, R.Y.: A study of boundary layer characteristics of turbomachine blade rows and their relation to over-all loss. ASME Trans., Series D - J. Basic Engineering, Vol. 82, No. 3, Sept. 1960, pp 589-592.
12. HODSON, H.P.: Boundary layer and loss measurements on the rotor of an axial flow turbine. ASME P 83-GT-4, 1983.
13. DEJC, M.E. & TROJANOVSKIJ; B.M.: Untersuchung und Berechnung axialer Turbinenstufen. Verlag Technik Berlin, 1973.
14. BRYNER, H.: Gegenwärtiger Stand in der Berechnung der Hinterkantenverluste einer Turbinenschaufel. VDI Berichte Nr 264, 1976.
15. SIEVERDING, C.H.; STANISLAS, M.; SNOECK, J.; The base pressure problem in transonic turbine cascades. ASME Trans., Series A - J. Engineering for Power, Vol. 102, No. 3, July 1980, pp 711-718, also VKI Preprint 1979-1.
16. PRUST, H.W. & HELON, R.H.: Effect of trailing edge geometry and thickness on the performance of certain turbine stator blading. NASA TN-D 6637, 1972.
17. DAVID, O.: Einfluss der Hinterkantenform einer Turbinenleitschaufel auf die Profilverluste. TH Aachen Mitt. 74-06, 1974.
18. SIEVERDING, C.H.: The influence of trailing edge ejection on the base pressure in transonic turbine cascades. ASME Trans., Series A - J. Engineering for Power, Vol. 105, No. 2, April 1978, pp 215-222. Also VKI Preprint 1981-25.

19. SIEVERDING, C.H.: On the influence of boundary layer state on vortex shedding from flat plates and turbine blades. von Karman Institute, unpublished.
20. GRAHAM, C.G. & KOST, F.H.: Shock boundary layer interaction on high turning turbine cascades. ASME P 79-GT-37, 1979.
21. ANSALDO, private communication.
22. SCHLICHTING, H.: Boundary layer theory. Seventh edition. McGraw-Hill, 1979.
23. DUNHAM, J.: A review of cascade data on secondary losses in turbines. J. Mech. Engrg. Sci., Vol. 12, 1970.
24. WOLF, H.: Die Randverluste in geraden Schaufelgittern. Wiss. Z. Techn. Hochschule Dresden, Bd. 10, Heft 2, 1961.
25. DUNHAM, J. & CAME, P.M.: Improvements to the Ainley-Mathieson method of turbine performance prediction. ASME Trans., Series A - J. Engineering for Power, Vol. 92, No. 3, July 1970, pp 252-256.
26. CHAUVIN, J.: Turbine cascade endwall losses. A review. In: "Secondary Flows in Turbomachines", VKI LS 72, Jan. 1975.
27. HULTSCH, M. & SAUER, H.: Sekundärströmungen in Beschaufelungen axialer Turbomaschinen. Maschinenbautechnik, Bd. 28, Heft 1, 1979.
28. GERSTEN, K.: Uber den Einfluss der Geschwindigkeitsverteilung in der Zuströmung auf die Sekundärströmung in geraden Schaufelgittern. Forschung auf dem Gebiet des Ingenierswesens. Bd 23, Heft 3, pp 95-101
29. KLEIN, A.: Untersuchungen über den Einfluss der Zuströmungrenzschicht auf die Sekundärströmung in den Beschaufelungen von Axialturbinen. Forsch. Ing., Bd 32, Nr 6, 1966/
30. CAME, P.M. Secondary loss measurements in a cascade of turbine blades. Inst. Mech. Engrs. C33/73.
31. WEGEL, S.: Strömungsuntersuchungen an Beschleunigungsgittern im Windkanal und in der Axialturbine. Dissertation TH Darmstadt 1970
32. SIEVERDING, C.H.: Secondary flows in straight and annular cascades. Presented at NATO Advanced Study Institute on Thermodynamics and Fluid Mechanics of Turbomachinery, Izmir (Cesme), Turkey, Sept. 1984.
33. GREGORY-SMITH, D.G.: Secondary flows and losses in axial flow turbines. ASME Trans., Series A - J. Engineering for Power, Vol. 104, No. 4, Oct. 1982, pp 819-822.
34. PEACOCK, R.F.: A review of turbomachinery tip gaps. Int. J. of Heat & Fluid Flow, Vol. 3, No. 4, 1982, pp 185-194.
35. BOLETIS, E.: Experimental research on secondary flows in annular turbine cascades at VKI. In: "Secondary Flows and Endwall Boundary Layers in Axial Turbomachines, VKI LS 1984-05.
36. HAAS, J.E. & KOFSKEY, M.G.: Cold-air performance of a 12.766 centimeter-tip-diameter axial-flow cooled turbine. III - Effect of rotor tip clearance on overall performance of a solid blade configuration. NASA TP 1032, 1977.
37. HUBERT, G.: Probleme der Sekundärströmungen in axialen Turbomaschinen. VDI-Forschungsheft 496, 1963.

38. BOOTH, T.C.; DODGE, P.R.; HEPWORTH, H.K.: Rotor-tip-leakage : Part I - Basic methodology. ASME P 81-GT-71, 1981.
39. ROEDER, A.: Experimentelle Bestimmung der Einzelverluste in einer einstufigen Versuchsturbine. Mitt. Inst. Therm. Turbomasch., ETH Zürich, No. 15, 1969.

ORGANIZATION OF DESIGN SYSTEMS FOR TURBOMACHINERY

George K. Serovy

Department of Mechanical Engineering
Iowa State University
Ames, Iowa 50011 U.S.A.

INTRODUCTION

Design systems for turbomachines include a series of steps in geometry selection. From each step a reduced number of more carefully defined geometries emerge. In successive levels of computation flow field analysis and performance prediction become more detailed, and experimental verification becomes more critical.

In each level of design or analysis by computation, two common elements are involved. First, a group of aerodynamic or hydrodynamic limits must be recognized and set numerically. Second, some means for integration and interaction of hub-to-tip and blade-to-blade flow field prediction must be established.

Figure 1 is a flow chart which follows a sequence typical of many axial-flow turbomachine design systems. Each organization tends to develop a unique design system based on its own choice of alternatives within each box. We will use Figure 1 as the basis for discussion of design of an example type of axial-flow turbomachine, a fan or compressor for a gas turbine engine application. A bibliography is included which gives background sources related to axial-flow compressor, pump and turbine cases.

DESIGN POINT VALUES--SYSTEM OPERATING REQUIREMENTS

Except for configurations designed for research alone, turbomachines are developed to meet engineering system requirements, and their interactions with other system components are of fundamental importance in determining success or failure in operation.

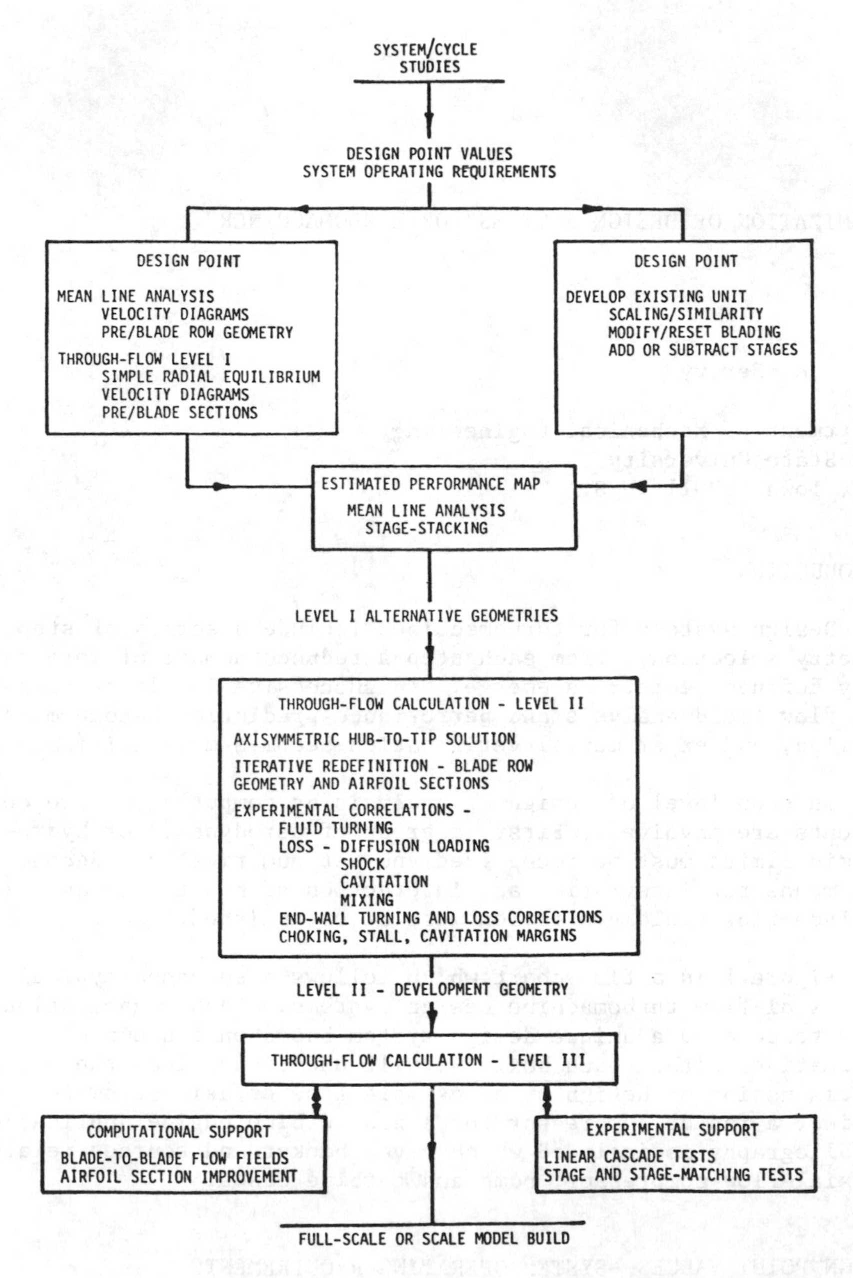

Figure 1. Structure of axial-flow turbomachine design system

No turbomachine design study should proceed without understanding and continuous review of the conditions imposed by system operation and physical arrangement.

Design objectives associated with a gas turbine engine compressor are listed in Figure 2. These objectives apply to both stationary and propulsion gas turbine components.

When a compressor for a gas turbine system is proposed a list of design operating point conditions might include

- working fluid state at compressor entrance including total temperature and total pressure
- tables or equations of state for working fluid properties
- entrance mass flow rate
- overall total-pressure ratio for compressor unit
- compressor efficiencies used in system analysis
- mass flow rates of working fluid subtracted from or added to entrance flow at points within the compressor with corresponding total pressure levels.

Limits on compressor dimensions, rotational speed and exit velocities might be given or might be the subject of negotiation between compressor designers and the designers of other system components. Entrance and exit geometry requirements are especially significant.

CONFIGURATION DEVELOPMENT BY SIMILARITY METHODS

Design and development of a new turbomachine configuration is always a long and expensive process. It has often been a much longer and expensive process than originally expected. Therefore scaling and geometric modification of existing configurations to match new application requirements is an attractive option. The usual similarity criteria are aerohydrodynamic in nature, and only refer indirectly to scaling of mechanical features of the configuration (e.g. shafts, bearings, seals, casing). It is necessary to remember that any departure from exact scaling in geometry (e.g. tip clearance, blade dimensions, surface finish) is a distortion in similitude, possibly influencing aerodynamic and aeromechanical behavior.

Figure 2. Design objectives for an axial-flow compressor in a gas turbine engine system

1. STABLE AERODYNAMIC AND AEROMECHANICAL CHARACTERISTICS, INCLUDING SUFFICIENT SURGE MARGIN AND NO EXCESSIVE BLADE OR DISK VIBRATIONS AT ALL POTENTIAL OPERATING CONDITIONS FOR THE SYSTEM INSTALLATION

2. ACCEPTABLE THERMODYNAMIC EFFICIENCY FOR ALL EXPECTED OPERATING CONDITIONS WITHOUT EXCESSIVE TIME-DEPENDENT DETERIORATION

3. ACCEPTABLE NOISE GENERATION CHARACTERISTICS

SUBJECT TO THE ABOVE CONSTRAINTS THE COMPRESSOR SHOULD BE DESIGNED FOR

4. HIGH MASS FLOW RATE PER UNIT FRONTAL AREA

5. HIGH AVERAGE PRESSURE RATIO PER STAGE

6. MINIMUM VALUES OF AXIAL STAGE LENGTH AND NUMBER OF BLADES USED IN EACH BLADE ROW

7. MINIMUM UNIT COMPLEXITY (VARIABLE GEOMETRY, BLEED PORTS)

NEW CONFIGURATION DESIGN AND DEVELOPMENT

Unless the alternative of geometric scaling is followed, axial-flow turbomachine design ordinarily includes all of the steps shown in Figure 1. The content of most of the sequential steps (boxes on the flow chart) has been discussed in other lectures and will be extensively reviewed in all of the following sessions. Only a few guidelines and references to additional ideas can be included here.

PRELIMINARY GEOMETRY SELECTION

Mean-Line Design Study

In the first step in configuration selection, a mean-line design study is usually made in order to determine a number of potentially satisfactory meridional flow path geometries. This would include the overall dimensions of a compressor or turbine and the number of stages required. Frequently, consideration of the use of axial or centrifugal/radial geometries enters this phase of design. At this point there is no good substitute for reference to what history has indicated to be possible, and for the reduction of variables to a minimum.

In an axial compressor, flow path dimensions are strongly influenced by through-flow velocity level and by assigned stage aerodynamic loading, blade velocities and attainable efficiency. Figure 3 shows the effect of average axial Mach number level and hub-tip radius ratio on equivalent mass flow rates through an annulus and applies specifically to inlet stage dimensions. Figures 4 and 5 show the general effect of diffusion loading on the pressure rise attainable in compressor and pump rotors. Figure 6 shows the levels of experimental performance achieved in compressor stage tests on a basis similar to Figure 4. An efficiency is assumed in Figure 4. Figure 6 represents measured single-stage performance for a series of NASA-designed research stages.

It is to be noted that not even rough preliminary stage dimensions can be estimated without some consideration of blade row solidity, blade numbers and aspect ratio.

Velocity Diagrams

Mean line studies should include consideration of varying degree of reaction, Mach number level, axial velocity × density product variation, and blade row diffusion scheduling. Reynolds numbers can be evaluated as soon as chord lengths are estimated. All of these items depend on the form of velocity diagram

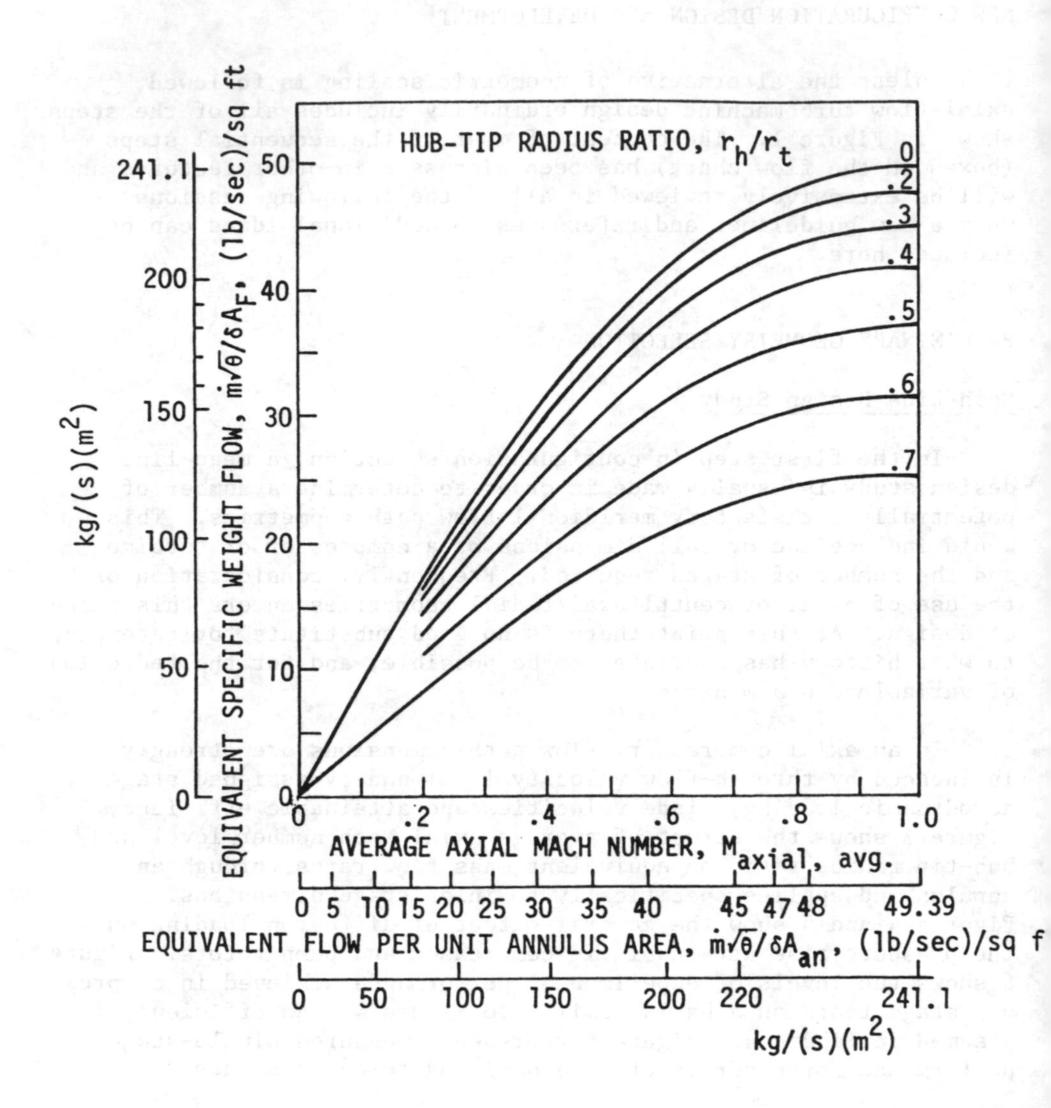

Figure 3. Effect of axial Mach number level on equivalent specific air flow rates through axial-flow turbomachine

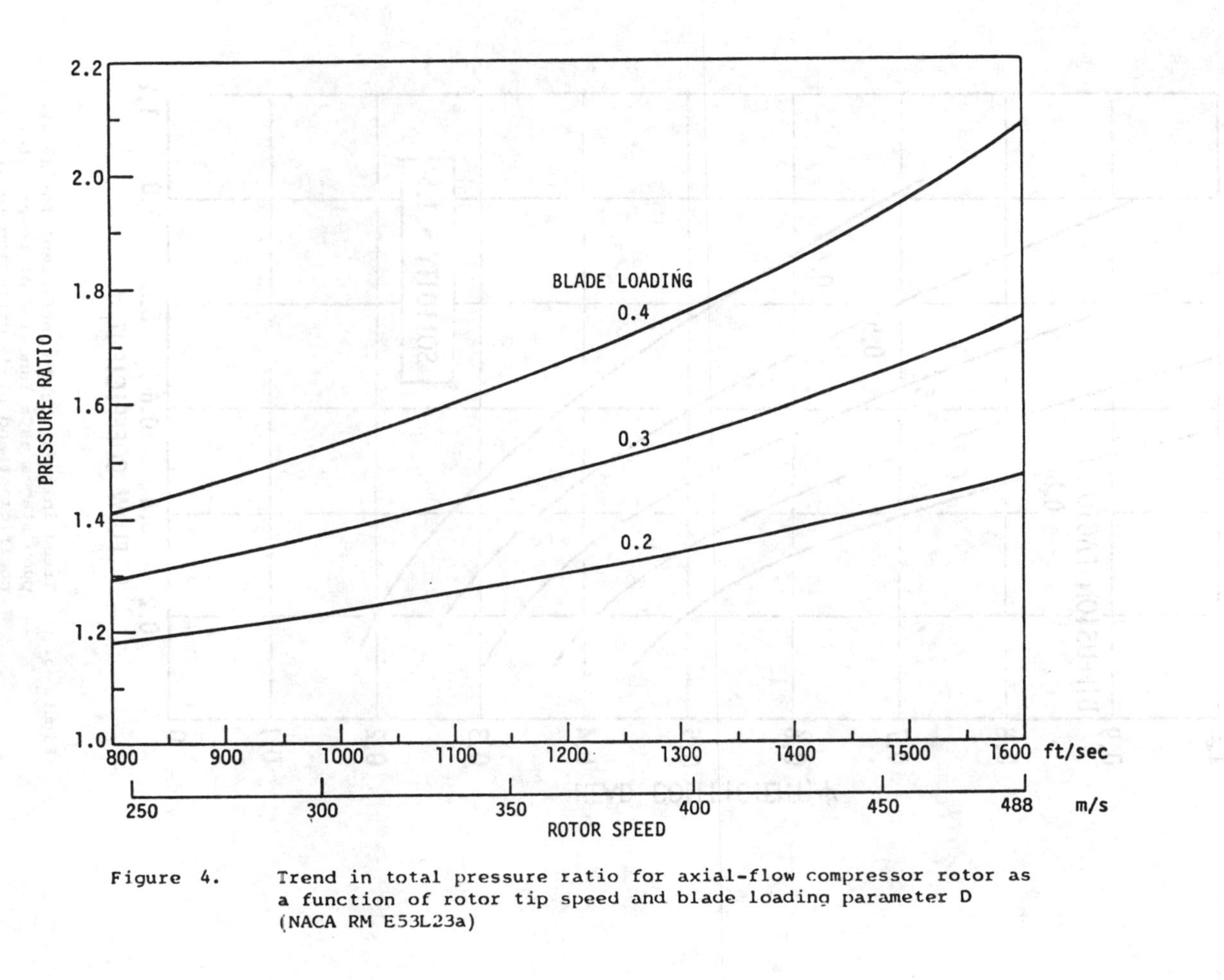

Figure 4. Trend in total pressure ratio for axial-flow compressor rotor as a function of rotor tip speed and blade loading parameter D (NACA RM E53L23a)

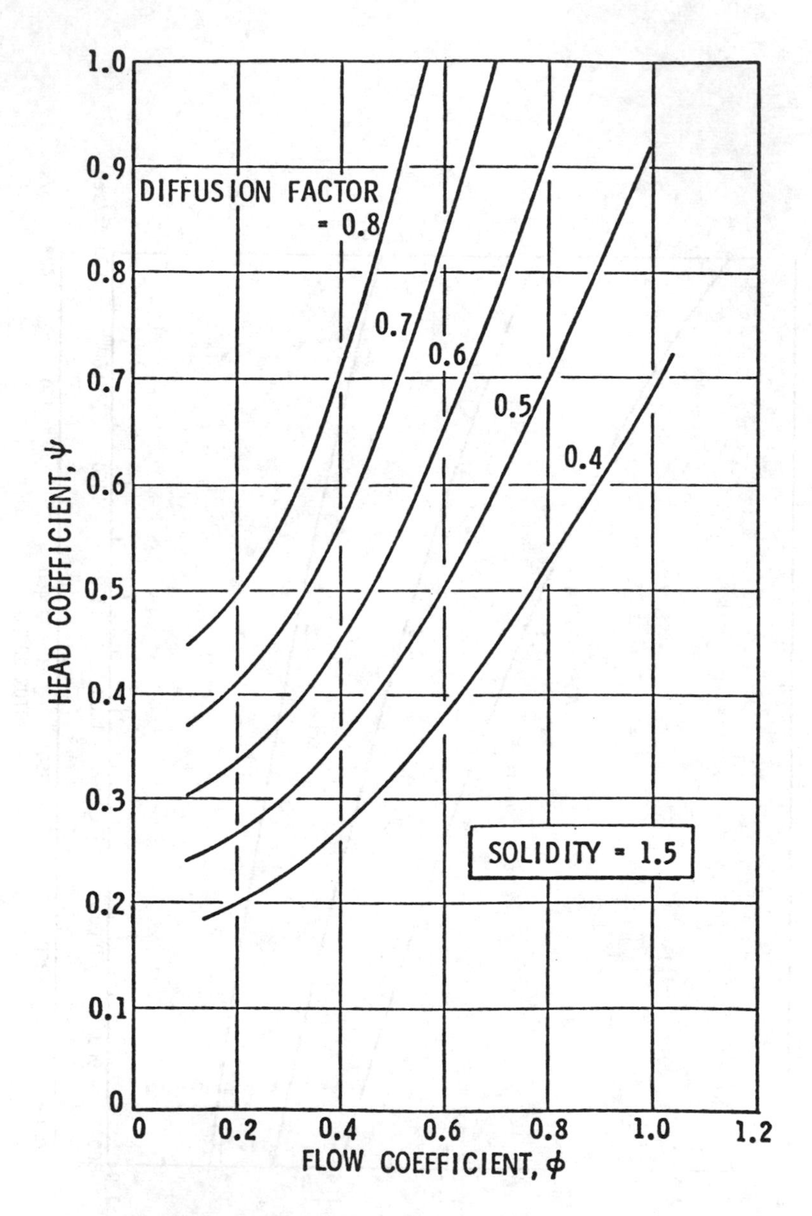

Figure 5. Trend in ideal head coefficient for axial-flow pump stages as a function of rotor inlet flow coefficient and blade diffusion loading parameter (NASA SP-304)

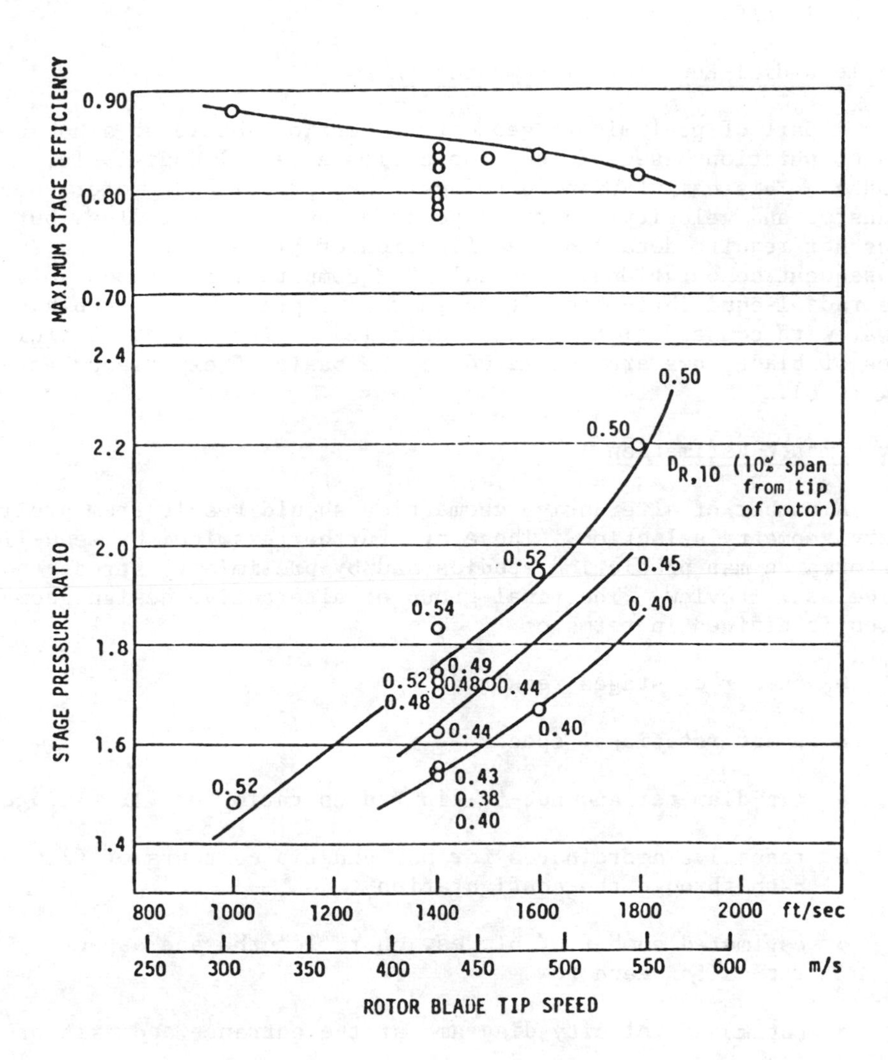

Figure 6. Measured trends in axial-flow compressor stage performance (D. M. Sandercock - NASA)

designated at the mean flow-path radius and on the variation in the radius itself.

Simple-Radial-Equilibrium Computation

A part of preliminary geometry selection should be a hub-to-tip computation based on isentropic simple radial equilibrium (ISRE). This computation introduces the variation in rotor energy transfer and velocity diagram type as a function of radius, but does not require detailed specification of blade sections as do subsequent computations. Typical ISRE computation methods solve the radial-equilibrium condition in radial planes between blade rows, with conservation of mass satisfied by iteration. Efficiencies of blade rows are estimated on the basis of experience (e.g. Figure 6).

Performance Estimation

A number of alternative geometries should result from preliminary geometry selection. These are further examined by mean-line performance map prediction studies and by preliminary stress and aeroelastic review. The final group of alternative design geometries is defined in terms of

- number of stages required
- shaft rotational speed
- tip diameter and hub-to-tip radius ratio for first stage
- tentative coordinates for hub and tip contours of flow path through the configuration
- estimated number of blades, chord lengths and aspect ratios for each row
- estimated velocity diagrams at the entrance and exit of each row
- estimated performance map.

Both the mean-line and ISRE flow path design and performance estimation methods utilize empirical and simplified allowances for blade row losses and for effective annulus area reduction (blockage) due to end-wall effects.

Preliminary Airfoil and Cascade Selection

For Level II through-flow calculations, enough information must be known about the geometry of individual blade rows so that reasonable estimates of loss distribution along the span can be made. In axial-flow turbomachines the necessary information includes

- number of blades in each row
- chord length distribution from hub to tip
- airfoil profile selection (e.g. effective camber angle, camber line shape, thickness distribution)
- stagger angle
- stacking line geometry.

Both correlated cascade data or blade-to-blade computation methods may be useful in selection and optimization of these geometric features of the candidate designs. However, at this point in the design process correlated data is probably the most reasonable alternative.

THROUGH-FLOW CALCULATION--LEVEL II

A very limited number of attractive configuration geometries should result from preliminary and Level I design. This number is reduced by detailed analysis using computational fluid mechanics codes based on the streamline-curvature duct flow and streamline-curvature through-flow flow field models. These are designated Level II codes. They almost always assume steady, axisymmetric (circumferentially-averaged) flow. At this level iterative redefinition of blade row geometries is called for along with distribution of losses and passage area blockage allowance along the span of each blade row and on stream surface approximations through each row.

The Level II through-flow codes are analysis codes. When used at the design operating point for a compressor or fan, cascade intersection (blade-to-blade) performance estimation includes fluid turning angle, shock losses, profile losses due to diffusion loading and secondary loss corrections where appropriate. Level II calculations are expensive to develop and expensive to use, but these costs are insignificant when related to the cost and time associated with full-scale testing.

THROUGH-FLOW CALCULATION--LEVEL III COMPUTATIONAL AND EXPERIMENTAL SUPPORT

Much of the content of this ASI lecture series has been descriptive of state-of-the-art work on the application of computational fluid mechanics and experimentation to the development of advanced turbomachinery. Integration of the methods described into future design systems is the homework assignment for the individuals who hear and discuss these lectures.

BIBLIOGRAPHY

PRELIMINARY GEOMETRY SELECTION

Mean-Line Design--Velocity Diagrams

LIEBLEIN, Seymour. Review of High-Performance Axial-Flow-Compressor Blade-Element Theory. NACA RM E53L22, 1954.

KLAPPROTH, John F. General Considerations of Mach Number Effects on Compressor Blade Design. NACA RM E53L23a, 1954.

ECK, Bruno. Fans. Oxford: Pergamon Press, 1973. (Translation of Ventilatoren, Fifth Edition. Berlin/Heidelberg/New York: Springer-Verlag, 1972.)

WHITNEY, Warren J. and STEWART, Warner L. Velocity Diagrams. IN Turbine Design and Application, Chapter 3. NASA SP-290, Vol. 1, 1972.

HORLOCK, J. H. Axial Flow Compressors. London: Butterworths, 1958. (Reprint with supplemental material, Huntington, New York: Krieger, 1973.)

HORLOCK, J. H. Axial Flow Turbines. London: Butterworths, 1966. (Reprint with corrections, Huntington, New York: Krieger, 1973.)

TRAUPEL, W. Thermische Turbomaschinen. Vol. I. Third Edition. Berlin/Heidelberg/New York: Springer-Verlag, 1977.

BALJE, O. E. Turbomachines. New York: John Wiley and Sons, 1981.

SCHWEITZER, J. K. and GARBEROGLIO, J. E. Maximum Loading Capability of Axial Flow Compressors. J. Aircraft. 21: 593-600. 1984.

Simple-Radial-Equilibrium Computation

BRYANS, A. C. and MILLER, M. L. Computer Program for Design of Multistage Axial-Flow Compressors. NASA CR-54530, 1967.

FLAGG, E. E. Analytical Procedure and Computer Program for Determining the Off-Design Performance of Axial-Flow Turbines. NASA CR-710, 1967.

KAVANAGH, Patrick and MILLER, Max J. Axial Flow Pump Design Digital Computer Program. Iowa State University. ERI-66300, 1970.

Performance Estimation

FINGER, H. B. and DUGAN, J. R., Jr. Analysis of Stage-Matching and Off-Design Performance of Multi-Stage Axial-Flow Compressors. NACA RM E52D07, 1952.

BENSER, W. A. Analysis of Part-Speed Operation for High-Pressure-Ratio Multistage Axial-Flow Compressors. NACA RM E53I15, 1953.

AINLEY, D. G. and MATHIESON, G. C. R. A Method of Performance Estimation for Axial-Flow Turbines. ARC Rep. and Memo 2974, 1957.

DOYLE, M. D. C. and DIXON, S. L. The Stacking of Compressor Stage Characteristics to Give an Overall Compressor Performance Map. Aero. Quarterly, 13, 1962: 349-367.

BULLOCK, Robert O. Analysis of Reynolds Number and Scale Effects on Performance of Turbomachinery. J. Eng. Power, Trans. ASME, Ser. A, 90, 1968: 149-156.

WASSELL, A. B. Reynolds Number Effects in Axial Compressors. J. Eng. Power, Trans. ASME, Ser. A, 90, 1968: 149-156.

DUNHAM, J. and CAME, P. M. Improvements to the Ainley-Mathieson Method of Turbine Performance Prediction. J. Eng. Power, Trans. ASME, Ser. A, 92, 1970: 252-256.

PAMPREEN, R. C. Small Turbomachinery Compressor and Fan Aerodynamics. J. Eng. Power, Trans. ASME, Ser. A, 95, 1973: 251-256.

SOUTHWICK, Robert D. A Stage-Stacking Simulation of Axial Flow Compressors with Variable Geometry. Aeronautical Systems Division, U.S. Air Force Systems Command. ASD-TR-74-38, 1974.

GRIEB, H., SCHILL, G. and GUMUCIO, R. A Semi-Empirical Method for the Determination of Multistage Axial Compressor Efficiency. ASME Paper 75-GT-11, 1975.

WALL, R. A. Axial-Flow Compressor Performance Prediction. IN Modern Prediction Methods for Turbomachine Performance, Paper 4. AGARD-LS-83, 1976.

HOWELL, A. R. and CALVERT, W. J. A New Stage Stacking Technique for Axial-Flow Compressor Performance Prediction. ASME Paper 78-GT-139, 1978.

OSTERWALDER, J. Efficiency Scale-Up for Hydraulic Turbomachines with Due Consideration of Surface Roughness. J. Hydraulic Research, 16, 1978: 55-76.

KOCH, C. C. Stalling Pressure Rise Capability of Axial Flow Compressor Stages. J. Eng. Power, Trans. ASME, 103, 1981: 645-656.

KACKER, S. C. and OKAPUU, U. A Mean Line Prediction Method for Axial Flow Turbine Efficiency. J. Eng. Power, Trans. ASME, 104, 1982: 111-119.

PRELIMINARY AIRFOIL AND CASCADE SELECTION--CASCADE PERFORMANCE ESTIMATION

LIEBLEIN, S. Analysis of Experimental Low-Speed Loss and Stall Characteristics of Two-Dimensional Compressor Blade Cascades. NACA RM E57A28, 1957.

MILLER, G. R., LEWIS, G. W., Jr. and HARTMANN, M. J. Shock Losses in Transonic Blade Rows. J. Eng. Power, Trans. ASME, Ser. A, 83, 1961: 235-242.

LIEBLEIN, Seymour. Experimental Flow in Two-Dimensional Cascades. IN Aerodynamic Design of Axial-Flow Compressors, Chapter VI. NASA SP-36, 1965.

ROBBINS, William H., JACKSON, Robert J. and LIEBLEIN, Seymour. Blade-Element Flow in Annular Cascades. IN Aerodynamic Design of Axial-Flow Compressors, Chapter VII. NASA SP-36, 1965.

DUNHAM, J. A Review of Cascade Data on Secondary Losses in Turbines. J. Mech. Eng. Sci., 12, 1970: 48-49.

DEJC, M. E. and TROJANOVSKIJ, B. M. Untersuchung und Berechnung axialer Turbinenstufen. Berlin (DDR): VEB Verlag Technik, 1974.

DENTON, J. D. A Survey and Comparison of Methods for Predicting the Profile Loss of Turbine Blades. Conference Proceedings of the IME, 3, 1973: 204-212.

STEWART, W. L. and GLASSMAN, A. J. Blade Design. IN Turbine Design and Application, Chapter 4. NASA SP-190, Vol. 2, 1973.

HUFFMAN, G. David and TRAMM, P. C. Airfoil Design for High Tip Speed Compressors. J. Aircraft, 11, 1974: 682-689.

WRIGHT, L. C. Blade Selection for a Modern Axial-Flow Compressor. IN Fluid Mechanics, Acoustics, and Design of Turbomachinery, Part II. U.S. NASA SP-304, 1974.

HORLOCK, J. H. and PERKINS, H. J. Annulus Wall Boundary Layers in Turbomachines. AGARDograph AGARD-AG-185, 1974.

LICHTFUSS, H.-J. and STARKEN, H. Supersonic Cascade Flow. IN Progress in Aerospace Science. New York: Pergamon Press Ltd., 15, 1974, 37-149.

KOCH, C. C. and SMITH, L. H., Jr. Loss Sources and Magnitudes in Axial-Flow Compressors. J. Eng. Power, Trans. ASME, Ser. A, 98, 1976: 411-424. (Discussion by N. L. Sanger in J. Eng. Power, 99, 1977: 197.)

GRAHL, Klaus G. Über den Stand der Kennfeldberechnung mehrstufiger Axialverdichter. Zeit. Flugwiss. Weltraumforsch., 1, 1977: 29-41.

SCHOLZ, N. Aerodynamik der Schaufelgitter. Vol. I. Karlsruhe: G. Braun, 1965. (Translated and revised by Dr.-Ing. A. Klein as Aerodynamics of Cascades. Advisory Group for Aerospace Research and Development, AGARDograph No. 220, 1977. NTIS Document AD-A051233.)

SEROVY, G. K. Deviation Angle/Turning Angle Prediction for Advanced Axial-Flow Compressor Blade Row Geometries. AFAPL-TR-77-81, 1978.

HAY, N., METCALFE, R. and REIZES, J. A. A Simple Method for the Selection of Fan Profiles. Proceedings of the IME, 192, 1978: 269-275.

PRINCE, D. C., Jr. Three-Dimensional Shock Structures for Transonic/Supersonic Compressor Rotors. J. Aircraft, 17, 1980: 28-37.

DeRUYCK, J. and HIRSCH, C. Investigations of an Axial Compressor End-Wall Boundary Layer Prediction Method. J. Eng. Power, Trans. ASME, 103, 1981: 20-33.

NATO Advisory Group for Aerospace Research and Development, Propulsion and Energetics Panel Working Group 12. Through Flow Calculations in Turbomachines. AGARD-AR-175, 1981.

WENNERSTROM, A. J. and PUTERBAUGH, S. L. A Three-Dimensional Model for the Prediction of Shock Losses in Compressor Blade Rows. J. Engr. for Gas Turbines and Power, Trans. ASME, 106, 1984: 295-299.

SULLEREY, R. K. and KUMAR, S. A Study of Axial Turbine Loss Models in a Streamline Curvature Computing System. J. Engr. for Gas Turbines and Power, Trans. ASME, 106, 1984: 591-597.

HOBBS, D. E. and WEINGOLD, H. D. Development of Controlled Diffusion Airfoils for Multistage Compressor Application. J. Engr. for Gas Turbines and Power, Trans. ASME, 106, 1984: 271-278.

THROUGH-FLOW CALCULATION--LEVEL II

CREVELING, H. F. and CARMODY, R. H. Axial Flow Compressor Design Computer Programs Incorporating Full Radial Equilibrium. Part I - Flow Path and Radial Distribution of Energy Specified (Program II). NASA CR-54532, 1968. Part II - Radial Distribution of Total Pressure and Flow Path or Axial Velocity Ratio Specified (Program III). NASA CR-54531, 1968.

CARTER, A. F., PLATT, M. and LENHERR, F. K. Analysis of Geometry and Design Point Performance of Axial-Flow Turbines. I - Development of the Analysis Method and the Loss Coefficient Correlation. NASA CR-1181, 1968.

PLATT, M. and CARTER, A. F. Analysis of Geometry and Design Point Performance of Axial-Flow Turbines. II - Computer Program. NASA CR-1187, 1968.

CARTER, A. F. and LENHERR, F. K. Analysis of Geometry and Design-Point Performance of Axial-Flow Turbines Using Specified Meridional Velocity Gradients. NASA CR-1456, 1969.

SEROVY, G. K., KAVANAGH, P., OKIISHI, T. H. and MILLER, M. J. Prediction of Overall and Blade-Element Performance for Axial-Flow Pump Configurations. NASA CR-2301, 1973.

HEARSEY, Richard M. A Revised Computer Program for Axial Compressor Design. Volume I: Theory, Description and User's Instructions. ARL TR 75-0001, Vol. I, 1975. Volume II: Program Listing and Program Use Example. ARL TR 75-0001, Vol. II, 1975.

DENTON, J. D. Throughflow Calculations for Transonic Axial Flow Turbines. IN Turbomachinery Developments in Steam and Gas Turbines. New York: ASME, 1977: 11-19.

WYSONG, R. R., PRINCE, T. C., LENAHAN, D. T. et al. Turbine Design System. AFAPL-TR-78-92, 1978.

HIRSCH, C. and WARZEE, G. An Integrated Quasi-3D Finite Element Calculation Program for Turbomachinery Flows. J. Eng. Power, Trans. ASME, 101, 1979: 141-148.

CROUSE, James E. and GORRELL, William T. Computer Program for Aerodynamic and Blading Design of Multistage Axial-Flow Compressors. NASA Tech. Paper 1946, AVRADCOM Tech. Rep. 80-C-21, 1981.

UCER, A. S., YEGEN, I. and DURMAZ, T. A Quasi-Three-Dimensional Finite Element Solution for Steady Compressible Flow Through Turbomachines. J. Eng. Power, Trans. ASME, 105, 1983: 536-542.

JENNIONS, I. K. and STOW, P. A Quasi-Three-Dimensional Turbomachinery Blade Design System: Part 1 - Throughflow Analysis. ASME Paper 84-GT-26. 1984. Part 2 - Computerized System. ASME Paper 84-GT-27. 1984.

COMPUTATIONAL AND EXPERIMENTAL SUPPORT

STONE, A. Effects of Stage Characteristics and Matching on Axial-Flow-Compressor Performance. Trans. ASME, 80, 1958: 1273-1293.

BROWN, L. E. and GROH, F. G. Use of Experimental Interstage Performance Data to Obtain Optimum Performance of Multistage Axial Compressors. J. Eng. Power, Trans. ASME, Ser. A, 84, 1962: 187-194.

MILLER, M. J., OKIISHI, T. H., SEROVY, G. K., SANDERCOCK, D. M., and BRITSCH, W. R. Summary of Design and Blade-Element Performance Data for 12 Axial-Flow Pump Rotor Configurations. NASA TN D-7074, 1973.

MIKOLAJCZAK, A. A., ARNOLDI, R. A., SNYDER, L. E. and STARGARDTER, H. Advances in Fan and Compressor Blade Flutter Analysis and Predictions. J. Aircraft, 12, 1975: 325-332.

DANFORTH, C. E. Distortion-Induced Vibration in Fan and Compressor Blading. J. Aircraft, 12, 1975: 216-225.

DANFORTH, C. E. Blade Vibration: Same Key Elements in Design Verification. J. Aircraft, 12, 1975: 333-342.

BROWN, L. E. The Use of Experimental Interstage Data for Matching the Performance of Axial Compressor Stages Having Variable-Setting-Angle Blade Rows. ASME Paper 76-GT-42, 1976.

HETHERINGTON, R. and MORITZ, R. R. Influence of Unsteady Flow Phenomena on the Design and Operation of Aero Engines. IN Unsteady Phenomena in Turbomachinery, Paper 2. AGARD-CP-177, 1976.

KOVACH, K. and GRIFFITHS, P. R. Engine Compression System Surge Line Evaluation Techniques. Paper presented at Third International Symposium on Air Breathing Engines. Munich, March 1976.

SCHAFFLER, A. Experimental and Analytical Investigation of the Effects of Reynolds Number and Blade Surface Roughness on Multistage Axial Flow Compressors. J. Eng. Power, Trans. ASME, 102, 1980: 5-13.

BECKER, B., KWASNIEWSKI, M. and VON SCHWERDTNER, O. Transonic Compressor Development for Large Industrial Gas Turbines. J. Eng. for Power, Trans. ASME, 105, 1983: 417-421.

CARCHEDI, F. and WOOD, G. R. Design and Development of a 12:1 Pressure Ratio Compressor for the Ruston 6-MW Gas Turbine. J. Engr. for Power, Trans. ASME, 104, 1982: 823-831.

WISLER, D. C. Loss Reduction in Axial-Flow Compressors Through Low-Speed Model Testing. ASME Paper 84-GT-184. 1984.

DESIGN LIMITS, BASIC PARAMETER SELECTION AND OPTIMIZATION METHODS IN TURBOMACHINERY DESIGN

Ennio Macchi

Department of Energetics, Politecnico di Milano

1 INTRODUCTION

The design procedure of a turbomachine is a very complex engineering operation, involving thermodynamic,aerodynamic, technological, structural and economic aspects. Although the interactions among these various aspects are very strong in an actual design, in this lecture reference will be made to an idealized design system, where the thermo and fluid-dynamic aspects can be treated independently, and the influence exerted on the design by the other aspects can be confined to some constraints, related either to manufacture limitations (for instance, a limit minimum value for the trailing edge thickness of a blade), or mechanical stresses (say, a limiting value to peripheral speed to contain the centrifugal stresses), or economics (say, a limit value to number of blades).

The aerodynamic design of a turbomachine can be looked as an iterative procedure made of various steps, in which the previously assumed (or computed) values are progressively modified according to the results of more refined flow calculations. A typical structure of the design system for an axial flow turbomachine is represented in fig. 1: as shown in the figure, the preliminary definition of basic turbomachinery parameters can be obtained: 1) by an independent procedure, which will be referred to in the following as "optimization", or 2) by a development of existing units by some of the procedures indicated in the figure (scaling, similarity, etc.). The various aspects related to the first procedure, i.e. the selection of the basic design parameters of a "new" machine, will be the sub-

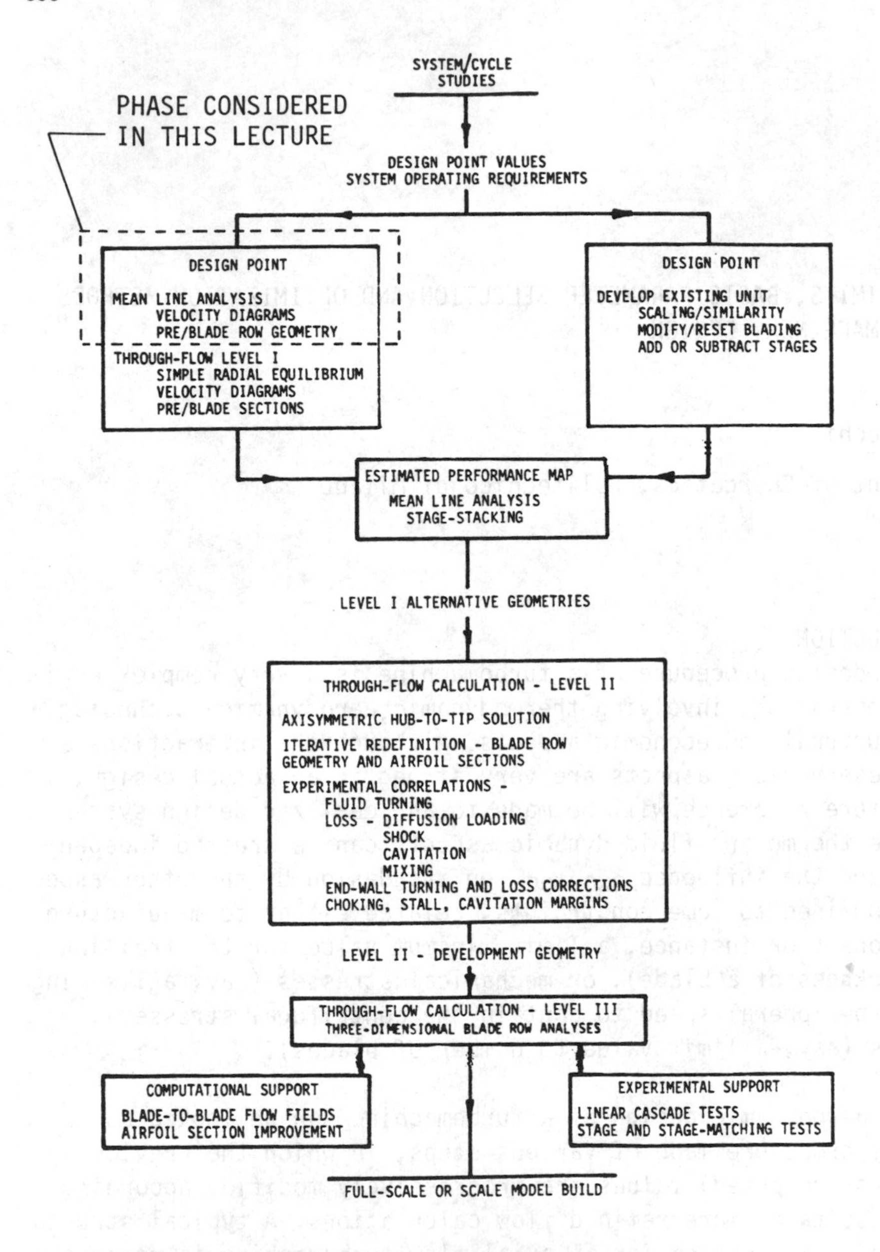

Fig. 1 Typical structure of a design system for axial-flow turbomachines, from (1).

ject of this lecture.

The analysis will be limited to the one- dimensional (or mean line) approach, which is just the very first step in the design of a turbomachine. However, the importance of this phase should not be underestimated: for instance, it is well known that the efficiency of a properly designed axial-flow turbine can be predicted with fair accuracy (say, 1 or 2%) by the adoption of simple mean line analysis methods which incorporate proper loss and flow angle correlations (by "proper"it is meant that the loss correlation must include the influence on loss coefficients of relevant geometric and kinematic variables). If, as it is the author's opinion, the above statement is correct, the important conclusion is reached that the efficiency of a turbine is set, within a narrow range, simply by the selection of the basic design parameters. All other design phases, though of great importance, cannot yield great efficiency improvements. Similar reasonings hold for other turbomachines.

Obviously, the above reasoning was made not to underestimate the importance of the role played by very accurate blade design procedures or flow analyses made nowadays possible by modern computers, but to stress that the first assumptions on turbomachine main characteristics, which are input data to subsequent calculations, are of great importance too. Hence, the adoption of sophisticated methods for their selection are justified.

The technical literature is surprisingly poor of indications on the selection procedure of the main design data of turbomachines. Possible explanations to this situation could be the following:most industrial machines are designed as development of previous units, and great part of design parameters are neither selected nor optimized, but simply assumed equal to the ones of existing models; "conventional" design parameters, based on experience, can be selected in many fields of application; last but not least, the subject looks relatively elementary, and therefore not very attractive for the researchers. Whichever the reason, the popularity of the few papers dealing on the subject (for instance, (2), (3) and (4))seems to demonstrate the need of reliable methods for the selection of design variables.

2 DEFINITION OF THE PROBLEM

For simplicity, explicit reference will be made in this section to the case of axial-flow turbine design. However, most reasonings hold for all other turbomachines.

2.1 Input Data

a. Thermodynamic properties of the working fluid (M, γ for perfect gas). If the fluid cannot be considered a perfect gas, approximate methods as the ones described in (5) can be used without significant errors.

b. Thermodynamic conditions at inlet (total temperature or enthalpy and total pressure) and outlet (static or total pressure).

c. Flow rate or power. Since the machine efficiency is generally not a priori known, an iterative procedure is required if power rather than flow rate is specified as input datum.

2.2 The Design Procedure

The design procedure can be looked at as an optimization problem. As in all optimization problems, four main points should be made clear: 1) definition of the function to be optimized; 2) independent variables; (3) constraints which limit the range of the search; and (4) relations between function, input data and variables.

2.2.1 Function to be optimized. The turbine efficiency is often chosen as the function to be optimized; other possible targets could be the turbine weight, or volume, or weighted combinations of various factors. The optimization results depend on the selected target: for instance, as shown in the example of Fig. 2 (from (6)), if the stage total-to-static efficiency is optimized ($\Phi_E = 0$), lower flow coefficients will be obtained than for stages optimized for total-to-total efficiency ($\Phi_E = 1$); the two solutions exhibit different velocity triangles and turbine dimensions.

2.2.2 Optimizing variables. The optimizing variables change according to design procedure. According to (7), a convenient set of variables i

- speed of revolution
- number of stages
- for each stage:
  - a number of parameters which specify the stage velocity diagrams, for instance: K_{is}, r^*, $(o/s)_1$, $(o/s)_2$, R_1/R_2
  - a number of parameters which specify the geometric variables of blanding necessary for calculating loss coefficients and angles, for instance: o_1, o_2, b_1, b_2

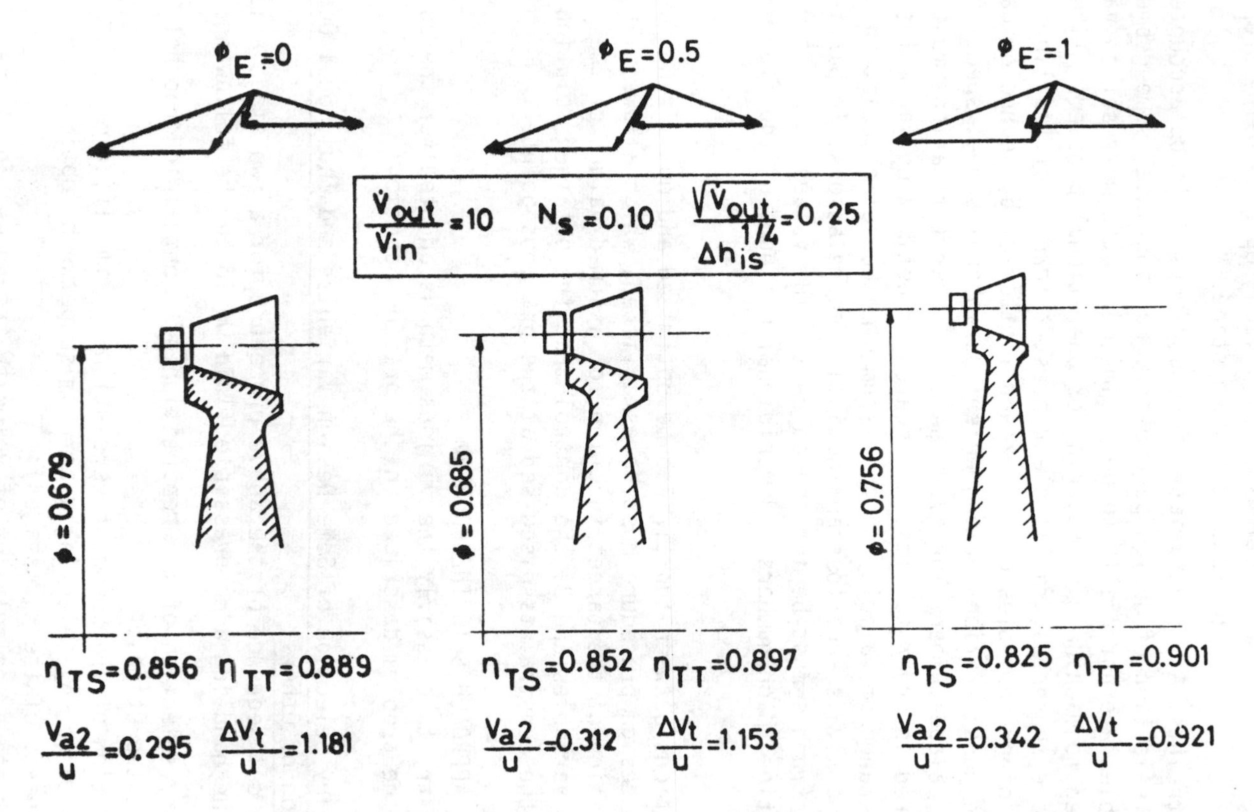

Fig. 2 Example of the influence of the assumed value for the utilization factor of leaving loss Φ_E on the results obtained in the optimization.

- the fraction of total isentropic enthalpy drop to be handled by the stage.

2.2.3 Constraints. A large variety of constraints is to be accounted for in the selection of design parameters. For instance, some geometric variables are not really to be optimized, but simply set to values related to machine manufacturing or operational problems, like: blade surface roughness, radial clearances, thickness of trailing edge. Some other variables cannot exceed limits set by technological or economic considerations, among which: peripheral speed, speed of revolution, blade number. Moreover, several other limitations must be considered in order to limit the search of optimizing variables within the range of validity of the assumed method of calculation (mean-line analysis) and loss and deviation correlations. In particular, constraints must be introduced on maximum values of blade hub/tip ratios, Mach numbers and meridional contours of the flow path.

2.2.4 Relations between function to be optimized and optimizing variables. A design procedure must be established, which allows: 1) the calculation of the target function for assumed values of the optimizing variables and 2) the control that the obtained solution is within the constraints discussed at the previous point.

3 POSSIBLE APPROACHES TO THE PROBLEM

Let's try to classify the various methods adopted for the solution of the problem described in the previous section:

3.1 Arbitrary Selection of Some Design Variables and Individual Optimization of Others.

A typical sequence of the above procedure for a two stage axial flow turbine handling a compressible fluid would be the following:

a) arbitrary selection of the repartition of enthalpic drop between the two stages;
b) optimization of the speed of revolution by simplified methods (for instance, by selecting a compromise between "optimum" specific speeds of the two stages);
c) for each stage, calculation of velocity triangles, for instance by:
   - arbitrary selection of the degree of reaction and consequent "optimization" of the peripheral speed of the stage;
   - arbitrary selection of flow coefficients at the exit of each blade row and consequent check of meridional contours of the

flow path;

d) for each blade row, determination of main blading characteristics, for instance by:

   - arbitrary selection of a blade dimension, say the axial chord;
   - optimization of blade solidity by approximate methods, say Zweifel (8) criterion or similar;
   - definition of all other blading geometric variables required for calculation of losses and deviations;

e) calculation of turbine performance;

f) individual optimization of some of the variables arbitrarily assumed in the previous points, by analysing their influence on the target function.

The main drawback of the above procedure or all other procedures based on individual optimization techniques, is the practical impossibility to account during the optimization process for the mutual influence exerted by the various parameters. For instance, let consider the two individual optimizations procedure quoted at point c): it is true that the head coefficient of a turbine stage is primarily a function of the degree of reaction; however, other variables influence the choice, as shown in fig. 3 (from (6)). Similarly, blade load criteria based on two-dimensional analyses cannot account for the influence of secondary losses (and therefore blade height) on the optimum blade spacing.

3.2 Use of Numerical Optimization Techniques

It was pointed out that the turbomachine design procedure can be considered as a problem of optimization of a multi-variable function, with a large number of constraints. Nowadays, numerical subroutines performing this duty properly are available on large computers even for complex designs. For instance, the author utilized this method for optimizing a large number of three stage axial flow turbines, as described in (7). The adopted optimizing variables, fixed input variables, variable input data and constraints are shown in tab. 1. A sample of variation of some optimizing variables and of the turbine efficiency during the numerical optimization procedure is given in fig. 4. About 15,000 iterations are required to optimize a three-stage turbine, (29 optimizing variables, 50 linear boundaries, 16 non linear boundaries), with an average CPU time of about 5' on UNIVAC 1100 computer.

For multi stage machines, since the various blade heights and

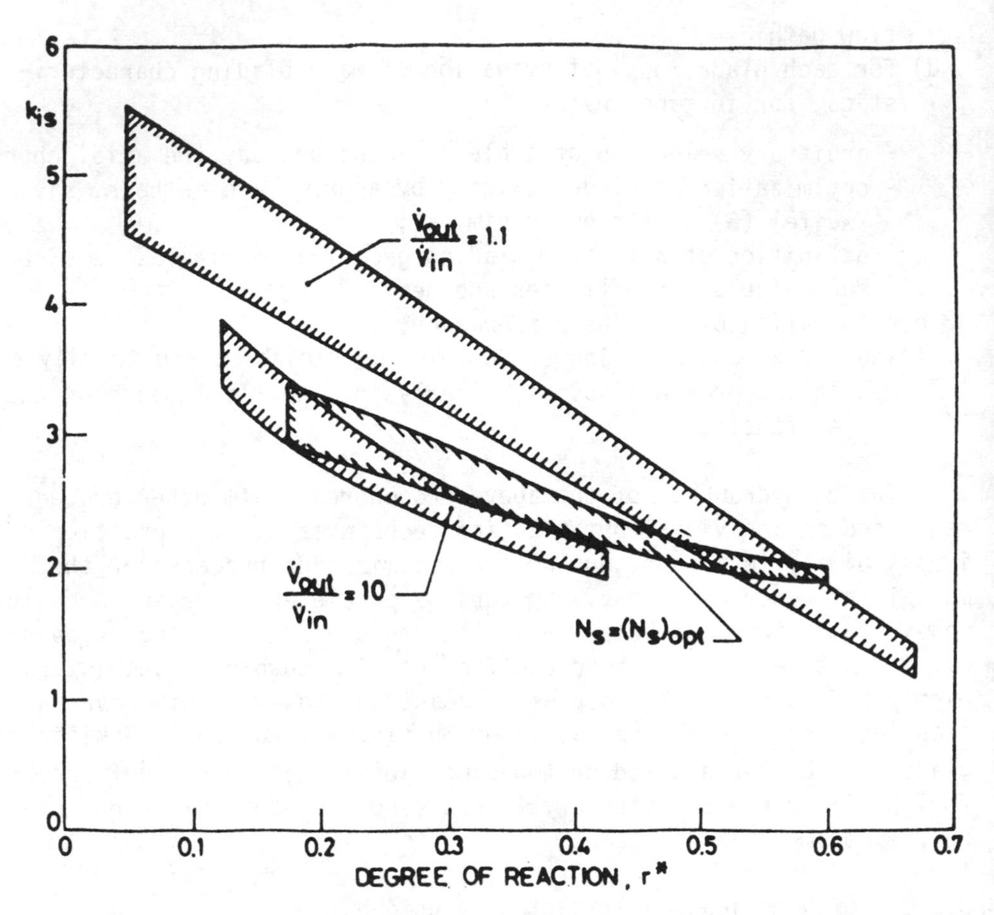

Fig. 3 Relationship between head coefficient and isentropic degree of reaction for optimized turbine stages of various characteristics.

diameters of the turbine stages are free to change during the design procedure, the duct between two succeeding stages can assume patterns that make difficult the recovery of kinetic energy. Hence, two modes of calculation can be considered:

1 constrained: a limit (± 30°) is set for the flare angle FL between rotor exit and the next stator; in this case, the losses occurring in the annulus from rotor discharge to stator inlet are computed according to the loss correlation method given in (9);
2 unconstrained: large variations of diameters between stages are accepted; a diffuser after each rotor, capable of recovering 50% of the kinetic energy ($\Phi_E = 0.5$) is stipulated, while the annulus losses are not accounted for.

1. Optimizing variables

for each stage: k_{is}, r^*, $(o/s)_1$, $(o_{min})_1$, b_1, $(o/s)_2$, o_2, b_2, R_1/R_2

for each stage, except the last: $(P_{TO}/P_2)_{ST}$

2. Fixed input variables

for all blade rows:

$(\alpha_o)_{ST\,1} = 90^o$

$Re = 5.10^5$

$k_s = 2.10^{-3}$ mm

δ_r = max (0.2 mm or R/1000)

t_n = max (0.2 mm or 0/10)

$b_{an\,1} = 0.2\ h_2$

$b_{an\,2} = h_2$

3. Variable input data

- Fluid thermodynamic properties (molecular mass, specific heat ratio)
- Inlet conditions (total pressure and temperature)
- Total to static enthalpy drop
- Mass flow rate
- Speed of revolution

4. Constraints

for each stage:

$0 < M_{W1} < 0.8$

$0 < M_{W2} < 1.4$

$-30^o < FL_1 < 30^o$

$-30^o < FL_2 < 30^o$

$0.001 < (h/D)_1, (h/D)_2 < 0.25$

$h_2/h_1 > 1.0$

$13^o < \sin^{-1}(o/s)_1 < 60^o$

$13^o < \sin^{-1}(o/s)_2 < 60^o$

$0 < (b/D)_1 < 0.25$

$0 < (b/D)_2 < 0.25$

$2\ o_1 < b_1 < 100$ mm

$2\ o_2 < b_2 < 100$ mm

1.5 mm $< o_{min}, o_1 < 100$ mm

1.5 mm $< o_2 < 100$ mm

$1.5 < k_{is} < 10$

$-0.01 < r^* < 0.9$

$1.0 < R_2/R_1 < 1.05$

$0 < u_1 < u_{max}$

$0 < u_2 < u_{max}$

$10 < z_1 < 100$

$10 < z_2 < 100$

For Z = 3

$$(1 + \frac{VR}{10})^{\gamma} < (\frac{P_{TO}}{P_2})_{ST\,1} \cdot (\frac{P_{TO}}{P_2})_{ST\,2} < (VR)^{\gamma}$$

For Z = 2

$$(1 + \frac{VR}{10})^{\gamma} < (\frac{P_{TO}}{P_2})_{ST\,1} < (VR)^{\gamma}$$

Table 1 Optimizing variables, fixed input variables, variable input data and constraints used in (7).

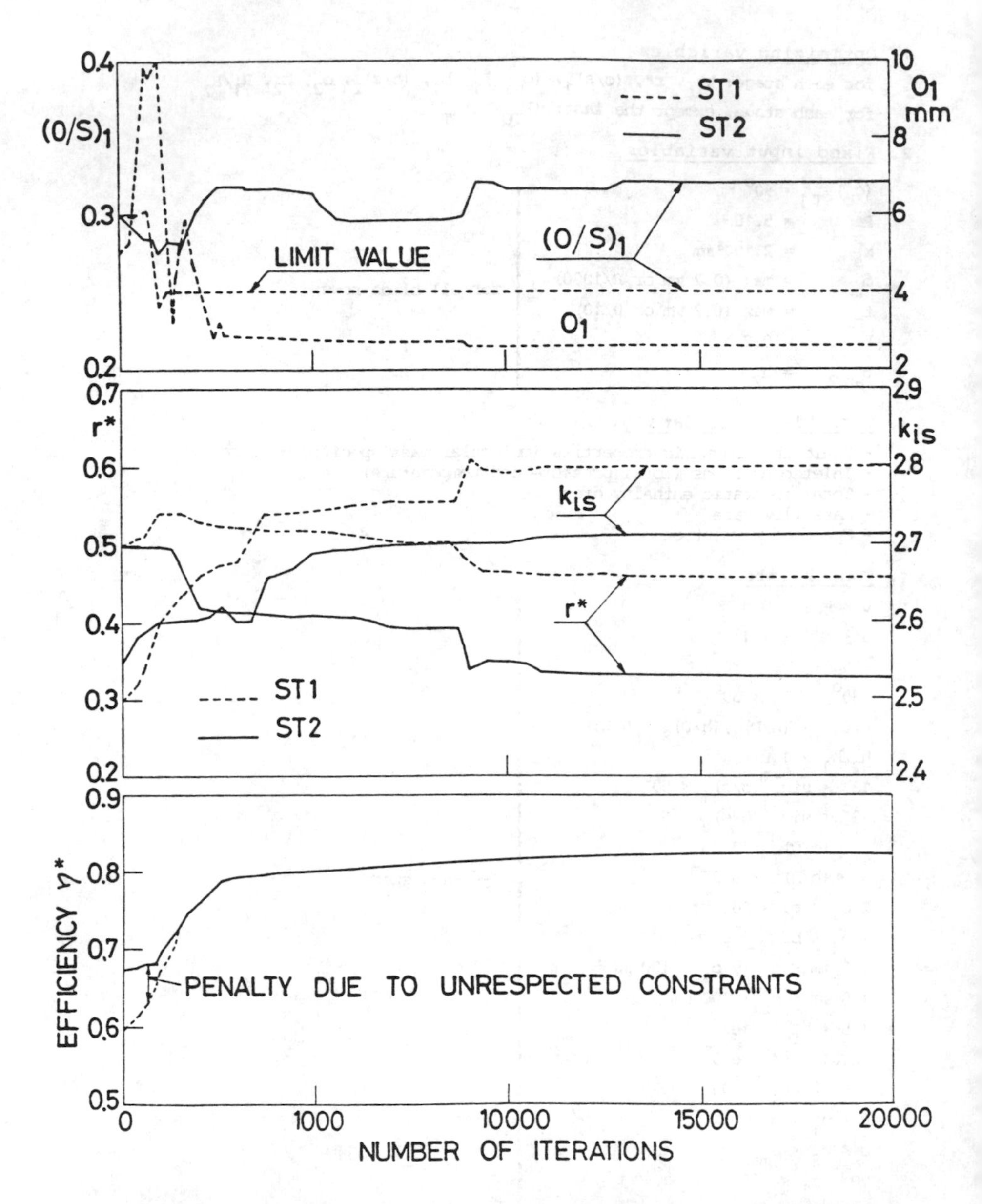

Fig. 4 Example of variation of the optimizing function (η^*) and of some optimizing variables during the iterative procedure of optimization. The initial solution does not satisfy all constraints, and the optimizing function is penalized.

Two examples of the results obtained with the two modes of operation are illustrated in Fig. 5a and b. For moderate expansion ra-

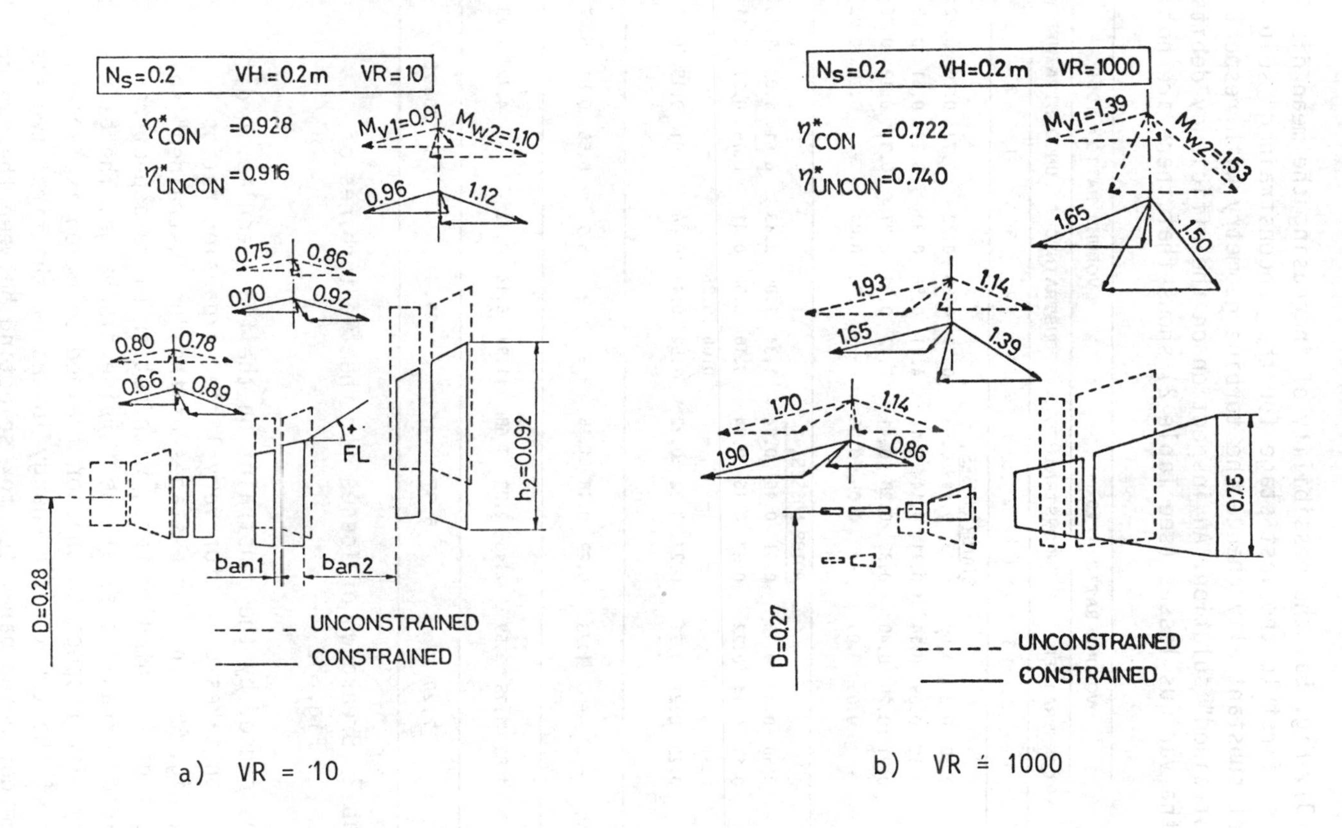

Fig. 5 Velocity triangles and meridional geometry of a three stage turbine optimized with two different approaches.

tios (VR = 10), Fig. 5a, the possibility of increasing the mean dia meter from the first to the last stage for the "unconstrained" solu tion does not substantially change the turbine geometry with respect to the "constrained" solution. An inspection on the efficiency debits related to the various losses (see Table 2) shows that there are not

	VOLUME RATIO = 10						VOLUME RATIO = 1000					
	CONSTRAINED			UNCONSTRAINED			CONSTRAINED			UNCONSTRAINED		
STAGE	1	2	3	1	2	3	1	2	3	1	2	3
				STATOR LOSSES, %								
PROFILE, $\Delta\eta_P$	0.30	0.31	0.60	0.49	0.44	0.63	0.79	0.88	0.32	0.72	0.76	0.32
SECONDARY, $\Delta\eta_S$	0.11	0.09	0.14	0.31	0.14	0.14	4.71	0.56	0.39	3.38	0.61	0.09
ANNULUS, $\Delta\eta_A$	0.21	0.20	0.40	0.32	0.28	0.41	0.42	0.31	0.37	0.31	0.42	0.26
LEAKAGE, $\Delta\eta_{CL}$	-	0.05	0.06	-	0.05	0.05	-	0.36	0.04	-	0.28	0.02
				ROTOR LOSSES, %								
PROFILE, $\Delta\eta_P$	0.50	0.45	0.73	0.51	0.46	0.84	1.34	1.86	2.43	6.61	1.02	1.30
SECONDARY, $\Delta\eta_S$	0.17	0.14	0.22	0.27	0.16	0.24	3.10	0.50	0.41	0.75	0.37	0.11
ANNULUS, $\Delta\eta_A$	0.31	0.33	-	-	-	-	0.06	0.26	-	-	-	-
LEAKAGE, $\Delta\eta_a$	0.23	0.21	0.21	0.27	0.22	0.22	1.52	0.41	0.06	2.01	0.15	0.06
KINETIC ENERGY, $\Delta\eta_K$	-	-	1.23	0.29	0.48	1.15	-	-	6.7	0.55	0.81	5.07
TOTAL	1.83	1.78	3.59	2.46	2.23	3.68	11.94	5.14	10.72	14.33	4.42	7.23
		7.20			8.37			27.81			25.98	

Tab 2 Breakdown of losses for the two turbines of Fig. 5.

disadvantages caused by the constraint on the flare angles. A dif ferent situation takes place for very large expansion ratios (VR = 1000), Fig. 5b : the three stages of the "unconstrained" solution exhibit different diameters, with better blade aspect ratio and a larger exhaust area at the turbine discharge. The gain in efficiency with respect to the constrained solution is larger than the sum of the two kinetic energy losses at the first two sta ges exit. The governing parameter for selecting between the two de sign modes is the expansion ratio: as shown in Fig. 6, only for high values of expansion ratio (VR>150÷200, corresponding to pressure ratios ranging between 200 for working fluids with complex molecu-

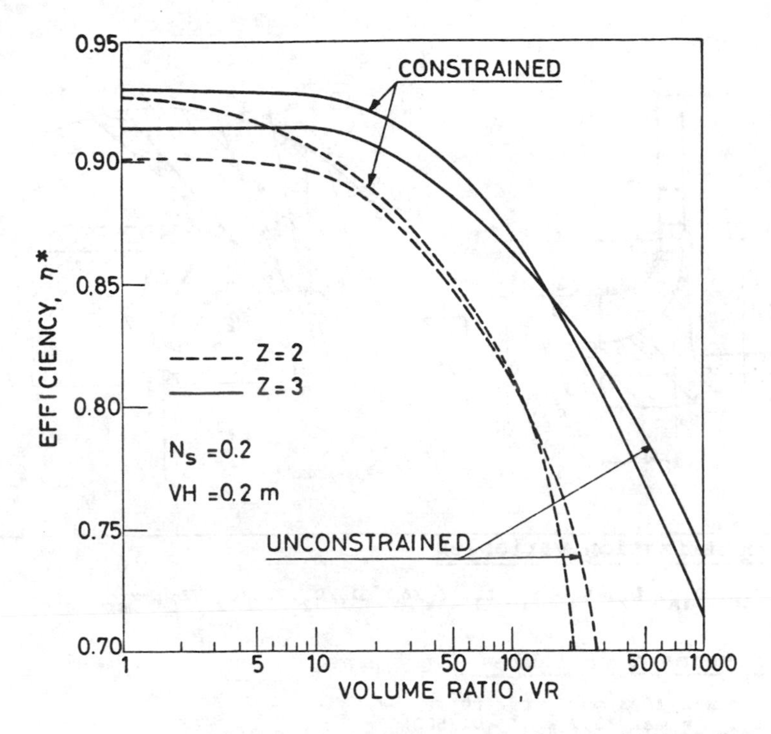

Fig. 6 Efficiency vs volume ratio variation for turbines optimized with and without a constraint in flaring angle.

le and 5000 for monoatomic gases), the unconstrained solution yields better results. For this reason, it is recommended to utilize the constrained mode of operation for conventional cases.

A similar procedure of contemporary optimization of design variables is described for centrifugal compressors in (10). In this case, 11 optimizing variables are necessary (tab. 3) and again a great number of constraints (43 as indicated in tab. 4) must be used. When proper loss and slip correlations are used, the program yields results in fair agreement with existing machines, as shown in fig. 7, from (11).

An advantage of these methods is their great flexibility, consequent to the very easy procedure of constraints introduction. The author used successfully the axial-flow turbine design programs described in (6) and (7) for the preliminary design of about 20 tur-

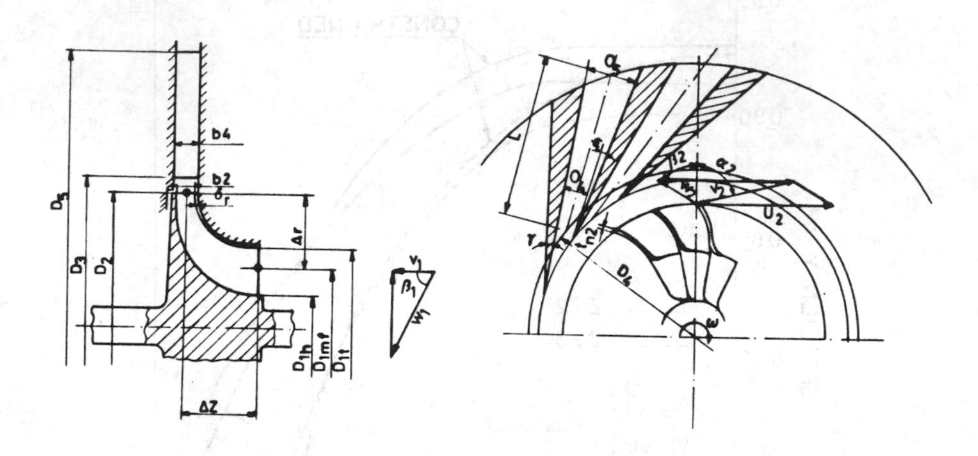

1. Optimization variables

   D_{1t}, D_{1h}, D_2, $\beta_2^{'}$, b_2, z_I, Δ_z/Δ_r, D_3/D_2, b_3/b_2, z_D, x_{sp}

2. Dependent variables

   δ = max (0,3 mm, $D_2/400$)
   t_{n1min} = max (0,2 mm, $D_2/500$)
   t_{n2min} = max (0,6 mm, $D_2/200$)
   t_{nt}/t_{nh} = 0,33

3. Design data

   - Fluid thermodynamic properties
   - Inlet conditions
   - Pressure ratio
   - Mass flow rate
   - Wheel alloy

Tab. 3 Nomenclature and optimization variables, dependent variables and design data used in (10).

bines for ORC engines (12), by introducing specific constraints suggested by various considerations. For instance, when it was required to utilize a limited number of blade profiles, it was simple to constrain the blade parameters to respect this condition.

3.3 Use of Similarity Parameters

The use of similarity parameters is very attractive in turbomachinery design, not only for the preliminary phase of the basic parameters selection, but also for the other steps of the

1. Geometric constraints

20°	<	β'_{1t}	< 70°	
0,4	<	D_{1t}/D_2	< 0,7	
0,3	<	D_{1h}/D_{1t}	< 0,7	
0,03m	<	D_2	< 2 m	
0°	<	β'_2	< 60°	
0,01	<	b_2/D_2	< 0,4	
8	<	z_I	< 60	
0,8	<	D_2/Dr	< 1,2	
0,4	<	X_{SP}	< 0,6	
1.05	<	D_3/D_2	< 1.2	
0,8	<	b_3/b_2	< 1,2	
0,2	<	b_4/O_4	< 7,0	
10	<	z_D	< 80	
4mm	<	O_4		
5°	<	γ	< 30°	(fig.1)
		D_5/D_2	< 2.0	

2. Fluidynamic constraints

$M_{w1t} < 1.4$ $\quad 0 < C_{pvl} < 0.4$

$M_{w1mf} < 0.9$ $\quad \varepsilon_{wake} < 0.8$

$W_{minI} > 0$ $\quad K_{bax} < 0.5$

$W_2/W_{1t} > 0.25$

$\lambda < 4 \ (\alpha_2 < 76^\circ)$

3. Mechanical stress constraints

$\sigma_A < \sigma_{lim\,A}$

$\sigma_B < \sigma_{lim\,B}$

$\sigma_C < \sigma_{lim\,C}$

Tab. 4 Geometric constraints, fluid-dynamic constraints and mechanical stress constraints.

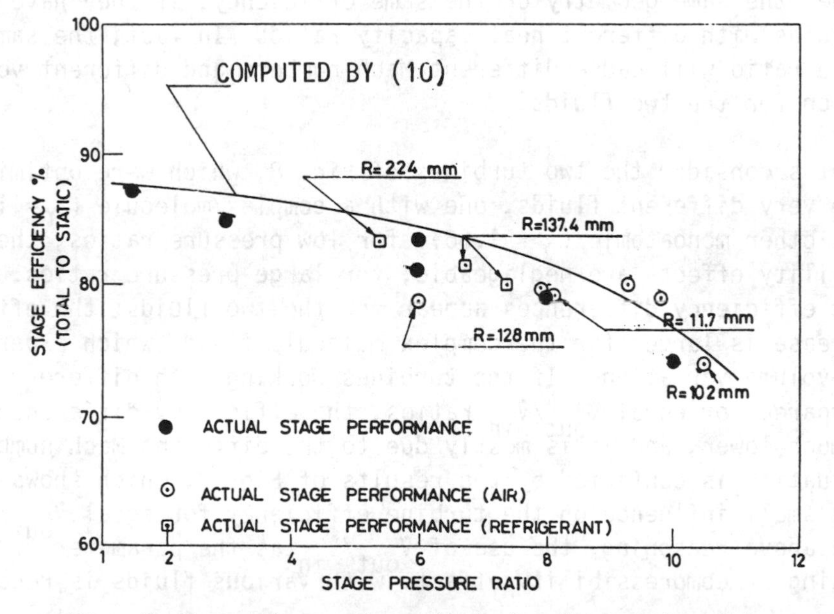

Fig. 7 Comparison of predicted and experimental results of advanced centrifugal compressors (from (11)).

design procedure. The basic idea is to transfer the results of the optimization on a particular machine to other machines. Let's consider the conditions to be satisfied to obtain a rigorous similarity between two machines:

1. Equal thermodynamic behaviour of the working fluid, i.e. either incompressible flows, or fluids with the same pressure ratio and the same heat capacity ratio.
2. Geometric similarity (including tip clearance, surface roughness, blade trailing edge thickness, etc.).
3. Same specific speed N_s and same specific diameter D_s.
4. Reynolds number effects are neglected.

The consequences of the non respect of the two first requirements are often underevaluated, and it is common the use of N_s, D_s plots to predict machine efficiency and/or to select design variables (2,3). Let's first discuss the first point: the thermodynamic properties of the working fluid affect the machine design and performance in two main ways: (1) loss coefficients are a function of Mach numbers; (2) the fluid volume variation during the expansion influences the geometry of the machine. Therefore, two optimized turbines having the same specific speed and, say, the same pressure ratio will not have either the same geometry or the same efficiency, if they have working fluids with different heat capacity ratios. In fact, the same pressure ratio will cause different Mach numbers and different volume variation for the two fluids.

Let's consider the two turbines of Fig. 8, which were optimized for two very different fluids, one with a complex molecule ($\gamma = 1.10$) and the other monoatomic ($\gamma = 1.66$): for low pressure ratios, the compressibility effects are negligeable; for large pressure ratios, important efficiency differences appear for the two fluids; the efficiency decrease is larger for the complex molecule fluid, which experiences larger volume variations. If the turbines working with different fluids are compared for equal $\dot{V}_{out}/\dot{V}_{in}$ ratios, the efficiency difference becomes much lower, and it is mostly due to the different Mach numbers. The situation is confirmed by the results of Fig. 9, which shows that γ has a small influence on the turbine efficiency for equal $\dot{V}_{out}/\dot{V}_{in}$. For the above reasoning, the use of $\dot{V}_{out}/\dot{V}_{in}$ as the parameter for accounting of compressibility effects with various fluids is recommended.

As far as the geometric similarity is concerned, it has been shown that several dimensional constraints must be introduced in

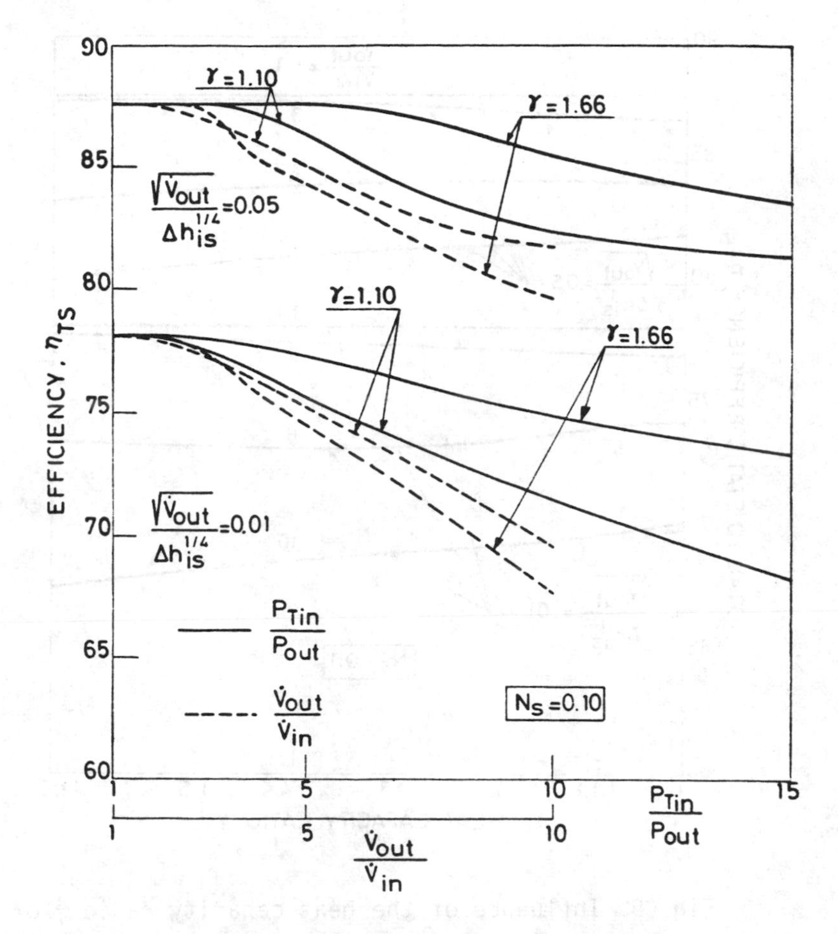

Fig. 8 Influence of the pressure ratio and of the volume flow rate variation on the efficiency of a turbine stage, for two values of γ and two values of turbine dimensions.

the analysis. Therefore, small turbines will behave differently from large turbines, because they have larger relative thickness, clearance, etc. According to the similarity rules, the parameter which accounts for the "size" of the considered turbine is $VH = \sqrt{\dot{V}_{out}}/\Delta h_{is}^{1/4}$. That is to say that similar turbines having the same $\sqrt{\dot{V}_{out}}/\Delta h_{is}^{1/4}$ have the same actual dimensions.

From the above discussion, it seems appropriate to state that the results obtained by optimizing a particular turbine can be transferred to other turbines having the same specific speed and specific diameter only if they have the same $\dot{V}_{out}/\dot{V}_{in}$ (similar compressi-

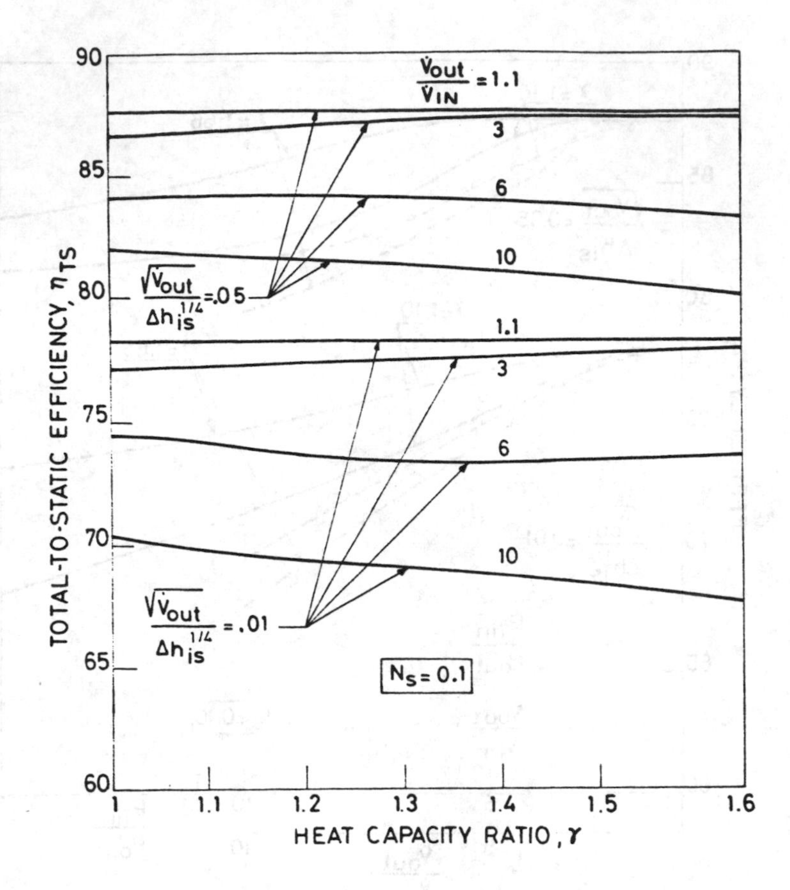

Fig. 9 Influence of the heat capacity ratio γ of the working fluid on the efficiency of a turbine stage, for two turbine dimensions; the comparison is made for constant volume flow rate ratios $\dot{V}_{out}/\dot{V}_{in}$.

bility effects) and $\sqrt{\dot{V}_{out}}/\Delta h_{is}^{1/4}$ ratios (geometric similarity) and if Reynolds number effects can be neglected.

The situation is illustrated in Fig. 10 (from (13)), where the results of turbine optimizations using different loss correlations are compared for various N_s and VH. It can be seen that both the compressibility parameter $\dot{V}$R (Fig. 10a) and the "size" parameter VH (Fig. 10b) exhibit a remarkable influence on the turbine efficiency. That is to say, the use of N_s - D_s plots is not warranted for turbines of small dimensions (say, VH < 0.10) or with important compressibility effects (VR>2). The errors on this regard can be much greater than the ones resulting from the adoption of different loss correlations.

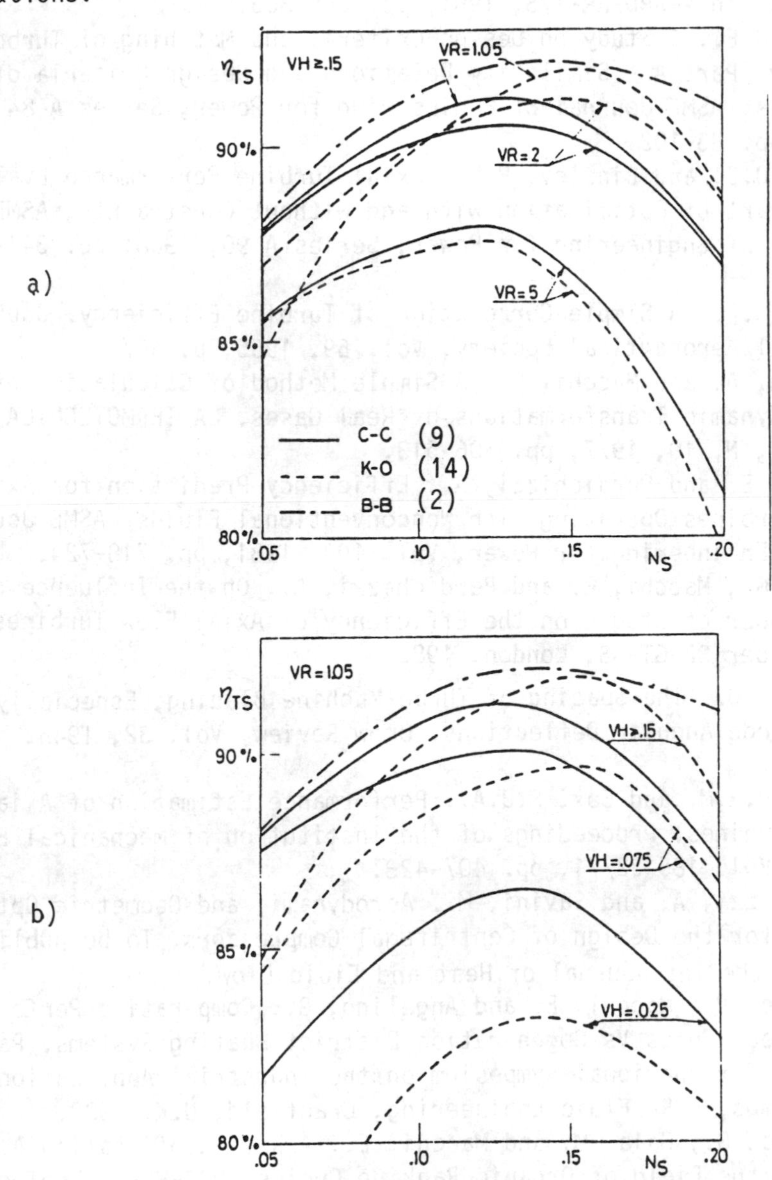

Fig. 10 Efficiency prediction for single stages axial flow turbines with different loss correlations.

a) Constant VR = 1.05
b) Constant VH = 0.15 m

REFERENCES

1. Serovy, G.. Axial-Flow Turbomachine Through-Flow Calculation Methods, in AGARD-AR-175, 1981, pp. 285-305.
2. Baljé, O.E.. A Study on Design Criteria and Matching of Turbomachinery: Part A - Similarity Relations and Design Criteria of Turbines. ASME Journal of Engineering for Power, Series A 84, 1962, pp. 83-102.
3. Baljé, O.E. and Binsley, R.L.. Axial Turbine Performance Evaluation, Part B: Optimization with and without Constraints, ASME Journal of Engineering for Power, Series A 90, 1968, pp. 341-348.
4. Smith, S.F.. A Simple Correlation of Turbine Efficiency. Journal of Royal Aeronautical Society, Vol. 69, 1965, p. 467.
5. Brentan, A. and Macchi, E.. A Simple Method of Calculation of Thermodynamic Transformations of Real Gases. LA TERMOTECNICA, Vol. 31, N. 10, 1977, pp. 506-513.
6. Macchi, E. and Perdichizzi, A.. Efficiency Prediction for Axial-Flow Turbines Operating with Nonconventional Fluids. ASME Journal of Engineering for Power, Vol. 103, 1981, pp. 718-724.
7. Lozza, G., Macchi, E. and Perdichizzi, A.. On the Influence of the Number of Stages on the Efficiency of Axial Flow Turbines, ASME paper 82-GT-43, London, 1982.
8. Zweifel, O.. The Spacing of Turbo-Machine Blading, Especially with Large Angular Deflections. Brow Review, Vol. 32, 1945, pp. 436-444.
9. Craig, H.R.M. and Cox, H.J.A.. Performance Estimation of Axial Flow Turbines. Proceedings of the Institution of Mechanical Engineers, Vol. 185 32/71, pp. 407-423.
10. Perdichizzi, A. and Savini, M.. Aerodynamic and Geometric Optimization for the Design of Centrifugal Compressors. To be published on the Int.Journal of Heat and Fluid Flow.
11. Centrone, P., Macchi, E. and Angelino, G.. Comparative Performance of Heat Pumps Vs Cogeneration District Heating Systems. Paper No. 51. International Symposium on the Industrial Application of Heat Pumps, BHRA Fluid Engineering, Cranfield, U.K. 1982.
12. Angelino, G., Gaia, M. and Macchi, E.. A Review of Italian Activity in the Field of Organic Rankine Cycles. ORC-HP-Technology, VDI, Zürich, 1984.
13. Lozza, G.. A Comparison between the Craig-Cox and the Kacker-Okapuu Methods of Turbine Performance Prediction. MECCANICA, Vol. 17, 1982, pp. 211-221.
14. Kacker,S.C. and Okapuu,U..A Mean Line Prediction Method for Axial Flow Turbine Efficiency.ASME paper n.81-GT-58, 1981.

LIST OF SYMBOLS

- for axial flow turbines:

b	axial chord, m
b_{an}	annulus length (see Fig. 5a), m
D	mean diameter, m
FL	flaring angle (see Fig. 5a), deg
h	blade height, m
k_{is}	stage head coefficient = $2(\Delta h_{is})_{ST}/u^2$
k_s	blade surface roughness, m
M	molecular mass
$M_{V,W}$	Mach number of the absolute and relative velocity
n	speed of revolution, rps
N_s	specific speed = $n\sqrt{\dot{V}_{out}}/\Delta h_{is}^{3/4}$
o	blade throat opening, m
o_{min}	blade critical throat, m
P	static pressure, Pa
P_T	total pressure, Pa
R	mean radius, m
r^*	isentropic degree of reaction
Re	Reynolds number, based on blade opening
s	blade spacing, m
t_n	trailing edge thickness, m
u	peripheral speed, at mean radius, m/s
UH	speed coefficient = $u/\sqrt{2\Delta h_{is}}$
W	relative velocity, m/s
V	absolute velocity, m/s
VH	size parameter = $\sqrt{\dot{V}_{out}}/\Delta h_{is}^{1/4}$, m
VR	volume ratio, $\dot{V}_{out}/V_{in}$
$\dot{V}_{in}$	volumetric flow rate, at turbine inlet total conditions, m^3/s
$\dot{V}_{out}$	volumetric flow rate, at turbine exit static conditions, calculated for isentropic process throughout the turbine, m^3/s
z	number of blades

Z	number of stages
α	absolute flow angle, deg
β	relative flow angle, deg
γ	heat capacity ratio
δ_r	radial clearance, m
Δh_{is}	total-to-static isentropic enthalpy drop relative to the whole turbine, J/kg
$\Delta\eta_A$	efficiency debit due to annulus losses
$\Delta\eta_{CL}$	efficiency debit due to leakages
$\Delta\eta_K$	efficiency debit due to leaving kinetic energy
$\Delta\eta_P$	efficiency debit due to profile losses
$\Delta\eta_S$	efficiency debit due to secondary losses
$\Delta\eta^*$	efficiency decrease due to lower than optimum UH
ϕ_E	utilization factor of leaving losses
η^*	overall efficiency, considering $\phi_E = 0.5$ on the last stage
π	total-to-static pressure ratio relative to the whole turbine

Subscripts

0	stator inlet
1	stator exit
2	rotor exit
ST	stage
ST 1	first stage
ST 2	second stage
ST 3	third stage
max	maximum value
min	minimum value
opt	optimized value

- for radial compressors:

a	speed of sound
A	geometric area
b	axial depth

Cp pressure recovery coefficient
D diameter
L diffuser length
Leu Eulerian work
K_b blockage factor
m mass flow rate
M_v absolute Mach number
M_w relative Mach number
n rotational speed, rps
N_s specific speed, $n \sqrt{\dot{V}_{in}}/\Delta h_{is}^{3/4}$
o diffuser throat width
p pressure
r isentropic degree of reaction
U tangential velocity
V absolute velocity
W relative velocity
t_n normal blade thickness
T temperature
$\dot{V}$ volume flow rate
X_{sp} % meridional length at splitter
Z number of blades
α absolute velocity angle
β relative velocity angle
β' geometric blade angle
δ clearance between impeller and shroud
Δh_{ext} external losses
Δh_{is} isentropic head
Δ_r impeller radial extent
ϕ flow coefficient = $(V_r/U)_2$
π pressure ratio
σ mechanical stress

Subscripts

ax	axial
D	diffuser
I	Impeller
mf	mean flow
OV	overall
r	radial
t	tip
tg	tangential
TS	total/static
TT	total/total
0	stagnation conditions
1	rotor inlet
2	rotor outlet
3	vaned diffuser inlet
4	diffuser throat
5	diffuser exit

DESIGN AND OPTIMIZATION OF CENTRIFUGAL COMPRESSORS

R. Van den Braembussche

von Karman Institute for Fluid Dynamics

1. INTRODUCTION

The increase of centrifugal compressor efficiency over the last decades has resulted in a much wider field of application. Its compactness, small weight and simplicity of the components allow to replace multistage axial compressors in an efficient way. The absence of mechanical friction, lower maintenance cost and high reliability makes it also superior to reciprocal compressors.

The real flow in a radial compressor is fully three dimensional, viscous and unsteady. Up to now there is no mathematical model which allows to predict the flow in such a machine, without neglecting some important aspects of the problem. In fact such a calculation is extremely difficult, both due to the complexity of the flow and the complexity of the geometry. Even when it existed, such a calculation would not be suited for a systematic investigation of flow in different geometries, as required by an optimization process, because it would be too long and too expensive.

Nevertheless, there is an urgent need for the designer to dispose of a calculation method, which not only allows to predict the main characteristics of the compressor but also to investigate the influence of the various parameters on the compressor performance.

The number of possible variables is so large that the optimization can only be performed by a stepwise procedure. Starting from the simplest possible description one will make extensive use of correlations and experience gained from previous designs. More elaborate methods can be split into two groups. One group describes the flow in a simplified way, but contains the possibility to account for important real flow phenomena, such as Mach number influence, boundary layer blockage and separation, losses, etc. Another group describes the flow in a rigourous way but neglects all viscous or unsteady flow effects. Methods which describe the real three dimensional viscous flow are now under development or applicable only to components.

In order to simplify the problem, one will discuss the design and optimization of impeller and diffuser separately in the next chapters. In a last chapter one will discuss impeller and diffuser matching in terms of range and stability. The definition of the different components used in this lecture is shown in figure 1.

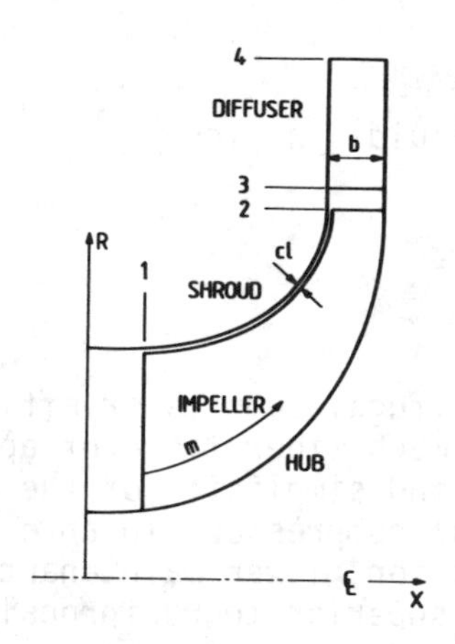

FIG. 1 - DEFINITION OF PARAMETERS

2. THE IMPELLER

The main purpose of the impeller is to perform part of the static pressure rise, and to provide the kinetic energy which will be transformed into pressure by the diffuser. This should be done with the highest possible efficiency and allow for a sufficient change in mass flow.

The attempts to correlate the efficiency of low and high pressure ratio centrifugal compressor with specific speed are quite controversial and have produced a large scatter in terms of maximum attainable efficiency. One will use here classical definition (using american units)

$$N_S = \frac{RPM \cdot Q^{1/2}}{H^{3/4}}$$

where Q = volume flow in cubic feet/sec
H = manometric head in ft.

It is obvious that the compressor efficiency depends on many more parameters than those used in the definition of specific speed. As stated by Rodgers [1], one should limit the correlation to impeller performances.

The specific speed correlation, shown in figure 2, is derived from data of Baljé [2] for low pressure ratio impellers with radial ending blades. It indicates the penalty in efficiency that can be expected when the optimum ratio between manometric head, rotational speed and volume flow is not respected.

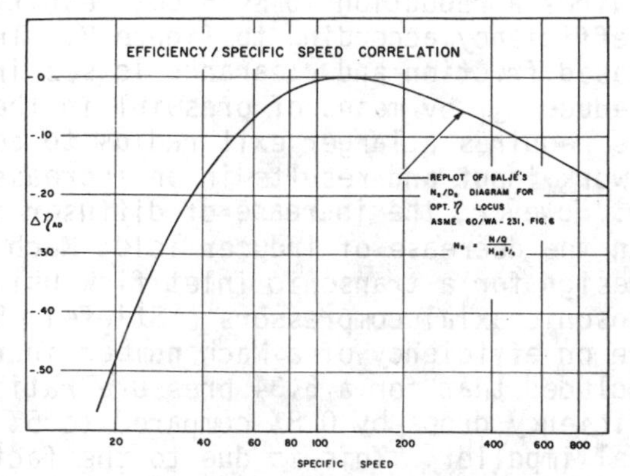

FIG. 2 - EFFICIENCY/SPECIFIC SPEED CORRELATION [4]

There are many reasons why impellers are not always designed at optimum N_S.

- RPM is sometimes imposed by the driving component or by mechanical limitations such as maximum RPM for industrial compressors.
- In multistage compressors the volume flow is decreasing in each stage and it is common practice to operate the first stage at higher than optinum N_S, and the last one at lower than optimum N_S in order to optimize the total efficiency [3].
- Higher specific speed results in smaller compressors rotating at higher RPM. This can be an advantage in aero applications because of the smaller weight or in turbocharger applications because of the smaller inertia resulting in faster accelerations.

At higher pressure ratios, the main problem arises from the increase of shroud inlet Mach number (M_{S1}) with pressure ratio (Fig. 3). At N_S = 100, sonic inlet velocity is reached at pressure ratio = 4. At pressure ratio = 10, M_{1S} will be 1.47 resulting in an efficiency drop due to inducer shock losses and premature flow separation.

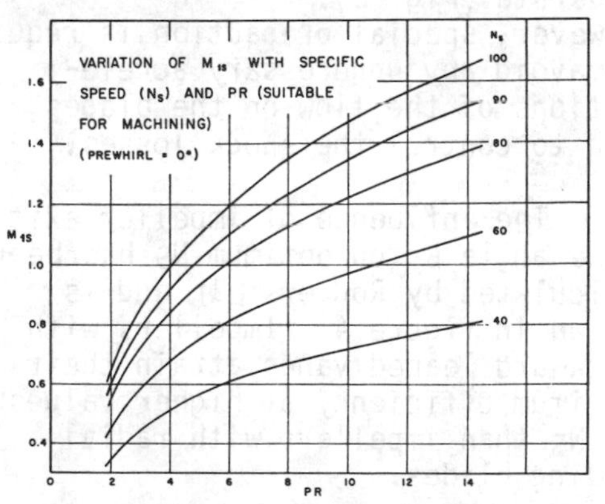

FIG. 3 - VARIATION OF M_{1S} WITH SPECIFIC SPEED AND PRESSURE RATIO [4]

Morris and Kenny [4] state three possible ways to overcome this problem :
- Reduce M_{S1} by designing for lower N_S. At pressure ratio 10 this requires a reduction to N_S = 60, resulting in a seven points drop in efficiency according to figure 2. This is mainly due to the increased friction and clearance losses in a longer impeller channel.
- Reduce M_{S1} by means of preswirl in the direction of rotation. This requires a larger exit radius to compensate for the reduction of work input and results in an increased diffuser inlet Mach number. However, the increase of diffuser inlet Mach number is smaller than the decrease of inducer inlet Mach number.
- Design for a transonic inlet flow using the experience gained in transonic axial compressors. Stiefel [5] has calculated the influence on efficiency of a Mach number increase from .85 to 1.34. He concluded that for a 6.34 pressure ratio radial impeller, the efficiency drops by 0.8% compared to 5% in a 1.5 pressure ratio axial impeller. This is due to the fact that the transonic inducer losses are compared to a much larger work input for the radial impeller.

The efficiency drop due to transonic flow effects is much smaller than the one resulting from a decrease in N_S. Although the situation in a centrifugal impeller is not fully comparable to an axial one because of the larger downstream influence in a longer and highly curved channel, the transonic flow losses have a much smaller influence.

It is recommended for a high pressure ratio impeller to design for an N_S value which is only slightly below the optimum because :
- this allows to reduce the relative inlet Mach number to an acceptable value (Fig. 3).
- For N_S values between 80 and 200, the efficiency drop is very moderate (Fig. 2).

However, special precaution is required for the inducer blade design to avoid any unnecessary accelerations of the flow on the blades, and to control the shock losses.

The influence of impeller exit vane angle β_2 on optimum N_S has been calculated by Rodgers [1] and is shown in figure 4. Impellers with backward leaned vanes attain their maximum efficiency at higher values of N_S than impellers with radial ending blades.

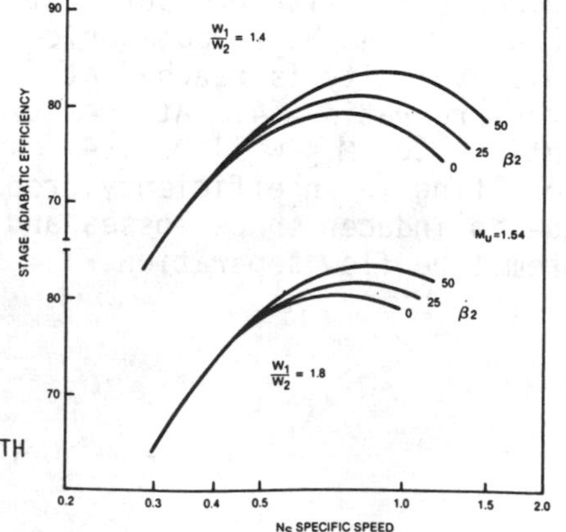

FIG. 4 - VARIATION OF OPTIMUM N_S WITH OUTLET ANGLE β_2 [1]

2.1 Inducer inlet design

Efficiency is not the only criterion in the design of a compressor. Most applications also require a sufficiently large range between surge and choking limit. Although inducer inlet geometry is not the only parameter defining range, it certainly has an important influence on it because it fixes the inlet Mach number and angle.

Based on low and high speed cascade data, Rodgers [6] has evaluated stalling and choking incidence at different inlet blade angles. These incidence limits are plotted in function of relative inlet Mach number in figure 5. The stable operating range decreases significantly with increasing Mach number and is strongly influenced by the blade inlet angle β_{1bl}. Smaller values of β_{1bl} give larger operating range because they allow for a larger stall incidence.

There are two ways to increase choking mass flow. One is by increasing the shroud radius, resulting in larger tip Mach numbers and smaller incidence range. The other one is by decreasing the inlet blade angle (β_{1bl}) resulting in larger o/t values (Fig. 6) and at the same time resulting in larger incidence range (Fig. 5).

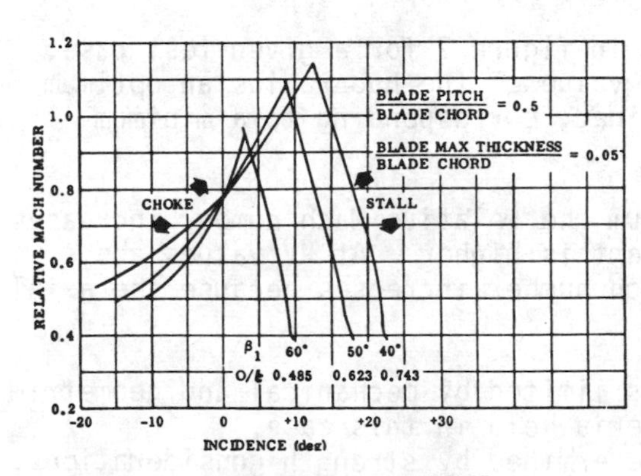

FIG. 5 - VARIATION OF INCIDENCE RANGE [6]

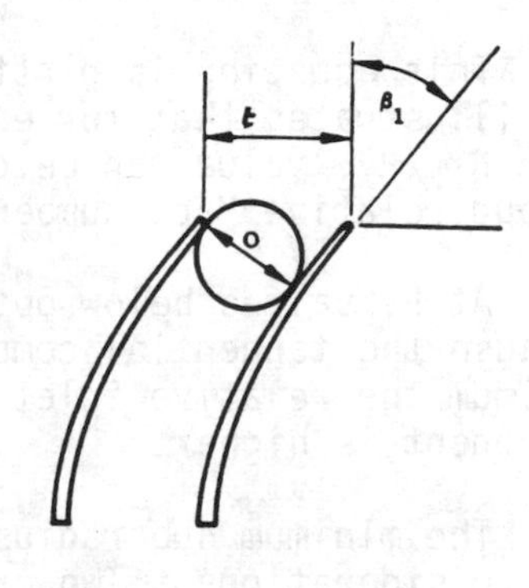

FIG. 6 - VARIATION OF THROAT SECTION [6]

A large amount of impeller losses are due to friction and are proportional to $\frac{\rho W_1^2}{2}$. Assuming that for a given geometry the same relative velocity deceleration can be achieved, one has interest in reducing the inlet Mach number to reduce channel friction losses. These considerations together with an eventual minimization of shock losses and the possibility to achieve a wider range are arguments to minimize the inducer inlet relative Mach number.

This Mach number is influenced by RPM, inlet hub and shroud radius and prerotation.

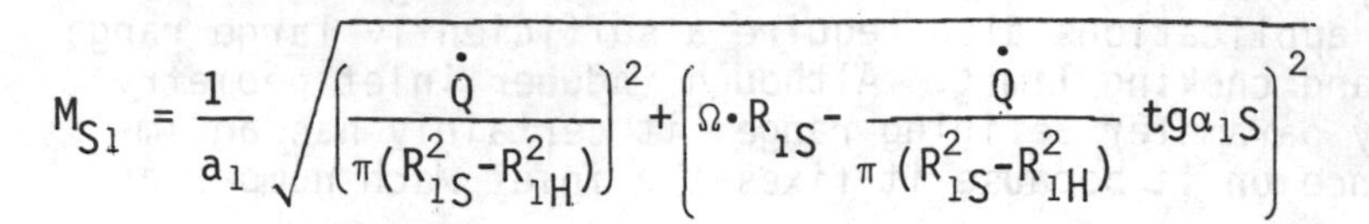

$$M_{S1} = \frac{1}{a_1}\sqrt{\left(\frac{\dot{Q}}{\pi(R_{1S}^2 - R_{1H}^2)}\right)^2 + \left(\Omega \cdot R_{1S} - \frac{\dot{Q}}{\pi(R_{1S}^2 - R_{1H}^2)}\, tg\alpha_{1S}\right)^2}$$

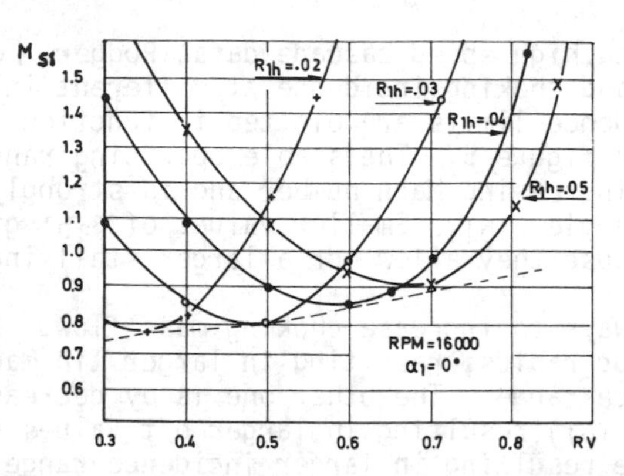

FIG. 7 - VARIATION OF SHROUD RELATIVE MACH NUMBER [17]

This equation is plotted in figure 7 for a given test case, and illustrates that for each value of the hub radius an optimum $RV = R_{1H}/R_{1S}$ value can be defined, corresponding to a minimum shroud relative Mach number.

At RV values below optimum the relative Mach number increases because the tangential component is higher. At RV values above optimum the relative inlet Mach number increases because the axial component is higher.

The minimum hub radius is limited by mechanical and geometrical considerations. Two criteria hold in this case.
- Minimum hub radius can be determined by strength considerations, which require a minimum diameter for the shaft to transmit torque or to avoid critical speed.
- Minimum hub radius can also be limited by the maximum blade blockage $k_{B_{max}}$ at the hub

$$R_{1H_{min}} = \frac{Z \cdot th/\cos\beta_{1bl}}{2\pi\, k_{B_{max}}}$$

z = blade number
th = blade thickness and defined from stress analysis

High values of blade blockage result in a flow acceleration and larger friction losses because of narrow channels. One way to reduce $R_1 H_{min}$ is by reducing the blade number at the inlet by means of splitter vanes.

2.2 Impeller outlet

The impeller outlet dimensions are in the first place defined by the total enthalpy rise required

$$\Delta H = U_2 \cdot V_{u2}$$

where Vu_2 is function of outlet flow angle (β_2), relative velocity (W_2) and peripheral velocity (U_2) as shown by the velocity triangle in figure 8. U_2 being defined by outlet radius and RPM, the problem reduces to the prediction of W_2 and β_2

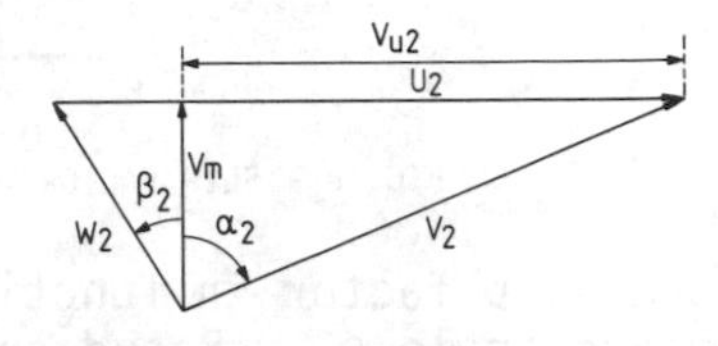

FIG. 8 - VELOCITY TRIANGLE

The flow leaving the impeller does not follow the blade direction but leans backwards over several degrees. This phenomenon, known as slip, is mainly due to the vorticity (-2Ω) of the relative flow, but is also strongly influenced by viscous effects.

Two definitions are commonly used to characterize this effect. The slip factor :

$$\sigma = Vu_2/U_2$$

is a direct function of blade outlet angle β_{2bl}.
The work reduction factor :

$$\mu = Vu_2/Vu_{2\infty}$$

where $Vu_{2\infty}$ is the fictitious tangential velocity that would exist if the flow followed the blade at impeller exit. This factor expresses better the reduction in work input due to the passage vortex.

Both definitions are the same for radial ending impellers because in that case $Vu_{2\infty} = U_2$. These factors are normally defined by an expression which is a combination of theoretical considerations and experimental observations.

Wiesner [7] made an extensive study of the different slip factor correlations and concluded that the expression developed by Busemann was the most generally applicable one. This expression is an exact calculation of the slip factor for incompressible, frictionless flow through an impeller with pure radial blades, at zero mass flow. The results are given in curves (Fig. 9) and define

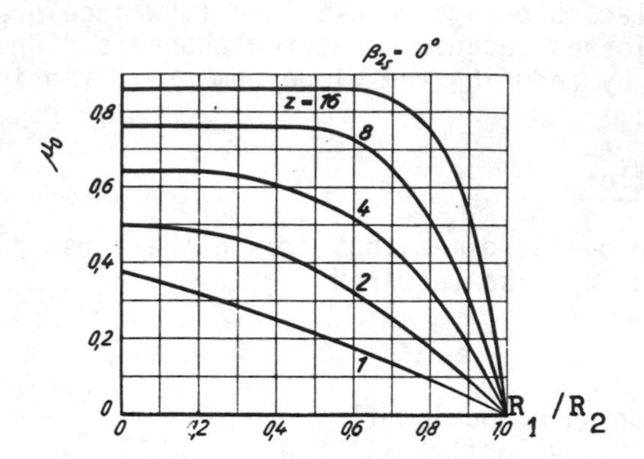

FIG. 9 - SLIP FACTOR IN FUNCTION OF R_1/R_2 AND BLADE NUMBER [8]

the slip factor in function of diameter ratio, blade number Z and blade angle β_2. Based on his comparisons with experimental data, Wiesner proposed a slightly different expression.

Although the agreement between measured and calculated slip factors is within ± 5%, this constitutes large errors on work predictions. Part of the scatter can be attributed to the interaction between impeller and vaned- or vaneless diffuser. But a systematic experimental study by Stiefel [8] revealed that μ depends also on the outlet flow coefficient ϕ_2, impeller shroud clearance and impeller efficiency.

The change of work reduction factor at off design has been evaluated by Traupel [9] who indicates an increase in μ for decreasing mass flow (Fig. 10). Eckert and Schnell [10] have

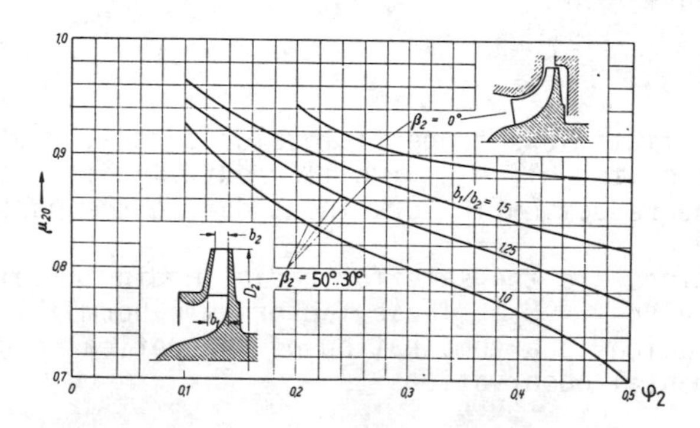

FIG. 10 - VARIATION OF SLIP FACTOR AT OFF DESIGN CONDITIONS [10]

evaluated the influence of viscous effects and flow separation and arrive at the following expression

$$\mu_{sep} = \frac{1}{1+(1-\varepsilon)\left(\frac{1}{\mu}-1\right)}$$

where ε is the relative width of the separated flow zone at imepller exit and μ is the value for non separated flows.

The value of the slip factor expresses only the work which has not been done and high or low slip factors have no direct effect on efficiency. However, the absence of more precise slip factor predictions is a major handicap in a more precise prediction of compressor overall performances.

The outlet relative velocity can be calculated from continuity, assuming uniform outlet flow.

$$W_2 = \frac{\dot{m}}{2\pi R_2 b_2 k_B \cos\beta_2 \rho_2}$$

where k_B is a blockage factor which accounts for blade thickness, and the calculation of ρ_2 requires the knowledge of static pressure and temperature

$$\rho_2 = \frac{P_2}{R_G T}$$

The static temperature rise in the rotor is given by :

$$C_p(T_2-T_1) = (U_2^2-U_1^2)/2 + (W_1^2-W_2^2)/2$$

The static pressure is given by

$$p_2 = p_2^{is} - \Delta p$$

where p_2^{is} is the static pressure, resulting from an isentropic compression.

$$p_2^{is} = p_1 \left(\frac{T_2}{T_1}\right)^{\kappa/(\kappa-1)}$$

and Δp are the pressure losses in the impeller.

The first part of the temperature is due to the change in radius from inlet to outlet, it is larger at the hub than at the shroud and independent of relative velocity and losses. The second part is due to the diffusion of the relative velocity in the blade passage and is larger at the shroud. This deceleration is analogue to the flow in a divergent channel and one will assume that the losses Δp are related to this diffusion. Both Vavra [11] and Moore [12] have evaluated these losses and compared to the relative velocity diffusion they obtained a wheel efficiency (ηw) of the order of .55 to .60. The low efficiency is not only due to diffusion and friction but is also influenced by inlet losses, shroud leakage, etc.

Similar to straight channel diffusers, the deceleration of the velocity is limited by separation. Dean [13] has stated that many centrifugal impellers operate sucessfully with a non negligible amount of flow separation and that the assumption of uniform exit flow is not realistic.

Because of the influence of Coriolis forces on the suction side boundary layer and the convex shape of the shroud, and because the deceleration is highest on the shroud contour, one can expect that the flow will separate in the shroud-suction side corner. This

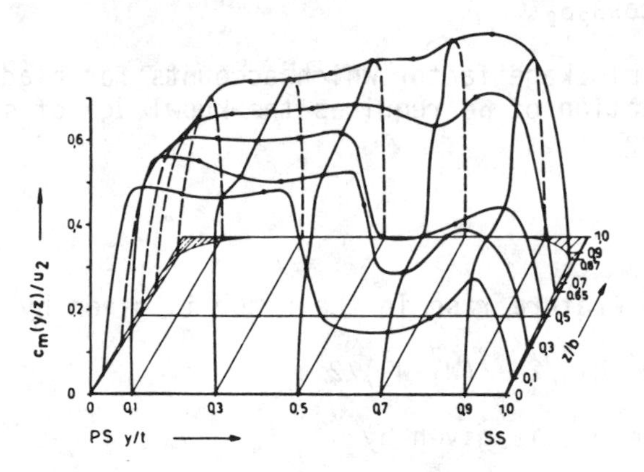

FIG. 11 - FLOW DISTORTION AT IMPELLER EXIT [14]

is confirmed by the measurements of Eckardt [14] (Fig. 11). The positive velocity component in the separated region is between 20% to 40% of the jet velocity and is due to the extensive secondary flows in the centrifugal impeller.

Critical values of impeller relative velocity ratio have been studied by Dallenbach [15]. He predicts flow separation when the local velocity over inlet velocity is less than .675 or .625, depending on the velocity gradient.

Maximum diffusion has also been evaluated by Young [16] based on experimental data. Assuming that downstream the separation point the Mach number remains constant in the jet, he observed a relation between actual diffusion ratio MR_2 and ideal diffusion ratio MR_i as shown in figure 12.

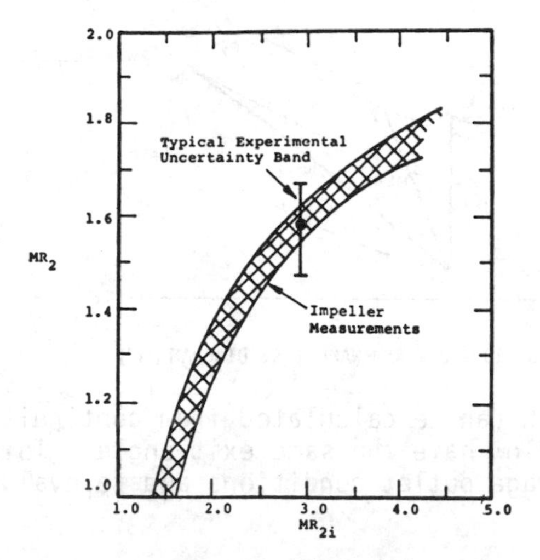

FIG. 12 - REAL FLOW DECELERATION/IDEAL FLOW DECELERATION [16]

$MR_2 = W_{1s}/W_{2jet}$

$MR_i = W_{1s}/W_{2i}$

where W_{2i} is the fictitious uniform outlet velocity for isentropic flow.

In Dean's theory the jet flow is assumed isentropic and all the losses are connected with the wake flow. This allows to calculate the static pressure and density in the jet in function of W_{2j}, because $\Delta pj = 0$. The static pressure in the wake can be evaluated from the static pressure in the jet p_{2j}, by means of the tangential equilibrium of pressure at the rotor exit [17]. Neglecting the blade curvature radius, the following approximated expression was obtained

$$p_{2w} = p_2 - \frac{\Omega \dot{m}}{z\, b_2}$$

where Ω is the angular velocity (rad/sec) of the impeller. The temperature in the wake is higher than the jet because the relative velocity is lower

$W_{2w} = \nu\, W_{2j} \qquad \nu = .2 \text{ to } .4$

The impeller outlet flow conditions calculated by this procedure are shown on the T.S. diagram of figure 13. From this figure one can see that the wake suffers an important entropy increase which results in heavy losses for the mean flow.

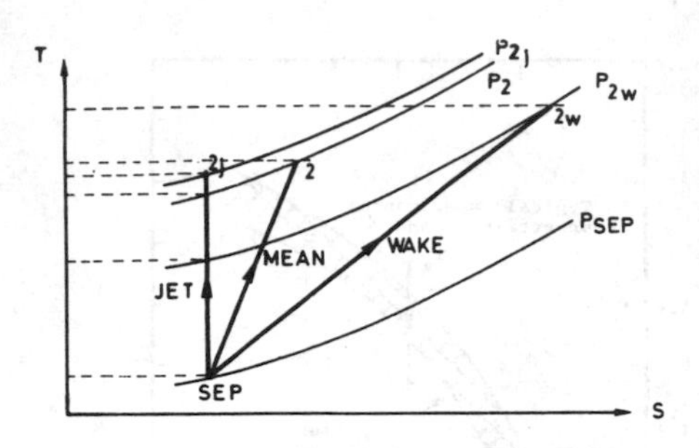

FIG. 13 - JET AND WAKE T.S. DIAGRAM [17]

The wake width can be calculated from continuity assuming that the wake and jet flow have the same exit angle. This allows to calculate the average outlet conditions and to evaluate the impeller efficiency.

The assumption of isentropic jet flow is somewhat optimistic because this does not account for the influence of channel geometry on friction and clearance losses. This is possible by introducing clearance and friction losses in the jet.

$$\Delta p_j = \frac{1}{2}(\omega_f + \omega_{c\ell})\,\rho_j\,W_j^2$$

ω_f can be evaluated in function of a friction coefficient, hydraulic length (L_H) and diameter (D_H) of the flow passage

$$\omega_f = 4.Cf.\ L_H/D_H$$

where Cf is function of Reynolds number and roughness. Based on an expression from Jansen [11] clearance losses can be evaluated by

$$\omega_{c\ell} = 2.4\ \frac{\delta_{c\ell}}{b_2}\cdot\left(1 - \frac{R_{1s}}{R_2}\right)\frac{U_2^2}{W_j^2}$$

Flow separation losses could be decreased by decreasing the impeller outlet width. However, this results in higher friction losses and clearance losses as shown in figure 14. The optimization of b_2 therefore requires a carefull balance between the different losses.

As one will see in the next chapter, wider diffusers are more efficient than narrow diffusers and the jet wake non uniformity at impeller outlet results in larger mixing losses at diffuser inlet. The optimization of b_2 also requires the knowledge of diffuser losses and is influenced by stability considerations.

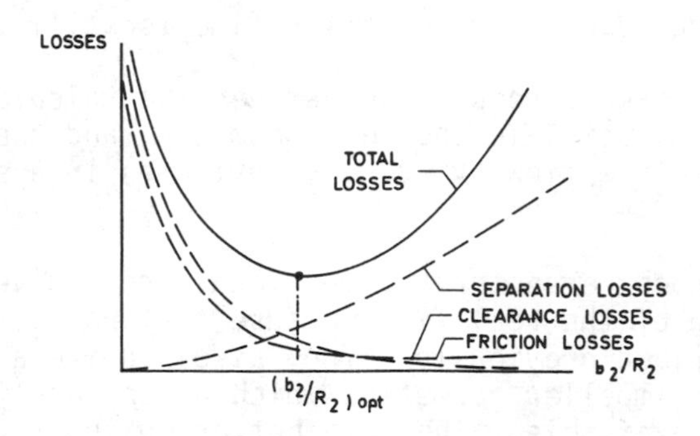

FIG. 14 - VARIATION OF IMPELLER LOSSES WITH OUTLET WIDTH

Larger backward lean angles also allow to decrease impeller diffusion. However, this results in a large outlet radius (to maintain the work constant) and therefore a small increase in friction losses. Besides a favorable effect on stability and range, one can estimate a 1 to 2 points in efficienty rise for a 10° increase of β_2.

1.3 Flow in the impeller

The previous method is highly based on correlations and does not fully account for the influence of impeller geometry on the flow. Results are therefore very approximated. A more reliable prediction is only possible with a better knowledge of the real velocity distribution.

Three dimensional solutions of the Navier-Stokes equations are very scarce and very time consuming. Walitt [19] needs 400 minutes on a CDC 7600 and estimates the time to solve a problem at 2 months. Although his method very nicely describes the

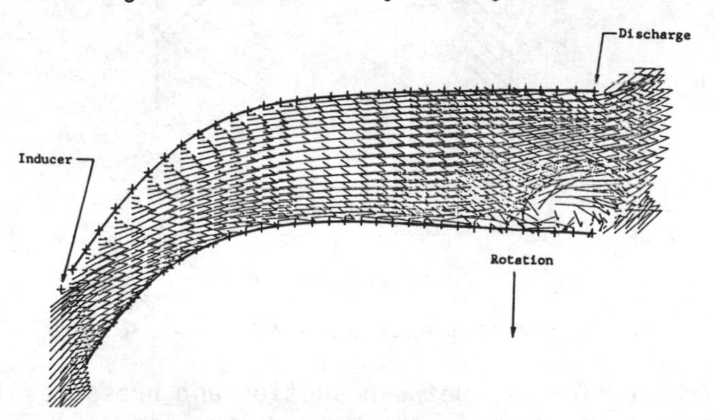

FIG. 15 - SOLUTION OF VISCOUS FLOW CALCULATION [19]

separated flow on the suction side, it is not yet suited for impeller optimization (Fig. 15). Other solution techniques [12] also give promising results, calculation time is still too long.

Other methods make a compromise between the calculation of a simplified flow (inviscid) in the real geometry, and the calculation of a more complete flow (viscous + inviscid) in a simplified geometry.

Inviscid methods, even three dimensional ones give a very simplified picture of the real flow because they neglect all viscous effects described previously. They also require a detailed description of the impeller geometry, which at the early stage of a design is not yet available. They are better suited for a final check of the detailed geometry. A comparative study of the different calculation methods for quasi 3D and full 3D, subsonic and transonic flow is made by Adler [20].

Methods which are better suited for an optimization procedure are combinations of viscous and inviscid calculations. They enable to obtain a more realistic description of the flow at relatively short computer times so that one can envisage an iterative use. Typical methods are those by Herbert [21] and Davies & Dussourd [22]. The mean relative velocity distribution between inlet and outlet is defined by a standard variation. Loading considerations based on blade angle and radius variation allow to

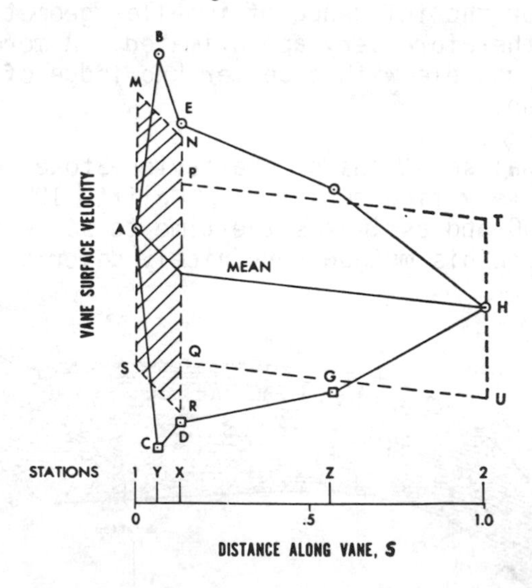

FIG. 16 - SIMPLIFIED SOLUTION OF BLADE TO BLADE FLOW [22]

calculate the difference between suction and pressure side velocity, which is then superposed on the mean value (Fig. 16). The blockage is estimated from correlations or by calculating the boundary layer growth along the channel walls, according to the method of

Stratford & Beavers [23] for non separated turbulent flow. The impeller exit velocity is then corrected in function of blockage and the procedure is repeated until convergence is achieved. The predicted impeller exit conditions are checked against slip factor predictions. An eventual disagreement in diffuser inlet flow angle is accounted for by applying a correction to the boundary layer calculation. The work input is complemented by
- disk friction losses,
- clearance losses or recirculation losses.

After comparison with the achieved pressure rise one can calculate the impeller isentropic efficiency.

However, these methods are still very approximated because of the prescribed velocity distribution and the very approximated boundary layer approximation. A more elaborated flow model has been proposed by Cipollone [24] and one will shortly discuss some of the main features.

This model starts with the standardization of the impeller such that the geometry can be analytically defined. The flow channels are defined by the intersection of the blade with hub and shroud contour. The intermediate points on the blades are defined by a straight line interpolation.

The meridional hub and shroud contour are approximated by a second degree Besier-Bernstein polynomial [25].

$$x(u) = u^2(x_0-2x_1+x_2) + 2u(x_1-x_0) + x_0$$

$$R(u) = u^2(R_0-2R_1+R_2) + 2u(R_1-R_0) + R_0$$

where :

x_0, R_0 are the coordinates of the leading edge
x_1, R_1 are the coordinates of the trailing edge
x_2, R_2 are the coordinates of point defined by the intersection of the tangent at leading and trailing edge (Fig. 17)
u is a parameter varying between 0 (leading edge) and 1 (trailing edge)

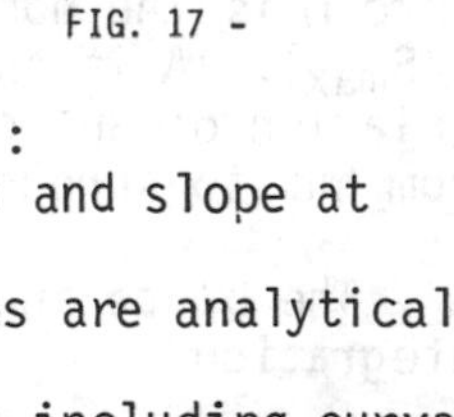

X_1, R_1
X_2, R_2
X_0, R_0

FIG. 17 -

The main advantages of such a polynomial are :
- the geometry is fully defined by the coordinates and slope at leading and trailing edge;
- arc length (s), local derivatives and coordinates are analytical functions of u and easy to calculate;
- continuous smooth variation of all the variables including curvature (this is not the case if a multiple circular arc is used).

This method has been checked on the Eckart impeller [26] and shows a good agreement (Fig. 18).

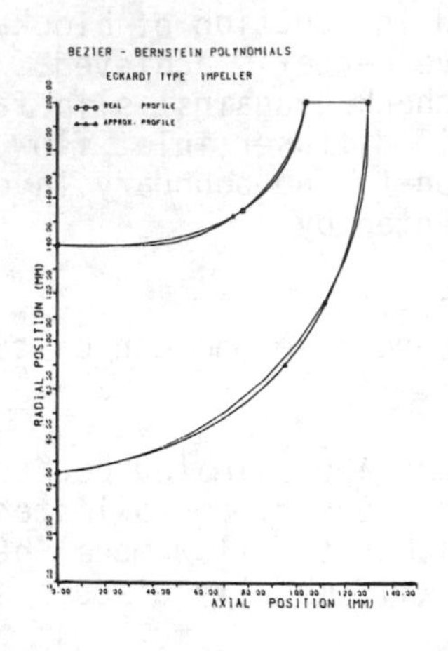

FIG. 18 - APPROXIMATION OF MERIDIONAL SHAPE

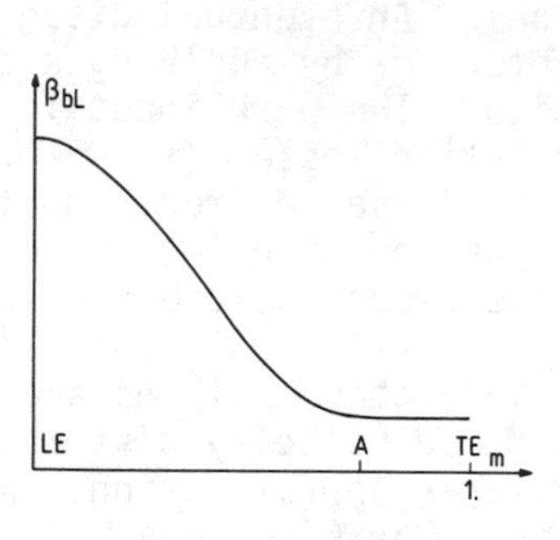

FIG. 19 - BLADE ANGLE DISTRIBUTION

The blade profiles are defined by a prescribed blade angle distribution β_{bl} along hub and shroud. Although they must represent the given leading edge and trailing edge blade angle, a lot of different distributions are possible. Each of them results in a different impeller. Part of the design optimization therefore consists of defining an optimum blade angle distribution, based on internal flow considerations.

One of the typical distributions, examined by Cipollone, is shown in figure 19

$$\beta_{bl}(m) = \frac{\beta_{1bl}+\beta_{2bl}}{2} + \frac{\beta_{1bl}-\beta_{2bl}}{2} \cos\pi \frac{m}{A} \qquad 0 < m < A$$

$$\beta_{bl}(m) = \beta_{2bl} \qquad A < m < 1$$

where m is the non dimensional length of the meridional contour (S/S_{max}). A is the non dimensional distance where the sinusoidal variation of β_{bl} changes into a linear one and will be different from hub to shroud.

The blade angular coordinate θ is defined by the following integration

$$\theta(s) = \int_{LE}^{TE} tg\beta_{bl} \cdot ds$$

The value of θ should not be too different from hub to shroud in order to avoid large lean angles for the blades and to obtain a machinable impeller.

Once the hub and shroud are defined, one can easily obtain a grid of intermediate points by linear interpolation along the quasi-orthogonals connecting equidistant points on hub and shroud (Fig. 20). These quasi streamlines define a set of axisymmetric streamsurfaces on which the velocity can be calculated.

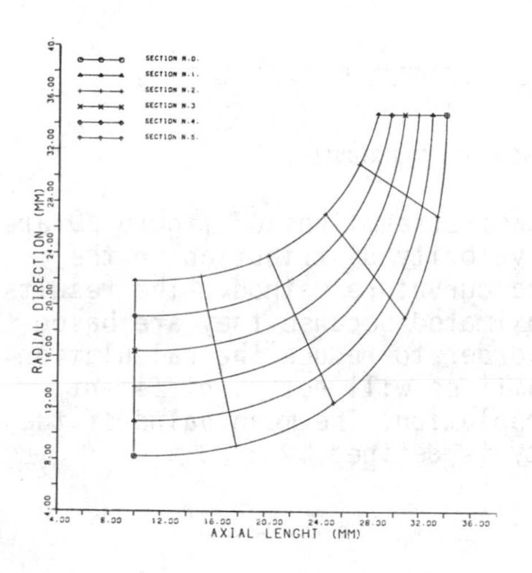

FIG. 20 - GRID GENERATION

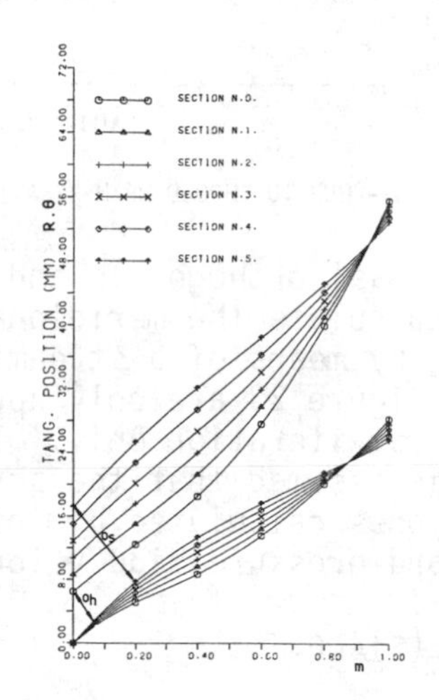

FIG. 21 - DEFINITION OF THROAT SECTION

The hub and shroud definition also allows to calculate the tangential coordinates of the blades (R.θ) in function of meridional distance (Fig. 21). This allows to define the position and width of the throat section on the different axisymmetric surfaces. This throat section is further corrected for blade thickness and boundary layer blockage. Assuming sonic velocity at the throat, one can calculate the maximum or choking mass flow in function of RPM. Typical results of such a calculation for a turbocharger impeller are compared with experimental data in figure 22.

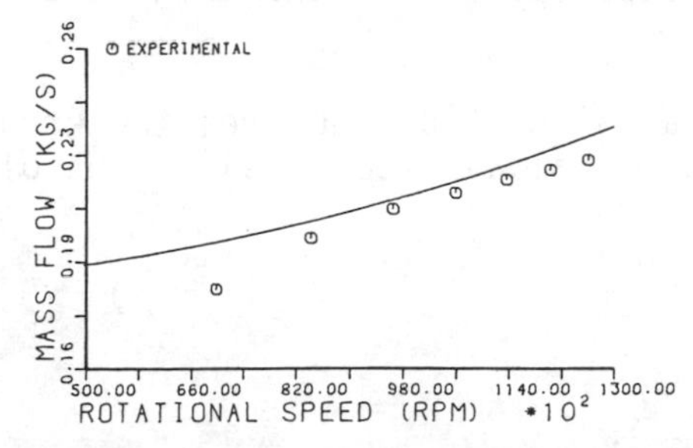

FIG. 22 - EXPERIMENTAL/PREDICTED CHOKING MASS FLOW

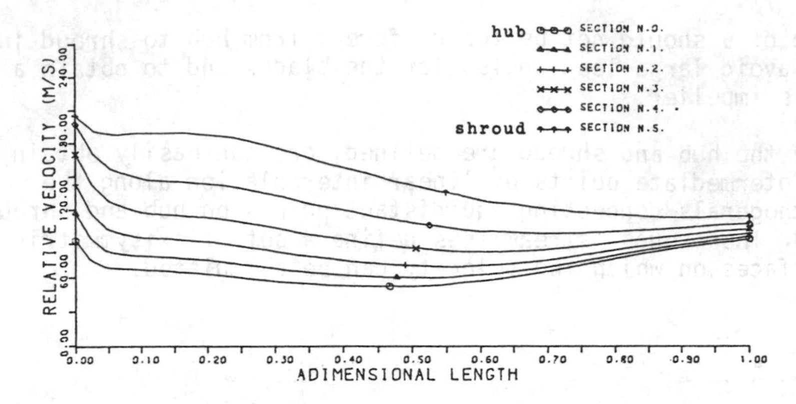

FIG. 23 - HUB TO SHROUD MEAN RELATIVE VELOCITY DISTRIBUTION

The quasi orthogonals and quasi streamlines of figure 20 are used to calculate the meridional velocity distribution in the impeller, by means of a streamline curvature method. The results shown in figure 23 are only approximated because they are based on the first iteration only. In order to reduce the calculation time it is assumed that the streamlines will not be different from the ones calculated by interpolation. The mean value of the suction and pressure side velocity is defined by

$$W(s) = W_m(s)/\cos\beta$$

The difference between suction and pressure side velocity on an axisymmetric surface is calculated from

$$\nabla_\wedge \overline{W} = -2\Omega$$

applied to a closed contour in the blade-to-blade plane. This results in

$$W_{ss}-W_{ps} = \cos\beta_{f\ell}\left(\frac{2\pi}{Z} - \frac{th}{R}\right)\frac{d}{ds}\left(\Omega R^2 - W_m R t_g \beta_{f\ell}\right)$$

This difference is superimposed on the mean value and shown in figure 24.

However, such a calculation requires the knowledge of the flow angle which is not always identical to the blade angle.

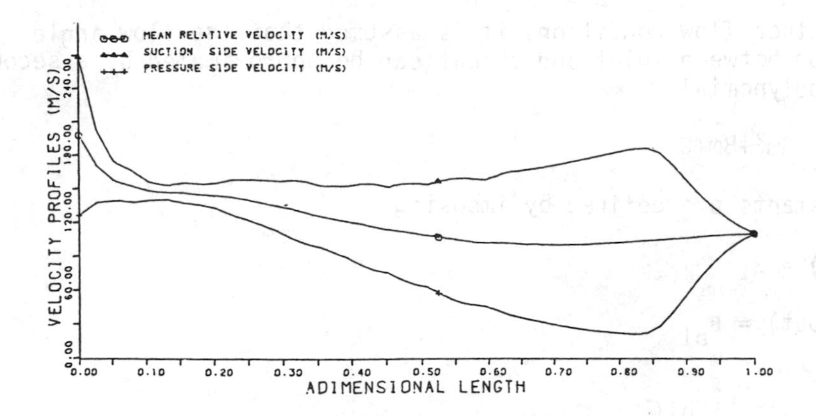

FIG. 24 - SHROUD RELATIVE VELOCITY VARIATION

At the trailing edge this difference is function of the slip factor. Downstream a critical value $m^{\star}$, the flow angle is approximated by a third degree polynomial

$$\beta_{f\ell}(m) = Am^3+Bm^2+Cm+D$$

The four constants are defined by the following conditions :

$$\beta_{f\ell}(m^{\star}) = \beta_{bl}(m^{\star})$$

$$\left.\frac{\partial\beta_{f\ell}}{\partial m}\right|_{m^{\star}} = \left.\frac{\partial\beta_{bl}}{\partial m}\right|_{m^{\star}}$$

This assures a zero and first order continuity between blade and flow angle at $m^{\star}$

$$\beta_{f\ell}(TE) = \beta_2(slip)$$

$$W_{ps}-W_{ss} = 0 \qquad \text{at trailing edge.}$$

This assures a correct flow angle and zero loading at the blade trailing edge.

The point $m^{\star}$ where the slip starts can be evaluated from a relation by Stanitz & Prian [27]

$$\ell n R^{\star}/R_2 = .71\ \frac{2\pi\cos\beta_{bl}(TE)}{Z}$$

At each RPM there is only one mass flow for which the flow angle at the leading edge equals the blade angle (zero incidence).

At all other flow conditions it is assumed that the flow angle variation between inlet ond throat can be approximated by a second degree polynomial

$$\beta_{f\ell}(m) = Am^2+Bm+C$$

The constants are defined by imposing

$$\beta_{f\ell}(m=0) = \beta_1$$

$$\beta_{f\ell}(\text{throat}) = \beta_{bl}$$

$$\left.\frac{\partial\beta_{f\ell}}{\partial m}\right|_{\text{throat}} = \left.\frac{\partial\beta_{bl}}{\partial m}\right|_{\text{throat}}$$

The variation of β used in the inviscid velocity distribution of figure 24 is shown in figure 25.

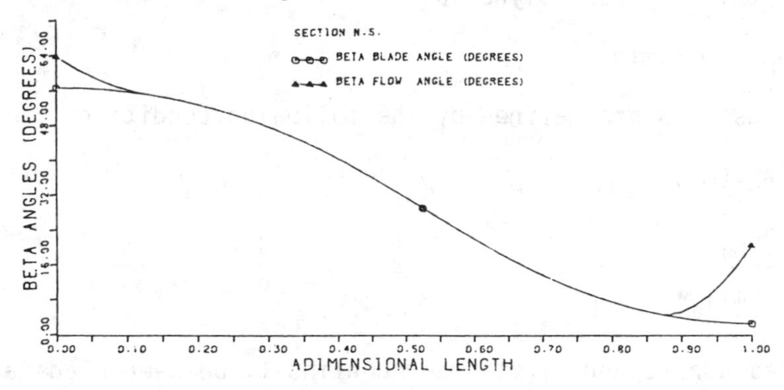

FIG. 25 - FLOW ANGLE AND BLADE ANGLE VARIATION

Viscous effects are introduced by the blockage predicted by a boundary layer calculation. However, ordinary flat plate boundary layer calculations are not suited for centrifugal impeller calculations.

- Due to the important radial velocity component, especially at the impeller exit, Coriolis forces are very large and cannot be neglected.
- The change from an axial inlet to a radial outlet, and the turning of the vanes result in important curvature effects, especially at the hub and shroud.

These effects are discussed in [28] and a simple modification of an integral boundary layer method is presented in [29].

Most centrifugal impellers have a strong deceleration on the pressure side and along the hub, and boundary layer methods which do not account for Coriolis or curvature will predict an early flow separation. Even when the boundary layer method predicts separation, curvature will prevent the development of separated flow zones, and it is acceptable to continue the calculations, keeping the blockage constant, until more favourable flow conditions occur. A limited amount of return flow can occur if the inviscid velocity becomes negative because of excessive blade loading due to the passage vortex -2Ω.

Coriolis effects do not allow to have large decelerations on the suction side at impeller exit. The calculated results of figure 26 correspond to the suction side velocity distribution of figure 24 and show that separation occurs because of the deceleration at the trailing edge.

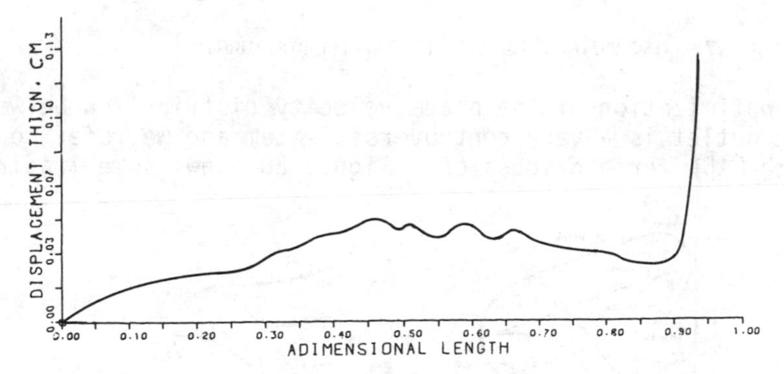

FIG. 26 - SHROUD SUCTION SIDE DISPLACEMENT THICKNESS

Boundary layer calculations, although they are only two dimensional, allow to estimate the boundary layer blockage in an impeller passage by integrating the boundary layer displacement thickness on suction and pressure side, hub and shroud. It accounts for the real deceleration in a given impeller and includes incidence effects. After the blade passage cross sections have been reduced in function of blockage, one can recalculate the velocity distribution and this iterative procedure is continued until convergence.

The influence of blockage on the velocity distribution is illustrated by the viscous and inviscid solution in a blade-to-blade plane close to the shroud (Fig. 27).

Downstream the separation point, the separated flow velocity distribution can be calculated assuming a constant velocity on the jet-wake intersection and a velocity ratio between wake and jet. This results in an additional blockage.

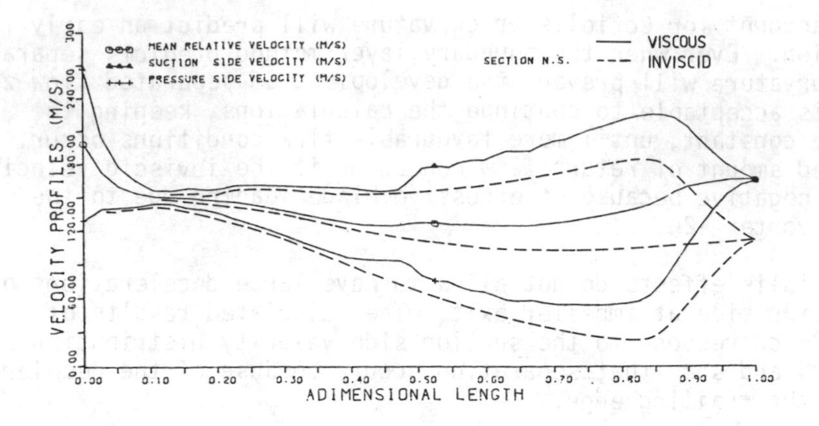

FIG. 27 - VISCOUS/INVISCID RELATIVE VELOCITY DISTRIBUTION

The optimization of the blade velocity distribution between inlet and outlet is a very controversial item and we refer to Dallenbach [15] for a discussion. Figure 28 shows some loading

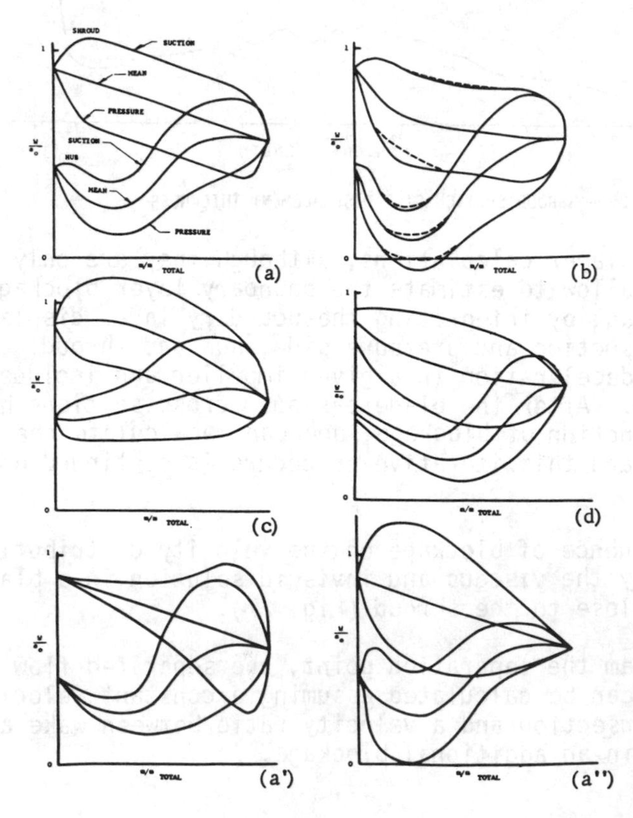

FIG. 28 - DIFFERENT TYPES OF RELATIVE VELOCITY DISTRIBUTIONS [15]

distributions at the hub and shroud of an impeller with radial bladed elements. They all have the same inlet and outlet relative velocity. Because of geometrical limitations of the blade shape, velocities along the shroud and hub cannot be chosen arbitrarily but are interconnected. Any modification of the shroud contour will have an influence at the hub.

Figure 28a shows a linear variation of the mean velocity at the shroud from inlet to outlet and a constant difference between suction and pressure side velocity. This results in an acceleration at the suction side leading edge. The corresponding hub velocity shows a rapid decrease of the velocity after the leading edge and a reacceleration towards the exit.

In order to avoid the shroud suction side acceleration, one could reduce the mean velocity more rapidly as shown in figure 28b. However, this results in a negative velocity at the hub pressure side and the blockage due to recirculation will influence the other velocities as indicated by the dashed lines.

In figure 28c the mean velocity at the shroud remains constant in the inlet region and decreases more rapidly towards the exit. This results in a slightly accelerating velocity at the hub and a larger overshoot at the shroud.

Figures 28a' and 28a" show velocity distributions with a linear variation of the mean velocity, such as 28a, but with a different loading distribution.

It has been shown that for a sufficiently high pressure rise separation cannot be avoided and will occur at $W/W_1 \leqslant .65$ [13]. Large pressure differences between pressure and suction side will result in higher secondary flows. A larger boundary layer thickness on the pressure side will also favour secondary flow because of Corilis forces. High loading at the inlet results in a velocity increase on the suction side which is unfavourable at high inlet Mach numbers and from the point of view of range. Friction losses are proportional to W^2 so that one should reduce the velocity as fast as possible.

Both 28a' and 28a" may be rejected for these reasons. A possible compromise is shown in figure 28d where the effect of loading is compensated at the inlet by a mean velocity reduction. The hub velocity deceleration is controlled by having the mimimum velocity further downstream.

Such a discussion is speculative and can only give an approximate direction in the impeller optimization.

3. VANELESS DIFFUSERS

The relative flow leaving the impeller is highly non uniform in tangential and axial direction because of the boundary layer development along the walls and because of secondary flow and separation in the impeller.

The circumferential non uniformity results in an unsteady absolute flow, with strong variations in velocity and flow direction (Fig. 29a). This unsteadiness favours the tangential uniformization of the flow.

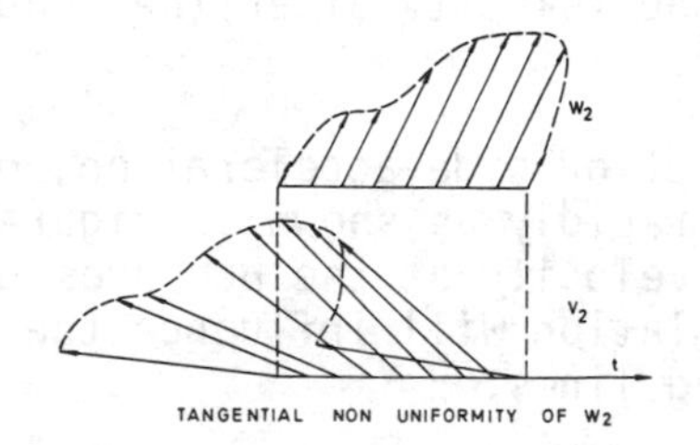

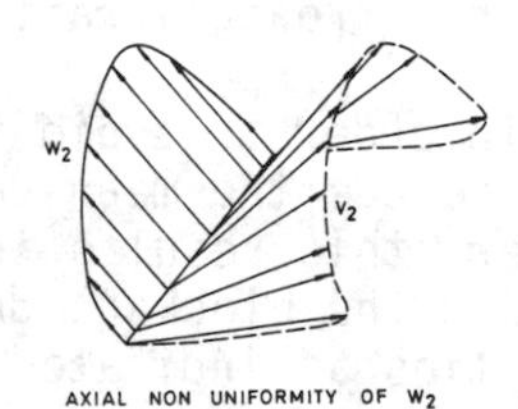

FIGS. 29a,b - TANGENTIAL AND AXIAL NON UNIFORMITY OF RELATIVE VELOCITY

The axial non uniformity of the relative flow is responsible for the skewness of the absolute flow (Fig. 29b). This axial non uniformity is continued in the diffuser boundary layers and can cause side wall separation and diffuser stall.

3.1 One dimensional calculation

The simplest method to calculate the flow in a vaneless diffuser is the one dimensional method of Stanitz [30], for radial and mixed flow diffusers. The flow is assumed uniform in axial and tangential direction. For adiabatic flow, the problem is solved by integrating the following equations in radial direction.

Continuity equation

$$b.R.\rho.V_m = b_2\ R_2\ \rho_2\ V_{m_2}$$

Energy equation

$$T + V^2/2C_p = T_2 + V_2^2/2C_p$$

Equation of state

$$P = \rho.R_G.T$$

Momentum equation in tangential and meridional direction.

$$F_u - \frac{V_m}{R} \cdot \frac{dRV_m}{dm} = 0$$

$$F_m + \frac{V_u^2}{R} \cdot \frac{dR}{dm} - V_m \frac{dV_m}{dm} - \frac{1}{\rho} \frac{dp}{dm} = 0$$

The tangential and meridional friction force is calculated by means of a friction coefficient Cf, which is function of Reynolds number and roughness

$$F_u = - Cf \frac{V.V_u}{b}$$

$$F_m = - Cf \frac{V.V_m}{b}$$

The method allows to calculate the streamlines, velocity, mean flow angle, static and total pressure distributions in the diffuser.

This method has been applied to a series of three vaneless diffusers to investigate the influence of diffuser width on total pressure losses.

$$\omega_D = \frac{P_{04}-P_{02}}{q_2}$$

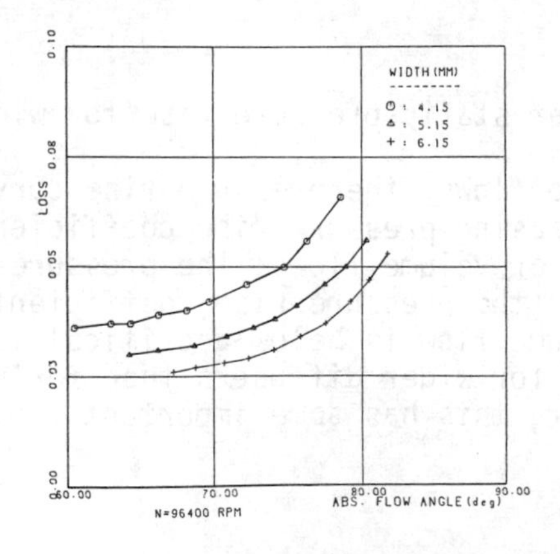

FIG. 30 - VARIATION OF DIFFUSER LOSSES WITH WIDTH

Results are plotted in figure 30 as a function of inlet flow angle. The three curves are calculated at a given impeller RPM for the same variation of mass flow in a turbocharger impeller with an exit width adapted to the diffuser width.

Wider diffusers show lower total pressure losses than narrow diffusers, and the losses increase with decreasing mass flow (larger values of α).

The variation of the static pressure rise coefficient with diffuser width and inlet conditions is shown in figure 31.

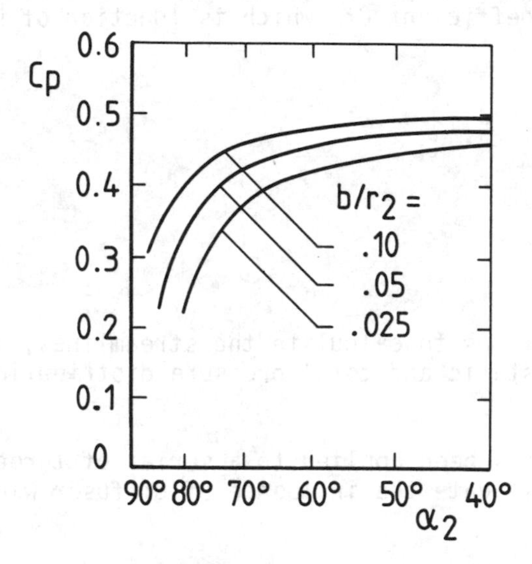

FIG. 31 - VARIATION OF STATIC PRESSURE RISE WITH WIDTH

$$C_p = \frac{P_4 - P_2}{q_2}$$

and indicates a larger static pressure rise for wider diffusers.

At larger volume flows, the pressure rise curve has a small positive slope (increasing pressure rise coefficient with increasing mass flow). At smaller volume flows, the pressure rise curve has a large positive slope (the pressure rise coefficient decreases rapidly when the volume flow is below a critical value. The critical value is smaller for wider diffusers than for narrow diffusers. As one will see later, this has some important consequences on diffuser stability.

3.2 Circumferential distortion

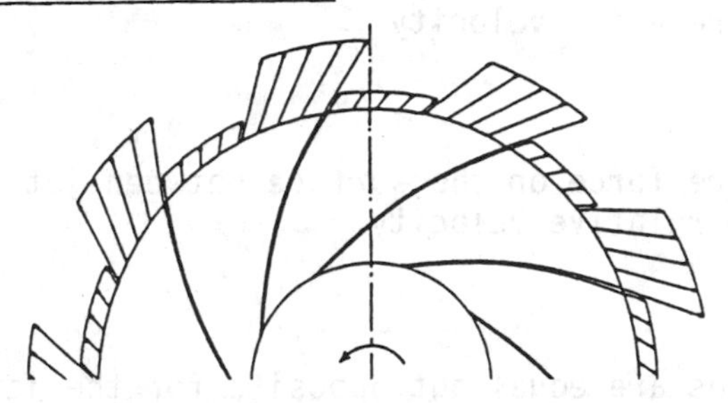

FIG. 32 - JET WAKE FLOW AT IMPELLER EXIT

The relative flow leaving the impeller has a strong circumferential distortion. A schematic representation of such a flow is shown in figure 32. The low velocity part accounts for the blade boundary layers, low energy fluid accumulated on the suction side by secondary flows and eventual flow separation. The high velocity part accounts for the inviscid flow core in the impeller.

Because of impeller rotation, this results in an unsteady absolute flow at the diffuser inlet. The absolute velocity vector changes in magnitude and direction as can be derived from the velocity triangle (Fig. 33). To avoid problems with the unsteady inertia terms, calculations of this jet and wake flow are performed in the relative plane rotating with the rotor. The complete calculation procedure is described in [31] and [17] and is similar to the one described in the previous chapter for uniform flow. The continuity equation, momentum equation in tangential and radial direction are now applied to the jet and wake separately. The total energy of jet and wake remains constant.

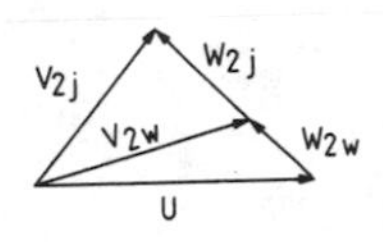

FIG. 33 -

The driving forces for the mixing process are :
- Side wall friction forces acting on the jet in the direction opposite to the absolute jet velocity V_j

$$C_f \frac{V_j^2}{2} 2\pi R dR$$

- Side wall friction force acting on the wake in the direction opposite to the absolute wake velocity V_w

$$C_f \frac{V_w^2}{2} 2\pi R dR$$

- friction force on the shear layer between jet and wake in the direction of the relative velocity

$$C_m(V_j-V_w)^2 \; z \; b \; dR$$

- jet-wake pressure force on the surface between jet and wake perpendicular to the relative velocity

$$\Delta p \; zb \; dR$$

The last two forces are equal but opposite for the jet and the wake and the last one is responsable for a reversible work exchange between jet and wake. The problem can be solved by adding the equation of state and geometrical relations concerning the velocity triangles.

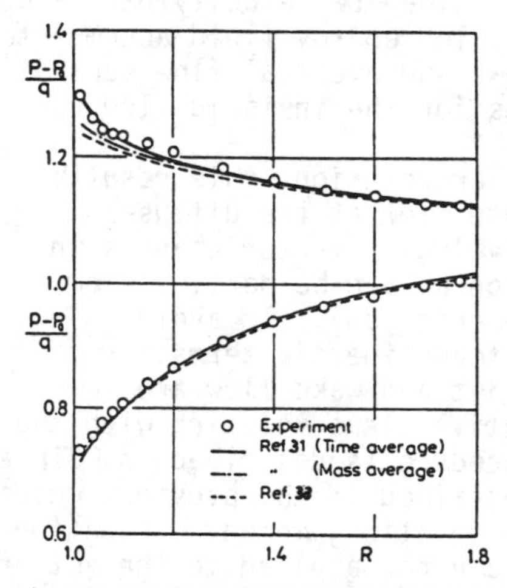

FIG. 34 - INFLUENCE OF MIXING ON DIFFUSER PERFORMANCES [34]

Typical results of such a calculation are compared with experimental data in figure 34. The large total pressure loss and static pressure rise at the diffuser inlet are due to the mixing process which is already completed at $R/R_2 = 1.2$. The downstream variation of total and static pressure are in agreement with uniform flow calculations.

This rapid mixing (when compared to axial compressor wakes) is due to the reversible work exchange between jet and wake.

The larger total pressure losses are due to increased friction on the diffuser walls. The velocity triangle (Fig. 33) shows that the wake absolute velocity has a larger tangential component and therefore also larger wall shear stresses. The radial component is smaller such that the mass flow per unit circumferential length will be smaller than in the jet. Higher losses are thus distributed on a smaller mass flow, resulting in a higher loss coefficient.

The shear stresses between jet and wake are relatively small because they are proportional to the square of a velocity difference $(W_j-W_w)^2$ and they apply to a surface which is much smaller than the diffuser side walls.

$$z\ b\ dR \ll 2\ RdR$$

One can therefore conclude that the "mixing losses" are mainly due to an increased friction on the side walls and not so much to mixing in the classical sense [32].

Johnston and Dean [30] proposed to approximate the mixing process by a sudden expansion at impeller exit and a uniform flow downstream in the diffuser. However, the agreement with experimental data is mainly due to a compensation of errors [34].

3.3 Axial distortion

The axial non uniformity of the relative velocity at the rotor exit results in a skewed velocity profile at the diffuser inlet. Contrarily to the circumferential perturbations this axial perturbation persists downstream in the vaneless diffuser because of the unfavourable interaction with the radial pressure gradient. The pressure gradient makes an angle α with the mean velocity vector, which maintains the three dimensional character of the boundary layer downstream in the diffuser.

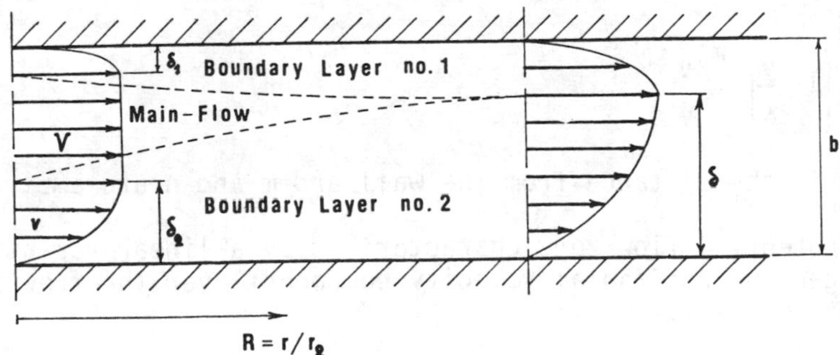

FIG. 35 - MODEL FOR 3D VISCOUS FLOW CALCULATION

The theoretical models of Jansen [35] improved by Senoo et al., [36] allow to calculate such a flow, assuming incompressible steady flow with no circumferential perturbations. The three dimensional velocity profile is represented by figure 35.

a) Two viscous zones corresponding to the three dimensional side wall boundary layers. They are characterized by the thickness δ and the skew factor ε

$$\varepsilon = tg\,\gamma$$

where γ is the angle between the limiting wall streamline and the main flow streamline (Fig 36a). The three dimensional velocity

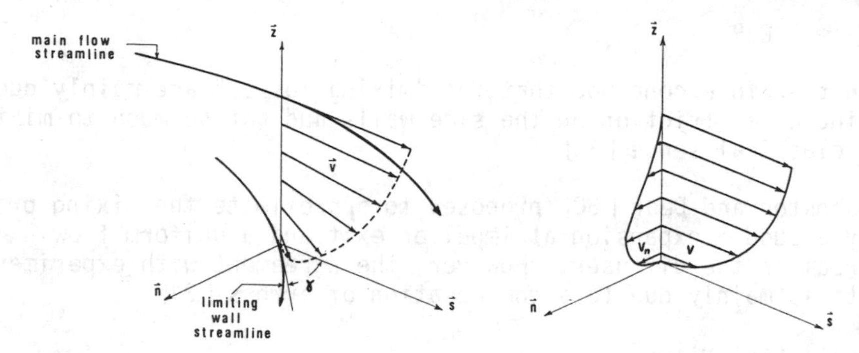

FIGS. 36a,b - 3D BOUNDARY LAYER DEFINITION

distribution is defined by a component in the main flow direction

$$\frac{v}{V} = \left(\frac{y}{\delta}\right)^{1/n}$$

and a cross flow component (Fig. 36b)

$$\frac{v_n}{V} = \varepsilon \left(1 - \frac{y}{\delta}\right)^m \frac{v}{V}$$

where y is the distance from the wall and m and n are empirically defined.

b) The potential flow zone characterized by a linear variation of the tangential and radial velocity component over the free section = $b-\delta_1-\delta_2$.

This problem requires the solution of a system of first order ordinary differential equations, consisting of the radial and tangential momentum equation for both boundary layers and core flow, and the continuity equation applied to the full diffuser width. Special precaution must be taken when the potential core disappears because the two boundary layers have merged together.

Typical results of such a calculation by Senoo [36] are shown in figure 37. Good agreement between experiments and theory is

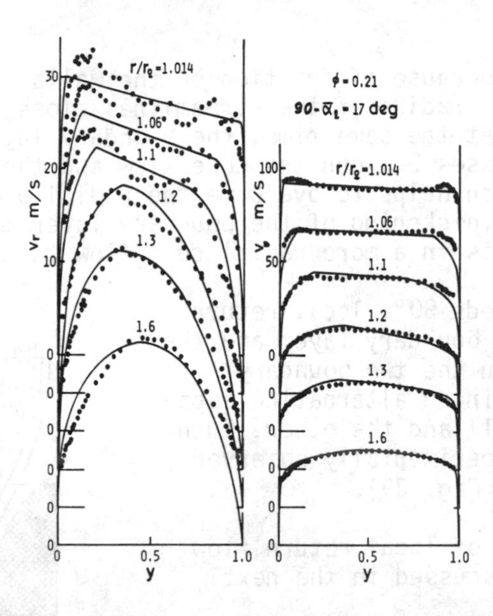

FIG. 37 - EXPERIMENTAL AND CALCULATED VELOCITY PROFILES [36]

obtained even in the case of highly distorted inlet flow.

The variation of flow parameters in function of radius is shown in figure 38. In this example the flow is not only axially

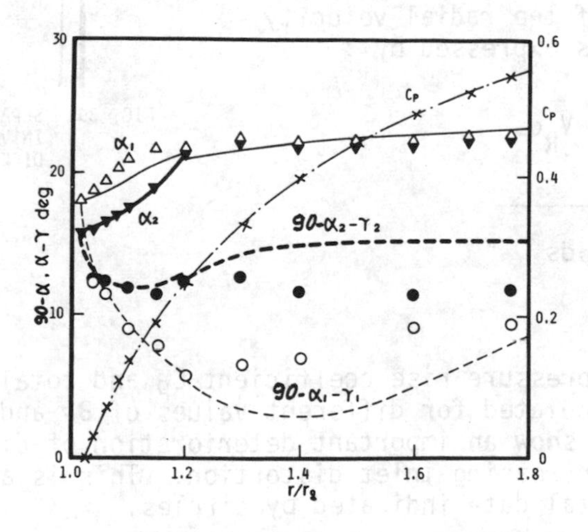

FIG. 38 - RESULTS OF DIFFUSER FLOW CALCULATION [36]

distorted, but also asymmetric and the evolution is different for both sides of the diffuser. The static pressure coefficient increases steadily from diffuser inlet to outlet. The tangential velocity component in the boundary layer is very large ($\approx U_2$) at diffuser inlet, and the corresponding centrifugal forces help to sustain the radial pressure gradient. However, this component reduces very fast because of friction on the walls, resulting in a smaller curvature radius of the streamlines close to the walls ($\alpha+\gamma$ increases). At the same time, the boundary layer gets thicker and the shear stresses between the core flow and the boundary layer is increasing which helps to overcome the radial pressure gradient ($\alpha+\gamma$ decreases). Thickening of the boundary layer means more blockage and results in a more radial core flow (α decreases).

When $\alpha+\gamma$ exceeds 90°, local return flow occurs in the boundary layer and the interaction between the two boundary layers can result in an alternating separation from one wall and the other. Such a flow has been experimentally observed by Rebernick [37] (Fig. 39).

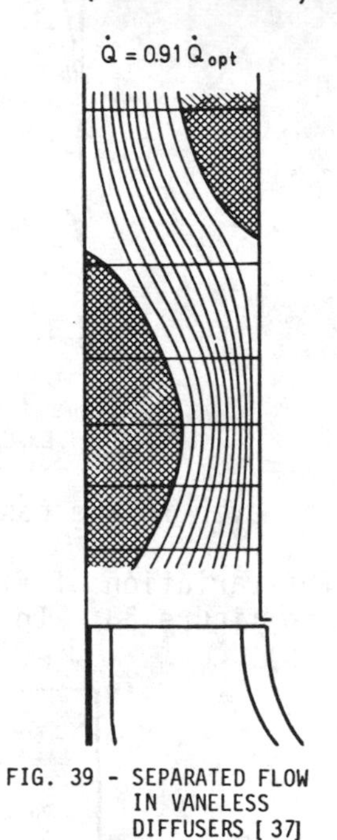

FIG. 39 - SEPARATED FLOW IN VANELESS DIFFUSERS [37]

The influence of local return flow on stability is discussed in the next chapter.

Senoo [38] has used this calculation method to evaluate the influence of inlet distortion on static pressure rise and total pressure losses. The inlet distortion is mainly due to the non uniformity of the radial velocity component and was expressed by :

$$Bf = \frac{\int_S \rho V_R^2 ds / \int_S \rho V_R ds}{\int_S \rho V_R ds / \int_S \rho ds}$$

The static pressure rise coefficient C_p and total pressure losses ω are calculated for different values of B_f and plotted on figure 40. They show an important deterioration of diffuser performances with increasing inlet distortion. This is also confirmed by the experimental data indicated by circles.

Large inlet distortion is mostly encountered in high specific speed impellers with wide diffusers. The better performances of wider diffusers, predicted by the one dimensional method are not always confirmed in the real cases because of inlet distortion.

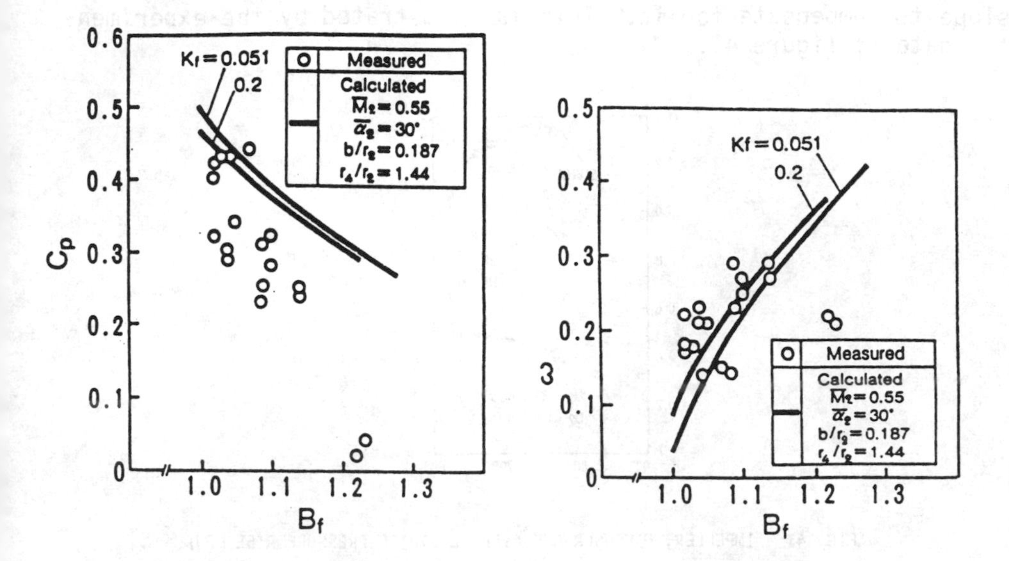

FIGS. 40a,b - INFLUENCE OF INLET DISTORTION ON DIFFUSER PERFORMANCES [38]

3.4 Stability of vaneless diffusers

When discussing stability, one must make a distinciton between rotating stall and surge [39].

Surge is a system instability depending on compressor and throttle characteristics and inlet and outlet geometry. It consists of large mass flow oscillations through the whole compressor and results in violent vibrations and an increasing noise level.

Theoretical studies by Emmons et al. [40], Taylor [41] and Greitzer [42] have revealed that surge will occur when the slope of the overall pressure rise curve exceeds a given positive value.

This value is a function geometry and throttle characteristics and decreases with increasing pressure ratio. A good approximation of this stability criterion is therefore

$$\frac{dP}{d\dot{m}} = 0$$

or surge will occur at maximum overall pressure rise.

The compressor overall pressure rise is the sum of impeller and diffuser pressure rise. The last one is shown in figure 31 and has a positive slope at all inlet flow angles. This means that stable operation is only possible if the impeller has a negative slope to compensate for it. This is illustrated by the experimental data of figure 41.

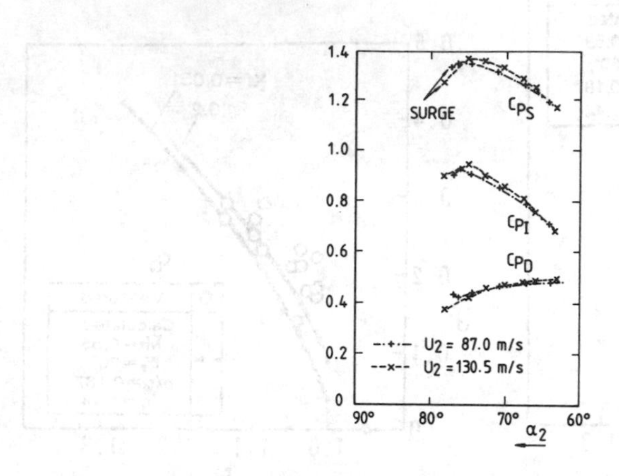

FIG. 41 - IMPELLER, DIFFUSER AND OVERALL STATIC PRESSURE RISE [39]

In a radial bladed impeller the negative slope will occur on the right side of the maximum efficiency point and only a small range of stable operations is possible. In a backward bladed impeller, the negative slope range of the impeller can be extended left of the maximum efficiency point because the work input increases with decreasing mass flow. At smaller mass flows the positive slope of the diffuser pressure rise curve (Fig. 31) becomes very large and cannot be compensated by the impeller pressure rise slope so that surge will occur anyway.

This critical value of α_2 is not well defined but decreases with decreasing diffuser width. It is therefore very doubtful if surge can be avoided by decreasing the diffuser width because this results in a decrease of α_2, similar to the critical value.

Rotating stall in a vaneless diffuser is a local instability of the diffuser flow and depends on local flow conditions. It consists of zones of low energy fluid, which rotate in the diffuser at subsynchronous speed. It does not influence very much the overall performances but creates unsteady forces. These forces create vibrations, which depending on geometry and flow conditions, can prevent a normal operation of the compressor.

Based on a stability analysis, Jansen [43] concluded that diffuser rotating stall will occur when there is local return flow in the diffuser. Using the method described in the previous chapter, Senoo et al. have calculated the inlet flow conditions for which there will be rotating stall, because of local return flow in the boundary layer. Results are summarized in figure 42

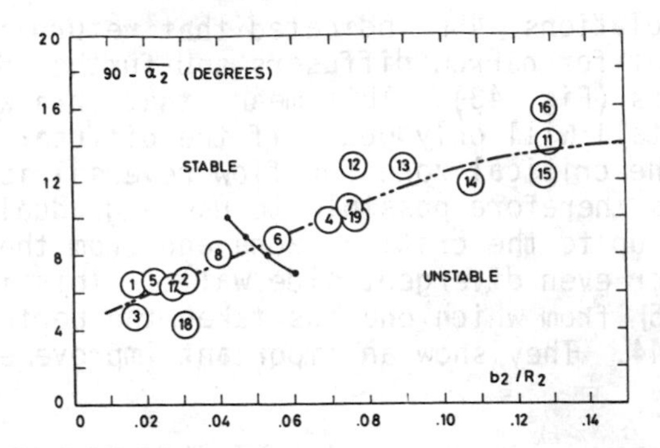

FIG. 42 - VARIATION OF CRITICAL INLET FLOW ANGLE WITH DIFFUSER WIDTH [39]

and show a dependence of the critical inlet flow angle α_C on relative diffuser width. Rotating stall will occur when the inlet flow angle becomes larger than the critical value.

The numbered circles in figure 42 represent experimental values of diffuser critical inlet flow angle, after they have been corrected for Reynolds number effects.

This critical inlet flow angle puts a limitation on maximum diffuser width. It was shown (Fig. 30) that, for a given mass flow, wider differences have lower losses but have a larger inlet flow angle. Increasing the diffuser width too much will result in rotating stall because the flow angle will become larger than the critical one. This is illustrated by the black points in figure 42, indicating a typical variation of diffuser inlet flow angle with diffuser width.

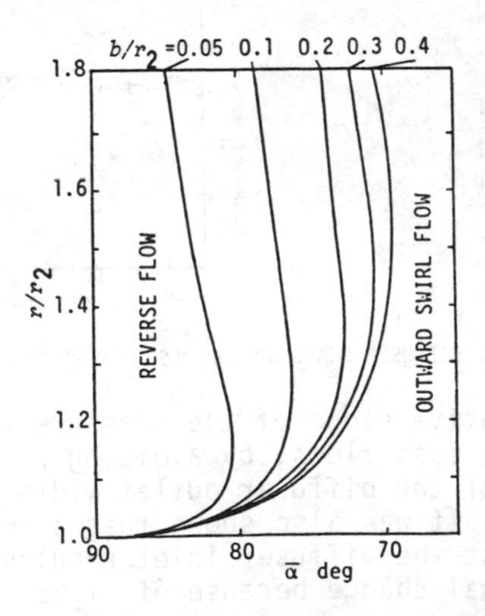

FIG. 43 - VARIATION OF CRITICAL INLET FLOW ANGLE WITH RADIUS RATIO [45]

Further calculations [45] indicated that return flow occurs closer to the inlet for narrow diffusers and further downstream for wider diffusers (Fig. 43). This means that in a wider diffuser, rotating stall will only occur if the diffuser is long enough, because the critical zone for flow reversal is further downstream. It is therefore possible to use a gradual decrease of diffuser width up to the critical zone and from there on, one can use parallel or even divergent side walls. This is clearly illustrated in [46] from which one has taken the performance curves of figure 44. They show an important improvement in surge

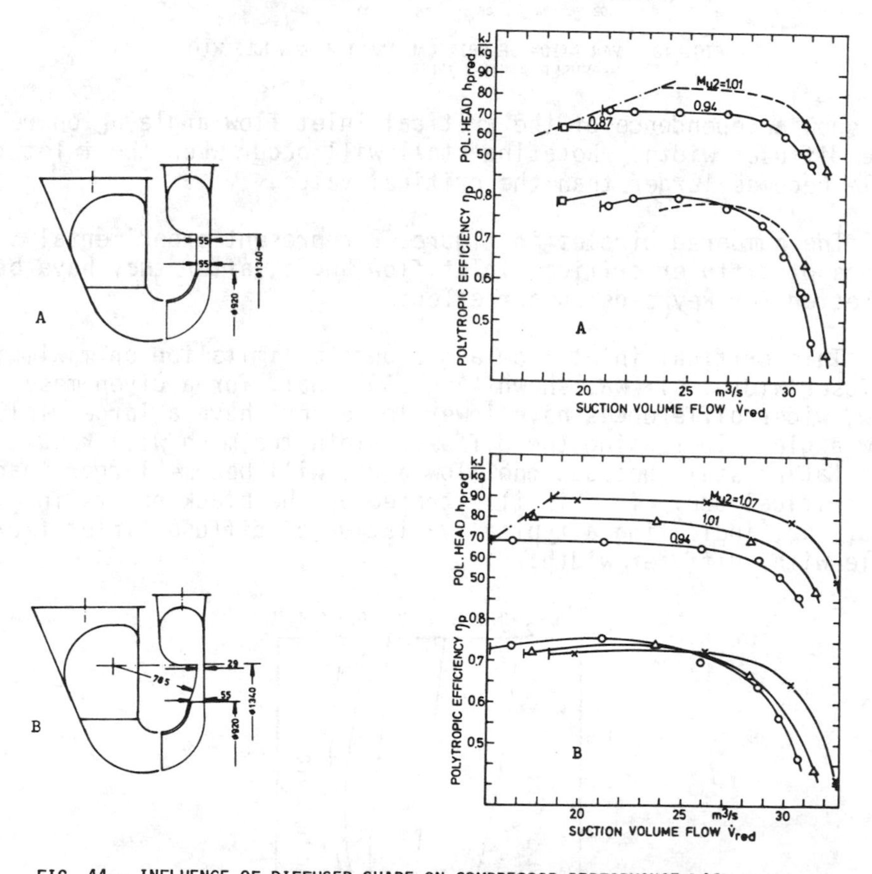

FIG. 44 - INFLUENCE OF DIFFUSER SHAPE ON COMPRESSOR PERFORMANCE [46]

limit because the negative slope of the pressure rise curve could be extended to smaller mass flows, by avoiding rotating stall. However, a reduction of the diffuser outlet width means also lower static pressure rise. It was also shown that a reduction of the diffuser width, right at the diffuser inlet, results in lower performances than a gradual change because of higher friction losses at larger mass flow.

4. VANED DIFFUSERS

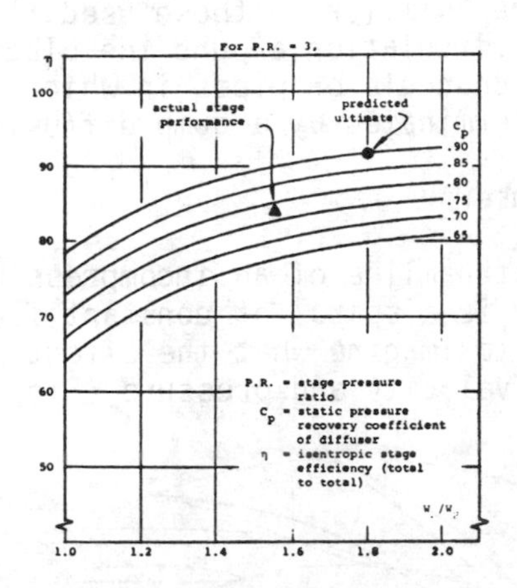

FIG. 45 - INFLUENCE OF DIFFUSER STATIC PRESSURE RISE ON EFFICIENCY[45]

The kinetic energy available at the diffuser inlet easily amounts to more than 50% of the total energy added by the impeller. An efficient transformation of this energy into pressure is therefore an important part of component design. The influence of diffuser static pressure rise on stage efficiency has been calculated by Dean [13] and is shown in figure 45. Increasing the diffuser static pressure rise coefficient by .1 results in an efficiency increase by 5 points.

A typical static pressure rise coefficient for a vaneless diffuser is about .5. This low value is due to the fact that vaneless diffusers accomplished diffusion by an inverse reduction of swirl flow with radius for conservation of angular momentum. The radial distance over which a sufficient reduction is achieved corresponds to a very long flow path because of the large inlet flow angles α. The flow in a vaneless diffuser with radius ratio 2, and inlet flow angle of 76°, will rotate over 400° before leaving the diffuser [30]. This results in high friction losses and the real pressure rise will be much lower than the ideal one.

$$C_{p_{real}} = C_{p_{ideal}} - \omega$$

One way to overcome this problem is by shortening the flow path over which the diffusion takes place by means of vanes. This also allows to create an outlet over inlet area ratio which is larger than R_4/R_2. Vanes have also a favourable effect on the three dimensional boundary layers because they create a pressure gradient which is better aligned with the main velocity vector.

Two types of vaned diffusers can be distinguished. One is composed of curved vanes similar to those used in cascades, and the flow is turned by the circulation around the blades. The other one consists of divergent channels or pipes in which the flow is decelerated, and sometimes terminated by a dump diffusion.

4.1 Curved vane diffusers

The undisturbed streamline of an incompressible inviscid flow in a vaneless diffuser is a spiral of constant flow angle α. It is therefore not easy to imagine what the effect of a geometry change will be on the velocity and pressure (Fig. 46)

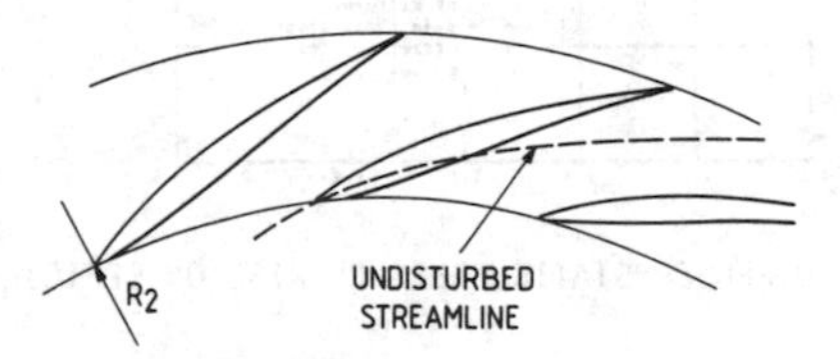

FIG. 46 - CURVED VANE DIFFUSER

A method which is well adapted to analyse such flows is the singularity method. The inlet flow conditions are easily generated by a source Q and vortex Γ at the center

$$Q = V_{m_2} \cdot 2\pi R_2$$

$$\Gamma = V_{u_2} \cdot 2\pi R_2$$

The influence of the blades is generated by vortices on the contour. Such a method becomes more complicated when the flow is compressible or where the walls are not parallel. It also neglects viscous effects and certainly secondary flow effects.

A second method is based on conformal transformation and makes use of experimental data available for cascades. This would make it possible

a) to achieve high diffuser efficiency;
b) provide off-design performances (stall and choke limit);
c) minimizing the development time.

The transformation is indeed very simple for the incompressible case

$$x = \ell n\ r$$

$$y = \theta \text{ (radians)}$$

The velocity components in the cascade and the diffuser are related by :

$W_u = W_y/R$

$W_R = W_x/R$

In case of compressible flow the transformation requires a correction in function of Mach number and is somewhat more complex.

Such a method has been used with success by Pampreen [47] using up to three rows of vanes. Similar work by Senoo et al. [48] for low solidity tandem cascades confirmed that a very large flow range between surge and choke can be achieved. This is due to the first blade row which at off design condition turns the flow to a direction which is optimum for the second row, in which the diffusion is achieved. The disadvantages are related to :
- the large number of vanes;
- blades are at very high stagger angles (α = 70°) for which no experimental data in cascades are available;
- the flow must be subsonic which requires sometimes a larger vaneless space.

4.2 Channel diffusers

Better suited to transonic flow (high pressure ratio compressors) are vaned diffusers in which the flow is diffused in divergent channels. These channels are defined by wedge type blades between parallel walls or by conical pipes (pipe diffusers)

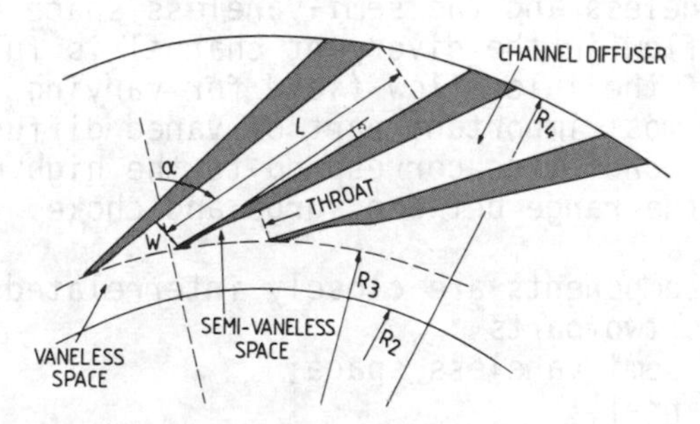

FIG. 47 - DEFINITION OF VANED ISLAND DIFFUSERS

They consist of three main elements :
a) a vaneless space just downstream of the rotor
b) the leading edge region or semi-vaneless space
c) the diffuser passage itself starting at the throat section and characterized by the outlet to throat area ratio.

At low exit pressure the throat section will be choked and the flow in the divergent channel will accelerate supersonically and then decelerate by shock. This results in flow separation and high losses. At increasing back pressure the shock will move forward. The shock will occur at lower Mach number and the corresponding

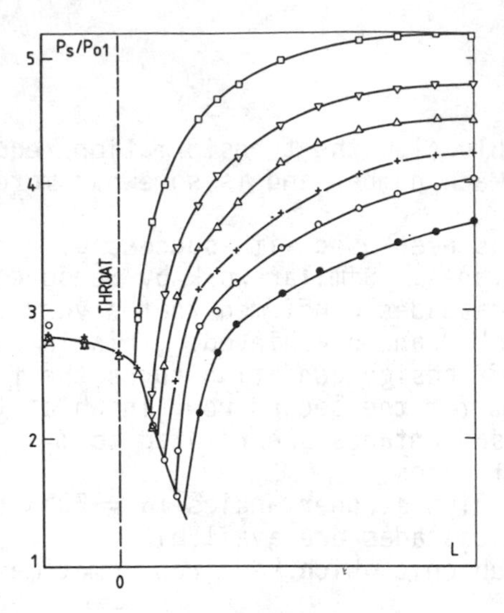

FIG. 48 - PRESSURE DISTRIBUTION IN DIFFUSER CHANNEL

losses will be lower. Up to now, the flow conditions upstream the throat are unchanged and the mass flow is constant (vertical performance curve).

The most interesting part is when the throat is unchoked. The flow in the vaneless and the semi-vaneless space changes with mass flow and the flow in the divergent channel is fully subsonic. The optimization of the inlet flow field for varying incidence and Mach number is the most important part of vaned diffuser design because these flow conditions correspond to the high efficiency points and define the range between surge and choke.

Although the components are closely interrelated the study is normally split into two parts :
- the vaneless and semi vaneless space;
- the divergent channel.
The interference between both components is expressed in terms of throat Mach number and blockage.

4.2.1 Vaneless and semi vaneless space. The flow in the vaneless and semi vaneless space is a very complex one. The non uniform unsteady flow leaving the impeller (see vaneless diffusers) is influenced by the diffuser vanes. The constant pressure lines change rapidly from almost circumferential to perpendicular on the velocity at the throat section. As shown by Krain [49] such a flow is very unsteady and the instantaneous flow conditions depend on the relative position between impeller and diffuser vanes.

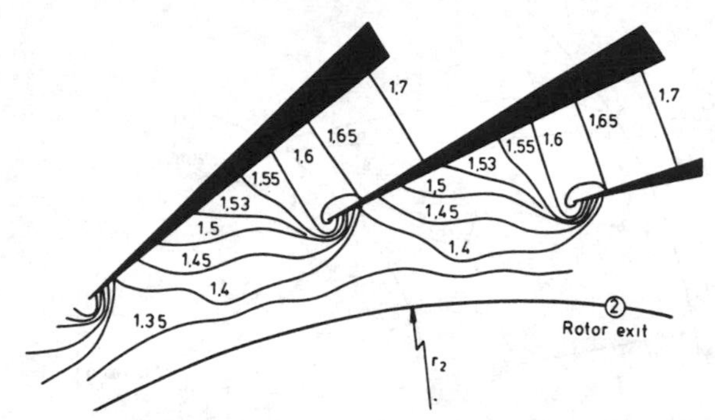

FIG. 49 - PRESSURE DISTRIBUTION IN VANELESS AND SEMI-VANELESS SPACE [49]

The geometry is defined by the impeller exit to the diffuser leading edge radius ratio (R_3/R_2), the tangent at the leading edge, blade number and blade suction side shape.

Three types of losses exist in this region :
a) Friction losses on the side walls and vane suction side. They include jet and wake mixing losses, as discussed for vaneless diffusers, and increase with radius ratio.
b) Shock and shock boundary layer interaction losses in the case of transonic flow. They are increasing with local Mach number and are therefore function of radius ratio and suction side shape.
c) Unsteady flow losses due to incidence changes on the vane when the flow is not yet fully mixed out at the leading edge.

Transonic and unsteady flow losses can be avoided by increasing the radius ratio R_3/R_2 upto values where the flow is subsonic and fully mixed out. This results in larger friction losses and throat blockage, which has a detremental effect on the pressure rise in the the downstream channel. A large vaneless space is also unstable as shown for vaneless diffusers.

No clear criterium is available to decide on optimum diffuser leading edge radius ratio. Normally a value of $R_3/R_2 = 1.05$ is accepted. This results in a moderate throat blockage but relatively high Mach number. This requires an optimization of the blade shape to minimize the shock losses. Came and Herbert [50] calculated that the maximum angular change between leading edge and throat section is defined by figure 50

$2\pi/Z'-\lambda-K$

λ is the vane leading edge angle for which the minimum value is defined by stress considerations.
K is the diffuser channel opening angle at the throat and the minimum values are defined at 1° to allow for a boundary layer growth.

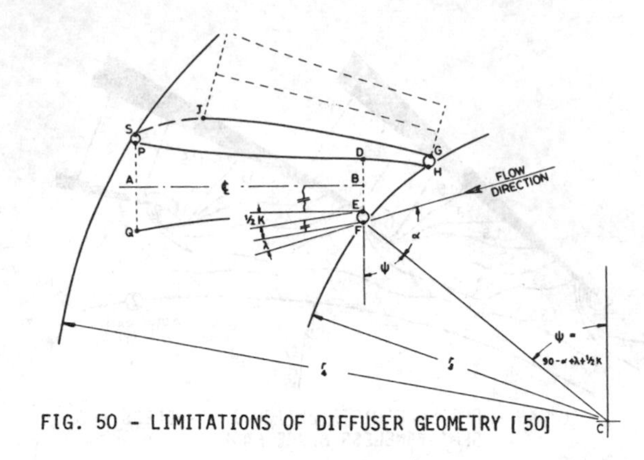

FIG. 50 - LIMITATIONS OF DIFFUSER GEOMETRY [50]

He uses a cubic contour to connect the leading edge and throat section in a smooth way.

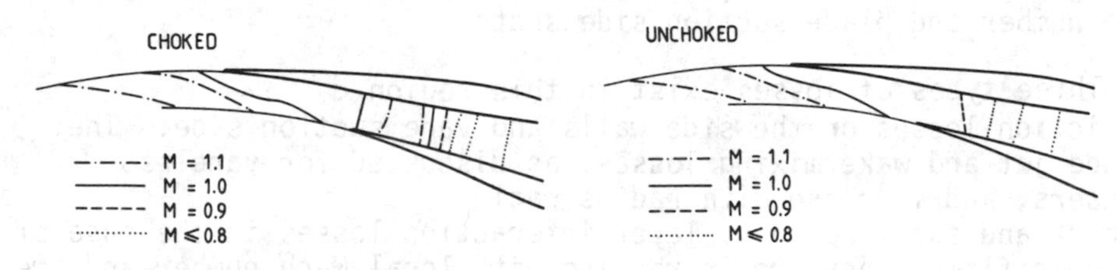

FIGS. 51a,b - CALCULATED ISOMACH LINES

The optimization of the diffuser inlet section is also possible by means of a transonic calculation method adapted for axisymmetric inlet conditions [51,52]. Typical results of such calculations are shown in figures 51. The uniform inlet conditions are defined from the average rotor outlet conditions. They allow to optimize the blade shape by assuming that the local Mach number is nowhere significantly increased above the leading edge value. As shown by Verdonk [51] such calculations require an iterative procedure in which the boundary layer blockage is calculated in function of the velocity distribution, and the local channel width is corrected in function of boundary layer blockage. This also allows to predict throat blockage and Mach number.

Kenny [53] has correlated throat blockage to the static pressure rise between vane leading edge and throat (Fig. 52) and shows an influence of inlet Mach number. The throat Mach number is calculated from continuity.

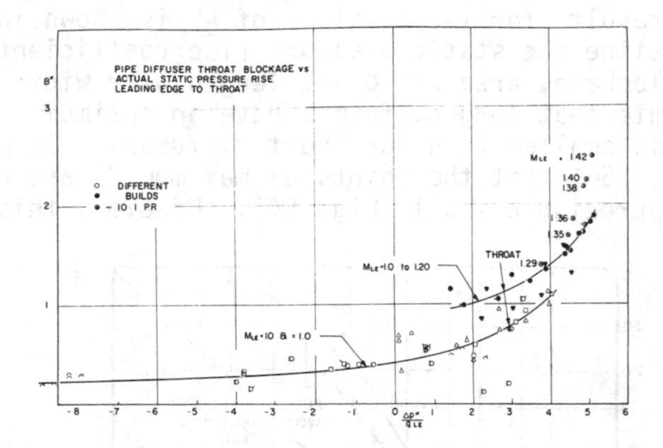

FIG. 52 - THROAT BLOCKAGE VS STATIC PRESSURE RISE [53]

4.2.2 Divergent channel. The flow in channel diffusers has been studied by Reneau et al. [54], Runstadler et al. [55] etc.

They found a dependence of pressure rise coefficient on :

- aspect ratio $AS = b/W$
- area ratio $AR = A_{out}/A^{\star}$
- length over width ratio $LWR = L/W$
- inlet Mach number $M^{\star}$
- inlet blockage $B^{\star}$
- Reynolds number
- cross sectional shape (circular or rectangular).

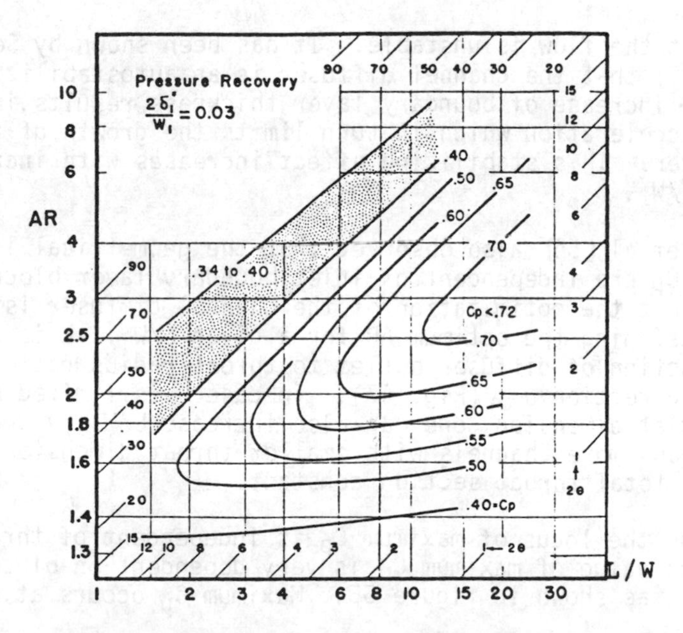

FIG. 53 - STATIC PRESSURE RISE VERSUS CHANNEL GEOMETRY [54]

A typical result for large values of AS is shown in figure 53 and allows to define the static pressure rise coefficient in function of inlet blockage, area ratio and length over width ratio. They also indicate that long diffusers have an optimum divergence angle 2θ which is smaller than for short diffusers. It was observed by Reneau et al. [54] that the points of maximum C_p are close to the conditions of appreciable stall (Fig. 54). However, this does

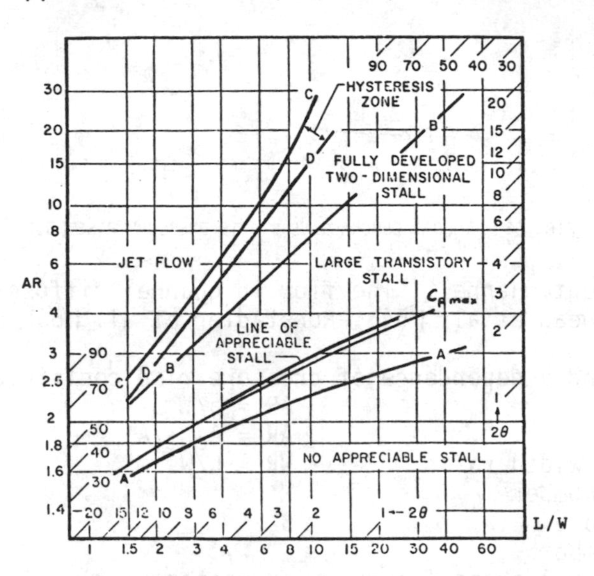

FIG. 54 - DIFFERENT FLOW CONDITIONS IN DIFFUSER CHANNEL [55]

not mean that the flow is unstable. It has been shown by Senoo and Nishi [56] that the channel diffuser is an autostabilizing system. Any increase of boundary layer thickness results in a local flow acceleration which in turn limits the growth of the boundary layer. This stabilizing effect increases with increasing values of δ^*/W_1.

Reneau et al [54] also observed that the geometrical locations for maximum C_p are independent of inlet boundary layer blockage. This means that the optimization of the channel diffuser is reduced to defining the optimum AR for a given L/W_1. L is a geometrical function of diffuser outlet to throat radius ratio and throat flow direction α_{th} (Fig. 47). In case L is limited by the diffuser outlet dimension, one can also increase L/W_1 by decreasing W_1. This means more channels with smaller throat dimension to maintain the total throat section constant.

Although the locus of maximum C_p is independent of throat blockage, the value of maximum C_p is very dependent on blockage and aspect ratio, as shown in figure 55. Maximum C_p occurs at AS = 1

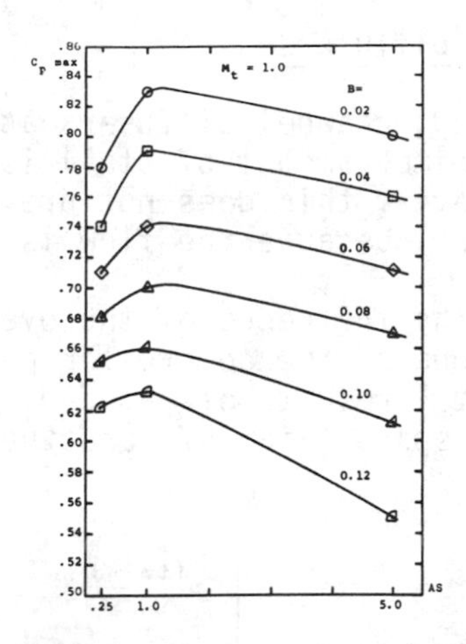

FIG. 55 - VARIATION OF MAXIMUM STATIC PRESSURE RISE [13]

and decreases rapidly for lower AS values. Larger AS values have less influence on maximum Cp, because optimum AS will occur further downstream because of the divergence of the diffuser channel.

Pipe diffusers show a higher static pressure rise than channel diffusers. This is mainly due to the lower friction losses because of a better ratio of wetted surface/cross section, and because the optimum aspect ratio is maintained up to the diffuser exit. Channel diffuser pressure recovery can be improved too, by using divergent side walls for the diffuser such that the optimum AS is conserved throughout the channel [50] .

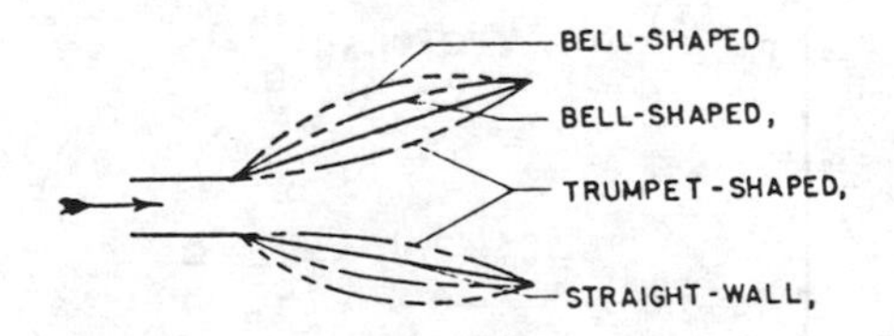

FIG. 56 - DIFFERENT SHAPES OF THE DIFFUSER CHANNEL [58]

Optimization of channel diffusers, by contouring the walls, has also been attempted (Fig. 56). Based on boundary layer considerations, Huo [57] concluded that losses are minimized when the flow is first decelerated to conditions close to separation, and then continued towards the exit with lower divergence. This corresponds to the bell-shaped diffuser tested by Carlson et al [58] , who showed a small improvement in doing so. However, this brings the flow close to separation, and a small change in inlet flow conditions can cause a large decrease of performance. Based on stability considerations, it was recommended [50] to use a trumpet shaped channel, because of the small influence on performances.

4.3 Stability of vaned diffusers

As already mentioned, channel diffusers achieve maximum static pressure rise when a limited amount of stall is present in the divergent channel. However, this does not prevent steady state operation of the diffuser, because the flow is auto-stabilizing [56].

More important is the influence of the overall diffuser static pressure rise on the slope of the compressor performance curve to avoid surge. Experimental results of Baghdadi [59] show the influence of the vaneless space (Fig. 57) and the divergent channel

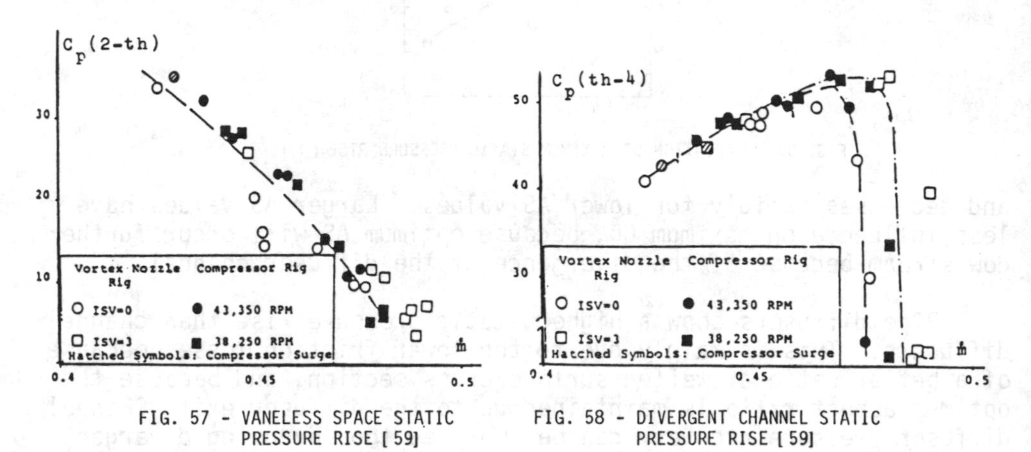

FIG. 57 - VANELESS SPACE STATIC PRESSURE RISE [59]

FIG. 58 - DIVERGENT CHANNEL STATIC PRESSURE RISE [59]

(Fig. 58) on the total static pressure rise in the diffuser (Fig. 59).

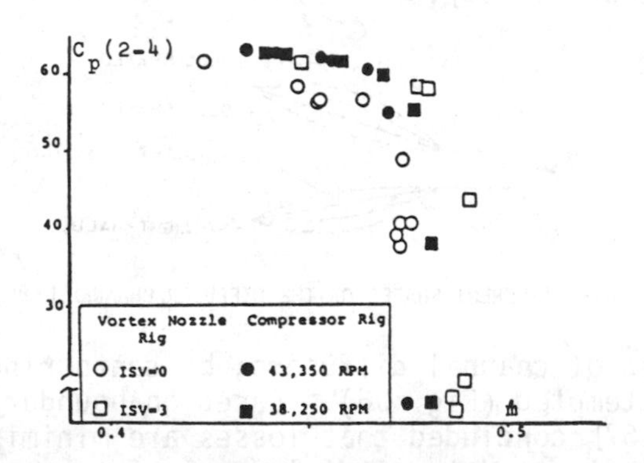

FIG. 59 - OVERALL STATIC PRESSURE RISE IN A VANED DIFFUSER [59]

The vaneless and semi vaneless space pressure rise curve show a negative slope ($\partial \Delta P_{2\text{-th}}/\partial \dot{m}<0$) and is therefore a stabilizing component. However, increasing static pressure rise in the semi vaneless space results in higher throat blockage and is therefore responsable for the positive slope of the channel static pressure rise ($\partial \Delta P_{th\text{-}4}/\partial \dot{m}>0$).

In the absence of any discontinuity in the performance curve, surge can be expected at the maximum of Δp_{2-4}. The main task in achieving maximum range between surge and choke is to increase the range with positive slope for Δp_{2}-th and at the same time, to control the throat blockage. This can be done by minimizing the pressure rise Δp_{2}-th at choking and by decreasing the mass flow where maximum Δp_{2}-th occurs.

When the diffuser inlet flow is subsonic, low static pressure rise and hence minimum throat blockage can be achieved at choking mass flow, because the flow accelerates to M = 1 at the throat. This results in a negative Δp_{2}-th and according to figure 52 in minimum blockage.

When the diffuser inlet flow is supersonic, deceleration to M = 1 at the throat is necessary, because this is the maximum Mach number at choking. Minimum blockage is therefore around 6 to 10% according to figure 52. The same figure has been replotted (Fig. 60) and shows that minimum throat blockage and static pressure rise are achieved at -2° to -4° negative incidence at the vane leading edge (Fig. 60).

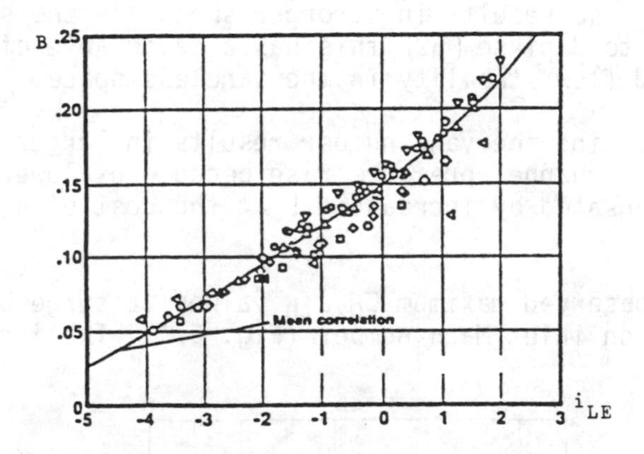

FIG. 60 - THROAT BLOCKAGE VERSUS INCIDENCE [60]

Decreasing the mass flow results in an increase of incidence to positive values. The corresponding increase in static pressure rise is limited by the stall incidence. The study of the detailed velocity distribution in the semi vaneless space, and the control of the local increase of Mach number, can help to increase the stall incidence and to decrease the surge mass flow. An experimental study is described by Bammert et al. [61].

A reduction of diffuser vane number is often mentioned as a possible method to decrease the surge mass flow. Although this rule is not always confirmed, there are some arguments to support it.

Came and Herbert [50] calculated a parameter

$$A_w/A_g = \frac{\text{wetted surface of the semi vaneless space}}{\text{geometrical throat section}}$$

which is proportional to hydraulic length over diameter. This value is increasing with decreasing number of vanes. It is reasonable to accept that larger values of A_w/A_g will allow a larger pressure rise similar to larger L/W_1 values in channel diffusers. A_w/A_g can also be increased by increasing the diffuser leading edge radius ratio (R_3/R_2). However, this does not have the same effect on surge limit because the stabilizing effect of the vanes is decreased.

Another argument is related to the change in suction side angle. Decreasing the number of vanes, requires a larger leading edge wedge angle to maintain the total throat section constant. This results in a more tangential suction side and lowers the flow incidence angle.

Fewer vanes also result in stronger shocks on the suction side. According to Japikse [62] this has a favourable effect on pressure rise and flow stability in the vaneless space.

However, reducing the vane number results in larger W_{th} values and a reduction in channel pressure rise because of lower L/W_1 values. This can be compensated by increasing L at the cost of a larger outlet diameter.

Kenny [63] observed maximum C_{p2-th} values at surge between .4 and .5 depending on inlet Mach number (Fig. 61). The increase of

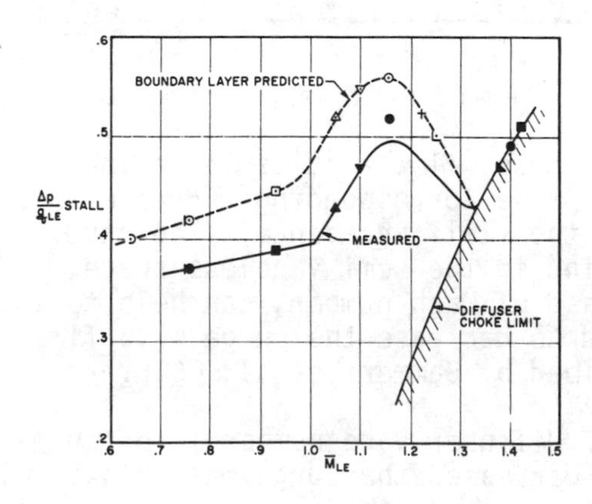

FIG. 61 - STALL STATIC PRESSURE RISE VERSUS INLET MACH NUMBER [63]

Δp at supersonic Mach numbers confirms the stabilizing effect of shocks mentioned by Japikse. Came & Herbert [50] observed surge at C_{p2-th} values between .4 and .5 (calculated by his off design performance program). They also state that much lower values can be observed if other diffusing components such as crossover and return channels are responsable for surge.

5. COMPRESSOR RANGE AND STABILITY

Compressor range is defined by the change in mass flow between surge and choking at each speed line.

The choking limit can be defined for each component (impeller-diffuser) separately. And the overall compressor choking limit is defined by the smallest choking mass flow of the components.

According to the general theory [42], surge occurs when the slope of the overall pressure rise curve exceeds a minimum positive value. Rotating or steady stall can proceed to surge, but they influence the surge limit mainly by changing the slope of the pressure rise curve.

The impeller-diffuser matching method of Yoshinaka [64] does not directly refer to previous theory but allows to gain some insight on how the range can be influenced by the different components. For this purpose he plots the compressor characteristic lines in function of inlet flow coefficient ϕ_1 (abscissa) and diffuser inlet flow angle α_2 (ordinate). Figure 62 shows such a performance map

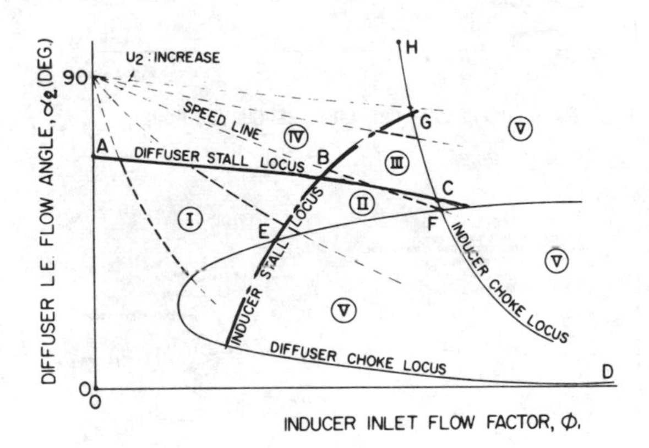

FIG. 62 - COMPRESSOR CHARACTERISTIC LINES [64]

on which the following lines have been plotted :
- Constant speed lines. They all converge at $\alpha_2 = 90°$ for zero mass flow. At high RPM they have higher α_2 values than at low RPM.
- The diffuser stall limit, defined by the critical static pressure rise criterium of Kenny [63].

- The diffuser choke limit is based on the assumption of sonic velocity at the throat section.
- The inducer stall limit, calculated from a critical value of static pressure rise coefficient between inducer leading edge and throat.
- The inducer choke limit, calculated from classical inducer choke predictions, assuming sonic velocity at the inducer throat.

The inducer choke and stall lines show a decreasing operation range with increasing RPM similar to figure 5. Also the diffuser shows a decreasing incidence range for increasing speed.

The experimental data of a high pressure ratio centrifugal compressor (Fig. 63) at zero preswirl are superposed on such a presentation and are shown in figure 64. The lines are

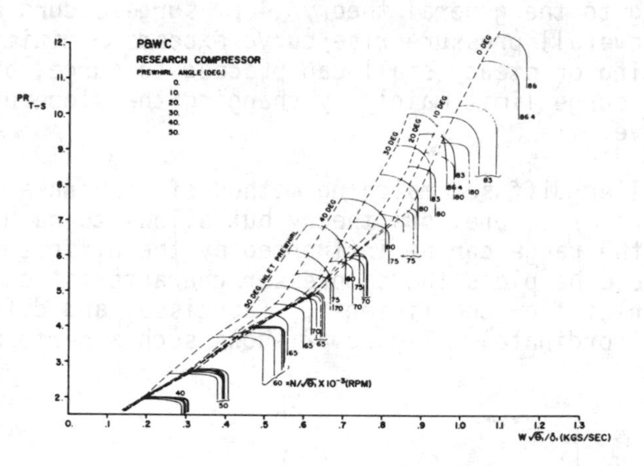

FIG. 63 - COMPRESSOR PERFORMANCE CURVES [64]

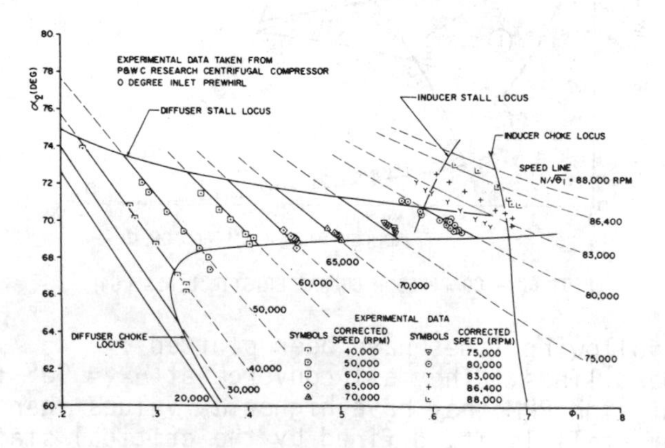

FIG. 64 - REPLOTTED PERFORMANCE CURVES (ZERO PREROTATION) [64]

predictions by means of previous criteria. The points are experimental data. This figure suggests that surge occurs when both impeller and diffuser are stalled. The diffuser tends to control compressor surge below design speed and the inducer controls surge at overspeed conditions. This observation is very well respected, except for the speed lines at 65000 and 75000 where the range is smaller than predicted.

Such a ϕ_1 versus α_2 diagram is a simple and useful tool to identify the effect of a component change on range limitations.

The use of a 40° prerotation at compressor inlet will shift the inducer stall and choke locus to smaller ϕ_1 values. As shown in figure 65 this results in an increased range at 60 000 to 75 000 RPM and is confirmed by experimental data.

A change in diffuser vane setting and throat section will shift the diffuser stall and choke limit to higher or lower α values and will have an unfavourable or favourable effect on range.

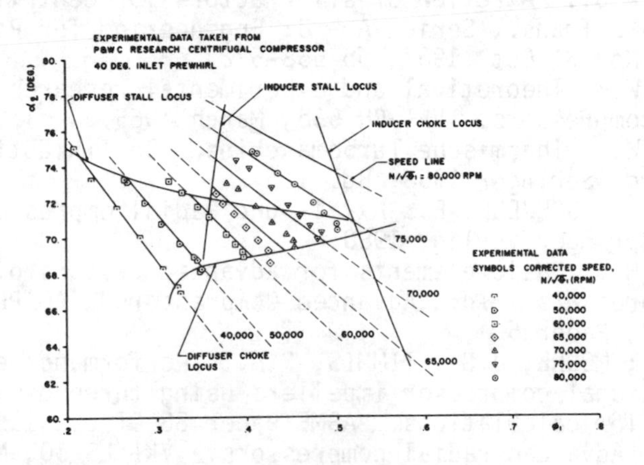

FIG. 65 - REPLOTTED PERFORMANCE CURVES (40° PREROTATION) [64]

Maximum range will be obtained for a given speed line, when the diffuser choke limit and inducer choke limit cross each other on the speed line.

However, one should keep in mind that all operating points above the diffuser stall locus are operating with a stalled diffuser which will have its consequences on efficiency.

REFERENCES

1. RODGERS, C.: Specific speed and efficiency of centrifugal impellers. Proceedings of the Symposium on "Performance Prediction of Centrifugal Pumps and Compressors", ASME Gas Turbine Confs., New Orleans, March 1980, pp 191-200.
2. BALJE, O.E.: A study on design criteria and matching of turbomachines: Part B - Compressor and pump performance and matching of turbocomponents. ASME Trans., Series A - J. Engineering for Power, Vol. 84, No. 1, Jan. 1962, pp 103-114.
3. BRYCE, C.A. et al.: Advanced two-stage compressor program - Design of inlet state. NASA CR 120 943, 1973.
4. MORRIS, R.E. & KENNY, D.P.: High pressure ratio centrifugal compressors for small gas turbine engines. In: "Helicopter Propulsion Systems", AGARD CP 31, June 1968, paper 6.
5. STIEFEL, W.: Experience in the development of radial compressors. In: "Advanced Radial Compressors", VKI LS 50, May 1972.
6. RODGERS, C.: Typical performance characteristics of gas turbine radial compressors. ASME Trans., Series A - J. Engineering for Power, Vol. 86, No. 2, April 1962, pp 161-175.
7. WIESNER, F.J.: A review of slip factors for centrifugal impellers. ASME Trans., Series A - J. Engineering for Power, Vol. 89, No. 3, Oct. 1967, pp 558-572.
8. STIEFEL, W.: Theoretical and experimental research on limit loading compressors. VKI CN 53b, March 1965.
9. TRAUPEL, W.: Thermische Turbomaschinen. Berlin/Göttigen/Heidelberg, Springer 1958, Bd. 1.
10. ECKERT, B. & SCHNELL, E.: Axial- und Radialkompressoren. Berlin, Springer Verlag, 1980.
11. VAVRA, M.H.: Basic elements for advanced designs of radial flow compressors. In: "Advanced Compressors", AGARD LS 39, May 1970, Paper 6.
12. MOORE, J.; MOORE, J.G.; TIMMIS, P.H.: Performance evaluation of centrifugal compressor impellers using three dimensional viscous flow calculations. ASME Paper 83 GT 63, 1983.
13. DEAN, R.: Advanced radial compressors. VKI LS 50, May 1972.
14. ECKARDT, D.: Detailed flow investigations within a high speed centrifugal compressor impeller. ASME Trans., Series I - J. Fluids Engineering, Vol. 98, No. 3, Sept. 1976, pp 390-402.
15. DALLENBACH, F.: The aerodynamic design and performance of centrifugal and mixed flow compressors. SAE Tech. Progres Series, Vol. 3, 1961, pp 2-30.
16. YOUNG, L.: Discussion on "Impeller stalling as influenced by diffusion limitations" by C. Rodgers. ASME Trans., Series I, J. Fluids Engineering, Vol. 99, No. 1, March 1977, pp 94-95.
17. FRIGNE, P. & VAN DEN BRAEMBUSSCHE, R.: One dimensional design of centrifugal compressors taking into account flow separation in the impeller. VKI TN 129, June 1979.

18. JANSEN, W.: A method for calculating the flow in a centrifugal impeller when entropy gradients are present. Inst. Mech. Engrs., Internal Aerodynamics (Turbomachinery), 1970, Paper 12.
19. WALITT, L.: Numerical analysis of the three dimensional viscous flow field in a centrifugal impeller. In: Centrifugal Compressors, Flow Phenomena and Performance, AGARD CP 282, Nov. 1980.
20. ADLER, D.: Status of centrifugal impeller internal aerodynamics. Part 1 - Inviscid flow prediction methods. ASME Trans., Series A, J. Engineering for Power, Vol. 102, No. 3, July 1980, pp 728-737.
21. HERBERT, M.V.: A method of centrifugal compressor performance prediction. Proc. ASME Symposium on "Performance Prediction of Centrifugal Pumps and Compressors", New Orlean, March 1980, pp 171-184.
22. DAVIES, R.C. & DUSSOURD, J.L.: A unified procedure for the calculation of off-design performance of radial turbomachinery. ASME P 70 GT 64, 1970.
23. STRATFORD, B.S. & BEAVERS, G.S.: The calculation of the compressible turbulent boundary layer in an arbitrary pressure gradient - a correlation of certain previous methods. ARC R&M 3207, 1959.
24. CIPOLLONE, R.: Viscous flow calculation in a centrifugal impeller. VKI PR 1984-24, June 1984.
25. CASEY, M.V.: A computational geometry for the blades and internal flow channels of centrifugal compressors. ASME Paper 82-GT-155, 1982.
26. MOORE, J.: Eckard's impeller - A ghost from ages past. U. of Cambridge, Dept. of Eng., CUED/A Turbo TR 83, 1976.
27. STANITZ, J.D. & PRIAN, V.D.: A rapid approximation method for determining velocity distribution of impeller blades of centrifugal compressors. NACA TN 2421, July 1951.
28. JOHNSON, J.P.: The effects of rotation on boundary layers in turbomachine rotors. In: "Fluid Mechanics, Acoustics and Design of Turbomachinery, NASA SP 304, 1974.
29. VAN DEN BRAEMBUSSCHE, R. & ZUNINO, P.: Correction for streamline curvature and Coriolis force in a boundary layer integral method. ASME Paper 81 GT 97, 1981.
30. STANITZ, J.D.: One dimensional compressible flow in vaneless diffusers of radial and mixed flow centrifugal compressors, including effects of friction, heat transfer and area change. NASA TN 2610, 1952.
31. DEAN, R.C. & SENOO, Y.: Rotating wakes in vaneless diffusers. ASME Trans., Series D - J. Basic Engineering, Vol. 82, No. 3, Sept. 1960, pp 563-574.
32. VAN DEN BRAEMBUSSCHE, R.: Flow modelling and aerodynamic design techniques for centrifugal compressors. In: "Industrial Centrifugal Compressors", VKI LS 95, Feb. 1977.
33. JOHNSTON, J.P. & DEAN, R.C.: Losses in vaneless diffusers of centrifugal compressors and pumps. ASME Trans., Series A - J. Engineering for Power, Vol. 88, No. 1, Jan. 1966, pp 49-62.

34. SENOO, Y. & ISHIDA, M.: Behaviour of severely asymmetric flow in a vaneless diffuser. ASME Trans., - Series A - J. Engineering for Power, Vol. 97, No. 3, July 1975, pp 375-387.
35. JANSEN, W.: Steady fluid flow in a radial vaneless diffuser. ASME Trans., Series D - J. Basic Engineering, Vol. 86, No. 3, Sept. 1964, pp 607-619.
36. SENOO, Y.; KINOSHITA, Y.; ISHIDA, M.: Asymmetric flow in vaneless diffusers of centrifugal blowers. ASME Trans., Series I, J. Fluids Engineering, Vol. 99, No. 1, March 1977, pp 104-114.
37. REBERNIK, B.: Investigation of induced vorticity in vaneless diffusers of radial flow pumps. Proc. 4th Conf., on Fluid Machinery, Budapest, Sept. 1972, pp 1129-1139.
38. SENOO, Y.: Vaneless diffusers. In "Flow in Centrifugal Compressors, VKI LS 1984-07, May 1984.
39. VAN DEN BRAEMBUSSCHE, R.: Surge and stall in centrifugal compressors. In "Flow in Centrifugal Compressors", VKI LS 1984-07, May 1984.
40. EMMONS, H.W.; PEARSON, C.E.; GRANT, H.P.: Compressor surge and stall propagation. ASME Trans., Vol. 77, No. 2, April 1955, pp 455-469.
41. TAYLOR, E.: The centrifugal compressors. In " Aerodynamics of Turbines and Compressors", Vol. X in "High Speed Aerodynamics and Jet Propulsion", Princeton Univ. Press, 1964, pp 553-586.
42. GREITZER, E.M.: The stability of pumping systems. The 1980 Freeman scholar lecture. ASME Trans., Series I - J. Fluids Engrg., Vol. 103, No. 2, June 1981, pp 193-242.
43. JANSEN, W.: Rotating stall in radial vaneless diffusers. ASME Trans., Series D - J. Basic Engineering, Vol. 86, No. 4, Dec. 1964, pp 750-758.
44. SENOO, Y. & KINOSHITA, Y.: Influence of inlet flow conditions and geometries of centrifugal vaneless diffusers on critical flow angles for reverse flow. ASME Trans., Series I - J. Fluids Engrg., Vol. 99, No. 1, March 1977, pp 98-103.
45. SENOO, Y. & KINOSHITA, Y.: Limits of rotating stall and stall in vaneless diffusers of centrifugal compressors. ASME Paper 78-GT-19, 1978.
46. LÜDTKE, K.: Aerodynamic tests on centrifugal process compressors. The influence of the vaneless diffuser shape. ASME Trans., Series A - J. Engineering for Power, Vol. 105, No. 4, Oct. 1983, pp 902-909.
47. PAMPREEN, R.C.: The use of cascade technology in centrifugal compressor vaned diffuser design. ASME Paper 72-GT-39, 1972.
48. SENOO, Y.; HAYAMI, H.; UEKI, M.: Low-solidity tandem-cascade diffusers for wide-flow-range centrifugal blowers. ASME Paper 83-GT-3, 1983
49. KRAIN, H.: A study on centrifugal impeller and diffuser flow. ASME Trans., Series A - J. Engineering for Power, Vol. 103, No. 4, Oct. 1981, pp 688-697.

50. CAME, P.M. & HERBERT, M.V.: Design and experimental performance of some high pressure ratio centrifugal compressors. In: "Centrifugal Compressors, Flow Phenomena and Performance", AGARD CP 282, Nov. 1980, paper 15.
51. VERDONK, G.: Vaned diffuser inlet flow conditions for a high pressure ratio centrifugal compressor. ASME Paper 78-GT-50, 1978.
52. TEIPEL, I. & WIEDERMANN, A.: The influence of different geometries for a vaned diffuser on the pressure distribution in a centrifugal compressor. ASME Paper 84-GT-68, 1984.
53. KENNY, D.: Supersonic radial diffusers. In "Advanced Compressors", AGARD LS 39, May 1970.
54. RENEAU, L.R.; JOHNSTON, J.P.; KLINE, S.J.: Performance and design of straight, two-dimensional diffusers. ASME Trans., Series D - J. Basic Engineering, Vol. 89, No. 1, March 1967, pp 141-150.
55. RUNSTADLER, P.W. & DEAN, R.C.: Straight channel diffuser performance at high inlet Mach numbers. ASME Trans., Series D - J. Basic Engineering, Vol. 91, No. 3, Sept. 1969, pp 397-422.
56. SENOO, Y. & NISHI, M.: Prediction of flow separation in a diffuser by a boundary layer calculation. ASME Trans., Series I - J. Fluids Engineering, Vol. 99, No. 2, June 1979, pp 379-389.
57. HUO, S.: Optimization based on boundary layer concept for compressible flow. ASME Paper 74-GT-63, 1974.
58. CARLSON, J.J.; JOHNSTON, J.P.; SAGI, C.J.: Effect of wall shape on flow regimes and performance in straight, two-dimensional diffusers. ASME Trans., Series D - J. Basic Engineering, Vol. 89, No. 1, March 1967, pp 151-159.
59. BAGHDADI, S.: The effect of rotor blade wakes on centrifugal compressor diffuser performance - A comparative experiment. ASME Trans., Series I - J. Basic Engineering, Vol. 99, No. 1, March 1977, pp 45-52.
60. KENNY, D.P.: A novel correlation of centrifugal compressor performance for off-design prediction. AIAA Paper 79-1159, 1979.
61. BAMMERT, K.; JANSEN, M.; RAUTENBERG, M.: The influence of the diffuser inlet shape on the performance of a centrifugal compressor stage. ASME Paper 83-GT-9, 1983.
62. JAPIKSE, D.: The influence of diffuser inlet pressure fields on the range and durability of centrifugal compressor stages. In: "Centrifugal Compressors, Flow Phenomena and Performance", AGARD CP 282, Nov. 1980, paper 13.
63. KENNY, D.P.: A comparison of the predicted and measured performance of high pressure ratio centrifugal compressor diffusers. In "Advanced Radial Compressors", LKI LS 50, May 1972.
64. YOSHINAKA, T.: Surge responsability and range characteristics of centrifugal compressors. Proc of Tokyo Joint Gas Turbine Congress, 1977, pp 381-390.

List of Symbols

a	speed of sound
AS	aspect ratio
AR	area ratio
B_f	diffuser inlet distortion factor
b	diffuser width
C_p	static pressure rise over inlet dynamic pressure specific heat coefficient
C_f	friction coefficient
D_H	hydraulic diameter
H	manometric head - total enthalpy
k_B	blockage factor
L_H	hydraulic length
$\underline{m}$	non dimensional length in meridional direction
$\dot{m}$	mass flow
M	Mach number
N_s	specific speed
O	impeller throat section
p	static pressure
PR	pressure ratio
q	dynamic pressure
Q	volume flow
R	radius
R_G	gas constant
RV	impeller inlet hub over shroud radius ratio
s	arc length
t	pitch
T	temperature
th	blade thickness perpendicular to camber line
U	peripheral velocity
V	absolute velocity
v	velocity component in the boundary layer
W	relative velocity or diffuser width
z	blade number
z'	diffuser vane number
α	absolute flow angle measured from radial direction
β	blade angle or relative flow angle
δ	boundary layer thickness
δ_{cl}	impeller shroud clearance
ε	relative wake width boundary layer skewness
θ	angular coordinate in peripheral direction half diffuser opening angle
κ	isentropic exponent
μ	work reduction factor
ν	wake over jet relative velocity ratio
ρ	density
σ	slip factor

ϕ flow coefficient V_m/U_2
ω loss coefficient
Ω rotational speed (rad/sec)

Subscripts

1 impeller inlet
2 impeller outlet
3 vaned diffuser leading edge
4 diffuser exit
∞ condition for ∞ blade number
bl refers to the blade
cl clearance
f friction
$f\ell$ refers to the flow
h at hub
j jet flow conditions
m meridional component
n perpendicular to the streamwise direction
PS pressure side
SS suction side
S at shroud
sep separation condition
th throat position
u peripheral component
w wake flow conditions

INCORPORATION OF VISCOUS - INVISCID INTERACTIONS IN TURBOMACHINERY DESIGN

Peter Stow

Chief Aerodynamic Scientist
Rolls-Royce
Derby
England

1. INTRODUCTION

The object of this paper is to discuss the incorporation of viscous - inviscid interactions in turbomachinery blade design. Quasi-three-dimensional flow on a general axisymmetric stream - surface is considered although some of the ideas will still apply to fully three-dimensional flows.

At the start of the design of blade sections the inlet and outlet conditions for the blade row, in terms of Mach number and whirl angle, will be specified from a through-flow analysis. As a consequence the overall sectional lift of the blade sections is prescribed and the designers freedom lies in choosing the space-chord ratio and the loading distribution from leading to trailing edges. The blade loading distribution will determine the surface boundary layer state (e.g. laminar or turbulent) and development and hence the loss produced by the blade. In many examples the designer does not have complete freedom over the blade geometry he can employ because of considerations such as blade cooling, leading edge erosion, stress levels etc. As a consequence the task of satisfying the through-flow conditions together with any geometric constraints and at the same time achieving acceptable values of loss, and possibly heat transfer rates, is often best achieved using a method with compatible mixed design and analysis modes. Examples will be presented later.

The main approach adopted in determining the loss characteristics of a blade is to use an inviscid method with a coupled boundary layer analysis, the latter being either an integral or finite difference method. Alternative means of coupling the two calculations as well as representing the boundary layer effects will be discussed. Results will be presented and the main limitations of the approach will be discussed. Fully viscous methods are under development aimed at removing these limitations and the main features of these methods will be discussed.

2. BOUNDARY LAYER ANALYSIS

Some of the main physical features that need to be described by a boundary layer method are as follows,

(i) Laminar flow
(ii) Laminar separation and reattachment
(iii) Start and end of transition
(iv) Turbulent flow
(v) Turbulent separation
(vi) Re-laminarization

Boundary layer methods fall into two main categories, integral and finite difference approaches. In the former, boundary layer parameters like momentum thickness θ and displacement thickness δ^* are determined from ordinary differential equations formed by integrating the boundary layer equations through the boundary layer. For example in the case of two-dimensional flow the momentum integral equation takes the form

$$\frac{d}{ds}\,[(\rho u^2)_\delta \theta] = \tau_w - (\rho u)_\delta\, \delta^* \frac{du_\delta}{ds}$$

where the co-ordinate system is shown in Fig.1 and where τ_w is the skin friction and the subscript δ refers to conditions at the edge of the boundary layer; see Appendix A for more details.

The advantage with this approach is that the solution of the equations is fast and this means that combined with an inviscid method that is itself fast an interactive viscous-inviscid design program can be produced. The integral equations, however, need to be supplemented by correlations from experiment for quantities like skin-friction, form factor and features such as laminar separation and reattachment, the start and end of transition, and re-laminarization also need to be described in terms of correlations.

With finite difference approaches the boundary layer equations are solved numerically. For example the momentum equation for two-dimensional flow takes the form, see Appendix A,

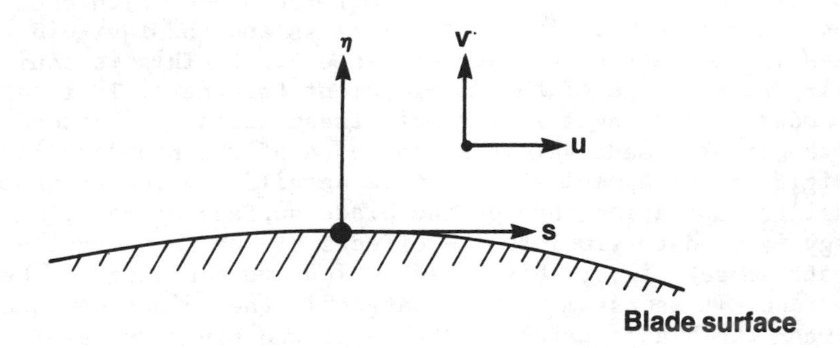

Fig. 1. Boundary layer coordinate system

$$\frac{\partial}{\partial s}(\rho u^2) + \frac{\partial}{\partial n}(\rho uv) = -\frac{\partial p}{\partial s} + \frac{\partial \tau}{\partial n}$$

where the s-n co-ordinate system is shown in Fig.1, u and v are the velocity components in the s and n direction and τ is the shear stress. As a consequence the solution time is longer than for integral methods, dependent largely on the accuracy of the procedures adopted. For turbulent flow the equations need to be supplemented with a turbulence model which will rely heavily on experimental data. Laminar separation usually involves adopting a "fix" of one sort or another in order to formally keep the method working but as the boundary layer equations breakdown locally in the region of separation the results need to be used with caution. With many methods correlations are still needed for determining the start and end of transition or re-laminarization. With methods adopting a turbulent kinetic energy equation then the means exists for trying to describe these phenomena using a differential equation, it is however early days to be able to comment quantitatively on the use of such approaches.

3. INVISCID - BOUNDARY LAYER COUPLING

There are three choices in deciding where to couple the solutions to the boundary layer and inviscid mainstream equations namely at the boundary layer edge, the edge of the boundary layer displacement thickness or the blade surface, see Murman and Bussing [1] for details.

In the main the effect of the boundary layer on the inviscid mainstream calculation is represented either by a displacement effect or by transpiration. In the former the blade is thickened by the boundary layer displacement thickness and the inviscid flow calculated in the reduced passage area see Fig.2. This is equivalent to coupling at the edge of the displacement thickness. It is assumed in this model that changes in the mainstream variables between the edge of the displacement thickness and edge of the boundary layer are negligible; see Appendix B for more details. In the transpiration model transpiration through the blade surface of mass, momentum and energy is used to simulate the effects of the boundary layer on the mainstream see Fig.3; this is equivalent to coupling at the blade surface. It is assumed that changes in the mainstream quantities between the blade surface and edge of the boundary layer are negligible; see Appendix B for more details. In the case of two-dimensional flow the transpiration velocity v_o normal to the blade surface is given by

$$(\rho v)_o = \frac{d}{ds} [(\rho u)_\delta \, \delta^*]$$

see Appendix B.

It can be shown from considerations of conservation of momentum and energy that the transpiration streamwise velocity u_o should be the local inviscid velocity i.e.

$$u_o = u_\delta$$

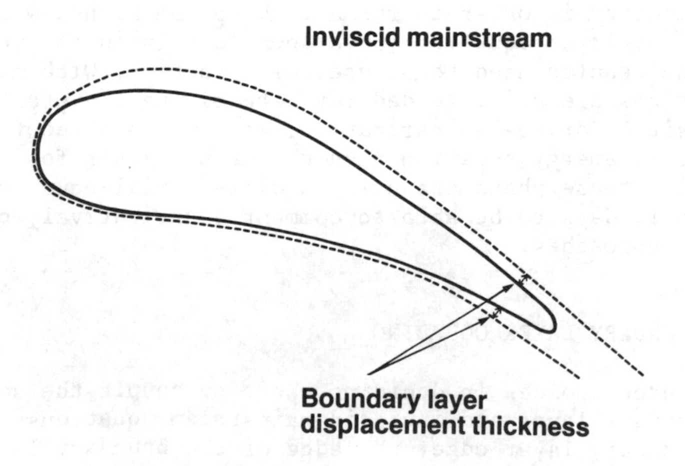

Fig. 2. Boundary layer displacement model

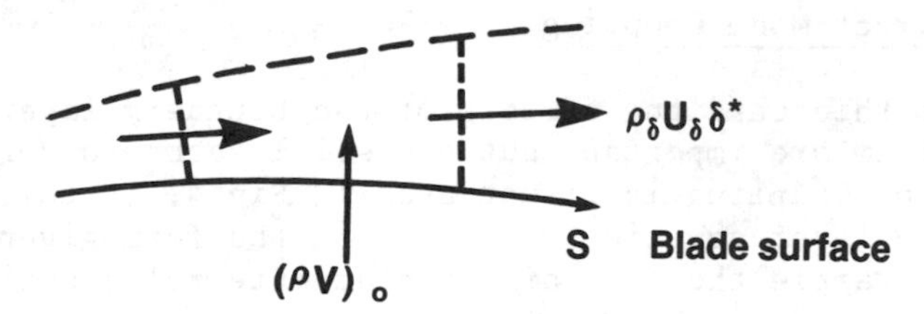

$$(\rho V)_o \delta s = (\rho_\delta U_\delta \delta^*)_{s+ds} - (\rho_\delta U_\delta \delta^*)_s$$

Fig. 3. Transpiration model

and that the transpiration total enthalpy H_o should also be the local inviscid value i.e.

$$H_o = H_\delta$$

see Appendix B.

In practice only small differences will arise between the two main methods of coupling the equations and the choice is governed mainly by features or properties of the inviscid method. For example, in the case of a streamline curvature method a displacement model is used since the edge of the displacement thickness now becomes the new effective blade streamline, see Stuart and Stow [2] and Wilkinson [3]. In a finite element approach where mesh generation is often quite expensive then a transpiration model is advisable, see Williams and Stow [4]. In time-marching methods, like that of Denton [5], either approach can easily be used, see for example the work of Haller [6] using a transpiration model and that of Calvert [7] using a displacement model.

In addition to considering how to represent the effects of the boundary layer on the mainstream flow one needs to consider how the two calculations should be iteratively coupled. In some cases the effects of the boundary layer on the mainstream flow will be so small that useful boundary layer information can be obtained from a single call to a boundary layer routine after the inviscid calculation is converged. In general, however, the boundary layer will have an effect on the inviscid flow in terms of blockage, deviation etc. and the two calculations must be iteratively coupled together. The type of coupling is determined in the main by the magnitude of the boundary layer effect and different modes of coupling have been developed to cater for this.

3.1 Direct Mode Coupling

In this case the effects of the boundary layer on the inviscid mainstream are important but are still second order. The direct manner of coupling is illustrated in Fig.4. In this case the boundary layer equations are used in the form given in Appendix A. So for example the form of momentum integral equation used is

$$\frac{d}{ds}\left[(\rho u^2)_\delta \theta\right] = \tau_w - (\rho u)_\delta H.\theta \frac{du_\delta}{ds}$$

and the entrainment equation is

$$\frac{d}{ds}\left[(\rho u)_\delta H_1 \theta\right] = (\rho u)_\delta C_E$$

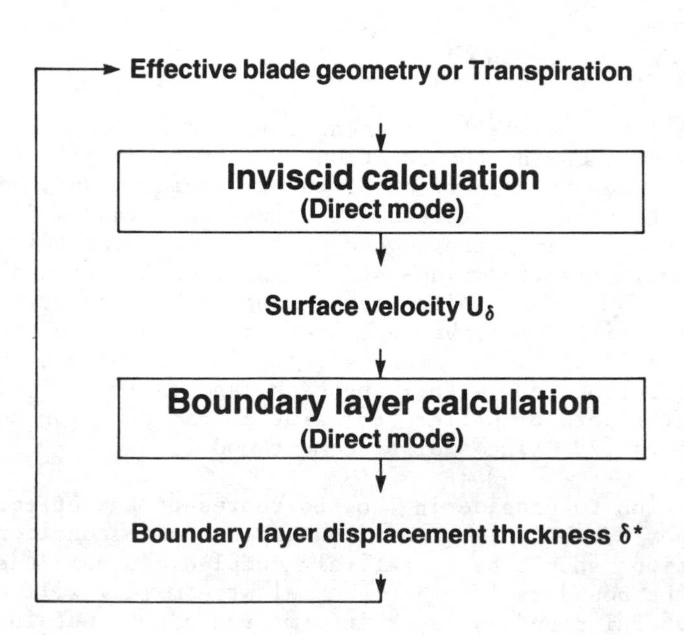

Fig. 4. Direct mode boundary layer coupling

The input to the boundary layer calculation is the inviscid velocity at the "effective" edge of the boundary layer (where this is either the edge of the displacement thickness for a displacement model or the blade surface for a transpiration model); the input to the inviscid calculation is the boundary layer displacement thickness used to determine either a new effective blade surface or to calculate transpiration quantities.

Figs 5a,b and c taken from Williams and Stow [4] illustrate results from a finite element velocity potential program using a transpiration boundary layer model. An integral boundary layer method is used based on Luxton and Young [8] for laminar flow and Green, Weeks and Brooman [9] for turbulent flow. In addition the method contains a laminar separation bubble calculation, a re-laminarization condition and a transitional treatment. It can be seen from Fig.5a that the surface Mach number agrees closely with the experimental results of Hodson [4a] including the leading edge region where a local spike occurs. Fig.5b shows comparisons of the momentum and displacement thicknesses of the suction surface boundary layer against experiment for the case of 4% free-stream turbulence. A laminar separation bubble is predicted near the leading edge followed by turbulent reattachment and then almost immediate re-laminarization of the boundary layer. Transition is predicted towards the back of the suction surface with the boundary layer being fully turbulent at the trailing edge. Fig.5c shows results for 1/2% free stream turbulence. In this case transition

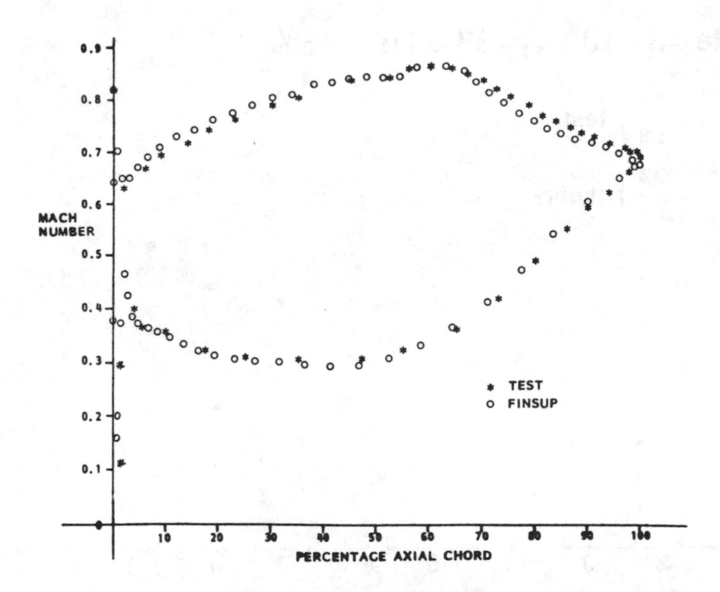

Fig. 5a. Comparison of predicted and measured mach number distribution around turbine blade tested in cascade

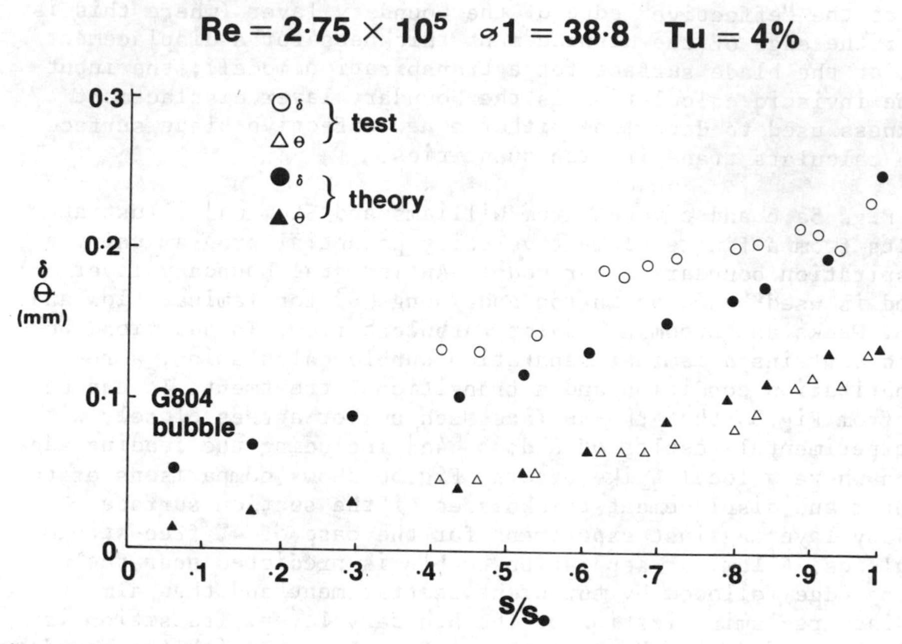

Fig. 5b. Suction surface boundary layer

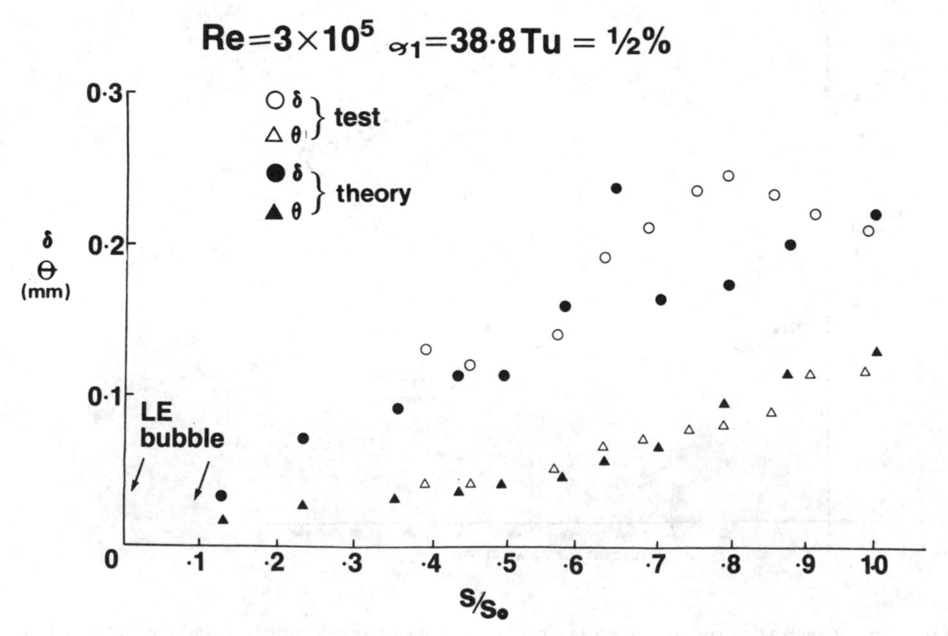

Fig. 5c. Suction surface boundary layer

is delayed and a second separation bubble occurs on the suction surface; this was found in the experiment. It can be seen that predictions up to the second separation are good but are less satisfactory for the displacement thickness after the separation indicating a short-coming in the bubble correlations. The momentum thickness is well predicted throughout this region.

This example also serves to illustrate the task of blade designer. In the case shown the leading edge velocity spike is relatively unimportant at design conditions but for other blades (or even this one at off-design conditions) it may be desirable to remove what might be a loss producing feature. In a pure analysis method this can be quite a tedious process guessing a modified geometry, re-analysing and then considering the resulting boundary layer. In a method with a mixed design and analysis mode the task can be made much easier. For example in the case considered the prescribed input velocity around the leading edge would have the spike removed, see Fig.6a. Fig.6b shows the resulting modified leading edge geometry as obtained from the design mode added to this particular method by Cedar and Stow [10]. Another example taken from reference [10] is shown in Fig.7a where boundary layer loss production for a supercritical controlled diffusion compressor blade is reduced using the design mode of the program, the resulting geometry being shown in Fig.7b. In both cases the lift of the blade is kept constant and only the loading distribution is changed. This illustrates how by coupling a boundary layer

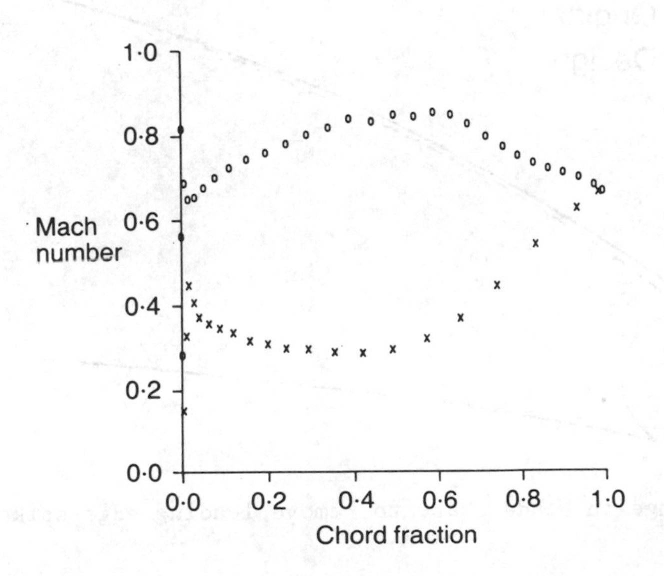

Fig. 6a (i). Analysis of turbine blade with leading edge spike

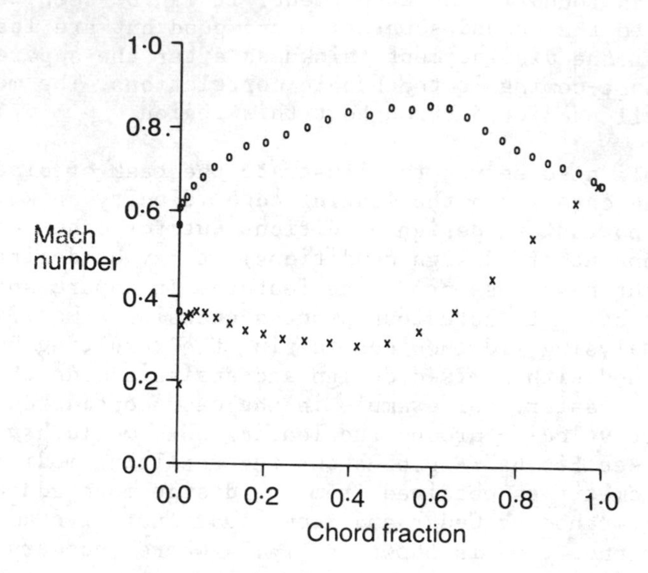

Fig. 6a (ii). Redesigned blade to remove leading edge spike

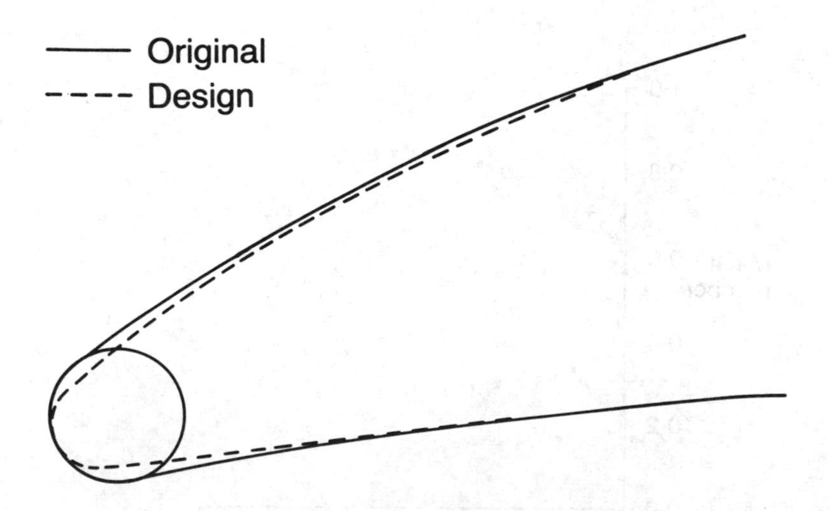

Fig. 6b. Change in blade shape to remove leading edge spike

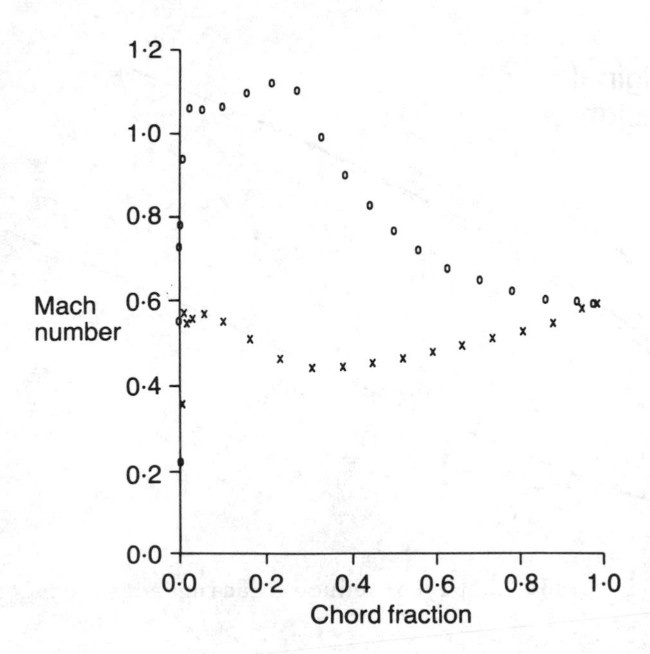

Fig. 7a (i). Mach number distribution around original BGK type blade

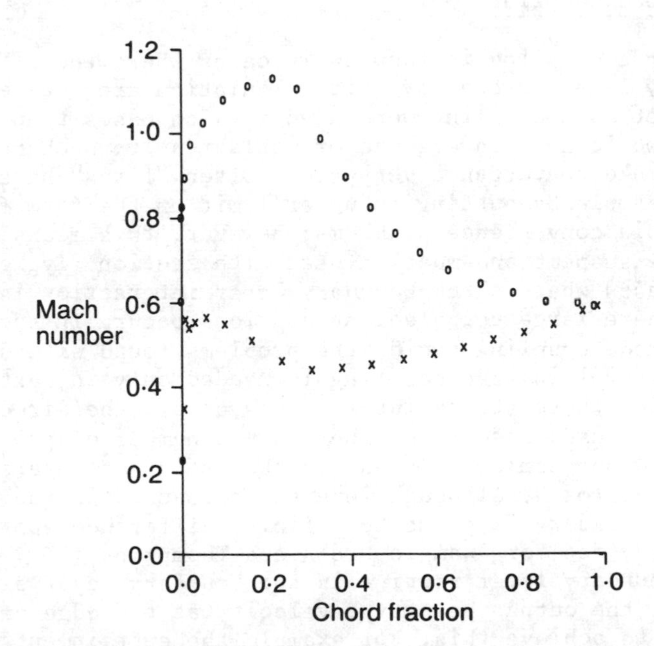

Fig. 7a (ii). Designed 'laid-back' type blade

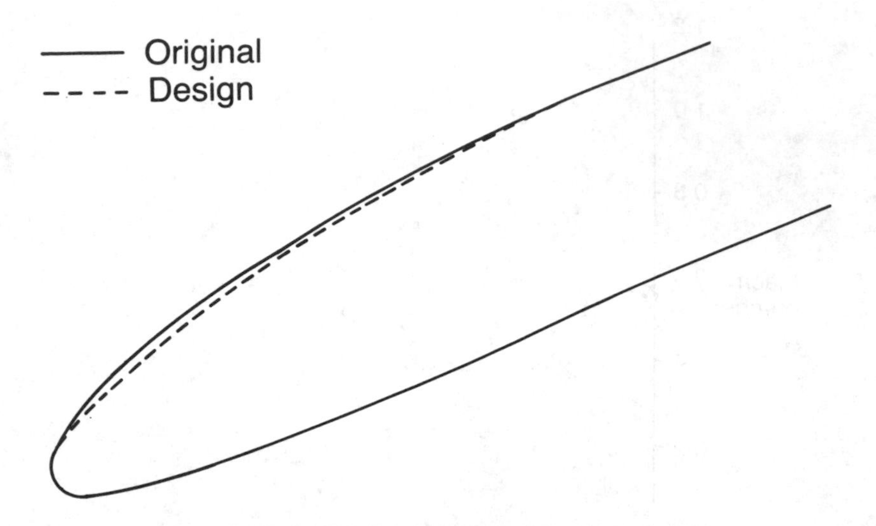

Fig. 7b. Change in blade shape to reduce leading edge acceleration

calculation to an inviscid method with a design mode a very powerful interactive design tool can be developed.

3.2 Inverse Mode Coupling

Inverse mode coupling is adopted in cases where the effects of the boundary layer on the inviscid calculation are not second order. If direct mode coupling were used in such cases then either the procedure would not convergence or would require such heavy damping as to make convergence very slow. Often "fixes" have to be adopted (for example by putting an upper limit on the form factor) in order to avoid convergence problems; however, this means that the results are suspect and must be used with caution. Typical examples are cases where shock boundary layer interaction is important or where large turbulent separations occur. In such cases inverse mode coupling avoids the problems found with direct coupling. In general inverse coupling is needed only in certain areas of the flow where strong interactions occur, the direct mode of coupling being used elsewhere. There are a number of procedures that can be used but common to these is the use of an inverse boundary layer approach. Although inverse approaches in the main adopt integral boundary layer methods finite difference approaches can also be used, see for example Drela and Thompkins [11]. The input to the boundary layer routine is the boundary layer displacement thickness, the output being the velocity at the edge of the boundary layer to achieve this. For example the entrainment equation is used in the form

$$\frac{\theta}{u_\delta}\frac{du_\delta}{ds} = \frac{1}{F_2}\left(\frac{d\delta^*}{ds} - F_1\right)$$

together with the standard momentum integral equation

$$\frac{d}{ds}[(\rho u^2)_\delta\theta] = \tau_w - (\rho u)_\delta\, H\theta\, \frac{du_\delta}{ds}$$

and the usual correlations hold.

How this is used in the inviscid calculation depends on the details of the method and whether a design mode exists for the method. In the case where such a mode does exist then full inverse coupling can be used, see Fig.8. The input to the inviscid calculation is the blade surface velocity in parts where the inverse boundary is applied, the output being the effective blade streamline or edge of the boundary layer thus giving the new displacement thickness.

Le Balleur [12] has proposed a semi-inverse approach where only an inverse boundary layer method is used. Direct mode input to the inviscid calculation in the form of boundary displacement

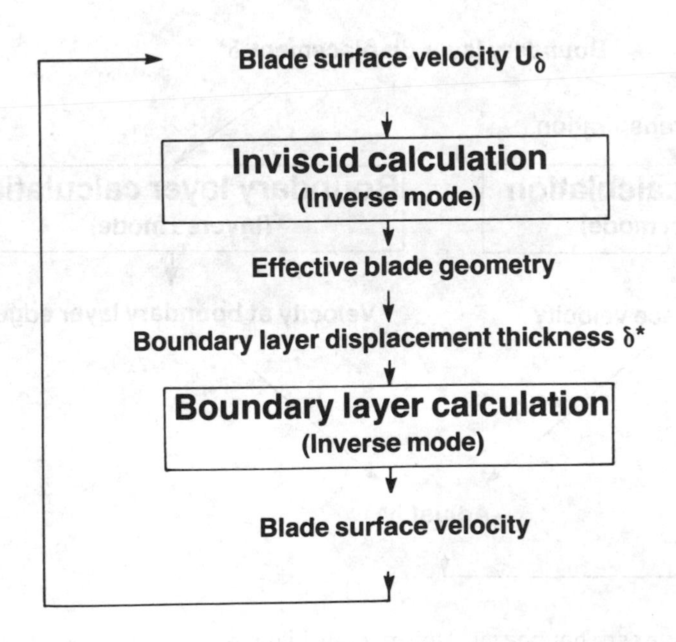

Fig. 8. Inverse mode boundary layer coupling

thickness or transpiration is used but now this is iteratively adjusted until the resulting surface velocity is equal to that from the inverse boundary layer calculation see Fig.9. A design mode type of approach can obviously be used to determine how to adjust the displacement thickness in order to achieve the velocity.

Calvert [7] has developed an inverse approach based on the Denton [5] inviscid time-marching method and an inverse integral boundary layer based on the lag-entrainment method of East, Smith and Merryman [13]. Fig.10 shows results from Calvert [14] for the case of a DFVLR transonic compressor designed for an inlet Mach number of 1.09. It can be seen that very good agreement with experiment is achieved for the distribution of surface pressure. Also shown in the figure is the result of an inviscid prediction indicating the importance of the boundary layer interaction in this case. Fig.11a from reference [14] shows results for a Detroit Diesel Allison compressor with an inlet Mach number of 1.46. It can be seen that very good agreement is found over the range of pressure ratios and the shock boundary layer interaction is obviously well modelled. Fig.11b gives an indication of the significance of the boundary layer in this example.

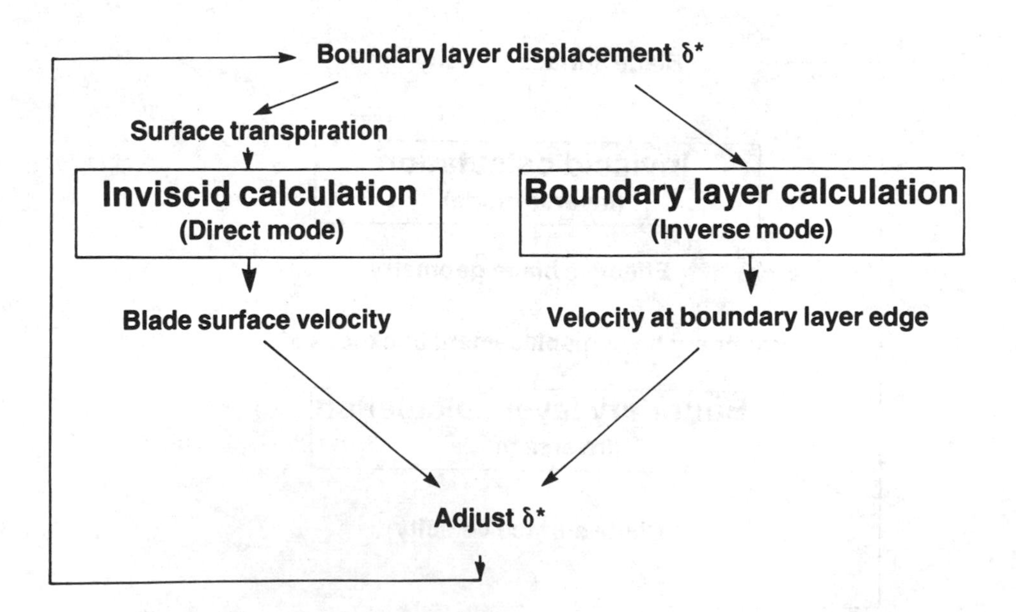

Fig. 9. Semi-inverse boundary layer coupling

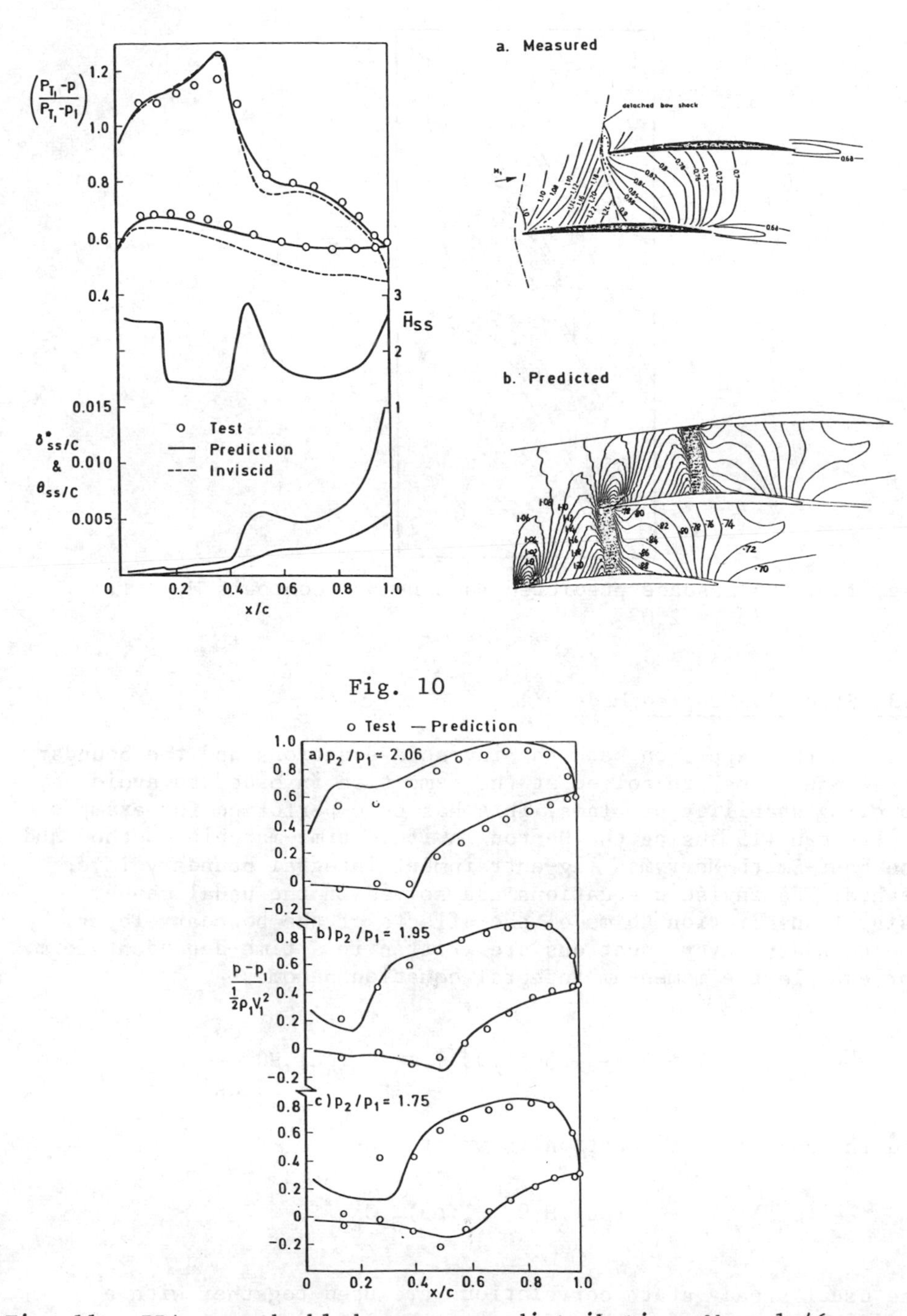

Fig. 10

Fig. 11a. DDA cascade blade pressure distributions $M_1 = 1.46$

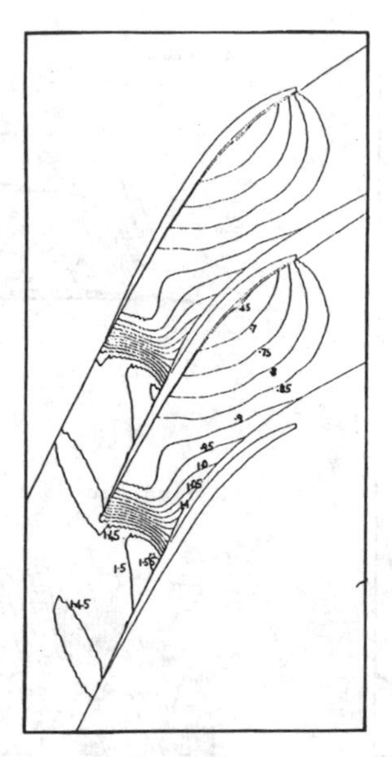

Fig. 11b. DDA cascade predicted Mach number contours $M_1 = 1.46$, $p_2/p_1 = 2.03$

3.3 Strongly Coupled Mode

In this approach both the inviscid equations and the boundary layer equations are solved at the same time in order to avoid coupling stability problems. This has been performed for example by Freeman [15] using the Denton inviscid time-marching method and the East-Smith-Merryman lag-entrainment integral boundary layer method. The inviscid equations are solved in the usual manner using transpiration to model the effects of the boundary layer. The boundary layer equations are written in a time-dependent form. For example the momentum integral equation becomes

$$\frac{\partial}{\partial t}((\rho u)_\delta \delta^*) = -\frac{\partial}{\partial s}[(\rho u^2)_\delta \theta] + \tau_w - (\rho u)_\delta H\theta \frac{du_\delta}{ds}$$

and the entrainment equation is written as

$$\frac{\partial}{\partial t}(\rho_\delta \delta) = -\frac{\partial}{\partial s}[(\rho u)_\delta H_1 \theta] - (\rho u)_\delta C_E$$

The usual steady state correlations are used together with a modified equation for the entrainment coefficient

$$\frac{1}{u_\delta}\frac{\partial C_E}{\partial t} = -\frac{\partial C_E}{\partial s} + K\,(C_{E_{eq}} - C_E)$$

These equations are solved using a simple explicit time-stepping scheme together with upwind differencing for the spatial derivatives.

The method has been applied by Freeman to the flow in fans involving strong shock boundary layer interactions; results similar to those of Calvert are found.

3.4 Mixing Loss Calculation

In calculating the loss of a blade it is important to include the effects of trailing edge mixing losses, arising from the base region, and wake mixing losses downstream. Boundary layer calculations can say little about the former but can be used to assess wake mixing by continuing the calculation downstream of the trailing edge for example in an integral calculation by dropping the skin friction and modifying the entrainment correlations. A simple one-dimensional mixing calculation, see for example Stewart [16], can then be used to calculate the overall mixing loss to uniform conditions. Base mixing losses are not included in the approach.

An alternative approach is to use the calculated inviscid and boundary layer conditions at the blade trailing edge, together with correlations for the base pressure, in a two-dimensional mixing calculation similar to that of Stewart, see Stow [17]. This solves the equations of continuity of mass, momentum and energy to give the fully mixed out uniform conditions for downstream of the blade row. It is found that the effects of non-uniformities across the trailing edge are important and should be included in the analysis.

4. FULLY VISCOUS ANALYSIS

The main reasons for interest in a fully viscous analysis are for the prediction of off-design losses, where laminar separation bubbles may occur and in some cases fairly extensive turbulent separation, and for the prediction of base flow mixing losses. These are two areas where a coupled inviscid-boundary layer analysis has limitations. This can be seen from the results shown in Fig.12 from Williams [18] for a typical LP turbine blade using the method presented in reference [4]. It can be seen that the predicted losses agree closely with measured values near to design conditions but become progressively less accurate as the off-design incidence increases; this is attributed to shortcomings in commonly adopted correlations for laminar separation and reattachment.

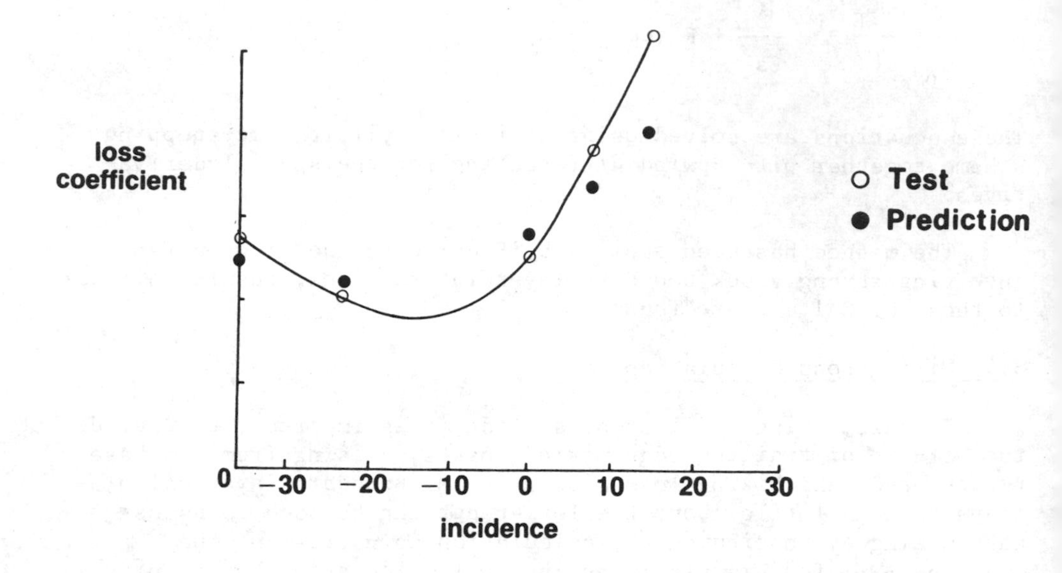

Fig. 12

In stating the above areas of interest it is important therefore that the full geometry with good grid resolution is used in a fully viscous analysis. Fig. 13a shows the type of grid being employed for example by Norton, Thompkins and Haimes [19] where O-type or C-type grids around the blade can be combined with a standard sheared blade-to-blade grid see Fig. 13b; it can be seen that good resolution can be achieved in both the leading edge region (important for off design analysis) and trailing edge region. It is also important with a fully viscous method that there are no restrictions with regard to handling regions of reverse flow for example at leading and trailing edges, in regions of separation etc. Another area for important consideration is the turbulence model needed to close the solution of the turbulent equations of motion. More care is often needed than in boundary layer methods especially for mixing length models since the edge of the boundary layer is in some cases difficult to determine accurately.

With a fully viscous analysis the Reynolds averaged Navier-Stokes equations are solved with no simplifying assumptions regarding streamwise derivatives as in a boundary layer approach; typical equations of motion for two-dimensional flow are given in Appendix C.

There are two main approaches being developed to solve the Reynolds averaged Navier-Stokes equations, pressure-correction

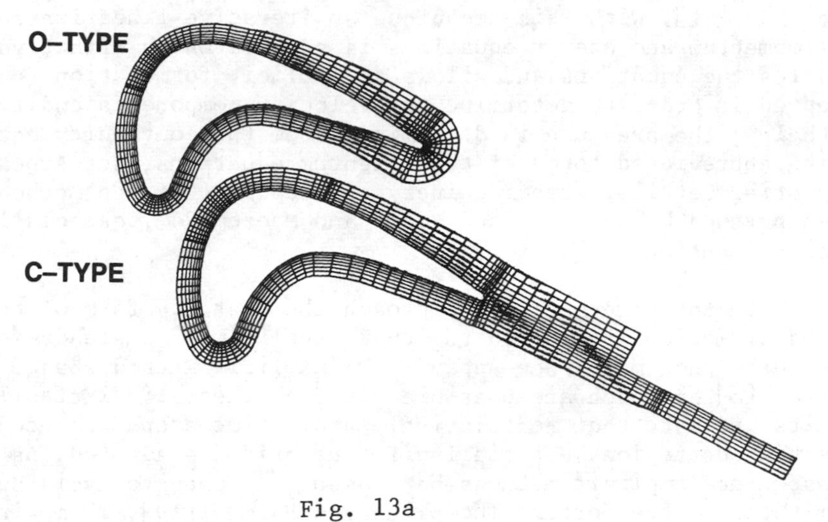

Fig. 13a

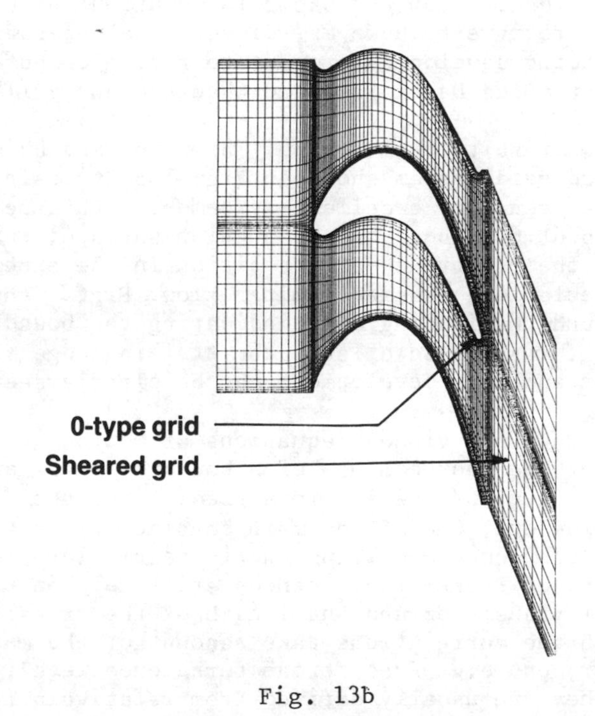

Fig. 13b

techniques and time-marching techniques. The pressure-correction method is based on the work of Caretto et.al. [20] and Spalding and Patankar [21]. With this technique an iterative linearization of the momentum and energy equations is adopted which effectively uncouples the equations and allows an implicit formulation to be adopted in order to determine the velocity components and total enthalpy; the pressure is determined from the continuity equation using abbreviated forms of the momentum equations, see Appendix C for brief details. Turbomachinery applications of the method have been pursued by Hah [23] and Moore and Moore [24], especially for three-dimensional flow.

With the time-marching approach the unsteady form of the equations of motion are solved together usually with a steady form for the turbulence model see Appendix D. Explicit methods based on the Denton [5] approach are possible although theoretical stability limits indicate that restrictively small time steps are needed in the viscous regions especially if fine grid are adopted. As a consequence implicit methods have been developed to avoid such limitations, see Norton, Thompkins and Haimes [19], Thompkins et.al. [25] and Dawes [26]. In this approach implicit correction techniques are adopted to solve the fully coupled finite difference equations very similar to a Newton-Raphson technique see Appendix D. Block matrix equations result from the basic formulation which can be fairly expensive to invert. However, differential operator splitting techniques are being developed in order to reduce computational times, see for examples Dawes [26] and Chaussee and Pulliam [27].

Fig. 14 shows results from Norton, Thompkins and Haimes [19] for the blade and grid system shown in Fig. 13b at design conditions. It can be seen that excellent agreement with experiment exists except about 60% chord on the suction surface; this is probably due to the occurrence of transition in the experiment whereas the calculations assumed laminar flow. Fig. 15 shows details of the flow around the leading edge indicating the boundary layer development. Fig. 16 shows details of the trailing edge flow where vortex formation and wake development can be clearly seen.

By solving the full viscous equations of motion it is possible to remove many of the shortcomings of a boundary layer analysis mentioned earlier e.g. off-design loss prediction, the detailed trailing edge flow etc. One of the main considerations for the future lies in turbulence modelling and in particular predicting the start and end of transition. Methods still rely on correlations for the start of transition and the length of the transitional region. Although the correlations take account of the main factors affecting the process e.g. free stream turbulence level, pressure gradient etc. they are usually derived from relatively few and simple experiments and are consequently somewhat restrictive. In

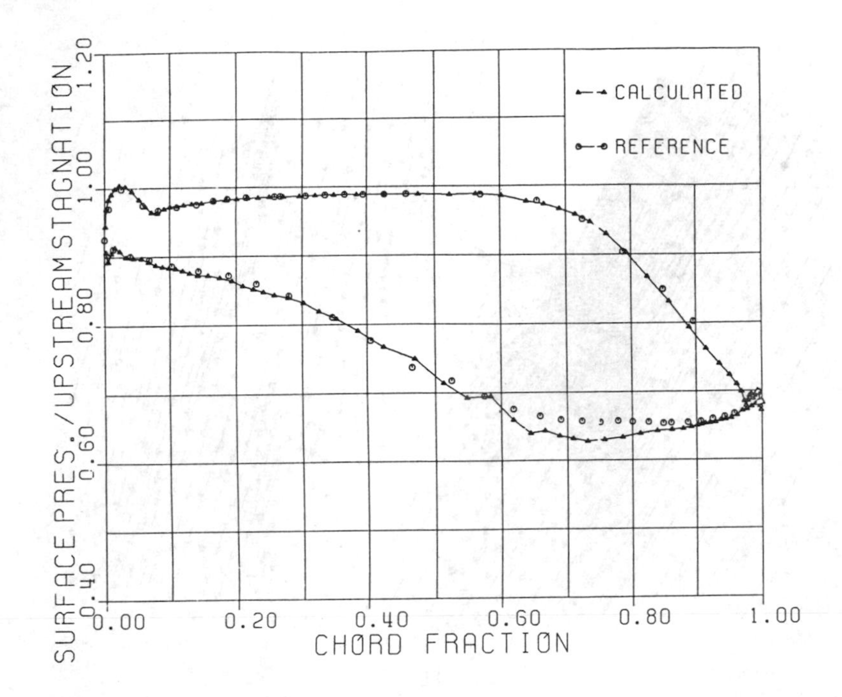

Fig. 14. Design Incidence

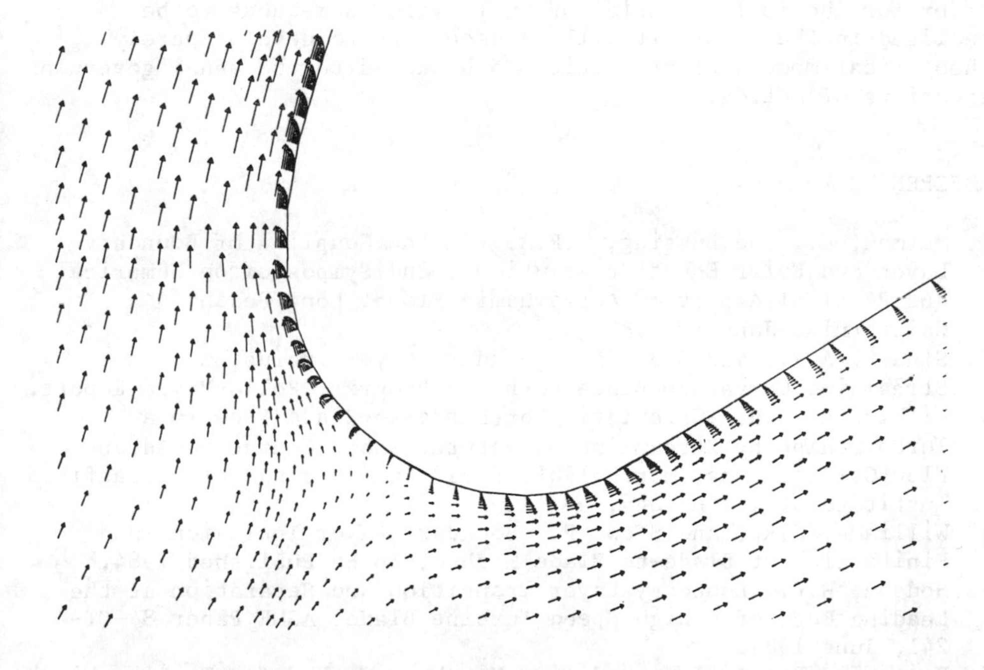

Fig. 15. Direction vector plot sheared + 0-type grid

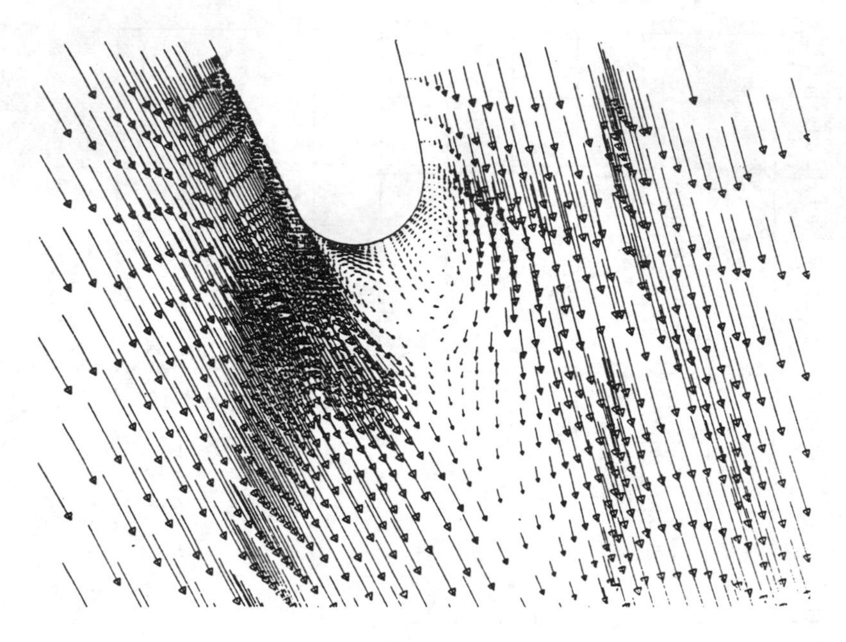

Fig. 16

order for the full potential of fully viscous methods to be realised in the future it will be necessary to develop purely theoretical models of transition to be added to the usual governing equations of motion.

REFERENCES

1. Murman, E., and Bussing, T.R.A. On the Coupling of Boundary Layer and Euler Equation Solutions. 2nd Symposium on Numerical and Physical Aspect of Aerodynamic Flows. Long Beach, California, January 1983.
2. Stuart, A.R., and Stow, P. Boundary Layer Inclusion is a Streamline Curvature Blade-to-Blade Program. Rolls-Royce Report.
3. Wilkinson, D.H. Calculation of Blade-to-Blade Flow in a Turbomachine by Streamline Curvature. Short Course on Advanced Flow Calculations, March 1981. Fluid Engineering Unit, Cranfield Institute of Technology.
4. Williams, C.R., and Stow, P. Boundary Layer Inclusion in a Finite Element Blade-to-Blade Method. To be Published 1984.

4a. Hodson, H.P. Boundary Layer Transition and Separation at the Leading Edge of a High Speed Turbine Blade. ASME Paper 84-GT-241, June 1984.

5. Denton, J.D. An Improved Time-Marching Method for Turbomachinery Flow Calculation. ASME 82-GT-239, 1982.

6. Haller, B.R. The Effects of Film Cooling Upon the Aerodynamic Performance of Transonic Turbine Blades. Ph.D. Thesis, University of Cambridge 1980.
7. Calvert, W.J. An Inviscid-Viscous Interaction Treatment to Predict the Blade-to-Blade Performance of Axial Compressors with Leading Edge Normal Shock Waves. ASME 82-GT-135.
8. Luxton, R.E., and Young, A.D. Skin Friction in the Compressible Laminar Boundary Layer with Heat Transfer and Pressure Gradient. Aeronautical Research Council Report 20336.
9. Green, J.E., Weeks, D.J., and Brooman, J.W.F. Prediction of Turbulent Boundary Layers and Wakes in Compressible Flow by a Lag-Entrainment Method. R.A.E. Tech. Report 72231.
10. Cedar, R.D., and Stow, P. A Compatible Mixed Design and Analysis Finite Element Method for the Design of Turbomachinery Blades. To be published International Journal of Numerical Methods in Fluids, 1984.
11. Drela, M., and Thompkins, W.T. A Study on Non-Unique Solutions of the Two-Dimensional Boundary Layer Equations at Laminar Separation and Reattachment Points. 2nd Symposium on Numerical and Physical Aspect of Aerodynamic Flows. Long Beach, California, January 1983.
12. Le Balleur, J.C. Numerical Viscous-Inviscid Interaction in Steady and Unsteady Flow. 2nd Symposium on Numerical and Physical Aspects of Aerodynamic Flows. Long Beach, California, January 1983.
13. East, L.F., Smith, P.D., and Merryman, P.J. Prediction of the Development of Separated Turbulent Boundary Layers by the Lag-Entraintment Method. RAE. Tech. Report 77046.
14. Calvert, W.J. Application of an Inviscid-Viscous Interaction Method to Transonic Compressor Cascades. AGARD PEP Slot Meeting on Viscous Effects in Turbomachines, Denmark, June 1983.
15. Freeman, C. A Viscous Inviscid Interaction Method for Calculating the Flow in Transonic Fan Blade Passages by the use of a Time Dependent Free Stream Calculation and a Time Dependent Boundary Layer Calculation. Published in "Computational Methods in Turbomachinery" Institute of Mechanical Engineers, Birmingham University, England. April 1984.
16. Stewart, W.L. Analysis of Two-Dimensional Compressible Flow Loss Characteristics Downstream of Turbomachine Blade Rows in Terms of Basic Boundary Layer Characteristics. NACA TN 3515, 1955.
17. Stow, P. A Two-Dimensional Mixing Loss Model. Rolls-Royce private communication.
18. Williams, C.R. Loss Predictions for LP Turbine Blades. Rolls-Royce private communication.
19. Norton, R.G., Thompkins, W.T., and Haimes, R. Implicit Finite Difference Schemes with Non-Simply Connected Grids - A Novel Approach. AIAA, 2nd Aerospace Sciences Meeting. Jan, 1984 Reno.

20. Caretto, L.S., Gosman, A.D., Patankar, S.V., and Spalding, D.B. Two Calculation Procedures for Steady Three-Dimensional Flow with Recirculation. Proceedings of the Third International Conference on Numerical Methods in Fluid Dynamics, Paris. Volume II, page 60, 1972.
21. Patankar, S.V., and Spalding, D.B. A Calculation Procedure for Heat, Mass and Momentum Transfer in Three-Dimensional Parabolic Flows. International Journal of Heat and Mass Transfer Volume 15, page 1787-1806, 1972.
22. Hah, C. A Navier-Stokes Analysis of Three-Dimensional Turbulent Flows Inside Turbine Blade Rows at Design and off-design Conditions. ASME 83-GT-40.
23. Moore, J.G., and Moore, J. Calculation of Horseshoe Vortex Flow without Numerical Mixing. ASME Paper No. 84-GT-241.
24. Connell, S.D., and Stow, P. The Pressure Correction Method. To be published in Computers and Fluids, 1984.
25. Thompkins, W.T., Tong, S.S., Bush, R.H., Usab, W.J., and Norton, R.J.G. Solution Procedures for Accurate Numerical Simulations of Flows in Turbomachinery Cascades. AIAA Paper 83-0257, 1983.
26. Dawes, W.N. Computation of Viscous Compressible Flow in Blade Cascades Using an Implicit Iterative Replacement Algorithm. Published in "Computational Methods in Turbomachinery". Institute of Mechanical Engineers, Birmingham University, England, April 1984.
27. Chaussee, D.S., and Pulliam, T.H. Two-Dimensional Inlet Simulations Using a Diagonal Implicit Algorithm. AIAA Journal, Volume 19, Number 2, 1981.

Appendix A. Two-Dimensional Boundary Layer Equations

Consider a local s-n coordinate system based on the blade surface, see Fig.1.

Then for two-dimensional flow, neglecting the effects of blade surface curvature and changes in stream tube height, the boundary layer equations are

$$\frac{\partial}{\partial s}(\rho u) + \frac{\partial}{\partial n}(\rho v) = 0 \tag{A.1}$$

$$\frac{\partial}{\partial s}(\rho u^2) + \frac{\partial}{\partial n}(\rho uv) = -\frac{\partial p}{\partial s} + \frac{\partial \tau}{\partial n} \tag{A.2}$$

$$\frac{\partial}{\partial s}(\rho uH) + \frac{\partial}{\partial n}(\rho vH) = -\frac{\partial q}{\partial n} + \frac{\partial}{\partial n}(u\tau) \tag{A.3}$$

where u and v are velocity components in the s and n directions, H is the total enthalpy, q is the heat flux and τ the shear stress. It should be noted that in a practical situation the effects of streamline radius, stream-tube height and blade rotation would be included.

Integral equations are obtained by integrating the above equations with respect to n from n=0 to n=δ, the edge of the boundary layer. Forexample, equation (A.1) gives

$$\frac{d}{ds}\int_o^\delta \rho u \, dn - (\rho u)_\delta \frac{d\delta}{ds} + (\rho v)_\delta = 0$$

where the subscript δ represents conditions at the edge of the boundary layer. This may be written in terms of displacement thickness δ^* as

$$\frac{d}{ds}[(\rho u)_\delta(\delta-\delta^*)] = (\rho u)_\delta \frac{d\delta}{ds} - (\rho v)_\delta = (\rho u)_\delta C_E \qquad (A.4)$$

where C_E is defined as the entrainment coefficient; equation (A.4) is referred to as the entrainment equation.

In a similar manner equation (A.2) can be shown to give the momentum integral equation

$$\frac{d}{ds}[(\rho u^2)_\delta \theta] = \tau_w - (\rho u)_\delta \delta^* \frac{du_\delta}{ds} \qquad (A.5)$$

where θ is the momentum thickness and τ_w the skin-friction and where changes in u_δ in the n direction are ignored.

Similarly from equation (A.3) an energy integral equation can be obtained

$$\frac{d}{ds}[(\rho uH)_\delta \theta_H] = q_w \qquad (A.6)$$

where q_w is the blade surface heat flux and θ_H the total enthalpy thickness of the boundary layer defined as

$$(\rho uH)_\delta \theta_H = \int_o^\delta \rho u \, (H-H_\delta) \, dn \qquad (A.7)$$

In deriving equation (A.6) H_δ is taken as constant in the s-direction.

It is usual to write equations (A.5) and (A.6) as

$$\frac{d}{ds}\,[(\rho u^2)_\delta \theta] = \tau_w - (\rho u)_\delta\, H.\theta\, \frac{du_\delta}{ds} \tag{A.8}$$

and

$$\frac{d}{ds}\,[(\rho u)_\delta\, H_1 \theta] = (\rho u)_\delta\, C_E \tag{A.9}$$

where

$$H = \delta^*/\theta$$

and

$$H_1 = (\delta - \delta^*)/\theta$$

Appendix B. Inviscid-Boundary Layer Coupling

In coupling separate inviscid and boundary layer calculations we use the inviscid solution outside of the boundary layer and the viscous solution inside the boundary layer. If we couple the calculations of any point other than the edge of the boundary layer then we need to determine boundary conditions for the inviscid calculation at that point such that the correct solution results at the edge of the boundary layer. We consider two cases coupling at the edge of the displacement thickness and at the blade surface. For simplicity the effects of streamline radius, stream-tube height and blade rotation are ignored although these can easily be included.

1. DISPLACEMENT MODEL

In the local s-n blade coordinate system, see Fig.1, the inviscid equations of motion are

$$\frac{\partial}{\partial s}(\rho u) + \frac{\partial}{\partial n}(\rho v) = 0 \tag{B.1}$$

$$\frac{\partial}{\partial s}(\rho u^2) + \frac{\partial}{\partial n}(\rho uv) = -\frac{\partial p}{\partial s} \tag{B.2}$$

$$\frac{\partial}{\partial s}(\rho uv) + \frac{\partial}{\partial n}(\rho v^2) = -\frac{\partial p}{\partial n} \tag{B.3}$$

$$\frac{\partial}{\partial s}(\rho uH) + \frac{\partial}{\partial n}(\rho vH) = 0 \tag{B.4}$$

where the effects of blade surface curvature are ignored. The boundary layer equations are

$$\frac{\partial}{\partial s}(\overline{\rho u}) + \frac{\partial}{\partial n}(\overline{\rho v}) = 0 \qquad (B.5)$$

$$\frac{\partial}{\partial s}(\overline{\rho u^2}) + \frac{\partial}{\partial n}(\overline{\rho uv}) = -\frac{\partial \bar{p}}{\partial s} + \frac{\partial \tau}{\partial n} \qquad (B.6)$$

$$\frac{\partial \bar{p}}{\partial n} = 0 \qquad (B.7)$$

$$\frac{\partial}{\partial s}(\overline{\rho u H}) + \frac{\partial}{\partial n}(\overline{\rho v H}) = -\frac{\partial q}{\partial n} + \frac{\partial}{\partial n}(\bar{u}\tau) \qquad (B.8)$$

where the '-' notation denotes a viscous value.

With the displacement model we require a boundary condition for the inviscid flow at $n=\delta^*$ such that at $n=\delta$ the inviscid and viscous solutions agree i.e.

$$u_\delta = \bar{u}_\delta;\ v_\delta = \bar{v}_\delta;\ H_\delta = \bar{H}_\delta;\ p_\delta = \bar{p}_\delta \quad \text{etc.} \qquad (B.9)$$

Integrating equation (B.1) with respect to n from $n=\delta^*$ to $n=\delta$ gives

$$\int_{\delta^*}^{\delta} \frac{\partial}{\partial s}(\rho u)\, dn + (\rho v)_\delta - (\rho v)_{\delta^*} = 0 \qquad (B.10)$$

Integrating equation (B.5) with respect to n from n=0 to $n=\delta$ gives

$$\int_{o}^{\delta} \frac{\partial}{\partial s}(\overline{\rho u})\, dn + (\overline{\rho v})_\delta = 0 \qquad (B.11)$$

From equations (B.10) and (B.11), using equation (B.9),

$$(\rho v)_{\delta^*} = \int_{\delta^*}^{\delta} \frac{\partial}{\partial s}(\rho u) dn - \int_{o}^{\delta} \frac{\partial}{\partial s}(\overline{\rho u})\, dn$$

$$= \frac{d}{ds}\int_{\delta^*}^{\delta} \rho u dn + (\rho u)_{\delta^*}\frac{d\delta^*}{ds} - \frac{d}{ds}\int_{o}^{\delta} \overline{\rho u}\, dn$$

$$= \frac{d}{ds}\int_{\delta^*}^{\delta} \rho u dn + (\rho u)_{\delta^*}\frac{d\delta^*}{ds} - \frac{d}{ds}\left[(\rho u)_\delta(\delta-\delta^*)\right]$$

If we ignore changes in the inviscid flow in the region $\delta^* < n < \delta$ then the above equation gives

$$v_{\delta^*} = u_{\delta^*} \frac{d\delta^*}{ds} \qquad (B.12)$$

which means that the displacement thickness is a streamline of the inviscid flow; this is the boundary condition that would be used in many calculations.

Applying the above procedure to equations (B.2) and (B.6) gives

$$(\rho uv)_{\delta^*} = \frac{d}{ds} [(\rho u^2)_{\delta^*}(\delta^*+\theta)] - \tau_w$$

Using the momentum integral equation, see Appendix A, gives

$$(\rho uv)_{\delta^*} = (\rho u^2)_{\delta^*} \frac{d\delta^*}{ds}$$

i.e. (B.13)

$$v_{\delta^*} = u_{\delta^*} \frac{d\delta^*}{ds}$$

Applying the procedure to equations (B.3) and (B.7) and using the energy integral equation results again in equation (B.12).

2. TRANSPIRATION MODEL

With this model we require boundary conditions at n=0 for the inviscid calculation such that equation (B.9) is satisfied.

Integrating equation (B.1) with respect to n from n=0 to δ gives

$$\int_o^\delta \frac{\partial}{\partial s} (\rho u) dn + (\rho v)_\delta - (\rho v)_o = 0$$

Using equations (B.11) and (B.9) gives

$$(\rho v)_o = \int_o^\delta \frac{\partial}{\partial s} (\rho u) dn - \int_o^\delta \frac{\partial}{\partial s} (\overline{\rho u}) dn = \frac{d}{ds} \int_o^\delta (\rho u - \overline{\rho u}) dn$$

or using the definition of displacement thickness

$$(\rho v)_o = \frac{d}{ds} [(\rho u)_\delta \delta^*] \qquad (B.14)$$

where variations in the inviscid flow in the region $0 < n < \delta$ are ignored.

In a similar manner equation (B.2) can be shown to give

$$(\rho uv)_o = u_\delta \frac{d}{ds} [(\rho u)_\delta \delta^*]$$

or

$$u_o = u_\delta \tag{B.15}$$

where changes in u_δ in the n-direction have been ignored.

It can also be shown that equation (B.3) gives

$$H_o = H_\delta \tag{B.16}$$

Consistent with the earlier assumptions we assume

$$p_o = p_\delta$$

so that ρ_o and v_o can be found from the above relationships.

Appendix C. Pressure-Correction Method

For two-dimensional flow the equations of conservation of mass, momentum and energy can be written in the conservative form

$$\frac{\partial \underline{F}}{\partial x} + \frac{\partial \underline{G}}{\partial y} = \frac{\partial \underline{R}}{\partial x} + \frac{\partial \underline{S}}{\partial y} \tag{C.1}$$

where

$$\underline{F} = \begin{vmatrix} \rho u \\ \rho u^2 + p \\ \rho uv \\ \rho u(E+p/\rho) \end{vmatrix} \qquad \underline{G} = \begin{vmatrix} \rho v \\ \rho uv \\ \rho v^2 + p \\ \rho v(E+p/\rho) \end{vmatrix} \tag{C.2}$$

$$\underline{R} = \begin{vmatrix} 0 \\ \tau_{xx} \\ \tau_{xy} \\ R_4 \end{vmatrix} \qquad \underline{S} = \begin{vmatrix} 0 \\ \tau_{xy} \\ \tau_{yy} \\ S_4 \end{vmatrix}$$

where

$$R_4 = u\tau_{xx} + v\tau_{xy} + k\frac{\partial T}{\partial x}$$

$$S_4 = u\tau_{xy} + v\tau_{yy} + k\frac{\partial T}{\partial y}$$

$$E = C_v T + \frac{1}{2}(u^2+v^2)$$

where u and v are velocity components in the x and y-directions, T is the static temperature and $\underline{\tau}$ is the shear stress.

The non-linear system of equations is first linearised in order to establish an iterative scheme for solution and then the individual equations integrated over suitable control volumes usually on a staggered grid, see Fig. C1 from Connel and Stow [24]. For example with the x-momentum equation the convective terms for the u-momentum are taken as known from the previous iteration. So referring to Fig. C1 the integrated equation becomes

$$u_P - a_N^u u_N - a_S^u u_S - a_E^u u_E - a_W^u u_W = Q_P^u(p_P - p_W) + S_P^u$$

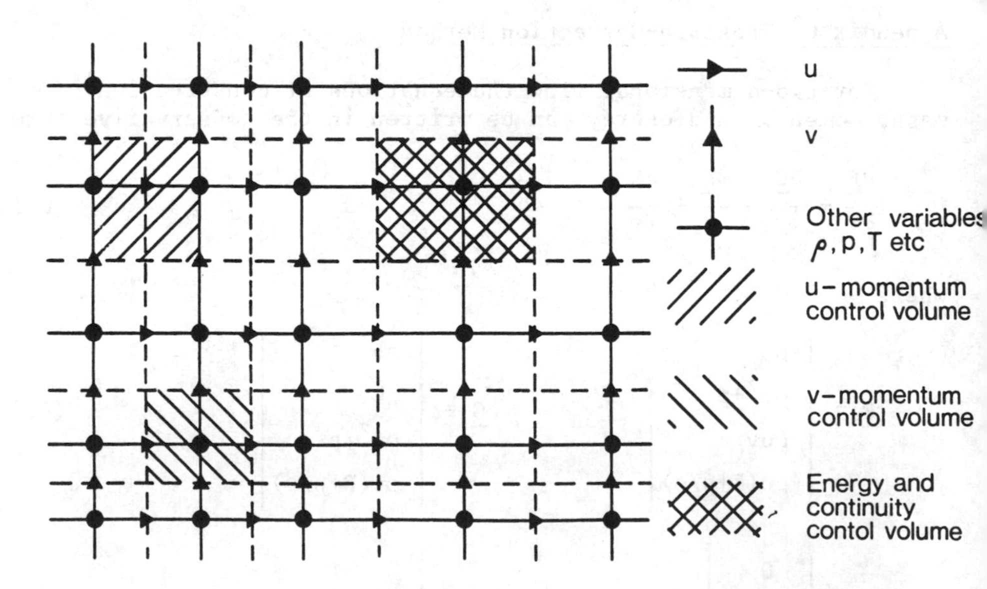

Fig. C1. Two dimensional staggered grid

where the suffix P refers to the point under consideration, N,S,E and W referring to points above, below to the right and left respectively; if higher order differencing is adopted then more points occur in the equation.

The resulting system of equations from the momentum and continuity equations can be written in the form

$$\underline{u} - A^u\underline{u} = \underline{S}^u - Q^u\underline{p} + \underline{K}^u \tag{C.3}$$

$$\underline{v} - A^v\underline{v} = \underline{S}^v - Q^v\underline{p} + \underline{K}^v \tag{C.4}$$

$$G_x\underline{u} + G_y\underline{v} = \underline{K} \tag{C.5}$$

where $\underline{u}$, $\underline{v}$ and $\underline{p}$ are the vectors of values at the points and where the $\underline{K}$ vectors arise from the boundary conditions. The matrices A^u, A^v, G_x, G_y, Q^u and Q^v and the vectors $\underline{S}^u$ and $\underline{S}^v$ depend on ρ, u and v and are taken as known from the previous iteration.

It is assumed in equations (C.3) and (C.4) that the pressure vector $\underline{p}$ is known from the previous iteration so that the equations can be solved for $\underline{u}$ and $\underline{v}$ respectively. A two-step alternating direction scheme is usually adopted in the solution procedure.

The pressure is up-dated from the continuity equation (C.5) using abreviated forms of the momentum equations (C.3) and (C.4). For example suppose using the coefficients determined by the previous iteration $\underline{u}^n$, $\underline{v}^n$ and $\underline{p}^n$, the solution to equations (C.3) and (C.4) is $\underline{u}^*$ and $\underline{v}^*$. In general these will not satisfy equation (C.5) and need to be corrected. We write

$$\underline{u}^{n+1} = \underline{u}^* + \delta\underline{u}$$

$$\underline{v}^{n+1} = \underline{v}^* + \delta\underline{v}$$

$$\underline{p}^{n+1} = \underline{p}^n + \delta\underline{p}$$

and require that as well as satisfying the momentum equations the continuity equation is also satisfied so that

$$G_x.\delta\underline{u} + G_y.\delta\underline{v} = -(G_x.\underline{u}^* + G_y.\underline{v}^*) + \underline{K} \tag{C.6}$$

From equations (C.3) and (C.4) we can see that

$$\delta \underline{u} - A^u \delta \underline{u} = -Q^u \delta \underline{p} \tag{C.7}$$

$$\delta \underline{v} - A^v \delta \underline{v} = -Q^v \delta \underline{p} \tag{C.8}$$

so that substituting into equation (C.6) gives the pressure correction equation.

$$[G_x (I-A^u)^{-1} Q^u + G_y (I-A^v)^{-1} Q^v] \delta \underline{p} = G_x \underline{u}^* + G_y \underline{v}^* - \underline{K} \tag{C.9}$$

Equation (C.9) is solved for the pressure correction and the $\underline{u}$ and $\underline{v}$ velocity vectors up-dated using equations (C.7) and (C.8).

In the original scheme by Caretto et.al.[20] and Patankar and Spalding [21] the A^u and A^v terms in equations (C.7) - (C.9) are ignored. Improved convergence rates have however been found by using extended pressure correction schemes based on the above analysis, see Connell and Stow [22].

For example using the approximation

$$(I-A^u)^{-1} \simeq I + A^u$$

$$(I-A^v)^{-1} \simeq I + A^v$$

equation (C.9) may be written in the form

$$[G_x Q^u + G_y Q^v] \delta \underline{p} = G_x (\underline{u}^* - A^u \delta \underline{u}) + G_y (\underline{v}^* - A^v \delta \underline{v}) - \underline{K}$$

The $\delta \underline{u}$ and $\delta \underline{v}$ terms on the right-hand side are added in an iterative manner repeating the process a number of times before updating the coefficient matrices A^u, A^v etc. Although more work is involved per iteration it is found that the number of iterations reduces to ensure a significant overall saving.

Appendix D. Implicit Time - Marching

The equations of motion are written in the form

$$\frac{\partial \underline{U}}{\partial t} + \frac{\partial \underline{F}}{\partial x} + \frac{\partial \underline{G}}{\partial y} = \frac{\partial \underline{R}}{\partial x} + \frac{\partial \underline{S}}{\partial y} \tag{D.1}$$

where

$$\underline{U} = \begin{bmatrix} \rho \\ \rho u \\ \rho v \\ \rho E \end{bmatrix}$$

and $\underline{F}$, $\underline{G}$, $\underline{R}$ and $\underline{S}$ are defined in Appendix C.

A general single step implicit scheme can be written in the form

$$\frac{1}{\delta t}[U^{n+1}-U^n] + \theta\left\{\frac{\partial \underline{F}}{\partial x} + \frac{\partial \underline{G}}{\partial y} - \frac{\partial \underline{R}}{\partial x} - \frac{\partial \underline{S}}{\partial y}\right\}^{n+1} + (1-\theta)\left\{\frac{\partial \underline{F}}{\partial x} + \frac{\partial \underline{G}}{\partial y} - \frac{\partial \underline{R}}{\partial x} - \frac{\partial \underline{S}}{\partial y}\right\}^{n} = 0 \qquad \text{(D.2)}$$

where $\theta=1$ gives the first order accurate backward Euler scheme and $\theta=1/2$ gives the second order Trapezoidal scheme and where the superscript denotes the time level.

We can write

$$\frac{\partial \underline{F}}{\partial x}^{n+1} = \frac{\partial \underline{F}}{\partial x}(\underline{U}^{n+1}) = \frac{\partial}{\partial x}\left[\underline{F}(\underline{U}^n) + \frac{\partial \underline{F}}{\partial \underline{U}}(\underline{U}^{n+1} - \underline{U}^n) + \ldots\right]$$

$$= \frac{\partial \underline{F}^n}{\partial x} + \frac{\partial}{\partial x}\left[A^n(\underline{U}^{n+1} - \underline{U}^n) + \ldots\right] \qquad \text{(D.3)}$$

where A is the Jacobian matrix

$$A = \frac{\partial \underline{F}}{\partial \underline{U}}$$

Consequently equation (D.2) may be written as

$$\frac{1}{\delta t}\,\delta\underline{U}^{n+1} + \theta\left\{\frac{\partial}{\partial x}(A^n\,\delta\underline{U}^{n+1}) + \frac{\partial}{\partial y}(B^n\,\delta\underline{U}^{n+1}) - \frac{\partial}{\partial x}(C^n\,\delta\underline{U}^{n+1}) - \frac{\partial}{\partial y}(D^n\,\delta\underline{U}^{n+1})\right\} + \left\{\frac{\partial \underline{F}}{\partial x} + \frac{\partial \underline{G}}{\partial y} - \frac{\partial \underline{R}}{\partial x} - \frac{\partial \underline{S}}{\partial y}\right\}^n = 0$$

or

$$[I + \theta\delta t\,(\frac{\partial}{\partial x} A^n + \frac{\partial}{\partial y} B^n - \frac{\partial}{\partial x} C^n - \frac{\partial}{\partial y} D^n)]\ \delta\underline{U}^{n+1} = \Delta\underline{U}^n \qquad (D.4)$$

where

$$\delta\underline{U}^{n+1} = \underline{U}^{n+1} - \underline{U}^n$$

$$\Delta\underline{U}^n = -\delta t\,[\frac{\partial\underline{F}}{\partial x} + \frac{\partial\underline{G}}{\partial y} - \frac{\partial\underline{R}}{\partial x} - \frac{\partial\underline{S}}{\partial y}]^n$$

and B, C and D are the Jacobians

$$B = \frac{\partial\underline{G}}{\partial\underline{U}}, \quad C = \frac{\partial\underline{R}}{\partial\underline{U}}, \quad D = \frac{\partial\underline{S}}{\partial\underline{U}}$$

It should be noted that the Jacobians A,B,C and D can be determined analytically, see Dawes [25].

The equation (D.4) can be written in the form of a two-step alternating direction procedure as

$$[I + \theta\delta t\,(\frac{\partial}{\partial x} A^n - \frac{\partial}{\partial x} C^n)]\ \delta\underline{U}^* = \Delta\underline{U}^n$$

and

$$[I + \theta\delta t\,(\frac{\partial}{\partial y} B^n - \frac{\partial}{\partial y} D^n)]\ \delta\underline{U}^{n+1} = \delta\underline{U}^* \qquad (D.5)$$

With the usual finite differencing employed for spatial derivatives (e.g. central differences) the above equations (D.5) give rise to block tri-diagonal matrices to solve whereas a block penta-diagonal scheme would have resulted from equation (D.4). The above equations are supplemented for turbulent flow usually with a steady state form of turbulence model e.g. mixing length, two equation model etc.

It is found that because of the coupling of the correction equations then convergence results in a relatively small number of iterations e.g. 300; there are also no stability restrictions on the time step δt than can be used. If one is interested in time accurate solutions then the same time step would be used over the whole domain otherwise a varying time step chosen from local considerations can be used to speed convergence to a steady state solution.

LIST OF SYMBOLS

C_E	entrainment coefficient
E	total energy
H	stagnation enthalpy,form factor
H1	modified form factor
n	normal to blade surface
p	static pressure
q	heat flux
s	streamwise direction
t	time
T	static temperature
U	velocity components
$\underline{U}$	state vector
δ	boundary layer thickness
δ^*	displacement thickness
θ	momentum thickness
θ_H	total enthalpy thickness
ρ	density
τ	skin friction

Subscripts

o	blade surface value
δ	edge of boundary layer
w	wall value
eq	equilibrium value
—	viscous value

Superscripts

n	iteration number

TURBOMACHINERY BLADE DESIGN USING ADVANCED CALCULATION METHODS

Peter Stow

Chief Aerodynamic Scientist
Rolls-Royce
Derby
England

1. QUASI-THREE-DIMENSIONAL BLADE DESIGN SYSTEM

1.1. Introduction

In the design of compressors and turbines one designs individual nozzles, rotors, or stages to perform specific tasks. So, for example, one might require a compressor stage to achieve a desired increase in total pressure or a turbine stage to achieve a desired work output, both with some lower limits on the efficiency to be achieved.

Components are designed with the aid of theoretical methods and computer programs. If one is to have confidence in achieving design targets in practice using these theoretical methods then it is essential that the mathematical model employed in the method adequately describes the physical processes involved. However, the flow in turbomachines is very complex. There are rotating and non-rotating components. The blade geometries are three-dimensional; the sectional geometry as well as the orientation (stagger) changes for hub to tip. There are boundary layers on the annulus walls and the blade surfaces, wakes from the trailing edges of the blades, over-tip leakage flows, cooling flows being ejected over the blade or annulus surfaces etc. Consequently the flow is unsteady, three-dimensional and has regions where viscous effects are important. The solution of the full equations of motion with the full boundary conditions represents a formidable task both from a computational and modelling point of view. Even for steady flows it is likely that for some years to come methods for solving three-dimensional problems will be of such a speed that only isolated blade rows or at best single stages can be analysed in the design times that are available; it should also be mentioned that full three-dimensional design is not envisaged as being an easy task.

Many design systems adopt the philosophy proposed by Wu [1], for the design of single blade rows, namely to tackle the full problem in two stages by calculating the flows on two intersecting families of stream-surfaces S1 and S2, see Figure 1. The S1 family of surfaces is essentially a set of blade-to-blade surfaces, the solution of the flow on each of these surfaces being referred to as the blade-to-blade calculation. The S2 family of stream-surfaces lie between the blades, the solution of the flow on each surface of this family being referred to as the through-flow calculation. Using Wu's general theory the complete three-dimensional flow field through a blade row could be obtained by adopting an iterative procedure to link the calculations for the flow on the two families of stream-surfaces, see for example Krimerman and Adler [2]. In general, however, this procedure has not been adopted. Instead the majority of the workers have used the philosophy of Wu's S1 and S2 stream-surfaces without adopting an iterative link. One commonly adopted procedure is to use only one S2 stream-surface through the blade row together with a number of S1 stream-surfaces; with this

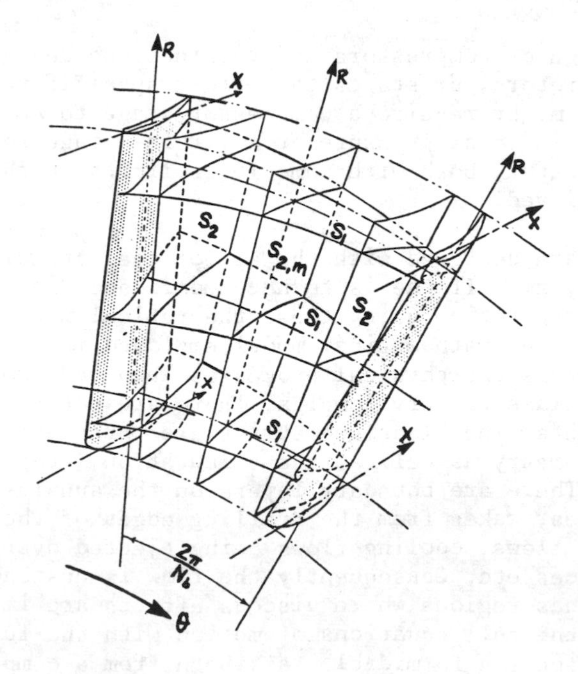

Fig. 1. S1, S2 stream surfaces

approach the stream-surfaces must by necessity be taken as axisymmetric. The procedure is often referred to as a quasi-three-dimensional design system.

Although, with this approach, the equations to be solved are simpler and the design task itself is made easier, by reducing the number of design variables (three-dimensional to two-dimensional), there are a number of short comings. It is found that certain flow features cannot be modelled and that design freedoms to control or exploit these features are consequently restricted. This point will be discussed later together with the need for a full three-dimensional design system.

In the following sections each stage in the quasi-three-dimensional design system is presented, see Figure 2. Also included are the ways in which three-dimensional effects and considerations can be included in the system.

1.2 Through-Flow Calculation

In the through-flow calculation the flow through a number of stages is usually considered. The flow is taken as steady in an absolute coordinate system for a non-rotating component and in a relative coordinate system for a rotating component. As a consequence the detailed effects of blade wakes and flow

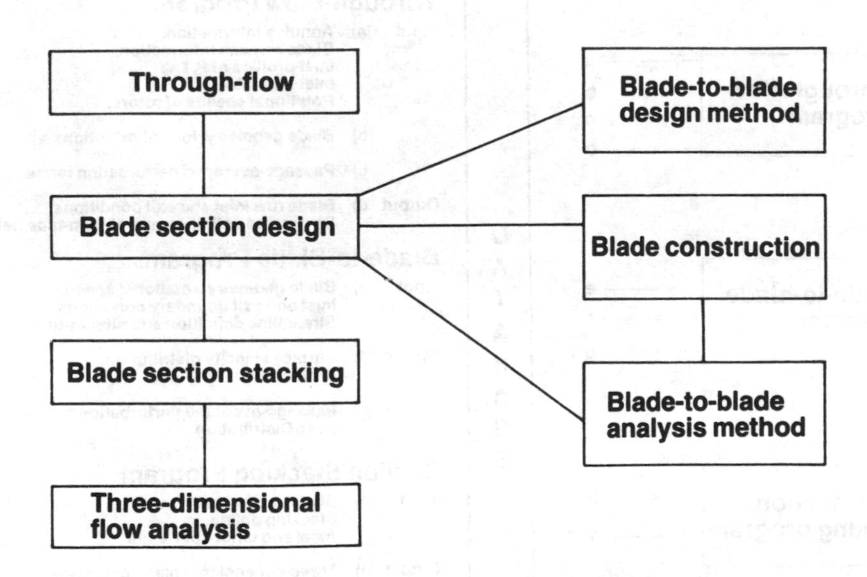

Fig. 2. Blade design system

unsteadiness are ignored. The three-dimensional equations of motion are usually reduced to a two-dimensional form either by assuming fully axisymmetric flow or by adopting a passage-averaging (or mean stream-sheet approach). In the former only calculations outside the blade rows are performed and assumptions need to be made as to how properties needed in the blade-to-blade calculation vary through the blade row. In the passage-averaging technique the three-dimensional equations are applied within the blade rows and integrated in the circumferential direction to produce two-dimensional equations for averaged quantities; in this way the effects of the blade geometry on the through-flow can be accounted for. Figure 3 illustrates the information flow in such a quasi-three-dimensional through-flow system, see Jennions and Stow [3] for more details and also Hirsch and Warzee [4].

Certain design targets, for example stage pressure ratio or stage work etc, are specified as boundary conditions in order to solve the through-flow equations. From the solution information is produced which defines the axisymmetric S1 stream-surfaces and the spacings of these (stream-tube heights) through the blade rows as well as the inlet and outlet flow conditions, i.e. velocities and whirl angles, to be used as boundary conditions for the blade design. It is found that the stream-tube height variation, and in

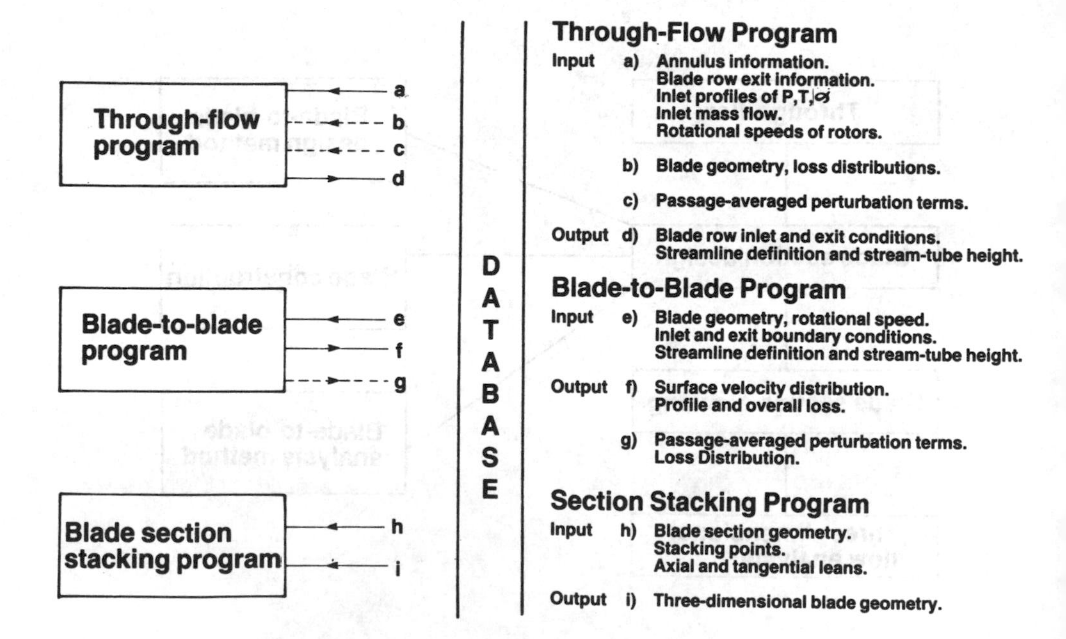

Fig. 3. Quasi-Three-Dimensional Through-Flow System

some cases the radius variation for a rotor, can have a very important effect on the blade section aerodynamics and hence the surface boundary layer development and consequent profile loss, see for example Figure 4 taken from Cedar and Stow [5] for a supercritical controlled diffusion compressor. In addition it is found that the inlet flow conditions to be used for blade design can be affected by the streamline definition through the blade row, see for example Figure 5 from Jennions [6] illustrating the effect for a fan. It is important therefore, in order to have confidence in ones design predictions or in ones ability to exploit design freedoms (e.g., annulus shape), that an accurate definition is produced from the through-flow calculation.

The stream-tube height and streamline radius variations through the blade row are affected by the blade geometry, i.e., blade blockage, turning and stack, as well as the annulus shape and consequently calculating stations within the blade rows are often essential. In the early stages the blade effects can be included in an approximate manner with the system depicted in Figure 3 being used in later stages.

In the quasi-three-dimensional approach the effects of the blade force, annulus shape etc. can be accounted for but the

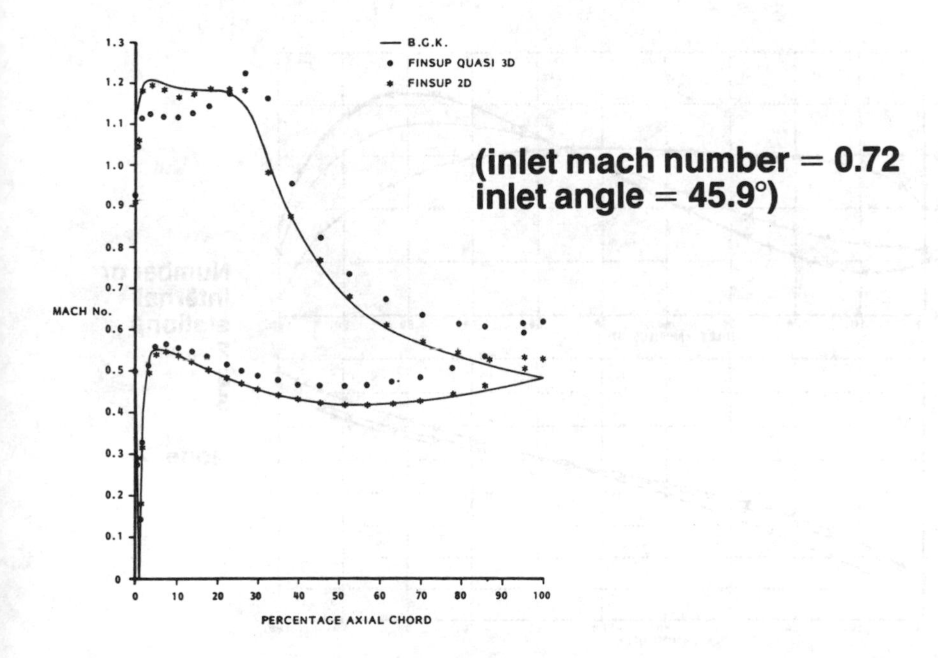

Fig. 4. Predicated mach number distribution around Bauer-Grabedian Korn test blade

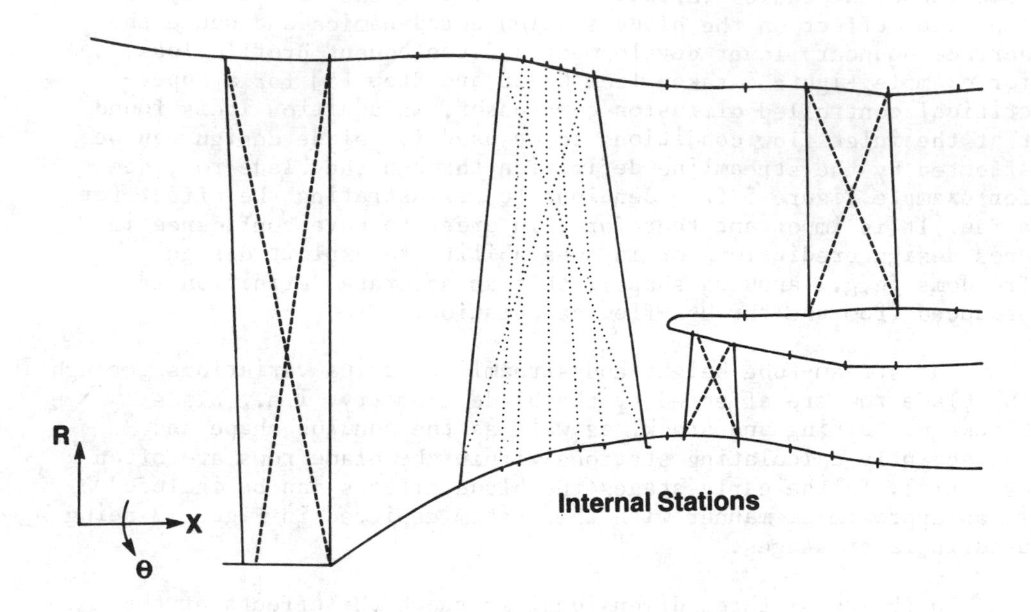

Fig. 5a. Q263 input generator program

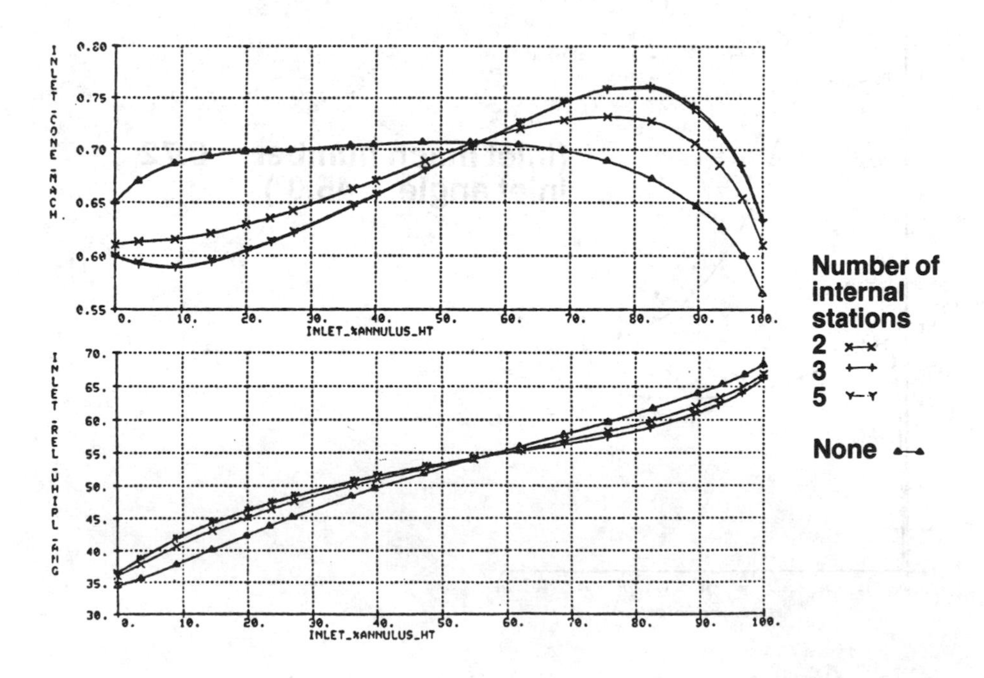

Fig. 5b. Q-3D through-flow effects

blade-to-blade stream-surfaces are assumed to be axisymmetric through the blade row. In a true three-dimensional flow, stream-surfaces, even if axisymmetric at inlet to a blade row, would become non-axisymmetric as they passed through the blade row under the influence of generated streamwise vorticity. This vorticity might be from inlet mainstream streamwise vorticity, inlet circumferential vorticity in the end-wall boundary layers, non-uniform spanwise blade loading or circulation etc. The magnitude of the stream-surface twisting through the blade row is governed by blade parameters like turning, blockage and stack.

The secondary flow generated by the streamwise vorticity serves to transport mass and momentum through the assumed axisymmetric stream-surfaces. Whereas the overall passage conservation of flow properties will be correct in a quasi-three-dimensional approach, the details may not be. This could affect the losses produced by the blade sections, the turning produced, the effectiveness of a cooling system (from the flow temperatures assumed) etc. If the secondary flow effects are large then detailed refinement in the design of blade sections will not be worthwhile and one needs to consider the full three-dimensional flow. It is advisable then that the secondary flow effects are considered in the early stages of blade design. This can often be achieved easily and conveniently using what is termed inviscid secondary flow theory, see for example Hawthorne [7], Lakshminarayana and Horlock [8], Marsh [9], and incorporating into or linking to a through-flow analysis as in Smith [10] and James [11]. This approach can give valuable information about secondary flows and the importance of the three-dimensional effects, see for example Figure 6 from James [11] on predicted outlet angle. It is also possible to use the approach in a design sense to calculate blade angles in the end-wall regions to reduce the effects of boundary layer over-turning.

1.3 Blade Section Design

In the blade-to-blade analysis, the flow on one isolated S1 stream-surface through a single blade row is considered at any one time. The flow is taken as steady in an absolute or relative coordinate system depending on whether a stator or rotor is being analysed. Information from the through-flow calculation is used in order to reduce the equations of motion to a two-dimensional form as well as to provide inlet and outlet boundary conditions for the design. In general the S1 stream-surface is taken as axisymmetric although if a more general linked S1-S2 calculation were adopted then this approximation could be avoided.

The inlet flow conditions are generally taken as uniform and steady so that the effects of wakes from upstream blade rows are not considered (at least in detail, their effects being included

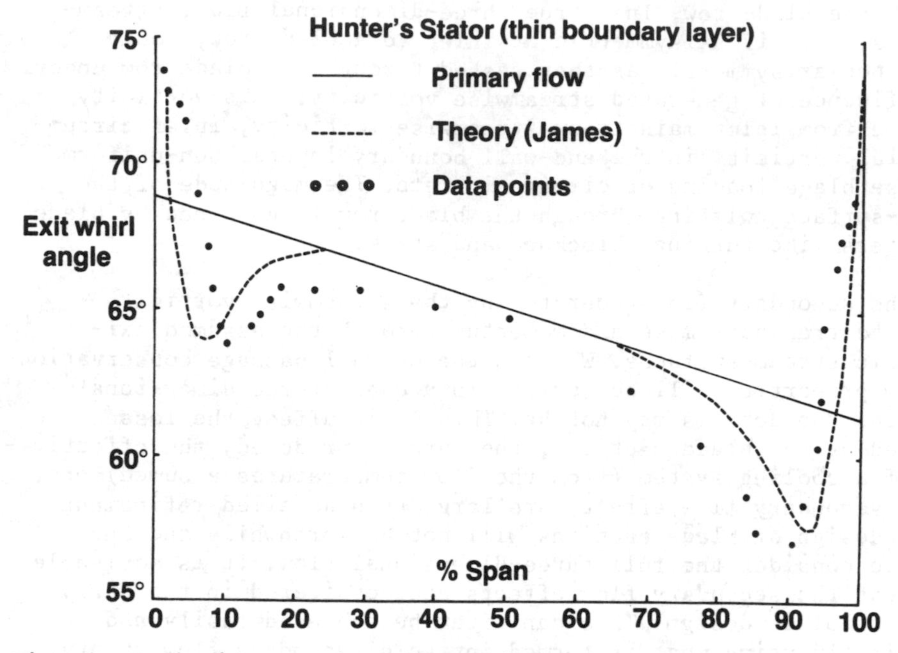

Fig. 6. Inviscid secondary flow whirl angles

in a mean sense since the prescribed loss for the upstream row determines the inlet flow conditions to be used).

Two main approaches are adopted for designing a blade section that satisfies the through-flow boundary conditions and any additional aerodynamic or geometric constraints that may be imposed, e.g., efficiency targets, blade cooling requirements. The two approaches are to use a design calculation or an analysis calculation.

1.3.1 Design Method

With a design method the blade surface velocity is prescribed and the method produces the geometry that will achieve this; surface boundary layers can be taken into account in arriving at the final geometry to be manufactured. Examples of such methods are Stanitz [12], Schmidt [13], Bauer, Garabedian and Korn [14]. In prescribing the surface velocity, boundary conditions and aerodynamic constraints, e.g., efficiency targets, absence of boundary layer separation, shock free transonic patches etc., can be taken into account either directly, in the calculation method or indirectly by the user of the method.

Figure 7 illustrates results from a method based on the work of Stanitz for the design of an H.P. turbine blade.

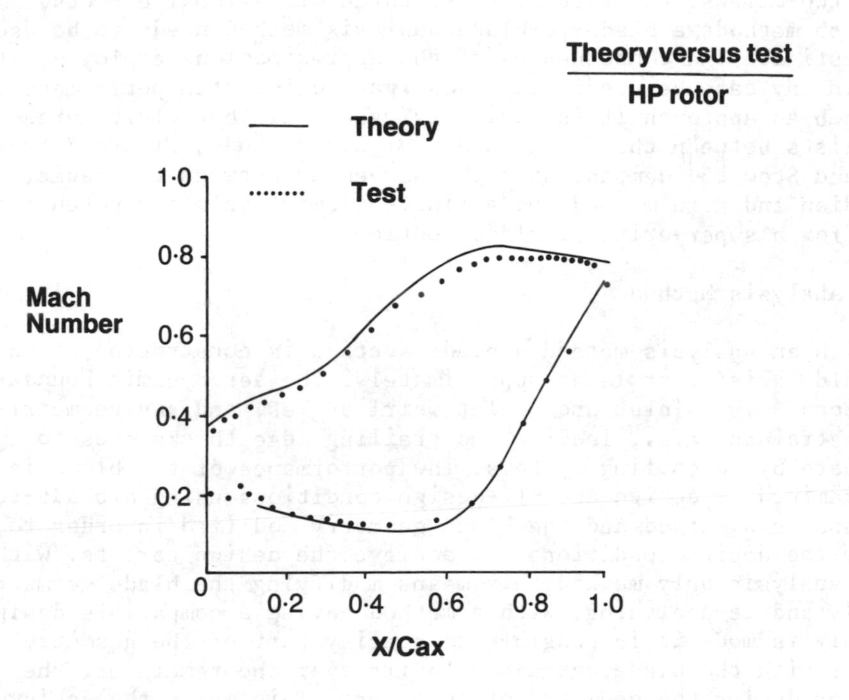

Fig. 7a. Design method

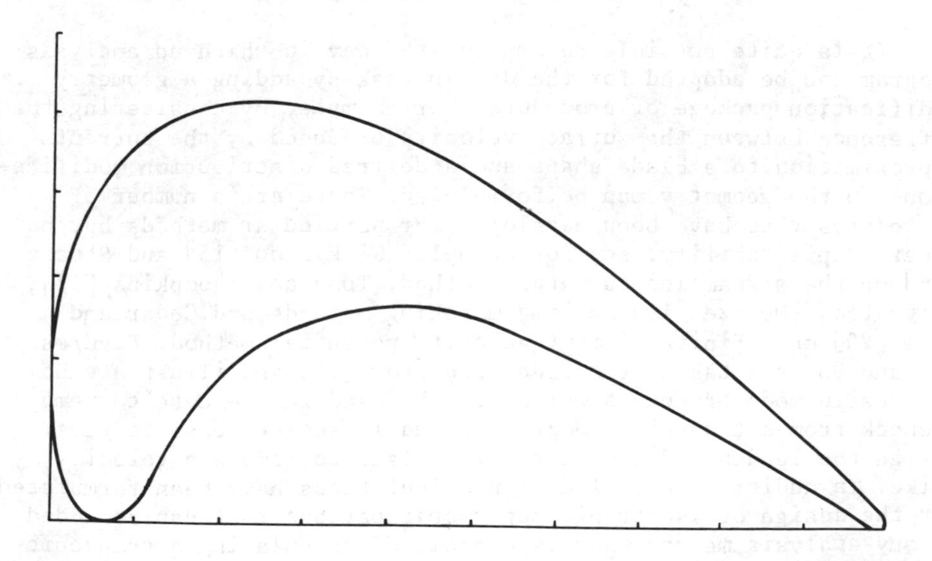

Fig. 7b. HP rotor

In many design methods approximations need to be adopted in the mathematical model in order to be able to solve the equations of motion e.g., the Bauer, Garabedian and Korn [14] method is purely two-dimensional with no quasi-three-dimensional effects. With such methods a blade-to-blade analysis method needs to be used to investigate the consequences of the approximations employed; it would in any case be needed to investigate off-design performance. With such an approach it is obviously essential that close agreement exists between the design and analysis methods; Figure 8 from Cedar and Stow [5] demonstrates the agreement between the Bauer, Garabedian and Korn method and a finite element velocity potential method for a super-critical blade section.

1.3.2 Analysis Method

With an analysis method a blade section is constructed so that it should satisfy, at least approximately, the aerodynamic boundary conditions, e.g., inlet and outlet whirl angles, and any geometrical constraints, e.g., leading and trailing edge thicknesses to accomodate blade cooling systems. The performance of the blade is then examined at design and off-design conditions using a blade-to-blade analysis method and the blade geometry modified in order to satisfy the design conditions and achieve the design targets. With a pure analysis only method this means modifying the blade geometry directly and re-analysing. With a method having a compatible design and analysis mode it is possible to specify part of the geometry together with the blade surface velocity over the remainder, the method producing the geometry of this part. This makes the achievement of mixed aerodynamic and geometric constraints easier.

It is quite possible to improve the way in which an analysis program can be adopted for the design task by adding a geometry modification package or procedure. For example, by considering the difference between the surface velocity produced by the current approximation to a blade shape and a desired distribution modifications to the geometry can be formulated. There are a number of procedures that have been developed for particular methods but have general applicability, see for examples Wilkinson [15] and Stuart [16] on the streamline curvature method, Tong and Thompkins [17], Paige [18], Meauze [19] on time-marching methods and Cedar and Stow [20] on a finite element velocity potential method. Figures 9a and 9b are taken from Cedar and Stow [20] and illustrate how the design mode of such a method can be used in one case to remove a shock from a transonic compressor and in another case to redesign the leading edge of a turbine blade to remove a velocity spike. In addition special design calculations have been formulated for the design of isentropic supersonic patches that can be added to any analysis method that is capable of calculating a transonic patch (in an accurate or an approximate manner), see for example Sobieczky, Yu, Fung and Seebass [21].

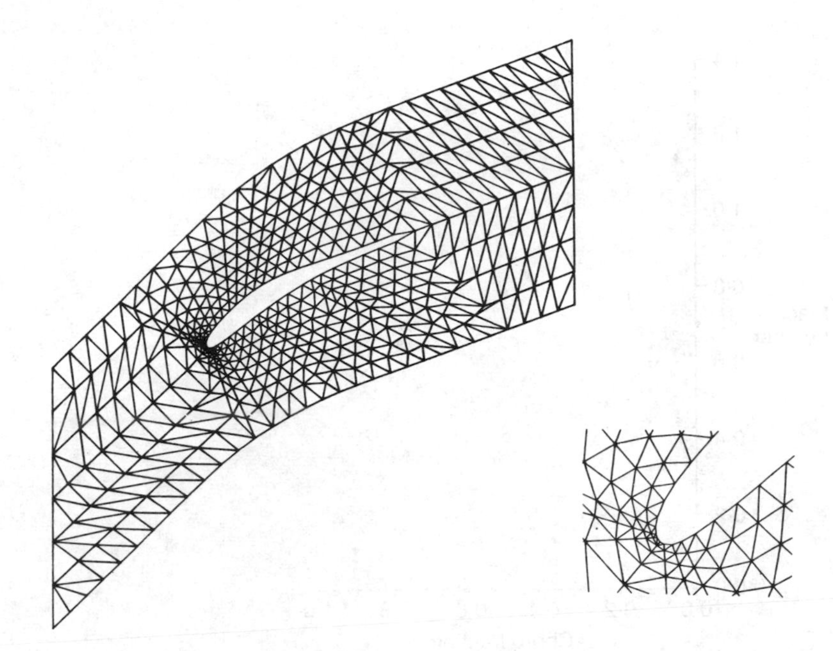

Fig. 8a. Finite element mesh used to analyse Bauer Garabedian Korn

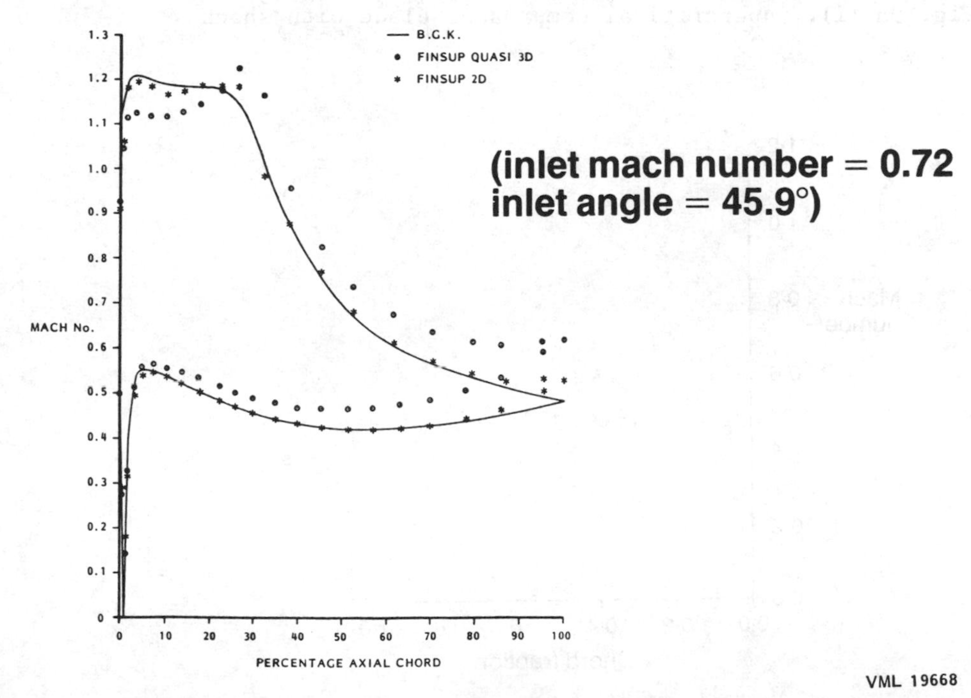

Fig. 8b. Predicted mach number distribution around Bauer-Garabedian Korn test blade

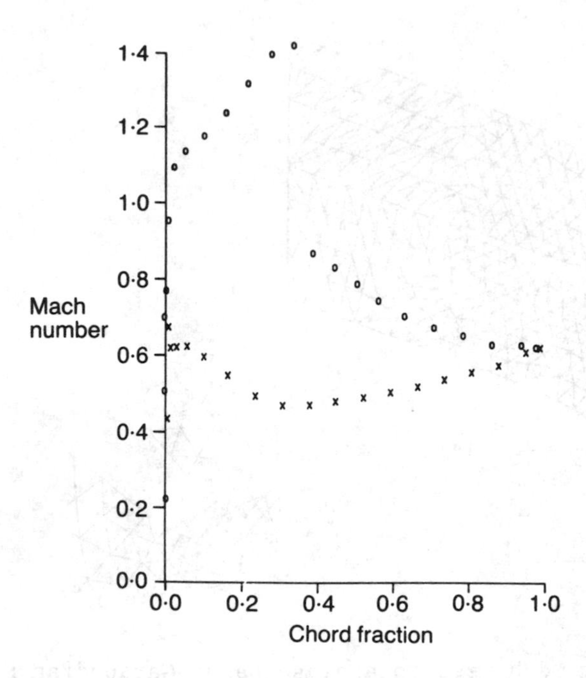

Fig. 9a (i). Supercritical compressor blade with shock

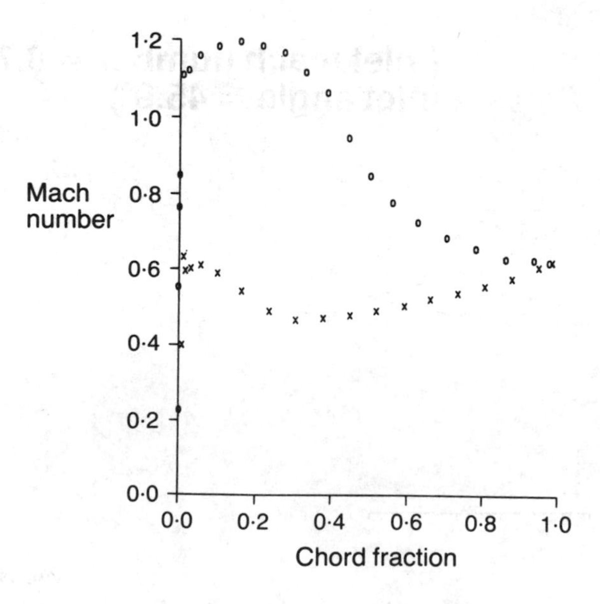

Fig. 9a (ii). Design mach number distribution

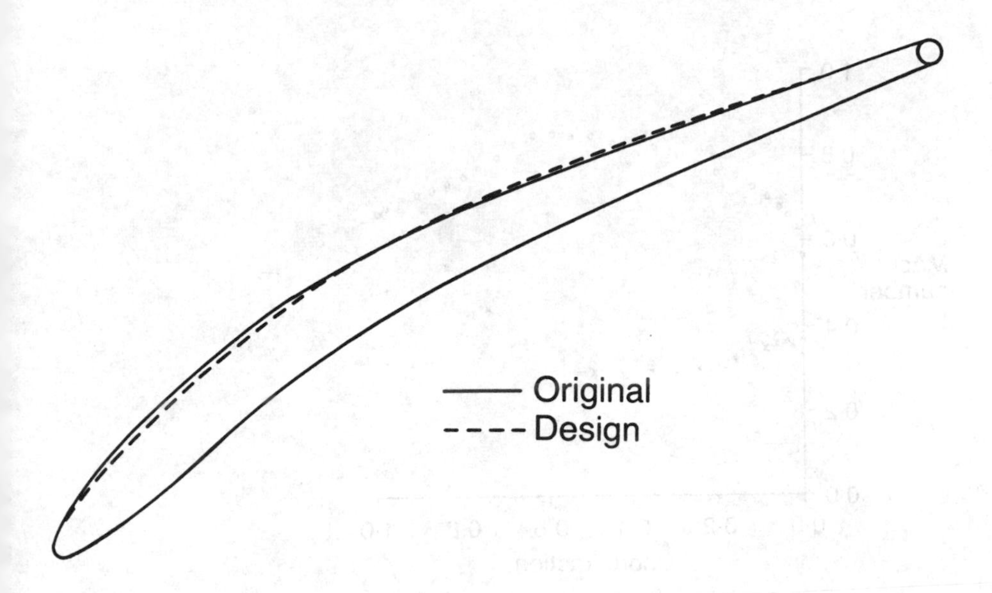

Fig. 9a (iii). Change in blade shape to remove shock

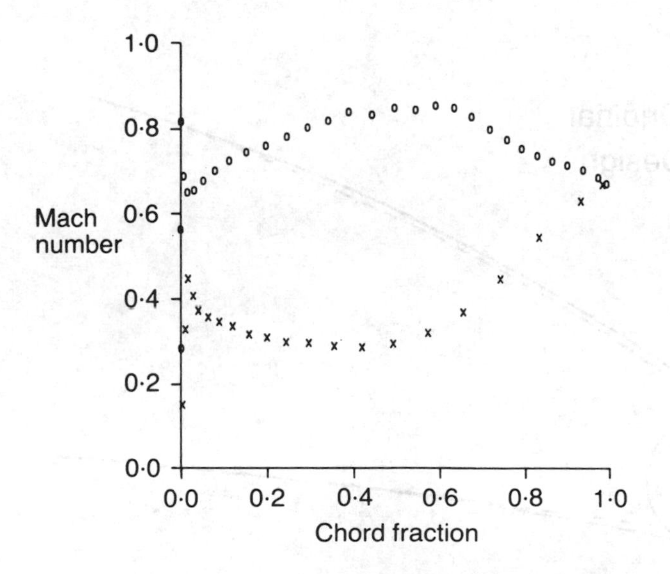

Fig. 9b (i). Analysis of turbine blade with leading edge spike

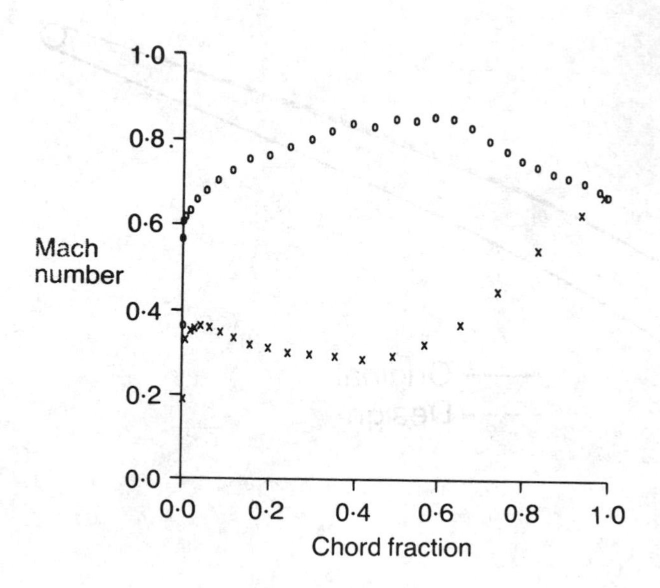

Fig. 9b (ii). Redesigned blade to remove leading edge spike

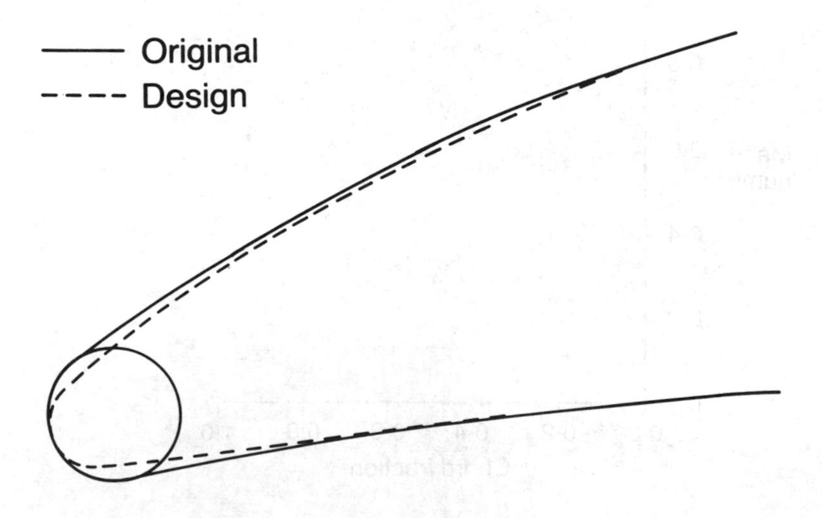

Fig. 9b (iii). Change in blade shape to remove leading edge spike

In some less demanding cases it may be that design targets can be achieved using blade sections belonging to a standard family e.g., double or multiple circular arc, NACA series etc. Boundary conditions together with experimentally derived correlations for blade loss and flow deviation against incidence are used to determine the parameters describing the family, e.g., space-chord, thickness-chord, stagger etc. In general, however, current efficiency demands mean that this approach is both conservative and restrictive but the technique of parameterising blades of a family is still useful. Using theoretical methods, e.g., a design method together with a surface boundary layer calculation, one can arrive at a new type of profile form. From this one attempts to describe the profile in terms of a number of parameters, these in turn being related to flow parameters or blade duty. With this procedure it is possible to save significant amounts of design time especially for compressors where many stages need to be designed. It can also be usefully employed in generating starting profiles in the design process.

1.3.3 Loss Considerations and Predictions

In order to predict the blade profile loss it is essential to have detailed and accurate surface velocity distributions at both design and off-design conditions.

Especially important is the leading edge region where with som e blades leading edge velocity spikes can occur especially at off-design conditions sufficient to cause laminar boundary layer separation bubbles. This was a region often neglected in the past, especially in design programs producing cusped leading edges where leading edge circles were often fitted with little consideration for the aerodynamic consequences. In some cases design of the blade leading edge can have an important effect on the profile loss (as long as the blade can be manufactured to the desired accuracy).

For each blade section to be designed the desired inlet and outlet flow conditions, and hence the overall lift of the section, are prescribed. The freedom in the design lies in the lift distribution which determines the surface boundary layer growth and hence the profile loss. Whereas in theory optimisation techniques could be coupled to design or analysis methods for blade design one still often resorts to 'stylised' distributions, e.g., aft-loaded, flat-topped etc., see Figure 10. Which type of distribution is preferable often depends on the flow conditions to be experienced by the blade e.g., Reynolds number, free stream turbulence etc. It is important, therefore, that boundary layer prediction methods, whether they are finite difference, finite element or integral, properly account for these effects. Other important aspects,

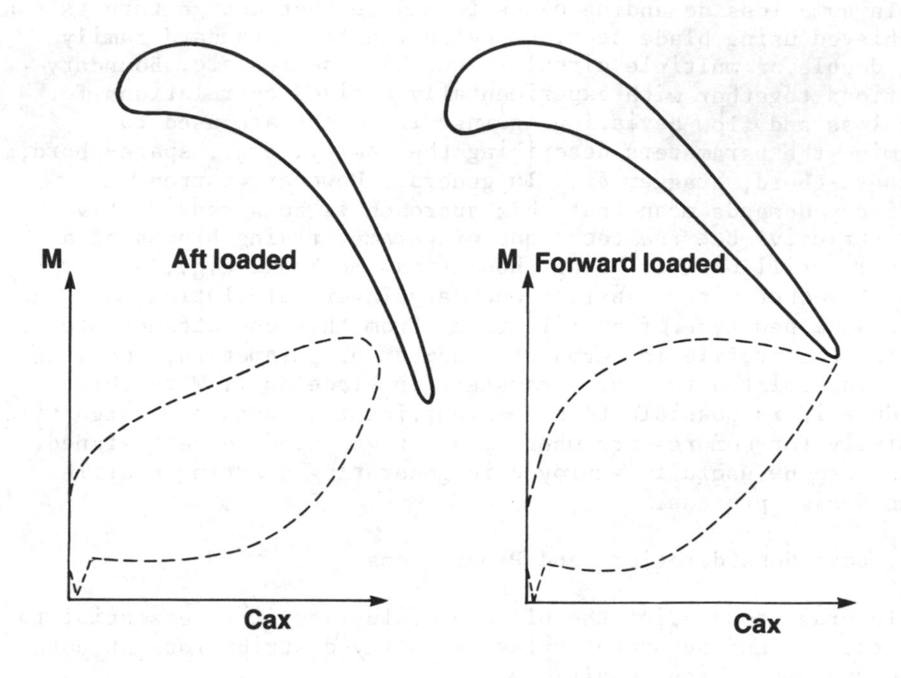

Fig. 10. Loading distribution

especially at off-design conditions, are the prediction of laminar separation bubbles and turbulent separation.

Although fully viscous blade-to-blade methods exist, see for example Dawes [22], Steger, Pulliam and Chima [23] and Thompkins et.al. [24], most workers have concentrated on using an inviscid mainstream calculation with a separate surface boundary layer calculation; often in the interests of computational speed the latter being an integral method. For some flows it is found that the boundary layer has only a minor influence on the mainstream flow and meaningful predictions can be made using a single call to the boundary layer calculation. For many flows, however, the boundary layer can have a significant influence on the flow and coupled mainstream-boundary layer calculations are essential. In cases where the boundary layer is relatively thin and can still be considered to have a second order effect, e.g., on many turbine blades or controlled diffusion compressor blades etc. then this coupling can be achieved using direct modes of each calculation, see Figure 11. However, in cases where the boundary layer plays a dominant role e.g., compressor blades with a significant turbulent separation, shock boundary layer interactions, then the direct mode of operation becomes unstable. One solution to the problem is

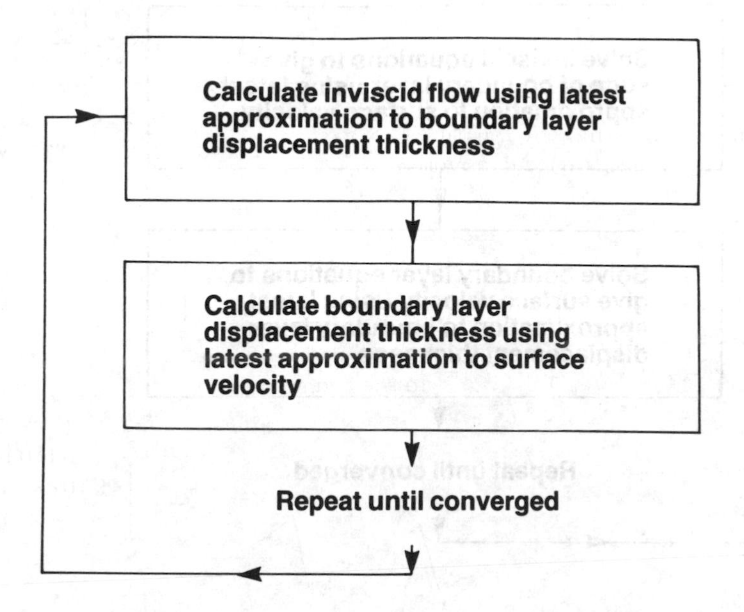

Fig. 11. Direct mode boundary layer

to adopt an inverse mode approach for both the mainstream and boundary layer calculations as developed for example by Calvert [25]; see also Figure 12 for details of the linking with the inverse approach. Another possibility is to solve the two sets of equations simultaneously in a more strongly coupled mode.

The above statements on coupling are irrespective of whether a finite difference or integral boundary layer calculation is used or whether a displacement or surface transpiration model is adopted to account for the effects of the boundary layer on the mainstream.

1.4 Blade Section Stacking

Using a number of S1 stream-surfaces from hub to tip, sections of the blade are designed; these are then stacked radially and circumferentially to form the full three-dimensional blade geometry, see Figure 13. In the stacking procedure stress, geometric and aerodynamic constraints are taken into account. It is only after this stage that the effects of the three-dimensional geometry on the flow can be assessed. This can be done in part using a quasi-three-dimensional through-flow analysis, see Section 1.2 and Figure 3, where the effects of blade geometry and blade passage variations in flow quantities are taken into account. As already

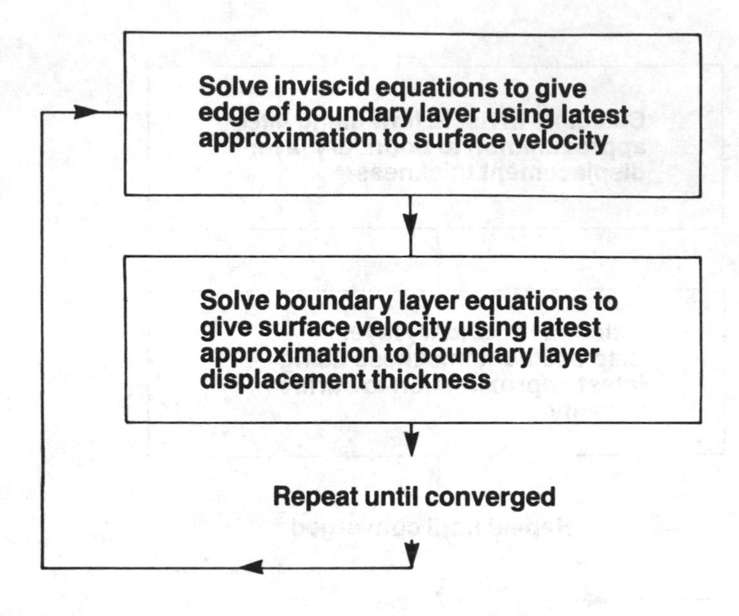

Fig. 12. Inverse mode boundary layer

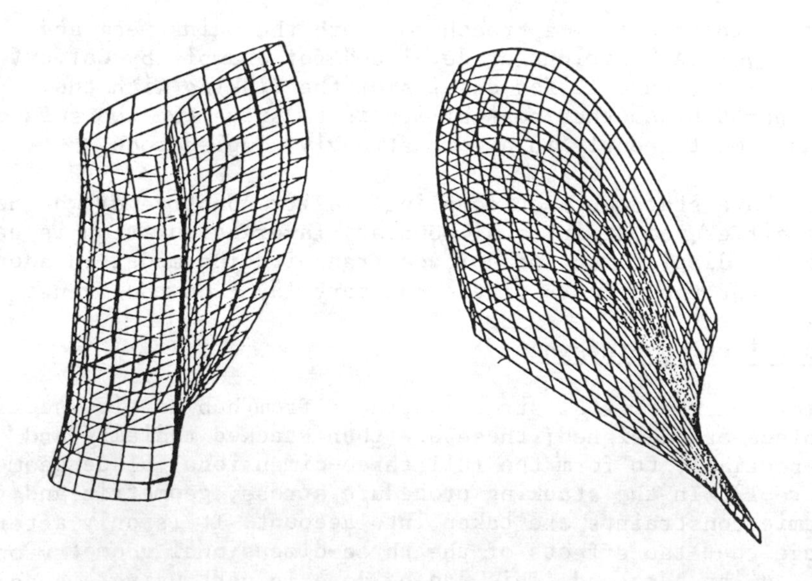

Fig. 13. Blade section stacking

mentioned the approximation adopted in the analysis is that the S1 blade-to-blade stream-surfaces are axisymmetric. In some cases this may be acceptable, in other cases twisting of the stream-surfaces, as indicated by a secondary flow analysis, may be significant and mean that a fully three-dimensional analysis is necessary.

Figures 14a and 14b taken from Jennions and Stow [3] show how such a linked system compares well with both experimental data and a fully three-dimensional analysis for the turbine geometry shown.

2. THREE-DIMENSIONAL PHENOMENA

It is worthwhile at this stage considering some of the three-dimensional aspects of turbomachinery flows that cannot be modelled by a quasi-three-dimensional system even one incorporating linking between the through-flow and blade-to-blade calculations.

As has already been mentioned in Section 1.2, in reality stream-surfaces are not axisymmetric through a blade row. This means that there will be transport of mass, momentum and energy through the blade-to-blade stream-surfaces used for designing. If these effects are large then they could affect a number of parameters. For example, temperature profiles and heat transfer rates that were used in designing blade cooling systems, the exit whirl angle distribution achieved by the row and the actual profile losses achieved. So that as well as affecting the performance of the designed row they could also affect that of the downstream row.

In the quasi-three-dimensional system blade sections are designed in isolation. There are, however, certain flow phenomena that cannot necessarily be analysed in this fashion e.g., transonic, choked or shocked flows. In the case of a shocked flow there is little reason to suppose in general that a shock predicted in a quasi-three-dimensional blade-to-blade analysis will form part of the true three-dimensional shock system. For a choked blade row there is little reason to suppose that, without certain constraints being built into the blade stack, a throat for a blade section forms part of the overall three-dimensional throat window for the passage, where annulus effects must be considered. Similar conclusions apply to boundary layer transition and laminar or turbulent separation.

The interaction of the blade with the end-wall boundary layer is a truly three-dimensional phenomenon. The end-wall boundary layer separates near the leading edge to form a horse-shoe vortex, see Figure 15, with one leg around the pressure surface and the other around the suction surface. The pressure surface leg tends to move across the passage towards the suction surface interacting

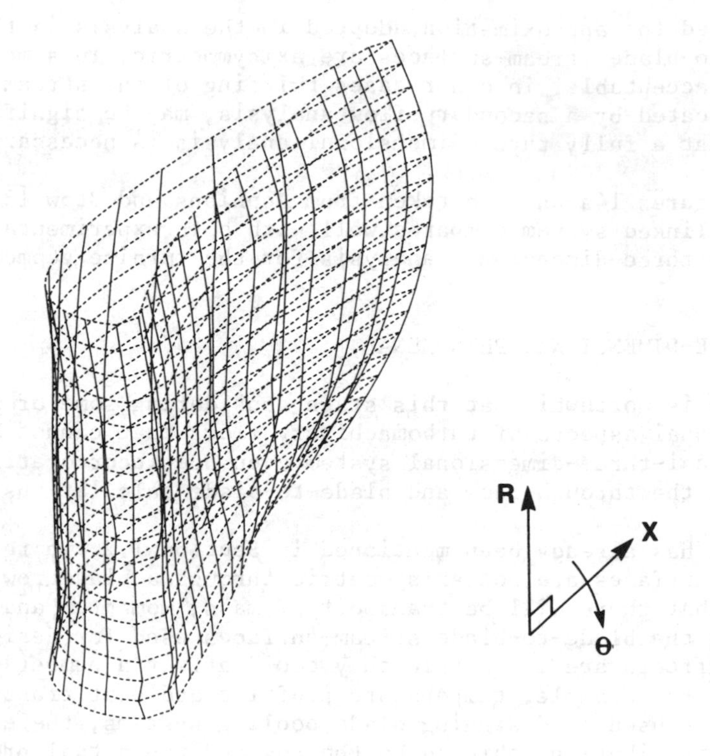

Fig. 14a. Vane geometry

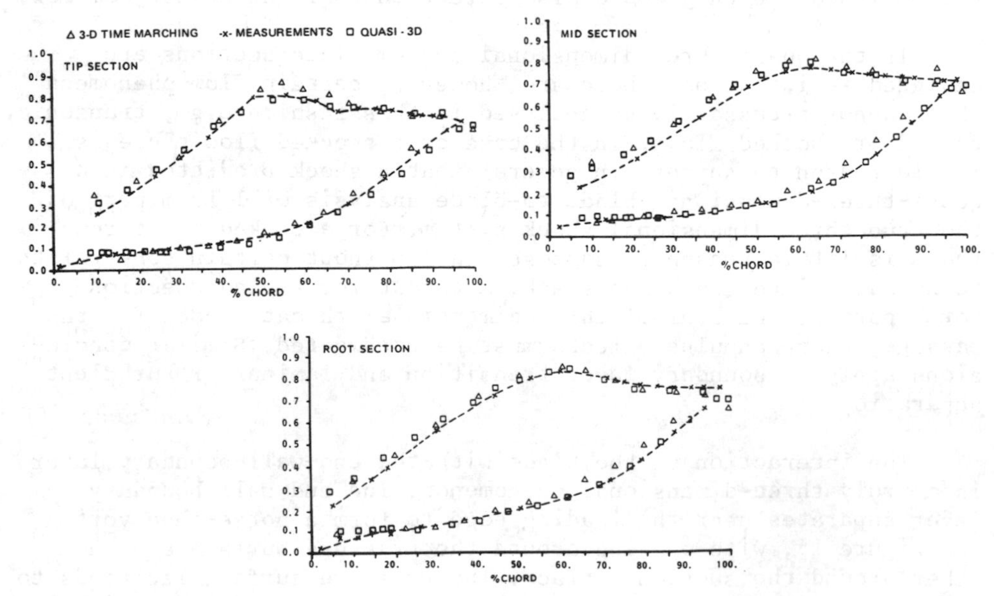

Fig. 14b. Surface Mach number comparison with fully 3D results

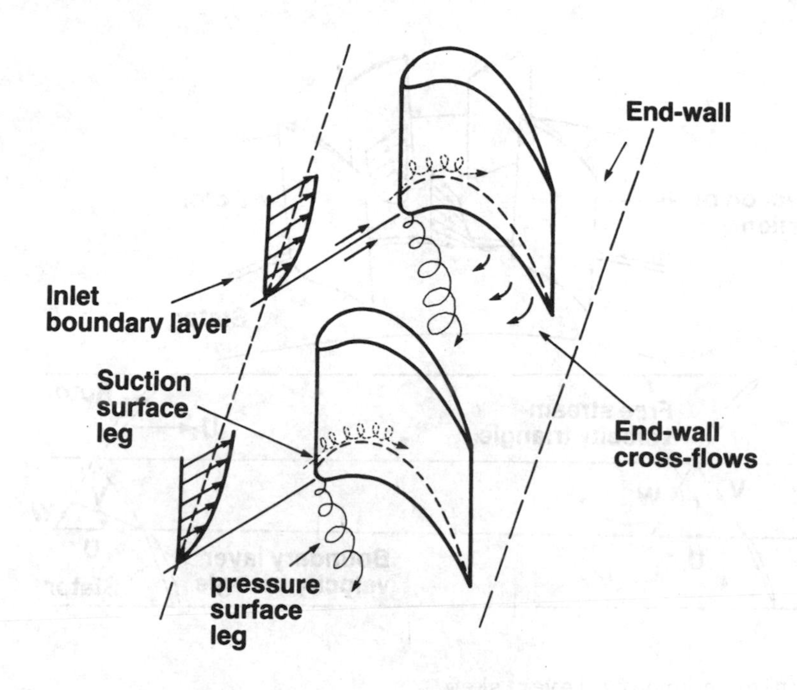

Fig. 15. Leading edge horse-shoe vortex

with the other leg and often lifting off the end-wall. Blade surface boundary layers are obviously affected by these vortices being drawn into new end-wall boundary layers feeding the vortices.

Even without the prescence of the leading edge vortices the flow characteristics of the end-wall boundary layers would, in general, be three-dimensional due to the action of the blade force. Outside of the boundary layer the blade force balances the momentum of the inviscid mainstream. Without special design of the blade ends the force will be approximately constant through the boundary layer and hence, because of the lower momentum in the boundary layer, the flow will follow a path of higher curvature and a cross-flow will be set-up in the boundary layer. This flow would be fed by flow down the pressure-surface and up the suction-surface of the blade.

The end-wall boundary layer is further complicated by the fact that the annulus has both moving and stationary parts. Consider for example the flow out of a rotor and into the downstream stator, see Figure 16. It can be seen that even in the case where the flow in the end-wall boundary layer from the rotor is collateral (i.e., there are no cross-flows and the relative velocity vector lies in

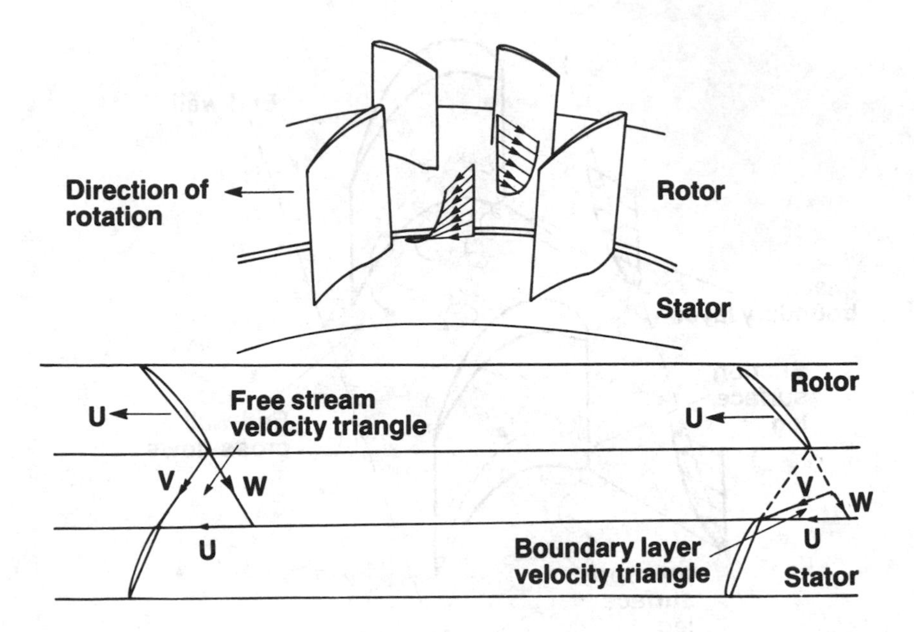

Fig. 16. Inlet boundary layer skew

a plane) then the absolute velocity vector through the boundary layer will have large cross flows. The velocity on the stator end-wall will be zero so that large cross-wise shear stresses are created which, given sufficient distance to act, will destroy the cross-flow and create a collateral boundary layer in the absolute frame. This effect can be enhanced or counteracted by the presence of the blade. By designing the blade ends to take this effect into account the end-wall boundary layer and the blade surface boundary layers can be affected.

Further complications exist at the tip of rotor blades. In the case of a shrouded rotor, see Figure 17 there will be some flow over the tip seals which will eventually mix in with the mainstream flow downstream of the rotor. In the case of an unshrouded rotor there is a leakage flow over the tip of the blade. The flow rolls up to form a tip leakage vortex, see Figure 18, with obvious effects on the blade surface boundary layers.

3. THREE-DIMENSIONAL FLOW MODEL

The engineering task is to design a three-dimensional blade passage including the end-wall regions. Basically we need to be able to model the effects on the flow of the geometry under our

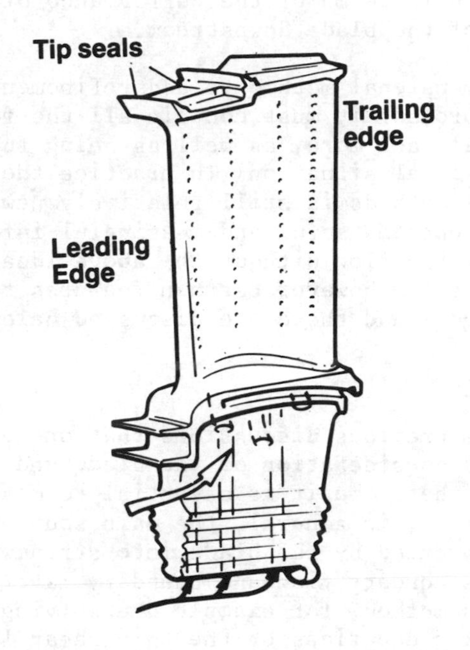

Fig. 17. Shrouded rotor

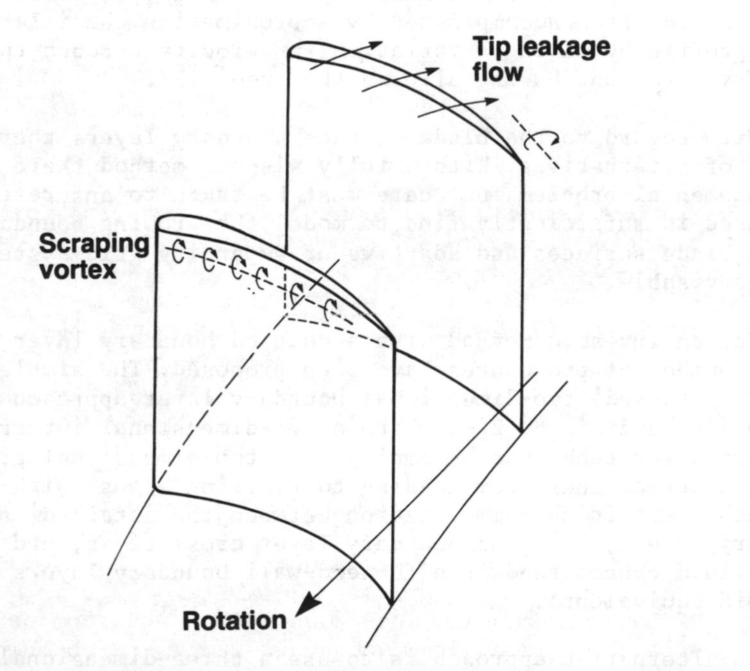

Fig. 18. Unshrouded rotor

control; this is both in terms of the performance of the individual blade row and that of the blade downstream.

For a three-dimensional method to add refinement to the two-dimensional design process it must contain all the features of the two dimensional model, and more, as well as being fully compatible. This however, is an ideal situation. In practice the application of three-dimensional methods is still relatively new and as a consequence one can obtain useful and meaningful information about the main features of the flow without the above ideal situation being achieved. There are however certain features that are essential for the model and these are discussed below.

3.1 Boundary Layers

We can see from previous discussions that one of the main areas of interest is co-sideration of the blade and end-wall boundary layers. Of these two it is essential to model the latter in some form as this is, in general, the main source of inlet vorticity, being converted by the blade into streamwise vorticity and giving rise to secondary flow and boundary layer over-turning. With a fully viscous method, for example one solving the Reynolds averaged Navier-Stokes equations or the thin shear layer form of these, then modelling presents no fundamental problems. With a purely inviscid method, however, approximations are needed and modelling is often accomplished by approximating the inlet boundary layer profile by having a variation in velocity through the boundary layer but having slip on the end-walls.

With regard to the blade surface boundary layers there are a number of alternatives. With a fully viscous method there are again no fundamental problems but care must be taken to ensure that the grid used is sufficiently fine to model the growing boundary layers on the blade surfaces and adaptive or expanding grid systems are often advisable.

For an inviscid method with a coupled boundary layer calculation a number of procedures have been proposed. The simplest is to adopt a 'stacked' two-dimensional boundary layer approach see Denton [26] and Singh [27], where a two-dimensional integral boundary layer technique is employed on two-dimensional grid lines or quasi-streamlines from leading to trailing edges. With this approach there is no communication between the 'stacked' surface boundary layers, i.e., no boundary layer cross flows, and boundary layer fluid cannot feed into the end-wall boundary layers (or inviscid equivalent).

An alternative approach is to use a three-dimensional surface boundary layer analysis, see for example Thompkins and Usab [28], although ideally with this approach the surface adopted should include blade and end-walls to facilitate communication between these regions. To date, direct modes of coupling have been adopted. In general this is quite adequate for relatively thin boundary layers but as in the case of two-dimensional flows it is likely that for thick or separated boundary layers an inverse approach or a directly coupled technique will be needed.

3.2 Losses

In using a three-dimensional method a designer's aim is to study the effects on the blade loss and turning of the changes he makes to the geometry under his control. In the absence of a loss calculation in the method he is left trying to assess this from the understanding of the flow he gains from the method, coupled possibly with experimental data from earlier designs. This is a difficult process demanding aerodynamic understanding and experience. The ideal situation is for the method to produce information regarding the losses, for example by adopting a viscous analysis. However, even with such an analysis one needs to be sure that pseudo or numerical viscous effects and losses do not swamp the true viscous features. Care is needed in the differencing of the differential equation, in the manner in which surface boundary conditions are applied, or the smoothing or other numerical techniques (e.g., upwinding) that may need to be adopted etc. It should be noted that these comments also apply to inviscid methods where spurious losses can arise. In addition to the numerical effects the loss will be governed by the turbulence model (including transition). However, in order to evaluate and develop adequate turbulence models the spurious numerical effects need to be eliminated by the use of fine grids in order to produce grid independent solutions where the viscous features are governed purely by the turbulence model; this is often difficult to achieve and expensive in computer time in a three-dimensional problem.

With a purely inviscid method it is often an advantage to be able to prescribe the overall loss and its distribution through the blade row. A similar situation occurs in two-dimensional blade-to-blade analysis methods but there the stream-tube height is often a convenient parameter to adopt for this purpose; in a three-dimensional analysis this is not possible. One way is to introduce a shear stress term into the momentum equations and relate this to the prescribed loss, see for example Jennions and Stow [3] and Denton [26]. With this loss model it is necessary to know the loss distribution through the blade row. This is a difficult task but a first approximation is to adopt a 'stacked' two-dimensional approach with the loss being specified as a function of the distance along quasi-streamlines from leading to trailing edges. These quasi-streamlines can be taken, for example, from the original through-flow streamlines, or grid lines can be used. In this way one can run a three-dimensional method to a pressure ratio defined by a through-flow calculation.

3.3 Blade Geometry

It is obviously important to model the effects of the blade geometry on the three-dimensional flow. In general, interpolation from designed sections will be necessary in the three-dimensional calculation. Interpolation along each blade section will usually present no problems if these are defined by a sufficient number of points. However, care needs to be taken in the hub to tip direction if there are rapid changes in the blade geometry, and sufficient sections need to be available for the interpolation to adequately represent these changes.

In many calculations approximations to the geometry at the leading and trailing edges are needed. At the leading edge these may be due to, for example, the grid system used, instability of the method for reversed flow, partially parabolic marching techniques employed etc. Often a cusped leading edge is employed which may be fixed or moveable (i.e., iteratively adjusted), solid or non-solid. In many cases this is a perfectly adequate treatment of the region as long as care is taken with conservation over the cusp and details of the flow are not required. In the end-wall boundary layer even though the stagnation point near the leading edge and reversed flow will not be modelled, the effects of the remaining blade geometry will be. In order to model the details, e.g., leading edge vortex, leading edge velocity spikes etc., then the full geometry needs to be included possibly with special consideration given to the grid in this area, depending on the method. In general simple grids have been adopted in three-dimensional programs based for example on 'stacked' blade-to-blade quasi-streamline, quasi-orthogonal grids, see Denton [29].

However, in the future it is likely that more complex grid systems will be investigated based for example on the two-dimensional work of Norton, Thompkins and Haimes [30], see Figure 19.

At the trailing edge the situation is similar. Again, in general, a cusped trailing edge is adopted, for example in an inviscid method to avoid an unrealistic stagnation region with the full geometry, or in a partially parabolic method to avoid regions of reverse flow. Generally the treatment is quite adequate but care is needed, as in two-dimensional calculations, for supersonic exit flows. This is an obvious area for future research in order to resolve the details of the flow.

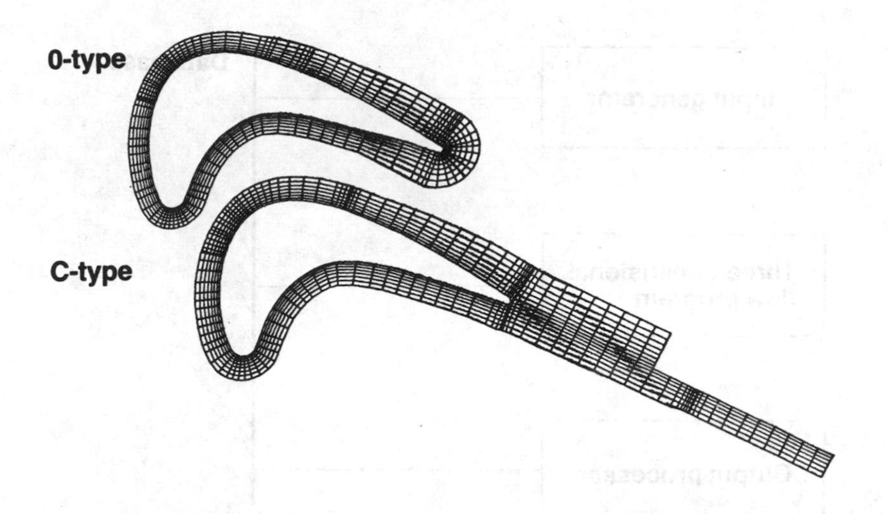

Fig. 19. Sample orthogonal O-type and C-type finite difference grids for turbine cascade

4. THREE-DIMENSIONAL AERODYNAMIC DESIGN SYSTEM

The basic essentials of a three-dimensional aerodynamic design system are shown in Figure 20, comprising of an input generator, core calculation and an output processor. If written in a universal form the input generator and output processor can be used with a number of core programs that may be needed to cover the range of design tasks e.g., transonic/supersonic fans, subsonic/transonic compressors and turbines. The linking of the system with the previously discussed quasi-three-dimensional design system is shown in Figure 21. The main idea with the system is to interactively generate the input data, have relatively long batch runs of the core program or schedule runs onto a super-computer, file the output onto a data base and then interactively process the output data.

4.1 Input Generator

This is shown in Figure 20 linking to a data-base containing the sectional blade geometry data and stacking information and so has a three-dimensional description of the stacked blade see Figure 22. Also available via the data base are the flow parameters from the through-flow analysis, either the true ones for engine conditions or those from a specially created analysis to cater for the particular blade under consideration. On the latter point, some calculations need a 'settling' distance between the input data

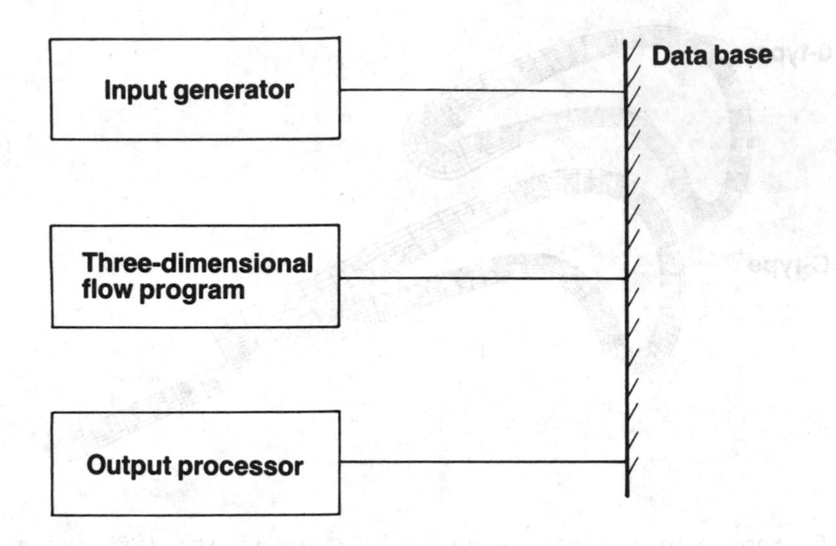

Fig. 20. Three-dimensional flow system

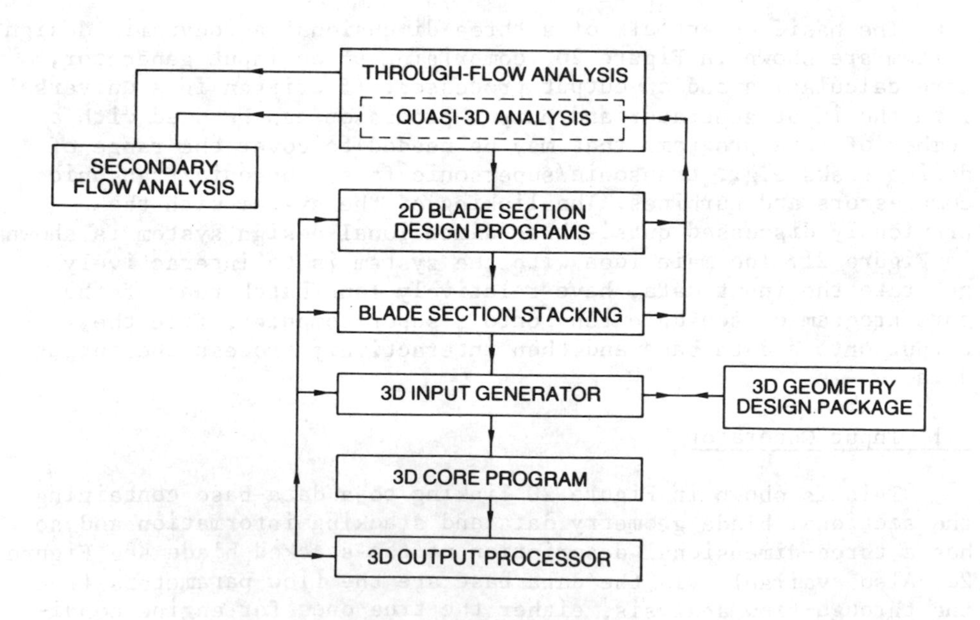

Fig. 21. Three-dimensional aerodynamic design system

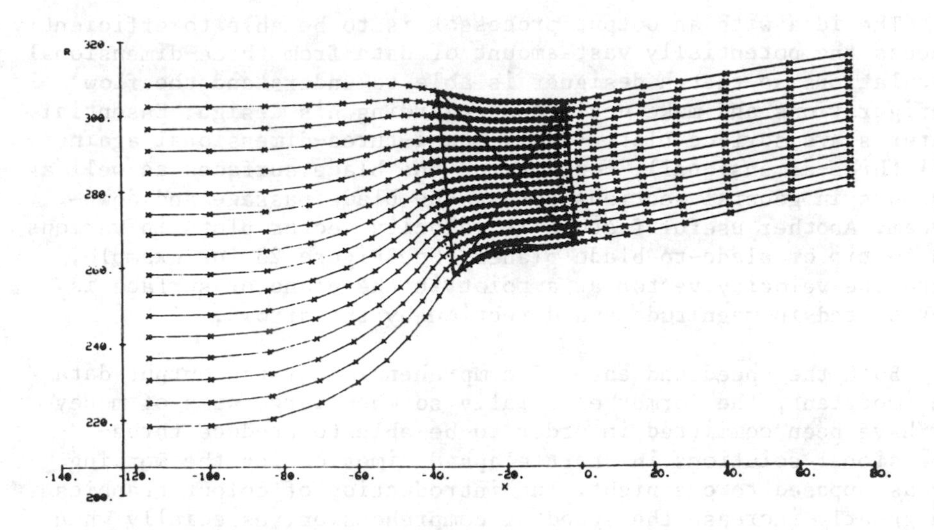

Fig. 22. Input generator

plane and the leading edge of the blade, thus necessitating a special 'cut-down' through-flow calculation.

An important feature of the input generator is the ability to modify the geometry or flow data in an interactive manner in order to study the effects on the three-dimensional flow. It is a simple matter to create loops back to the standard aerodynamic design system to facilitate this, see Figure 21. So, for example, a loop back to the stacking program enables the designer to alter the stack of the blades without altering the geometry of the individual sections. This procedure, although useful, is obviously restrictive in the freedom it allows to alter the blade force effects on the blade and end-wall boundary layers. The other loop back is to blade section design programs although this is obviously a longer loop back.

In the longer term a speedier and less restrictive procedure is to have a geometry design package linked as shown in Figure 21 where the blade and end-wall surfaces can be modified without re-entering the quasi-three-dimensional system. The idea of storing and manipulating surfaces is a versatile approach. As part of this, or alternatively linked to it, is the idea of generating an initial geometry for each blade section from a parameterised form of the profile together with stacking information.

4.2 Output Processor

The idea with an output processor is to be able to efficiently process the potentially vast amount of data from three-dimensional calculations so that a designer is able to understand the flow configurations and make decisions regarding his design. Essential features are surface plots (e.g., quasi-three-dimensional against full three-dimensional), contours on the blade surfaces as well as contours in general sections across the blade passage and downstream. Another useful feature is velocity vector plots in various hub to tip or blade-to-blade planes. See Figure 23 for example, where the velocity vector at a point in the plane or surface is represented in magnitude and direction by the arrows.

Both the speed and ease of comprehension of the output data are important, the former especially so when large sums of money may have been committed in order to be able to produce three-dimensional solutions in short elapsed times during the working day as opposed to overnight. The introduction of colour graphics can greatly increase the speed of comprehension, especially when applied to contour plots. Animation in the future is also likely to aid the interpretation of streamline plots and improve understanding of the three-dimensional aerodynamics involved.

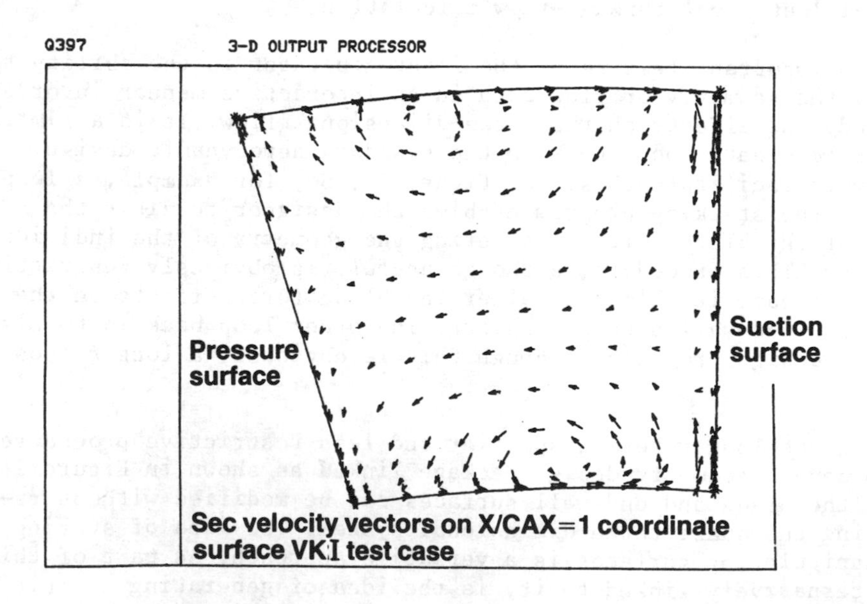

Fig. 23. Output processor

4.3 Core Programs

The quality of the output data is obviously governed by the core program. The aim with this is to produce accurate solutions in the shortest possible computer times in order to be able to reduce design times or to allow full optimisation of the design in the time available. Often this means a compromise between the fineness of the mesh used (which directly affects the accuracy) and the computer time involved. Introduction of super-computers such as a CRAY, CYBER, or dedicated super-minis with array processors, can improve this situation, but thought needs to be given to the data processing side of the operation i.e., the link to the design system.

In the following two sections some of the main points presented earlier concerning inviscid and viscous calculations are discussed. Although reference is given to some particular methods no attempt is made to review the various ones available or the alternative techniques adopted.

4.3.1 Inviscid Calculations

As stated earlier the inlet end-wall boundary layer must be approximated using a wall slip velocity; often this involves a compromise between the magnitude of the slip velocity (and hence the inlet vorticity that results) and the stability of the inviscid calculation.

One of the problems discussed earlier with not having loss predicted by the calculation is that a designer must study the details of the flow produced and exercise judgement with regard to the efficiency expected. Introducing a coupled surface boundary layer calculation can obviously go some way to helping in this direction but the additional losses created by the leading edge horse-shoe vortex and new end-wall boundary layers cannot be accounted for. It is found in some cases, for example turbines, that very useful results on the effects of blade stack and annulus geometry can be obtained without a coupled boundary layer calculation. In the case of transonic fans, however, the inclusion of the boundary layer blockage effects is essential in obtaining the correct shock position.

There are a number of methods available, finite difference/area and finite element. The most commonly adopted technique in the industry is explicit time-marching, see for example Denton [29] and Thompkins [31]. This is a very versatile approach in that it can handle subsonic, supersonic and transonic flows and capture shock waves that are present in the flow. The explicit nature of the method means that the time-step must be limited to ensure stability and hence convergence to a steady state solution; this is true whether a point-wise finite difference or integrated finite area technique is used. The maximum time-step is governed by the finite difference technique adopted and can be quite different for different methods. The upper limit is governed in theory by the Courant-Friedrichs-Lewy condition although in practice this limit is often reduced by the non-linear effects of the problem. If only steady state solutions are of interest then the time-step can be allowed to vary over the domain, being governed by local rather than global stability considerations. This can greatly improve the speed of the calculation. Another means of improving the convergence rate is the multi-grid technique, see for example Denton [29], where the solution is up-dated using Macro elements (composed of a number of small elements) in addition to the smaller elements; this is particularly easy to implement in a finite area approach. Although the stability condition seems at first sight quite restrictive with explicit time-marching, the method, by its very nature, is easily adapted to the architecture of parallel processing machines.

A sample calculation using a time-marching method based on the Denton approach has already been shown in Figure 14b where it is seen that good predictions of surface Mach numbers are obtained for the complex multi-lean turbine geometry shown in Figure 14a. In this calculation no account was taken of the inlet end-wall boundary layer.

Figure 24 shows results obtained by Connell [32] from the 3D Denton program (see Denton [29]) for a VKI low speed annular turbine, see Sieverding, Van Hove and Boletis [33]; in this case the inlet boundary layer was simulated.

It can be seen that very good agreement with experiment is obtained especially with grid refinement in the boundary layer.

4.3.2 Viscous Calculation

With a viscous method, based for example on the full Reynolds averaged Navier-Stokes equations or the thin shear layer form losses can be calculated. The loss predicted will be influenced by the truncation error of the numerical solution which in turn is governed by the mesh size used and any numerical techniques that need to be introduced to obtain a stable calculating procedure e.g., upwinding, smoothing etc. The other factor influencing the loss is the turbulence model adopted, including the treatment of transition.

One can obtain very useful results concerning end-wall and blade boundary layers using a relatively simple isotropic mixing

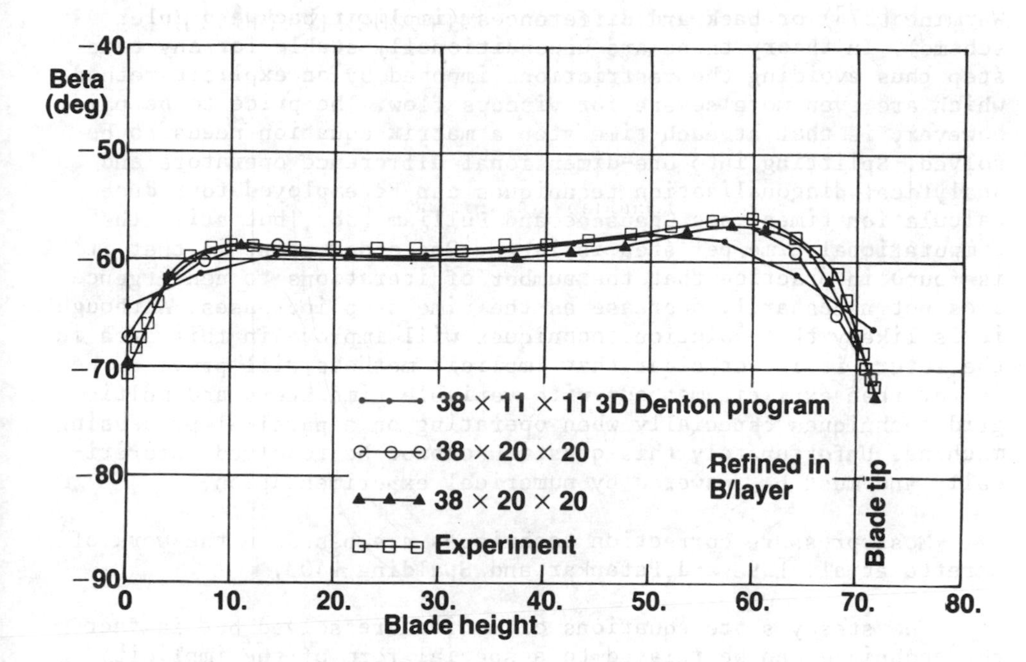

Fig. 24. -Beta angle distribution: X/CAX = 0.9

length model in the boundary layer with a free stream mixing length to represent the effects of free stream turbulence. There is, however still much work to be done on evaluating and developing turbulence models for turbomachinery applications, for example in the areas of three-dimensional transition, free stream turbulence effects, trailing edge and wake regions, skewed inlet boundary layers etc. It should be borne in mind that in order to develop the turbulence model one needs to obtain solutions free from numerical viscous effects i.e., grid independent solutions. This is something not easily achieved in three-dimensions except for simple problems because of the computer time needed. As solution techniques are improved and computer speeds increase this situation will change and higher order turbulence models can be employed.

There are a number of methods available and under development, the most commonly employed being based on time-marching see for example Pulliam and Steger [34], pressure correction techniques, see for example Moore and Moore [35], and mixed potential stream-like function approach (i.e., an irrotational rotational part) see Dodge [36].

With the time-marching approach implicit methods have been developed based on central differences in time (see Beam and

Warming [37]) or backward differences (implicit backward Euler scheme). In theory these are unconditionally stable for any time step thus avoiding the restrictions imposed by an explicit method, which are even more severe for viscous flow. The price to be paid, however, is that at each time step a matrix equation needs to be solved. Splitting into one-dimensional difference operators and analytical diagonalisation techniques can be employed to reduce calculation times, see Chaussee and Pulliam [38], but still the computational time per step is large. One other point is that it is found in practice that the number of iterations to convergence does not necessarily decrease as the time step increases. Although it is likely that solution techniques will improve in this area in the future it is not clear that implicit methods will prove to be faster than explicit methods with variable time steps and multi-grid techniques especially when operating on a parallel processing machine. Unfortunately this question cannot be resolved theoretically and must be answered by numerical experimentation.

Most pressure correction techniques are based on the work of Caretto et.al. [39] and Patankar and Spalding [40].

The steady state equations of motion are solved but in fact the technique can be related to a special form of the implicit time-marching technique where the momentum equations are uncoupled, see Connell [41]. Many workers have adopted what is termed a partially parabolic technique, see for example Moore and Moore [35], where in the viscous terms in the equations of motion second derivatives in the stream-wise direction are neglected, thus allowing a marching technique to be employed in this direction. The equations obtained are in fact the thin shear layer form of the Navier-Stokes equations. Often an additional restriction is that the stream-wise velocity must be positive (i.e. there can be no reverse flow in the stream-wise direction).

However, re-circulation in cross-stream planes can be accounted for, the equations being elliptic in these planes. In addition the pressure-correction equation is fully elliptic.

The partially parabolic technique is generally quite adequate and very useful results can be obtained even around blade leading edges. Where significant areas of streamwise reverse flow are expected or for example details around blade leading edges are required then a fully elliptic treatment can be adopted retaining the previously ignored viscous terms.

The pressure-correction technique although developed mainly for viscous flows can be applied to inviscid flows. In fact a valuable technique employed by Moore and Moore [35] is to first solve the inviscid form of the equations outside of the boundary layer regions in order to obtain a good first approximation to the pressure field economically. This is then used for the more expensive viscous calculation with fine grids in the boundary layer regions, the idea being that the inviscid pressure field will not change greatly under the effects of viscosity and thus save overall computing time. There are a number of other techniques that can be used to improve convergence rates, for example acceleration techniques together with improved and extended pressure correction schemes, see Connell [41]. Also likely to lead to further improvements are iterative multi-grid techniques.

Results from the partially parabolic method of Moore and Moore obtained by Birch [42] have already been shown in Figure 21 for the VKI low speed annular turbine; further results are given in Figure 25 showing that very good agreement with the experimental data is found with the marching procedure adopted.

Figure 26 shows predictions for the same cascade with a skewed inlet boundary layer on the inner wall, see Boletis, Sieverding and Van Hove [43] for details of the experiment. It can be seen that there is good agreement with the experimental results through the cascade. Also included in the figure are results from a VKI 3D inviscid time marching program. Figure 27 shows the secondary flow velocity vectors near the trailing edge indicating the effect of the inlet skew on the inner wall.

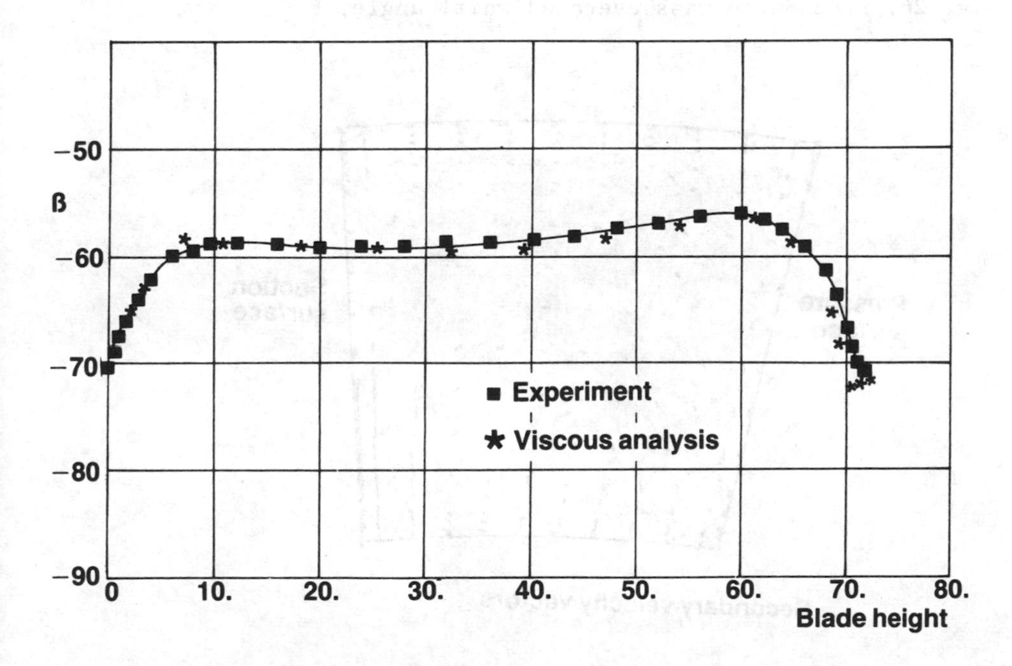

Fig. 25. Beta angle distribution: X/CAX = 0.9

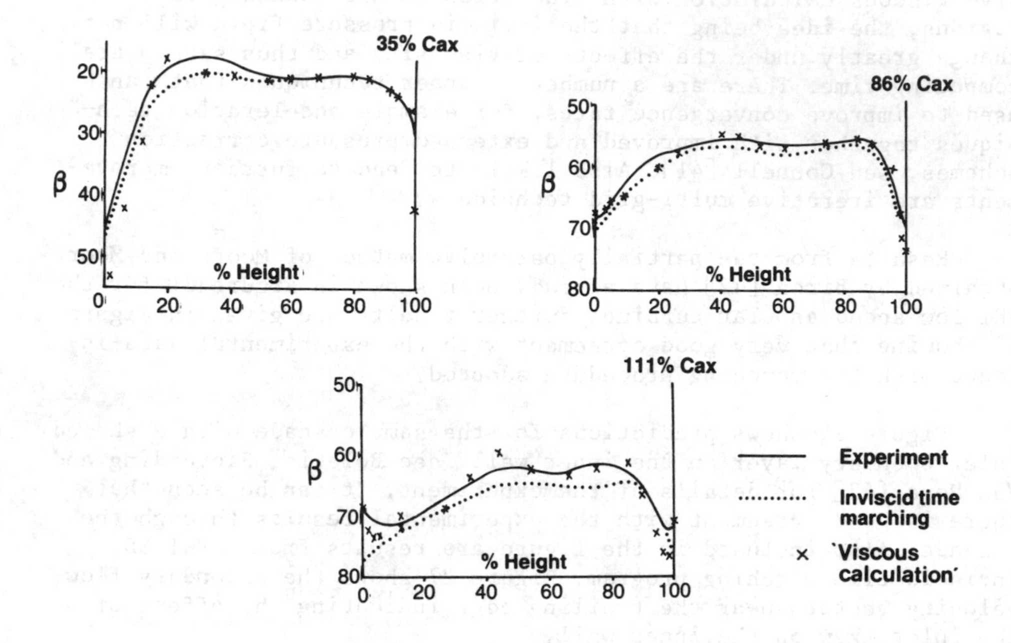

Fig. 26. Pitch-wise mass averaged whirl angle, $\bar{B}$

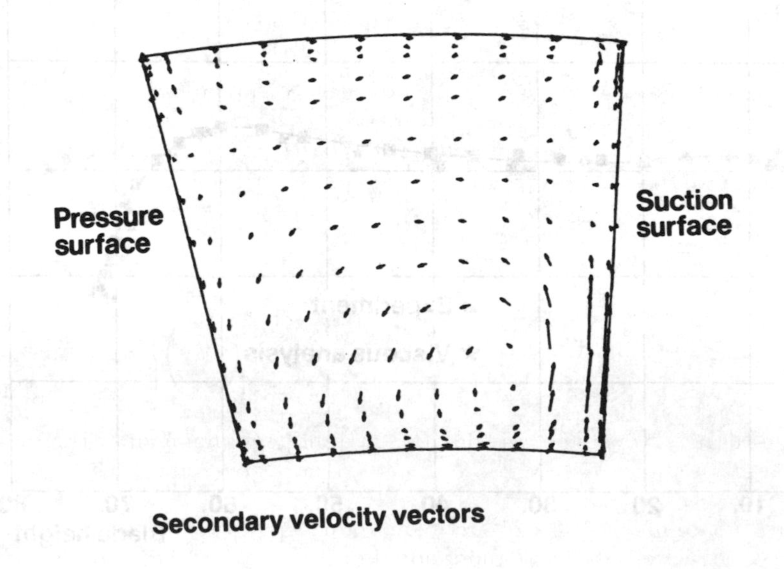

Fig. 27. 86% chord

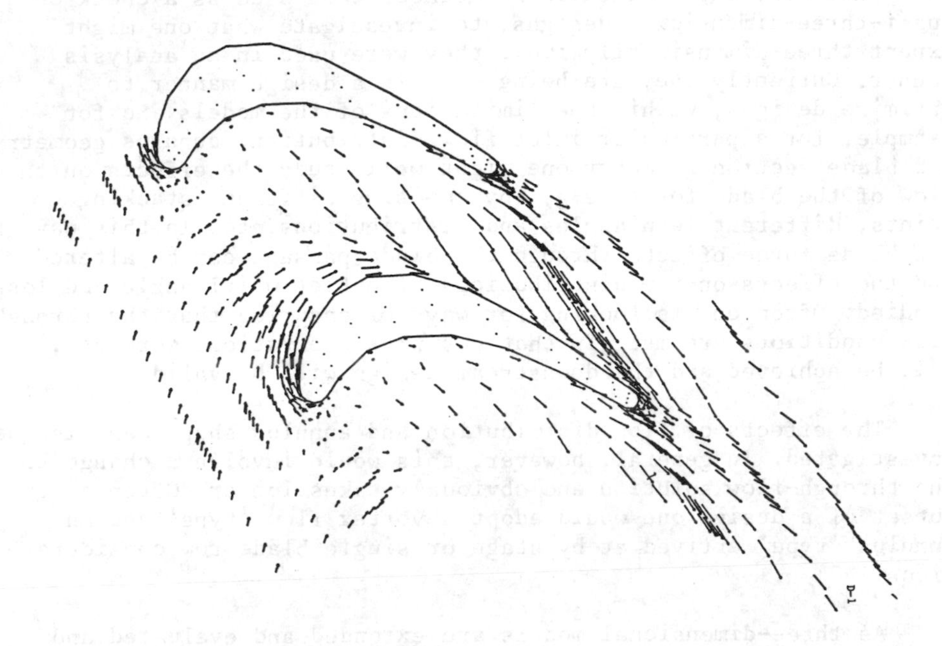

Fig. 28. Velocity vectors near end-wall

Figure 28 shows predicted velocity vectors near the end-wall of a linear turbine cascade similar to that of Langston [44] from an elliptic version of the Moore and Moore viscous method, see Moore and Moore [45]; details of the leading edge flow and horseshoe vortex formation can be clearly seen.

5. CURRENT AND FUTURE APPLICATIONS

It is still relatively early days in the use of three-dimensional methods in blade design at least compared with two-dimensional methods. Whereas a vast body of experience, knowledge and experimental data exists for the latter this is not yet the case for the former. Evaluation of a method is needed before design usage can take place so that the limitations of the model adopted are fully appreciated. At that stage one can start to exploit the design freedoms e.g., blade and annulus geometry, flow distributions (both leading to trailing edge and hub to tip) that are adequately represented by the model. In order to fully evaluate a model detailed experimental data is needed within the blade passage both from cascades and engine test rigs. This means that theoretical and experimental work needs to go hand-in-hand and often requires the development of new experimental as well as theoretical techniques.

Initially three-dimensional methods were used as a check on quasi-three-dimensional designs, to investigate what one might expect three-dimensionally i.e., they were used in an analysis manner. Currently they are being used in a design manner to optimise designs, within the limitations of the models. So for example, for a particular inlet flow distribution, annulus geometry and blade section geometry one might well study the effects on the flow of the blade forces e.g., by choosing different stacking points, different lean angles and distributions etc. In this way the blade force effects through the blade passage can be altered and the effects on the distributions of outlet whirl angle and loss studied. Often one is looking for ways of ensuring that the through-flow conditions are met. In that way pressure ratios, work etc. will be achieved and the downstream design will be valid.

The effects of flow distribution and annulus shape can also be investigated. In general, however, this would involve a change to the through-flow solution and obviously takes longer. Often at the outset of a design one would adopt a vortex flow 'type' and an annulus 'type' arrived at by stage or single blade row considerations.

As three-dimensional models are extended and evaluated and more understanding is gained then there will be further exploitation of three-dimensional design freedoms. Areas where effort is likely to be directed are the 'interaction' regions e.g., the blade ends, over tip leakage flows, trailing edge regions and base flows.

As more is learned about blade interaction effects then the more one can start to design individual blade rows or stages with the interaction taken into account.

Whereas currently it is usual to consider only design conditions in any detail in a three-dimensional design, as elapsed computer times become shorter and models are developed, then off-design analysis can be considered and a design optimised for the full range of operating conditions.

6. FUTURE RESEARCH

There are a number of areas where research is needed in the future both on the computational and mathematical modelling sides.

6.1 Computational Work

(i) Method reliability/ease of use
(ii) Speed of existing methods

(iii) Development/adaptation of methods for parallel processing machines
(iv) Multi-grid techniques
(v) Smoothing techniques especially in the region of shocks
(vi) Investigation of ways of imposing inlet and outlet boundary conditions to reduce 'settling' distances
(vii) Optimum computational grid

6.2. Mathematical Modelling Work

(i) Trailing edge region and wake development
(ii) Leading edge region
Development of blade surface boundary layers,
Horse-shoe vortices
(iii) Over-tip leakage region
(iv) Turbulence modelling, including transitional flows
(v) Stage calculations including blade row interactions, steady and unsteady
(vi) Film cooling modelling
(vii) Off-design analysis
(viii) Three-dimensional 'design calculations'.

REFERENCES

1. Wu, C.H. A General Theory of Three-Dimensional Flow in Subsonic and Supersonic Turbomachines of Axial, Radial and Mixed Flow Types. Trans ASME pages 1363-1380, November 1952.
2. Krimmerman, Y., and Adler, D. The Complete Three-Dimensional Calculation of the Compressible Flow Field in Turbo Impellers. Journal Mechanical Engineering Sciences, Volume 20, Number 3, pages 149-158, 1978.
3. Jennions, I.K., and Stow, P. A Quasi Three-Dimensional Turbomachinery Blade Design System. Part I - Through-Flow Analysis ASME 84-GT-26. PART II - Computerised System ASME 84-GT-27.
4. Hirsch, C.H., and Warzee, G. An Integrated Quasi-3D Finite Element Calculation Program for Turbomachinery Flows. Trans ASME, Journal Engineering for Power, Paper Number 78-GT-56, pages 1-8, 1978.
5. Cedar, R.D., and Stow, P. The Addition of Quasi Three-Dimensional Terms into a Finite Element Method for Transonic Turbomachinery Blade-to-Blade Flows. Int. J. for Num. Methods in Fluids to be published 1984.
6. Jennions, I.K. Rolls-Royce Private Communication, 1982.
7. Hawthorne, W.R. Rotational Flow Through Cascades. Part 1: The Components of Vorticity. Q. Jnl. Mech. Appl. Math., Number 8, part 3, pages 266-279, 1955.
8. Lakshminarayana, B., and Horlock, J.H. Generalised Expressions for Secondary Vorticity using Intrinsic Coordinates. Journal Fluid Mechanics, Volume 59, Part 1, pages 97-115, 1973.

9. Came, P.M., and Marsh, H. Secondary Flow in Cascades - Two Simple Derivations for the Components of Vorticity. Journal Mechanical Engineering Sciences 16, 1974.
10. Smith, L.H. Jr. Secondary Flow in Axial Flow Turbomachinery. Trans ASME 77, 1955.
11. James, P.W. Rolls-Royce Private Communication.
12. Stanitz, J. Design of Two-Dimensional Channels with Prescribed Velocity. Distributions Along the Channel Walls. NACA Report 115, 1952.
13. Schmidt, E. Computation of Supercritical Compressor and Turbine Cascades with a Design Method for Transonic Flows. ASME Paper Number 79-GT-30, 1979.
14. Bauer, F.,Garabedian, P., and Korn, D. Super-Critical Wing Sections, Volume I, II, III. Springer-Verlag, New York, 1972, 1975, 1977.
15. Wilkinson, D.H. Calculation of Blade-to-Blade Flow in a Turbomachine by Streamline Curvature. Short course on Advanced Flow Calculations, 23-27 March 1981. Fluid Engineering Unit, Cranfield Institute of Technology.
16. Stuart, A.R. Rolls-Royce Private Communication.
17. Tong, S.S. and Thompkins, W.T. jr. A Design Calculation Procedure for Shock-Free or Strong Passage Shock Turbomachinery Cascades. ASME 82-GT-220, 1982.
18. Paige, R.W. The Aerodynamic Design of Transonic Turbine Blades. Ph.D. Thesis Cambridge University, 1983.
19. Meauze, G. Methode de Calcul Aerodynamique Inverse Pseudo-Instationnaire. Library Translation 2048, 1980. Royal Aircraft Establishment.
20. Cedar, R.D., and Stow, P. A Compatible Mixed Design and Analysis. Finite Element Method for the Design of Turbomachinery Blades. To be published in Int. J. for Num. Methods in Fluids, 1984.
21. Sobieczky, H., Yu, N.J., Fung, K.Y., and Seebass, A.R. New Method for Designing Shock-Free Transonic Configurations. AIAA, Journal Volume 17, Number 7, pages 722-729, 1979.
22. Dawes, W.N. Computation of Viscous Compressible Flow in Blade Cascade using an Implicit Iterative Replacement Algorithm. Published in "Computational Methods in Turbomachinery". Inst., Mech. Engs. Birmingham University, England, April 1984.
23. Steger, J.L.,Pulliam, T.H., and Chima, R.V. An Implicit Finite Difference Code for Inviscid and Viscous Cascade Flow. AIAA 13th Fluid and Plasma Dynamics Conference, July 1980.
24. Thompkins, W.T., Tong, S.S., Bush, R.H., Usab, W.J., and Norton, R.J.G. Solution Procedures for Accurate Numerical Simulations of Flow in Turbomachinery Cascades. AIAA Paper 83-0257, 1983.
25. Calvert, W.J. An Inviscid-Viscous Interaction Treatment to Predict the Blade-to-Blade Performance of Axial Compressors with Leading Edge Normal Shock Waves. ASME Paper 82-GT-135.
26. Denton, J.D. Cambridge University Private Communication.

27. Singh, U.K. A Computation and Comparison with Measurements of Transonic Flow in an Axial Compressor Stage with Shock and Boundary Layer Interaction. ASME 81-Gr/GT-5, 1981.
28. Thompkins, W.T., and Usab, W.J. A Quasi-Three-Dimensional Blade Surface Boundary Layer Analysis for Rotating Blade Rows. ASME 81-GT-126, 1981.
29. Denton, J.D. An Improved Time Marching Method for Turbomachinery Flow Calculation. ASME 82-GT-239, 1982.
30. Norton, R.J.G., Thompkins, W.T., and Haimes, R. Implicit Finite Difference Schemes with Non-Simply Connected Grids. A Novel Approach. AIAA 22nd Aerospace Sciences Meeting Jan. 1984 Revs.
31. Thompkins, W.T. A Fortran Program for Calculating Three-Dimensional Inviscid Rotational Flows with Shock Waves in Axial Compressor Blade Rows. MIT G.T. and PDL Report No 162, 1981.
32. Connell, S.D. Rolls-Royce Private Communication.
33. Sieverding, C.H., Van Hove, W., and Boletis, E. Experimental Study of the Three-dimensional Flow Field in an Annular Turbine Nozzle Guide Vane. ASME 83-GT-120, 1983.
34. Pulliam, T.H., and Steger, J.L. Implicit Finite-Difference Simulations of Three-Dimensional Compressible Flow. AIAA Journal, Volume 18, No. 2, 1980.
35. Moore, J., and Moore, J.G. Calculations of Three-Dimensional Viscous Flow and Wake Development in a Centrifugal Impeller. Trans ASME, Journal of Engineering for Power. Volume 103, April 1981, pages 367-372.
36. Dodge, P.R. Numerical Method for 2D and 3D Viscous Flows. AIAA Journal, Volume 15, Number 7, 1977.
37. Beam, R.M., and Warming, R.F. An Implicit Finite-Difference Algorithm for Hyperbolic Systems in Conservation Law Form. Journal of Computational Physics. Volume 22, pages 87-110,1976.
38. Chaussee, D.S., and Pulliam, T.H. Two-Dimensional Inlet Simulation using a Diagonal Implicit Algorithm. AIAA Journal, Volume 19, Number 2, 1981.
39. Caretto, L.S., Gosman, A.D., Patankar, S.V., and Spalding, D.B. Two Calculation Procedures for Steady Three-Dimensional Flow with Recirculation. Proceedings of the Third International Conference on Numerical Methods in Fluid Dynamics, Paris, Volume II, page 60, 1972.
40. Patankar, S.V., and Spalding, D.B. A Calculation Procedure for Heat, Mass and Momentum Transfer in Three-Dimensional Parabolic Flows. International Journal Heat and Mass Transfer. Volume 15, page 1787-1806, 1972.
41. Connell, S.D. Numerical Solution of the Equations of Viscous Flow. Ph.D. Thesis, Nottingham University, 1983.
42. Birch, N.T. Rolls-Royce Private Communication.

43. Boletis, E., Sieverding, C.H., and Van Hove, W. Effects of a skewed inlet end wall boundary layer on the 3D flow field in an annular turbine cascade. AGARD PEP 61st Meeting on Viscous Effects in Turbomachines, Denmark, 1-3 June 1983.
44. Langston. Crossflows in a turbine cascade passage. Trans. ASME, J. of Eng. for Power. Vol. 102, pp. 866-874, 1980.
45. Moore, J.G., and Moore, J. Calculation of Horseshoe Vortex Flow without Numerical Mixing. ASME Paper No. 84-GT-241.

UNSTEADY EFFECTS

AN INTRODUCTION TO UNSTEADY FLOW IN TURBOMACHINES

Edward M. Greitzer

Massachusetts Institute of Technology,
Cambridge, Massachusetts, U.S.A.

1. INTRODUCTION

In this lecture we will examine some of the fluid dynamic phenomena that are associated with unsteady flow in turbomachines. It will be seen that there are several different sources of this unsteadiness, and that which of these is most important will depend on which aspect of the overall performance of the turbomachine is being examined.

It should be stressed at the outset that this lecture is intended to be only an *introduction* to the subject and to provide an appreciation for the basic phenomena and relevant fluid mechanic concepts, rather than to be an exhaustive review. Several recent reviews on the general topic of unsteady flows are listed in the References, and these can be consulted for more detailed descriptions of the specific applications and of the many calculation procedures that have been developed for these flows.

The topics to be covered are:

1. The Inherent Unsteadiness of Turbomachinery Flows
2. Sources of Flow Unsteadiness in Turbomachines
3. Relevant Parameters
4. Illustrations of Unsteady (Inviscid) Flows
5. Introductory Discussion of Unsteady Viscous Flows

6. Examples of Unsteady Effects in Turbomachines

   a. Uniform Inlet Flows

   b. Non-uniform Inlet Flows

In the lecture we will not consider aeroelastic problems such as flutter and/or resonant stresses, nor acoustic phenomena in turbomachines. It should be noted, however, that these are also areas in which there is a strong interest in unsteady flow.

2. THE INHERENT UNSTEADINESS OF TURBOMACHINERY FLOWS

The Unsteadiness Paradox

To introduce the topic of unsteady flow in turbomachines, we can examine an apparent paradox, which has been described by Dean [1]*. Consider the flow through an adiabatic, frictionless (i.e. reversibly operating) turbomachine, as shown in Figure 1. At inlet and outlet of the device, and at the location where the work is transferred (by means of a shaft, say), the conditions are such that the flow can be regarded as steady. In addition we restrict ourselves to situations in which the overall (average) state of the fluid *within* the control volume is not changing with time (this is actually the most important situation for turbomachinery applications) although the states of individual fluid particles passing through the device may be altered during their transit. Under these conditions we know from the energy equation for steady flow that the relation between the stagnation enthalpy, h_T, at inlet and

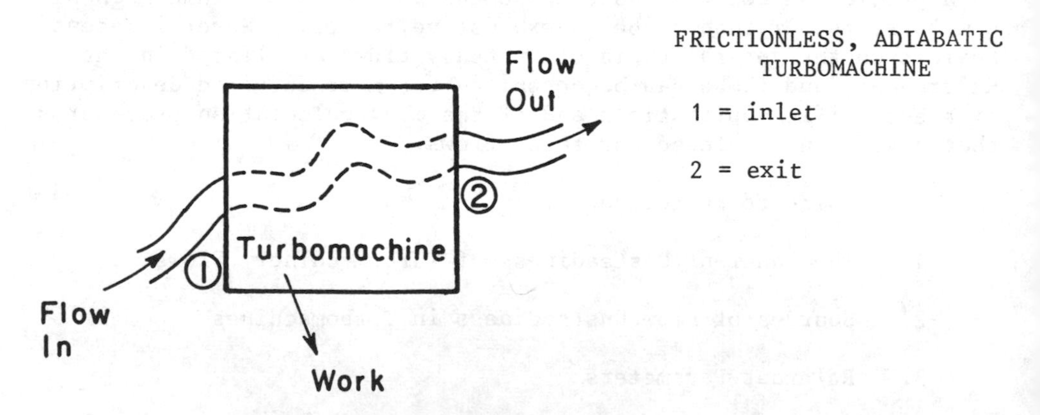

Figure 1. Flow through a frictionless, adiabatic, turbomachine.

*Numbers denote references at end of text.

outlet, and the rate of work done per unit mass flow is

$$h_{T_2} - h_{T_1} = - \left\{ \begin{array}{l} \text{rate of work done} \\ \text{by turbomachine} \\ \text{per unit mass flow rate} \end{array} \right\} \qquad (1)$$

In this equation, h_T, the stagnation enthalpy, is given by

$$h_T = h + \frac{1}{2} C^2 .$$

where h is enthalpy and C is the magnitude of the velocity.

We can now try to analyze this process in another way, by applying the "usual form" of the compressible Bernoulli equation (i.e. the first integral of the moment equation). Along a streamline through the machine (shown dotted in the Figure) the one-dimensional momentum equation is

$$- \frac{1}{\rho} dP = CdC \qquad (2)$$

where C is the velocity, P is static pressure, and ρ is density. For small changes in state we can write

$$dh = Tds + \frac{1}{\rho} dP \qquad (3)$$

Since our turbomachine is adiabatic and frictionless, the entropy change, ds, along a streamline is zero and Equation (2) becomes:

$$dh = - CdC \qquad (4)$$

This can be integrated to yield:

$$h + \frac{1}{2} C^2 = \text{constant along a streamline}$$

Hence,

$$h_{T_2} = h_{T_1} ,$$

so that the turbomachine does no work!

The source of this apparent inconsistency lies in writing the Bernoulli equation as a *steady flow* equation through the machine. In fact, the flow *inside* the device is unsteady and we are not justified in using the momentum equation without accounting for flow unsteadiness. If we include the unsteady terms, the equation for the pressure gradient along an *instantaneous* streamline is:

$$\frac{\partial C}{\partial t} + C\frac{\partial C}{\partial \ell} = -\frac{1}{\rho}\frac{\partial P}{\partial \ell} \tag{5}$$

where ℓ measures the distance along a streamline. Eq. (5) can be written as

$$\frac{\partial C}{\partial t} = -\frac{\partial}{\partial \ell}(h + C^2/2)$$

or

$$\frac{\partial C}{\partial t} = -\frac{\partial h_T}{\partial \ell} \tag{6}$$

This shows that the stagnation enthalpy must, in fact, change along a streamline when the flow is unsteady.

Multiplying Eq. (6) by C, and making use of Eq. (3) we find:

$$\frac{\partial}{\partial t}(h + C^2/2) + C\frac{\partial}{\partial \ell}(h + C^2/2) = \frac{1}{\rho}\frac{\partial P}{\partial t} \tag{7}$$

or

$$\boxed{\frac{Dh_T}{Dt} = \frac{1}{\rho}\frac{\partial P}{\partial t}} \tag{8}$$

where

$$\frac{D}{Dt} = \underbrace{\frac{\partial}{\partial t}}_{\text{rate of change at a point}} + \underbrace{C\frac{\partial}{\partial \ell}}_{\text{rate of change along streamline}}$$

$\frac{D}{Dt}$ is thus the rate of change following a fluid particle.
Hence, for a given particle in an isentropic flow, *the stagnation enthalpy can change only if the flow is unsteady.*

Let us now examine a physical situation illustrating this point - an axial compressor rotor as shown below.
The pressure field associated with the blades is such that pressure increases from the suction surface (S) to the pressure surface (P). This pressure field moves with the blades and hence an observer sitting at the *fixed point*, x, would see a pressure that varied with time, as sketched below

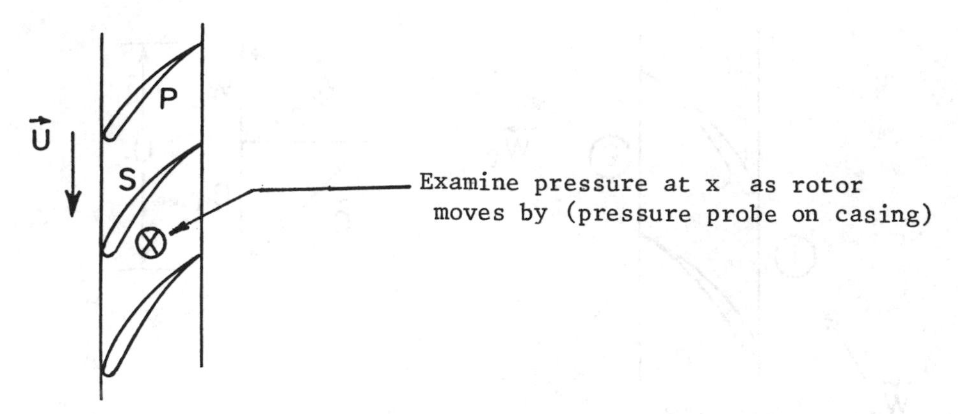

Figure 2 Axial compressor rotor

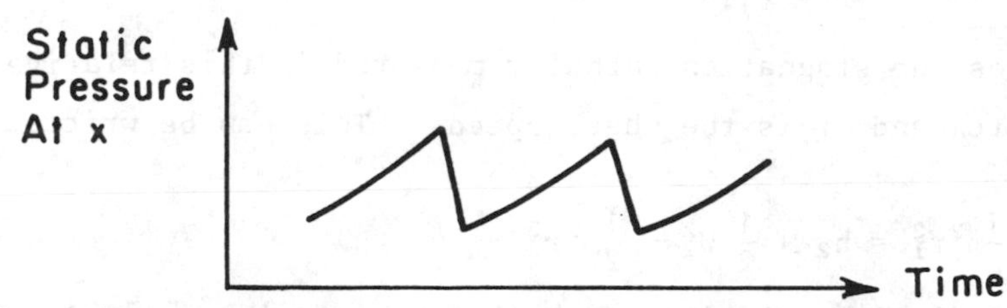

Figure 3. Unsteady pressure as measured on compressor casing.

Particles passing through the rotor will thus see positive $\frac{\partial P}{\partial t}$, and hence $\frac{Dh_T}{Dt}$ will also be positive. (For a turbine the opposite would be true.

In general, the computation of changes in stagnation enthalpy using Eq. (8) can be complicated, since one must know the flow field in detail. Some basic discussions of this topic are given by Preston [2] and Horlock and Daneshyar [3]. If one is mainly interested in the average value of such changes, therefore, an alternative is to work in terms of a coordinate system which is fixed to the individual blades. For such coordinate systems the flow is steady.

The stagnation enthalpy change across a rotor is also derived here in this manner, for a flow which is uniform at inlet and outlet.

Consider a rotor. In the blade fixed system

$$h'_{T_1} - \frac{1}{2}(\omega r_1^2)^2 = h'_{T_2} - \frac{1}{2}(\omega r_2^2)^2$$

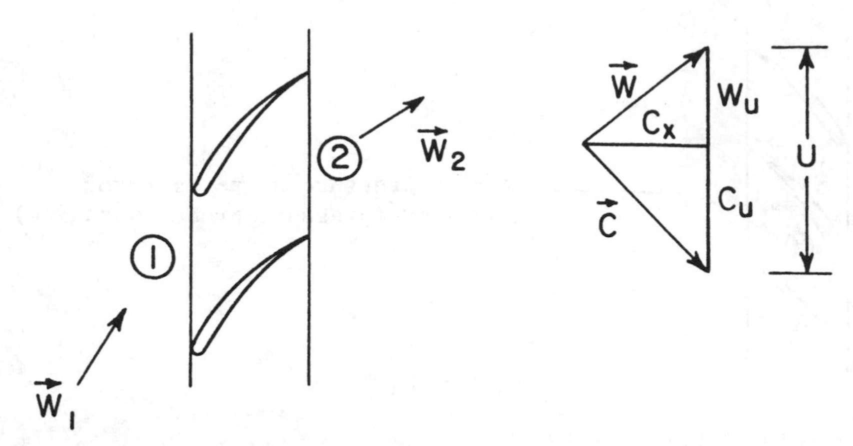

Figure 4 Rotor and velocity triangles.

where h_T' denotes the stagnation enthalpy measured in this relative coordinate system and ωr is the wheel speed. This can be written as

$$h_1 + \frac{1}{2} W_1^2 \frac{1}{2}\omega^2 r_1^2 = h_2 + \frac{1}{2} W_2^2 - \frac{1}{2}\omega^2 r_2^2$$

where W is the relative velocity. Using the velocity triangle, we have

$$\underline{h_1 + \frac{1}{2} (C_{x_1}^2 + C_{u_1}^2 + C_{r_1}^2)} - \omega r_1 C_{u_1} = \underline{h_2 + \frac{1}{2} (C_{x_2}^2 + C_{u_2}^2 + C_{r_2}^2)} - \omega r_2 C_{u_2}$$

The underlined terms are the stagnation enthalpy as measured in the absolute (engine fixed) reference frame.

Hence,

$$h_{T_2} - h_{T_1} = (r_2 C_{u_2} - r_1 C_{u_1}) \qquad (10)$$

which is one form of the well known Euler turbine equation.

3. SOURCE OF UNSTEADY FLOW IN TURBOMACHINES

We can list below some of the different kinds of flow unsteadiness encountered in turbomachines [4]. Also shown are typical time and length scales associated with these flows. Turbulence is listed here for reference, but it will not be considered in this lecture.

Types of Unsteady Flow	"Extent"	Typical Time Scale (sec)
unsteadiness due to:		
1) Turbulence	< 1 chord	
2) Wakes (rotor/stator)	1 chord	10^{-4} sec
3) Upstream potential field interactions due to blades	1 chord (gap)	10^{-4} sec
4) Inlet distortion	circumference (radius)	5×10^{-3} sec
5) Rotating stall	circumference (radius)	10×10^{-3} sec
6) Surge	length of compressor	10^{-1} sec

(Typical row assumed to have 50 airfoils, gap/chord ratio ≅ 1, 12,000 rpm.)

4. RELEVANT PARAMETERS IN UNSTEADY FLOWS

Consider a fluid mechanic device (i.e., an airfoil, a diffuser, a compressor blade passage, etc) which experiences a time varying flow. In particular, for purposes of illustration, consider the time varying flow to be of the form $e^{i\omega t}$. Thus, the time scale associated with this unsteadiness is $1/\omega$, i.e. significant changes occur in a time of the order of $1/\omega$.

There is another time scale in the problem as well. This is the time for fluid particle transport through the device. If the length is L and the throughflow velocity is C, this time is given by L/C.

The change in local flow quantities during the passage of the particle is dependent on the ratio of the two times, which is the quantity

$$\boxed{\Omega = \frac{\omega L}{C}} \, .$$

This is known as the *reduced frequency*. (It is sometimes defined using L/2 as the relevant length - hence, one would have $\Omega = \frac{\omega L}{2C}$.) The size of this parameter is a measure of the relative importance of unsteady effects compared to quasi-steady effects. Thus, as a rough guide, one has:

$\Omega << 1$ unsteady effects small - quasi-steady flow

$\Omega >> 1$ unsteady effects dominate

$\Omega \sim 1$ both unsteady and quasi-steady effects important

Many turbomachinery flows are characterized by values of Ω of order of unity.

Example 1: Diffusing Channel in Unsteady Flow

As an illustration of the use of the reduced frequency, we can examine a very simple example of unsteady flow. This is a diffusing passage subjected to a time varying inlet total pressure. This might be considered an elementary model of a compressor rotor blade passage (which is analogous to a diffuser) moving through a spatially non-uniform inlet total pressure. The basic situation is shown at the left where we see a diffuser with stations 1 at inlet and 2 at exit. As stated, we specify that at the inlet there is a perturbation in total pressure, which we take to be of the form $e^{i\omega t}$. At the exit we assume that the *static* pressure is constant. The coordinate ℓ measures the distance along the diffuser and L is the length of the diffuser.

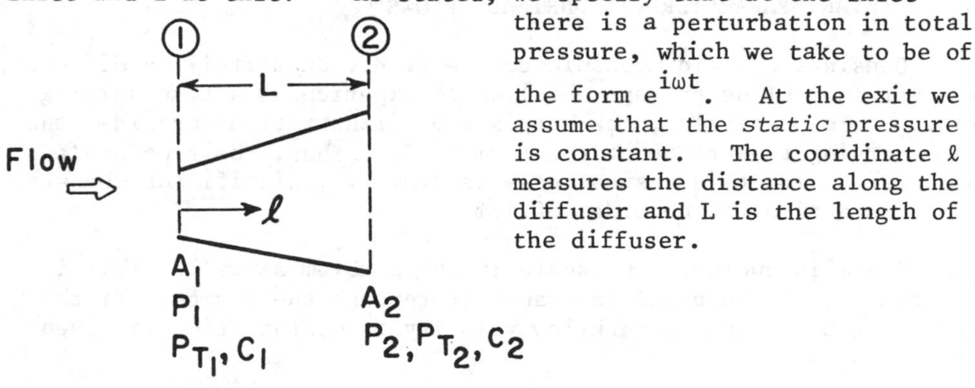

Figure 5. Unsteady flow in a diffuser passage.

We assume that all flow quantities are composed of a mean and a small perturbation so that we can adopt a linearized description of the problem. This shows the basic fluid mechanics without undue complexity. The mean flow quantities are denoted by overbars ($\bar{}$) and the perturbations by δ. The inlet total pressure will thus be written as

$$\delta P_{T_1} = \varepsilon e^{i\omega t},$$

say, where ε is the amplitude of the perturbation. In addition, again in keeping with the idea of eliminating undue complexity, we take the flow to be *inviscid* and *incompressible*.

The momentum equation for the diffusing passage can be written as

$$\frac{\partial C}{\partial t} + C\frac{\partial C}{\partial \ell} = -\frac{1}{\rho}\frac{\partial P}{\partial \ell} \tag{11}$$

Integrating this from 1 to 2 yields:

$$\int_1^2 \frac{\partial C}{\partial t}\, d\ell = \left.\frac{P}{\rho} + \frac{C^2}{2}\right|_2^1 \tag{12}$$

Let us examine the term on the left hand side. The continuity equation for the passage is

$$\begin{aligned} CA &= \text{constant} \\ &= C_1A_1 \end{aligned} \tag{13}$$

where A is the local flow-through area and is a function of the distance along the passage, ℓ. A_1 is the flow-through area at inlet. Using the continuity equation we can write the time derivative term as

$$\begin{aligned} \int_1^2 \frac{\partial C}{\partial t}\, d\ell &= \frac{dC_1}{dt}\int_1^2 \frac{A_1}{A}\, d\ell \\ &= \mathcal{L}\frac{dC_1}{dt} \end{aligned} \tag{14}$$

where $\mathcal{L}$ is an effective length of the diffuser and is a function of the diffuser geometry only. (As an example, if we had a linear variation of area with length, i.e., $A = A_1 + \left[\frac{A_1 - A_2}{L}\right]\ell$, we would find

$$\mathcal{L} = \left\{\frac{L[\ln(A_2/A_1)]}{[(A_2/A_1) - 1]}\right\}).$$

The integral of the momentum equation can thus be written as

$$\mathcal{L}\frac{dC_1}{dt} = \frac{P_{T1}}{\rho} - \frac{P_2}{\rho} - \frac{C_2^2}{2} \tag{15}$$

We now make use of the small perturbation concept and neglect products of perturbation quantities. In addition we note that for a time dependence of the form $e^{i\omega t}$ the time derivative term can be written as

$$\mathcal{L}\frac{dC}{dt} = i\omega\mathcal{L}\delta C_1 \tag{16}$$

Since in a steady flow the total pressure is the same at inlet and exit, the equation for the perturbation quantities becomes

$$L \frac{dC}{dt} = \frac{\delta P_{T_1}}{\rho} + \bar{C}_2 \ \delta C_2, \tag{17}$$

where we have used the condition of constant static pressure at the diffuser exit to put $\delta P_2 = 0$.

A quantity of interest is the inlet velocity variation, δC_1. We can again make use of the continuity equation to express the exit velocity in terms of the inlet velocities so that the former may be eliminated:

$$\bar{C}_2 = (A_1/A_2)\ \bar{C}_1$$

and

$$\delta C_2 = (A_1/A_2)\ \delta C_1$$

This leads to a final expression for the inlet velocity perturbation in terms of the imposed inlet total pressure non-uniformity δP_{T_1}. This is

$$\frac{\delta C_1}{(\delta P_{T_1}/\rho \bar{C}_1)} = \frac{1}{\left[(A_1/A_2)^2 + i\Omega\right]} \tag{18}$$

where the reduced frequency Ω is defined as

$$\Omega = \frac{\omega L}{\bar{C}_1} \ .$$

This relation is plotted on the next page. What is presented in the figure are the real and imaginary parts of $[\delta C_1/(\delta P_{T_1}/\rho \bar{C}_1)]$ as a function of reduced frequency Ω, for the case $A_2/A_1 = \sqrt{2}$, which is a condition representative of a typical turbomachinery application. For each value of Ω, a vector drawn from the origin to the curve represents the quantity $[\delta C_1/(\delta P_{T_1}/\rho \bar{C}_1)]$ in both magnitude and phase.

Several things are apparent from the curve:

1) At very low reduced frequency the non-dimensional velocity perturbation is very close to the steady-state value (2.0). It is almost in phase with the total pressure perturbation and its magnitude is near the steady-state value.

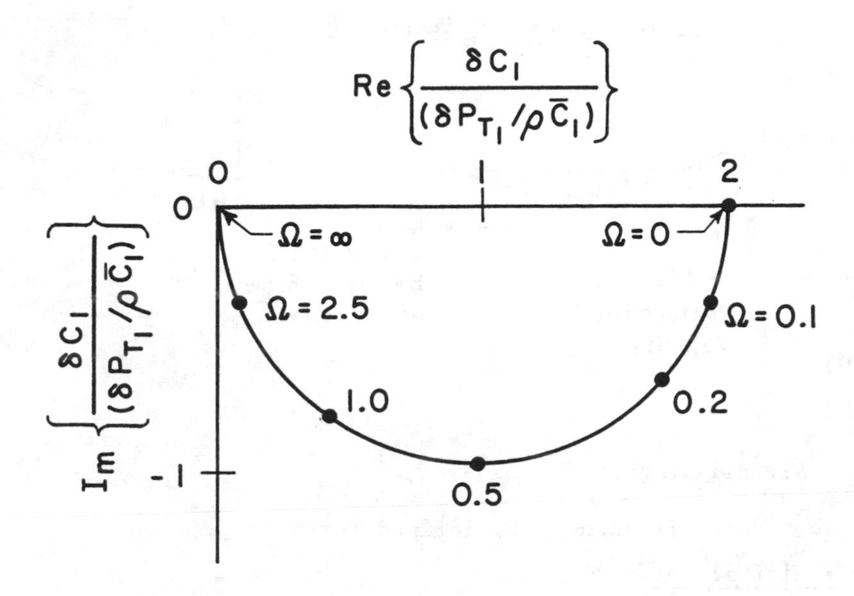

Figure 6. Diffuser inlet velocity perturbation; $A_2/A_1 = \sqrt{2}$

2) At high reduced frequency there is a phase difference of approximately $\pi/2$ between the velocity and total pressure perturbations, and the amplitude of the velocity non-uniformity is markedly reduced. In this situation, the local accelerations dominate over the quasi-steady type of convective acceleration terms.

3) The reduced frequency is an important parameter in unsteady flows.

Example 2: Unsteady Flow over a Flat Plate Airfoil

The above example illustrates some of the features of unsteady internal flows. We can briefly mention another example which is probably the most well known unsteady flow situation; the so-called Sears problem of an airfoil encountering a vertical gust [5]. The relevant geometry is shown below. A vertical velocity perturbation (gust) is convected by an isolated airfoil. The gust is of the form:

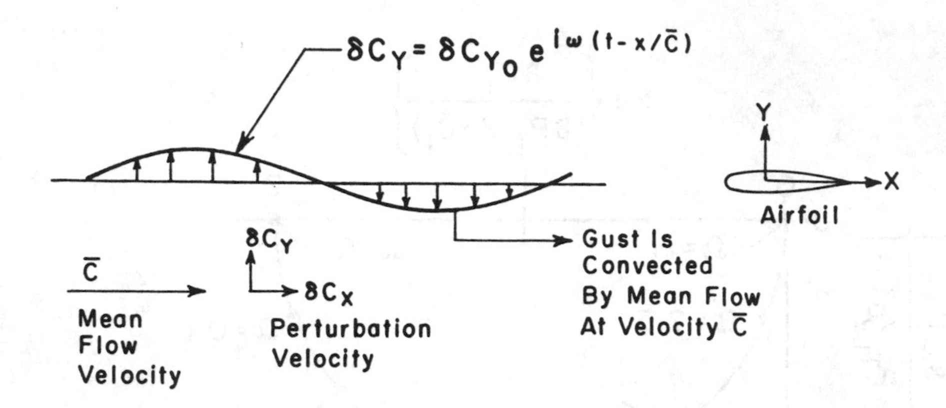

Figure 7. Airfoil in gust

where Ω, the reduced frequency, is defined *for this problem* as

$$\Omega = \frac{\omega \cdot \text{airfoil chord}}{2\bar{C}}$$

Due to changes in the angle of attack that are associated with gust passage, the circulation about the airfoil changes with time so that vorticity is shed into the airfoil wake. The lift of the airfoil also changes with time, and it is this quantity that we wish to find.

Note that as applied to this and other similar problems the reduced frequency has the following physical interpretation. The wavelength of the gust can be written as:

$$\text{Gust wavelength} = \omega/2\pi\ \bar{C}.$$

Thus the ratio of the airfoil chord to the gust wavelength, (which can be regarded as the change in phase of the gust along the airfoil) is just

$$\frac{\text{airfoil chord}}{\text{gust wavelength}} = \frac{\text{airfoil chord x } \omega}{2\pi C} = \frac{\Omega}{\pi}$$

The reduced frequency can thus be viewed as a measure of the wavelength of a disturbance compared to the relevant length of the airfoil. For a disturbance of wavelength long compared with the airfoil chord we have a situation where the blade is embedded in a flow that is essentially slowly varying but spatially uniform and hence the response is very similar to the steady-state response at that *instantaneous* angle of attack and inlet dynamic pressure. Also, the vorticity that is shed into the wake can be regarded as being "far away" (in this quasi-steady limit) so that it does not affect the flow around the airfoil. However, for higher reduced frequency one again would expect considerable departures from the quasi-steady response.

For the method of analysis the reader is referred to the original paper. The main conclusions are:

1) The non-dimensionalized lift is a strong function of the reduced frequency.

2) The basic form of the relation between lift, angle of attack, and velocity C (for a flat plate airfoil) is no longer valid in an unsteady flow. The expression for the non-dimensionalized lift is in fact given by the Sears function which looks as sketched in Fig.8. The magnitude and phase of the lift are shown by a vector from the origin to the curve. It can be seen that, as in the previous example, at very low reduced frequency the lift is quite close to the steady state value, but as one moves towards higher values of reduced frequency the behaviour becomes drastically different from the steady state siutation.

The Sears problem is for the isolated airfoil at zero mean angle of attack. However, the extension to the problem of the unsteady flow in cascades has been studied by many different investigators. For flat plates, subsonic flow is treated by Smith [6], and supersonic flow by Verdon [7,8] and Ni [9]. Finite thickness and turning have been examined by Verdon and Caspar,e.g.[10]. It should be stressed that these are only selected examples and the interested reader can consult the collections of papers listed in the References for more information.

Parameters Characterizing the Importance of Compressibility in Unsteady Flows

The example just given refers to an incompressible flow, but we have not addressed the question of just when turbomachine flows can be considered as incompressible. One condition for this in a steady flow is that the relevant Mach number, M, defined as M = C/a, where C is velocity and a is the speed of sound, is small. This is very often enough to characterize when a *steady* flow can be regarded as incompressible, but one must examine the situation in a bit more depth to see what additional parameters are required for this conclusion in an *unsteady* flow. A basic discussion of this topic is due to Lighthill[11] whose arguments we follow here. (See also Batchelor [12] for discussion of this point).

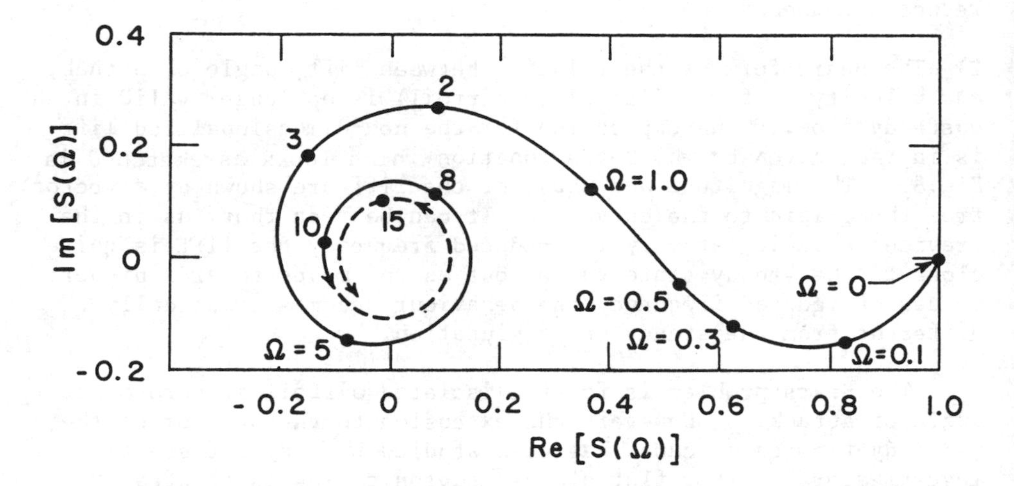

Figure 8. The Sears Function; Fluctuating Lift Due to Gust (Flat plate airfoil, zero mean angle of attack) Lift = $[\pi\rho.\text{chord}.\ \bar{C}\ \delta C y_0 e^{i\omega t}]\ S(\Omega)$ where $S(\Omega)$ is plotted above.

As a starting point, we consider the continuity equation, which can be written as [12]

$$\nabla \cdot \vec{C} = -\frac{1}{\rho}\frac{D\rho}{Dt} \tag{19}$$

The fluid can be regarded as incompressible if the magnitudes of the separate terms on the righthand side are much smaller than those on the left hand side. Thus, let us examine the sizes of these terms in an unsteady flow, again concentrating on the situation where the flow is periodic with radian frequency ω. Under this condition the magnitude of the density fluctuations will be

$$\frac{1}{\rho}\frac{D\rho}{Dt} \sim \omega \frac{\delta\rho}{\bar{\rho}} \sim \omega \frac{\delta P}{\bar{P}} \tag{20}$$

where $\delta\rho$ and δP are the perturbations in density and pressure associated with the fluctuations *at the frequency* ω, and $\bar{\rho}$ and $\bar{P}$ are the mean or ambient levels of these quantities. If δC is a typical perturbation velocity associated with the periodic flow,

$$\delta P \sim \bar{\rho}\omega L \delta C \tag{21}$$

where L is the relevant length of our fluid mechanic device. [Note that there will also be terms of order $\bar{\rho}C\delta C$ in δP, but that these will not invalidate the conditions under which the flow can be regarded as incompressible if those directly due to the accelerations seen in (21) do not]. We thus have that the right hand side of equation (19) is of magnitude given by

$$\frac{1}{\rho}\frac{D\rho}{Dt} \sim \frac{\rho\omega^2 L\delta C}{\bar{P}} \ .$$

The magnitude of the individual terms on the left hand side of (22) is $\delta C/L$ where, again, δC refers to the fluctuations at frequency ω. The condition that the flow can be viewed as incompressible is thus

$$\frac{\delta C}{L} << \frac{\bar{\rho}\omega^2 \delta C L}{\bar{P}}$$

or

$$\frac{\omega^2 L^2}{\bar{a}^2} << 1. \tag{23}$$

Therefore not only does the Mach number have to be small for the assumption of incompressibility to be valid, but the above parameter must be small as well. Another way of stating this is that L must be small compared to the "Radian wavelength", a/ω, of a sound wave of frequency ω [11]. Thus if this wavelength is long compared to L, then the flow can be considered incompressible. The parameter appearing in (23) can be written as

$$\frac{\omega L}{a} = \frac{\omega L}{C}\frac{C}{a} = \Omega M \tag{24}$$

so that the foregoing requirement is that the product of reduced frequency and Mach number be small. This can become important in applications where Mach numbers can be substantially below unity, but reduced frequencies are high enough so that the flow must be considered as compressible.

We have now looked at some of the basic ideas inherent in the description of unsteady flow. The main points that have been illustrated so far can be summarized as:

1) In an unsteady flow the total pressure (or the stagnation enthalpy) is not constant. It can vary along a streamline, it can vary along a particle path, and it can vary with time.

2) The reduced frequency, which can be regarded roughly as a measure of "how unsteady the flow is", is an important parameter. A large change in reduced frequency can change the overall character of the flows substantially.

3) Relations between flow quantities that are useful in a steady state (or quasi-steady) situation may not be valid in an unsteady flow. As an example of this, we found that along a streamline in the diffuser, even though the flow was inviscid, the total pressure was not constant. In addition, the simple correspondence between the lift of an isolated airfoil and the angle of attack which holds in a steady flow, is not valid in an unsteady flow.

4) The conditions for unsteady flows to be considered incompressible require not only that the Mach number is small, but that the product of reduced frequency and Mach number is small.

As a final note to this section we can point out that the reduced frequency which characterizes most of the list sources of unsteadiness will be of order unity so that unsteady effects are important. This can perhaps be readily seen in the blade/wake interactions but may not be so evident for the lowest frequency disturbances, those associated with compressor surge. The key concept to recognize is that under surge conditions even though the reduced frequency associated with a blade passage is quite small (and thus the *blades* can be performing quasi-steadily), the more relevant length scale for the motion is the length of the compressor.

Thus, as in the diffuser example, for the frequencies quoted the acceleration of fluid in the compressor duct gives rise to pressure differences that are comparable to those which are associated with quasi-steady effects, and, as we will see further in the lecture on surge, the unsteadiness must be taken into account in the attempt to formulate an accurate quantitative description of the phenomena.

5. ILLUSTRATIONS OF UNSTEADY (INVISCID) TURBOMACHINERY FLOWS

Total Pressure Changes in Unsteady Flow in Turbomachines

Another useful example which illustrates some additional features of unsteady flow was given initially by Preston [2], who examined the unsteady flow due to a moving row of bound vortices. This could serve as a basic model for a rotor blade row which moves relative to a stationary observer.

The flow is taken as inviscid, incompressible and irrotational, and the momentum equation is thus,

$$\frac{\partial \vec{C}}{\partial t} + \nabla(C^2/2) = -\frac{1}{\rho} \nabla P \tag{25}$$

or
$$\frac{\partial \vec{C}}{\partial t} = -\frac{1}{\rho} \nabla P_T \tag{26}$$

Further, $\vec{C}$ can be written as the gradient of a potential, ϕ,

$$\vec{C} = \nabla\phi.$$

Eq.(2) can therefore be integrated to give*

$$\frac{\partial \phi}{\partial t} + \frac{P_T}{\rho} = f(t) \tag{27}$$

This equation will form the basis of the analysis.

The specific configuration to be investigated is illustrated in Figure 9, which shows a row of bound vortices (representing the circulation which exists around the blades of a compressor rotor, for example). The vortices have a circulation of Γ in the counterclockwise direction, a spacing h, and they move to the left with a velocity U. We can make use of two coordinate systems, a fixed one, denoted by x,y, and a moving one denoted by x', y' which are also shown in the Figure. The moving coordinate system is tied to the row of bound vortices so that it too moves to the left with a velocity U.

Consider a point A which is on the y axis of the fixed system and define the time, t, so that at time t = 0 the fixed system and the moving system coincide. In terms of the moving coordinate system this point has coordinates

$$x' = x + Ut, \quad y' = y \tag{28}$$

*f(t) is a function of time only

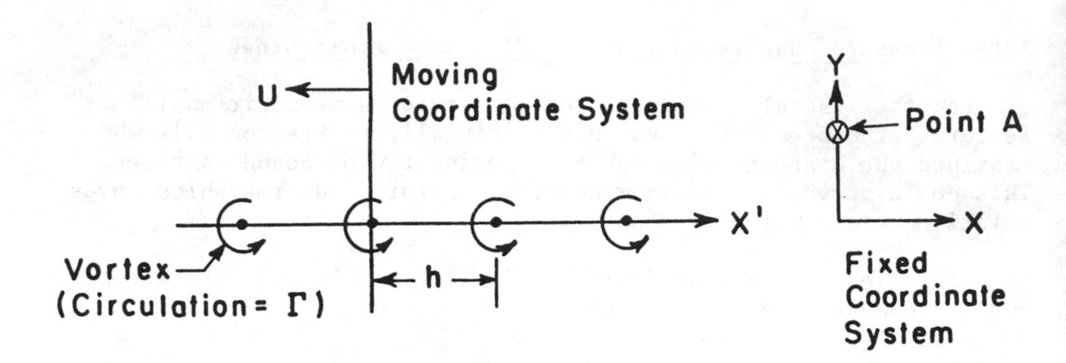

Figure 9.

Now in the moving coordinate system the flow is steady. In this coordinate system the velocity potential ϕ, for a row of vortices along the x' axis is given by Lamb[13] as

$$\phi = \frac{\Gamma}{2\pi} \tan^{-1} \left[\cot\left(\frac{\pi x'}{h}\right) \tanh\left(\frac{\pi y'}{h}\right)\right] \tag{29}$$

The velocity components as seen in the moving system are:

$$u' = -\frac{\Gamma}{2h} \frac{\sinh(2\pi y'/h)}{\cosh(2\pi y'/h) - \cos(2\pi x'/h)} \tag{30a}$$

$$v' = \frac{\Gamma}{2h} \frac{\sin(2\pi x'/h)}{\cosh(2\pi y'/h) - \cos(2\pi x'/h)} \tag{30b}$$

Now consider the conditions at the fixed point. Eq. (27) can be written as

$$\rho \frac{\partial \phi}{\partial t} + P_T = P_0 \tag{31}$$

where P_0 is a constant, because the motion is steady at $y = \pm \infty$. The quantity $\partial\phi/\partial t$ is the time rate of change of the velocity potential as seen by a fixed observer at point A. This can be related to the spatial rate of change of the velocity potential in the moving coordiante system, by

$$\left.\frac{\partial \phi}{\partial t}\right]_{\substack{\text{seen in}\\ \text{fixed system}}} = U \left.\frac{\partial \phi}{\partial x'}\right]_{\substack{\text{seen in}\\ \text{moving system}}} \tag{32}$$

The velocity u' is the velocity at point A(as measured by an observer in the moving coordinate system) at the time t. Since the coordinates of the point A (as measured by an observer in the moving system) are $x' = Ut$, $y' = y$, the velocity can be obtained by substituting these values in the formula for u' given in Eq. (39a). This gives (using (31) and (32))

$$P_T - P_0 = U \frac{\Gamma}{2h} \left[\frac{\sinh (2\pi y/h)}{\cosh (2\pi y/h) - \cos (2\pi Ut/h)} \right] \tag{33}$$

which is an expression for the instantaneous total pressure as measured by a stationary observer at point A.

Note that at $y = -\infty$ and $+\infty$ respectively the total pressures are found to be

$$\left. P_T \right)_{-\infty} = P_0 - \rho \frac{U\Gamma}{2h} \tag{34a}$$

$$\left. P_T \right)_{+\infty} = P_0 + \rho \frac{U\Gamma}{2h} \tag{34b}$$

Therefore the change in total pressure across the blade row is

$$\frac{\Delta P_T}{p} = \rho \frac{U\Gamma}{h} \tag{35}$$

Since the change in "tangential" (i.e. x) velocity from $-\infty$ to $+\infty$ is Γ/h, Eq. (35) can be seen to be the change in P_T that is given by the Euler turbine equation applied to this incompressible, inviscid flow.

Of more interest to us are the instantaneous variations in total pressure. These are shown in two different ways in Figures 10 and 11. First, the variations in total pressure have been plotted versus the vertical (y) location of point A for different non-dimensional times during the passage of the row of vortices. This is shown in Figure 10. The total time taken for the row to move one vortex spacing is h/U and this has been used to make the time non-dimensional. The position variable that is used is $(2\pi y/h)$

Several features can be seen in Figure 10. First it can be noted that the total instantaneous total pressure variations near the vortex row can be substantial, becoming as large or larger than the mean total pressure change across the row. However as one moves away from the blade row to a distance of $y/h = 1/2$, say, the fluctuations have decreased to roughly 10% of the total pressure change through the row, and at $y/h = 1.0$ they are less than 1%. A high response total pressure probe could therefore see a strong fluctuating pressure if it were placed in close enough proximity to this row, but would see an essentially steady flow if it were a blade gap away.

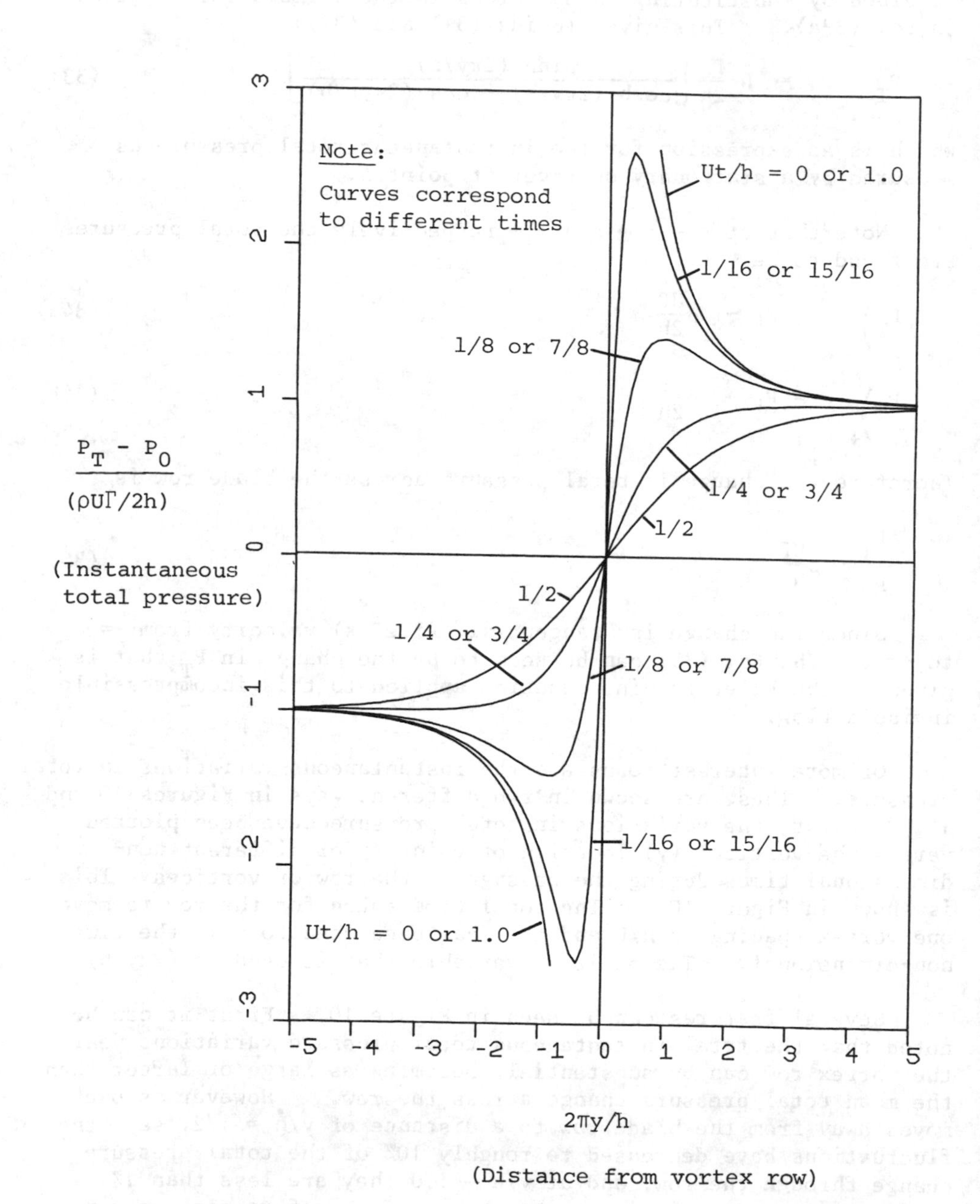

Figure 10. Instantaneous Total Pressure [2].

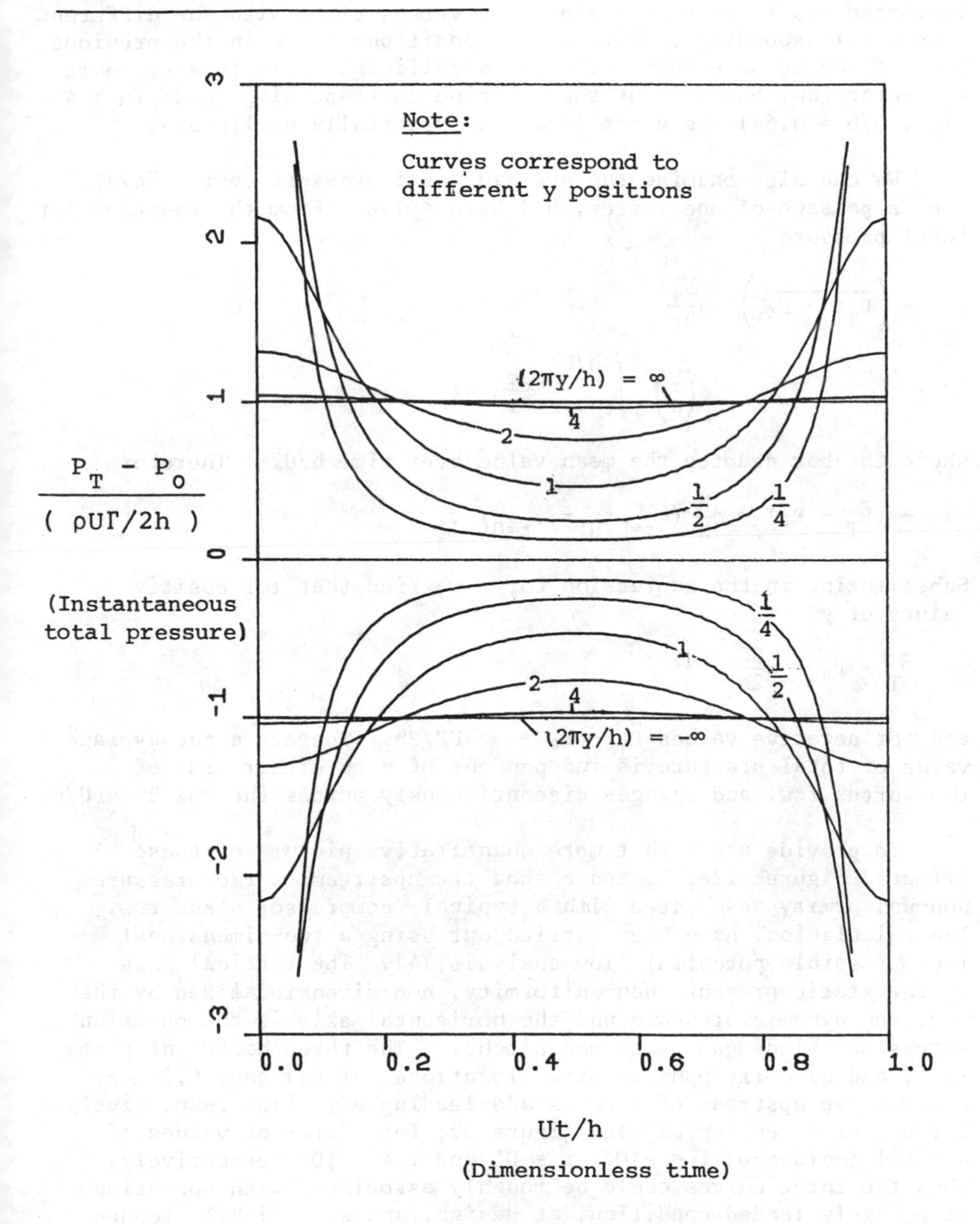

Figure 11. Instantaneous Total Pressure [2].

Figure 11 is a crossplot of Figure 10 and shows the instantaneous total pressure plotted versus time, with the different curves corresponding to different y positions. As in the previous plot it can be seen that there are significant variations close to the rotor row, but that at the location corresponding to $2\pi y/h = 4$ (i.e. $y/h = 0.64$) the variations are essentially negligible.

We can also examine the average total pressure over a "cycle" i.e. a passage of one vortex, $0 < Ut/h < 1.0$. From the equation for total pressure

$$-\left(\overline{P_T - P_0}\right) = \rho\overline{\frac{\partial\phi}{\partial t}}$$

$$= \rho\left(\frac{U}{h}\right)\int_0^{h/U}\frac{\partial\phi}{\partial t}\,dt$$

where the bar denotes the mean value over time h/U. Therefore

$$-\left(\bar{P}_T - P_0\right) = \rho\frac{U}{h}\left(\phi_{t=h/U} - \phi_{t=0}\right) .$$

Substituting in the expression for ϕ we find that for positive values of y

$$\overline{P_T} - P_0 = \rho\frac{U\Gamma}{2h}$$

and for negative values $\bar{P}_T - P_0 = -\rho U\Gamma/2h$. Therefore the average value of total pressure is independent of y on either side of the vortex row, and changes discontinuously across the row by $\rho\Gamma U/h$.

To provide a somewhat more quantitative picture of these effects, Figures 12a, b, and c show the upstream *static pressure* non-uniformity associated with a typical compressor blade row. The calculations have been carried out using a two-dimensional incompressible potential flow analysis[14]. The vertical axis is the static pressure non-uniformity, non-dimensionalized by the upstream dynamic pressure and the horizontal axis is the position across one blade gap (i.e. one pitch). The three different plots (a,b, and c) correspond to axial locations, of 0.1 gap, 0.2 gap, and 0.4 gap upstream of the cascade leading edge line respectively. The different curves on each figure are for different values of airfoil incidence: i = +10°, i = 0° and i = - 10° respectively. Thus the three curves could be roughly associated with operation at a highly loaded condition, at design, and at a lightly loaded condition.

It can be seen that the variations in static pressure are considerable near the blade, but decay rapidly as one moves upstream on the order of one-half the blade gap. In addition, however, it is important to note that the amplitude of the non-uniformity increases substantially as the incidence angle (i.e. blade loading) is increased. Probes placed upstream

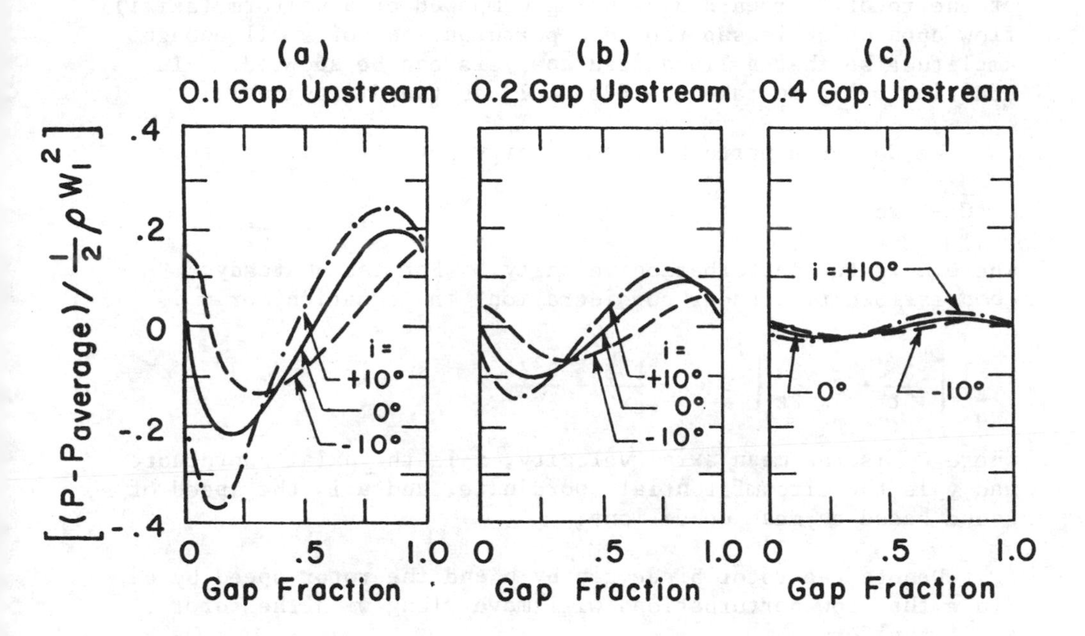

Figure 12. Static pressure variations upstream of "typical" compressor blade row (incompressible flow).

of a rotor can thus see a strongly fluctuating static pressure (as well as flow angle) and will thus also experience large variation in total pressure if they are in close proximity to a rotor, and this effect will be increased as the blade loading is increased.

Effects of Compressibility on the Unsteady Flow Field

In the discussion until now we have dealt with incompressible flow. This has been done in view of the desire to present the introduction to unsteady effects in the most basic manner. However, one should not lose sight of the fact that, in many flows of interest, compressibility plays an important role. It is thus appropriate to present a brief discussion of the effects of compressibility on some of the foregoing examples.

In particular, let us examine the upstream potential field of a blade row in a compressible flow. We can pose the question as follows: suppose that at some axial location (x = 0) we have an axial velocity perturbation due to the rotor. This is unsteady in the absolute reference frame. We wish to know how this perturbation decays with distance as a function of Mach number.

To analyze this example in a simple manner the flow upstream of the rotor is regarded as being composed of a uniform (axial) flow upon which is superposed a perturbation of small enough amplitude so that a linearized analysis can be adopted. In addition we again take the flowfield as two-dimensional.

We define a perturbation velocity potential ϕ such that

$$\vec{C} = \nabla\phi$$

where $\vec{C}$ is the perturbation velocity. For the unsteady, *compressible* flow under consideration, the equation for ϕ is [15]:

$$\frac{1}{\bar{a}^2}\left[\frac{\partial}{\partial t} + \bar{C}_x \frac{\partial}{\partial x}\right]^2 \phi = \frac{\partial^2\phi}{\partial x^2} + \frac{\partial^2\phi}{\partial y^2} \tag{36}$$

where $\bar{C}_x$ is the mean axial velocity, x is the axial coordinate and y is the circumferential coordinate, and $\bar{a}$ is the speed of sound based on mean conditions.

Denote the rotor blade gap by h and the rotor speed by U. Since the flow perturbations will move along with the rotor ϕ is of the form

$$\phi \sim f(x)\, e^{2\pi i[y/h - Ut/h]} \tag{37}$$

Defining a Mach number based on the mean axial velocity as M_x $(= \bar{C}_x/\bar{a})$ and a Mach number based on the blade speed as $M_B (= U/\bar{a})$ and substituting (37) into (36) the equation for f(x) is

$$(1 - M_x^2)\frac{d^2f}{dx} + \left(\frac{4\pi i}{h}\right)\left(M_x M_B\right)\frac{df}{dx} + \left(\frac{2\pi}{h}\right)^2 [M_B^2 - 1]f = 0 \tag{38}$$

The solution for ϕ is thus

$$\phi = A \exp\left\{\left(\frac{2\pi x}{h}\right)\left[\frac{-iM_x M_B + (1-M_x{}^2 - M_B{}^2)^{1/2}}{(1 - M_x{}^2)}\right]\right\} \tag{39}$$

where A is an unknown constant which is set by the amplitude of the velocity perturbation at x = 0.

What is of primary interest here is the change in the rate of decay of the velocity (and consequently the pressure) perturbation as the Mach numbers are increased. In making this comparison *we will hold the ratio of axial Mach number to Mach number based on rotor speed constant.* Thus our example corresponds approximately to a situation in which we examine the flow non-uniformities caused by a given compressor rotor as the rotor speed is increased. Specifically, let us choose the value of $M_x/M_B = 0.5$, which corresponds to a value of $C_x/U = 0.5$, typical of modern compressors. Note that holding M_x/M_B constant is equivalent to keeping the reduced frequency, based on a length L and the axial velocity, $\bar{C}_x$, constant. Thus although the reduced frequency does not increase with blade speed in this example, the product of reduced frequency and Mach number does.

To illustrate the changes that can be expected as the blade Mach number is increased, Figure 13 shows the decay of the axial velocity profile for several rotor Mach numbers. The vertical axis is the amplitude of the axial velocity non-uniformity normalized by its value at $x = 0$, and the horizontal axis is the upstream position, non-dimensionalized by the blade gap, h.

Several trends are apparent from this figure. First, as might be expected, for low Mach numbers ($M_B \leq 0.5$) we recover essentially the same results that were derived for the incompressible case. However, as the blade (and axial) Mach numbers increase, the upstream influence can be seen to extend considerably further than in the incompressible case, as shown by the results for $M_B = 0.8$. In addition, for high enough Mach numbers ($M_B \geq 0.9$) this simple analysis predicts that *for the situation under consideration* ($M_x/M_B = 0.5$, two-dimensional isentropic flow) there is no decay of the upstream velocity and pressure perturbation, but rather upstream propagation of these perturbations. This occurs when the quantity in the square root in Eq. (39) becomes negative. This is the condition at which acoustic waves are no longer "cut off" but can propagate upstream. It is to be emphasized that the foregoing analysis does not describe situations in which there are large amplitude velocity and pressure non-uniformities (such as shock waves) or three-dimensional effects. However, the main point, which is that unsteady interactions can become more significant due to the influence of compressibility, will also be true in these latter situations.

As an example of this Figures 14 and 15 show data from a transonic rotor operating at a tip relative Mach number of 1.4 at design conditions[17]. The data was recorded using high response pressure transducers mounted in the casing of the rotor. Figure 14 shows the pressure rise due to the inlet shock for three different operating conditions, all at the design speed. These are "near stall" (N/S), "design" (D), and "near choke" (N/C). The vertical axis shows the *shock pressure rise*, non-dimensionalized by the inlet absolute total pressure, and the horizontal axis is the distance upstream of the rotor, non-dimensionalized by the

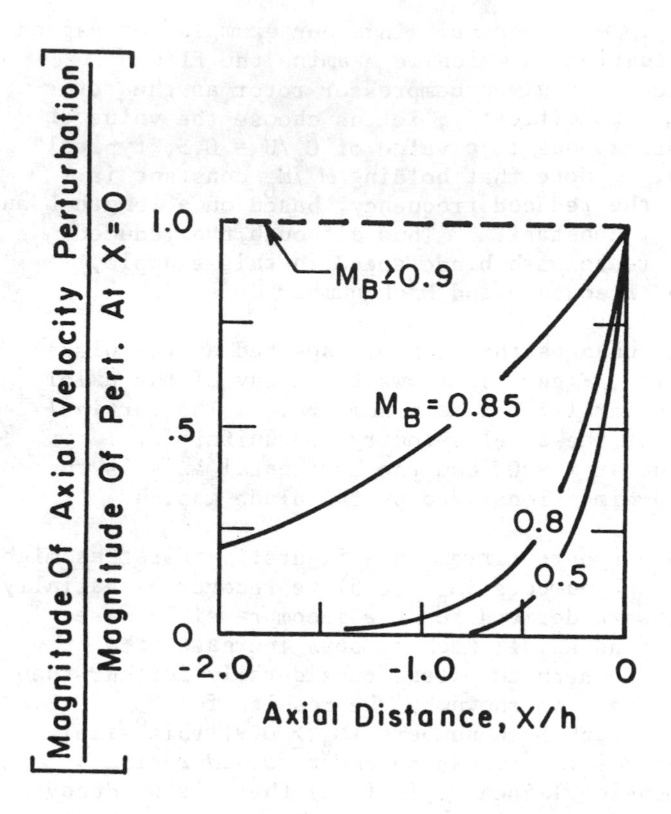

Figure 13. Upstream decay of axial velocity perturbation due to rotor ($M_x = 0.5\ M_B$)

blade gap at the 95% span position. It can be seen that as with the subsonic case, there is an increased effect as the blade loading is increased. However the upstream influence is now much more persistent. This example shows the increased potential for unsteady interactions with high Mach number stages.

A more detailed picture of the flow in this rotor is provided by Figure 15 which shows the static pressure isobars, referenced to the absolute inlet total pressure, for this rotor at the near stall conditions. It can be seen that the unsteadiness due to the rotor is appreciable even at the farthest upstream data location.

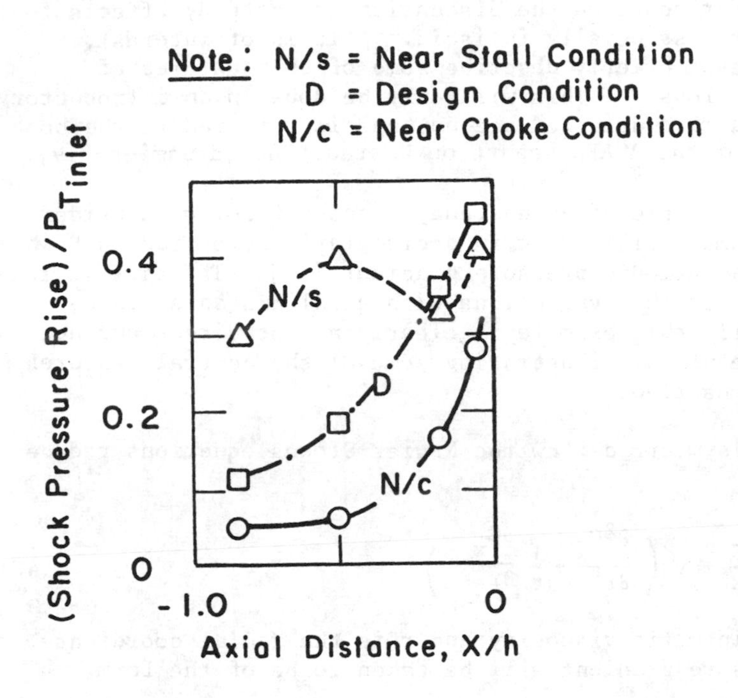

Figure 14. Shock pressure rise upstream of transonic rotor, [after Weyer and Hungenberg [17]].

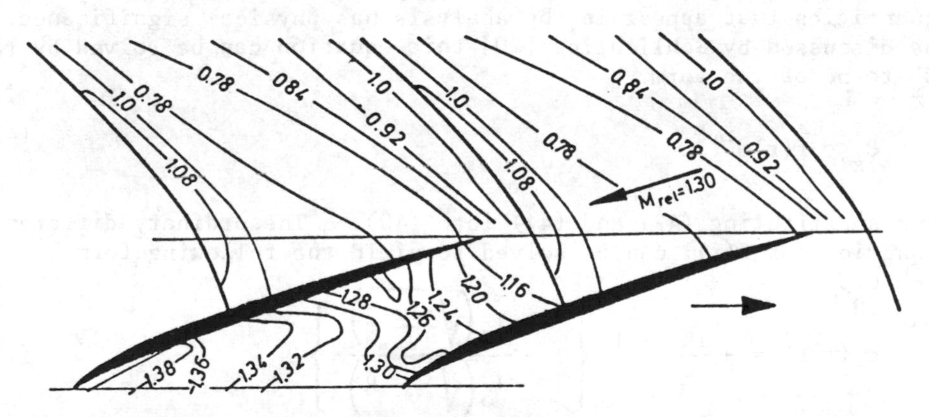

Figure 15. Isobars of (wall static pressure/inlet absolute total pressure) high back pressure ("near stall" point) [after Weyer and Hungenberg [17]].

Introductory Discussion of Unsteady Viscous Flows

We have so far confined the discussion of unsteady effects to phenomena that are essentially inviscid. It is of interest, however, to at least briefly describe some of the features of unsteady viscous flows. This will only be done in an introductory manner, and for a more detailed account one is referred to the book by Telionis [18] or the AGARD Report on Unsteady Aerodynamics [19].

As a simple example of an unsteady viscous flow, we consider the laminar, incompressible flow in a circular pipe of radius R which is subjected to a periodic pressure gradient [20]. The pipe is taken to be long enough so that variations with axial distance can be neglected. While this example is clearly not of direct engineering import, it is useful for illustrating some of the central features of unsteady viscous flow.

For this axisymmetric flow the Navier-Stokes equations reduce to

$$\frac{\partial C_x}{\partial t} = -\frac{1}{\rho}\frac{\partial P}{\partial x} + \nu\left(\frac{\partial^2 C_x}{\partial r^2} + \frac{1}{r}\frac{\partial C_x}{\partial r}\right) \tag{40}$$

where ν is the kinematic viscosity and r is the radial coordinate. The imposed pressure gradient will be taken to be of the form

$$-\frac{1}{\rho}\frac{\partial P}{\partial x} = k\, e^{i\omega t} \tag{41}$$

where it is again understood that only the real part of the complex quantities that appear in the analysis has physical significance. As discussed by Schlichting [20] this equation can be solved by taking C_x to be of the form

$$C_x = f(r)e^{i\omega t} \tag{42}$$

and substituting (42) and (41) into (40). The ordinary differential equation for f(r), can be solved to yield the following form for C_x.

$$c_x(r,t) = -\frac{iK}{\omega}\, e^{i\omega t}\left\{1 - \frac{J_o\left(\sqrt{\frac{-i\omega}{\nu}}\, r\right)}{J_o\left(\sqrt{\frac{-i\omega}{\nu}}\, R\right)}\right\} \tag{43}$$

In (43), J_o is the Bessel function of the first kind, of zero order. It can be seen from this solution that there is a new non-dimensional parameter, $(\sqrt{\omega/\nu})R$, that appears, which we will discuss in more detail below.

To see some of the central ideas connected with these unsteady flows, it is useful to examine two limiting cases of low and high frequency oscillations. Consider the former first. In the case of low frequency oscillations ($\sqrt{\frac{\omega}{\nu}}R \ll 1$) one can use the series expansion for the Bessel function to yield an approximate form for C_x. This is:

Low frequency approximation:
$$C_x = \frac{K}{4\nu}(R^2 - r^2)e^{i\omega t} \tag{44}$$

Thus for low frequency oscillations the velocity distribution is in phase with the pressure gradient and has a radial distribution that is the same as the (parabolic) steady state distribution for fully developed laminar flow in a circular pipe. In other words, as one might expect, at very low frequency, we see a quasi-steady viscous flow with the instantaneous flow at any time being that corresponding to the pressure gradient which occurs at that time.

For high frequencies, the asymptotic expansion of J_o for large arguments yields:

High frequency approximation :
$$C_x = -\frac{iKe^{i\omega t}}{\omega}\left\{1 - \sqrt{\frac{R}{r}}\exp\left[-(1+i)\sqrt{\frac{\omega}{2\nu}}(R-r)\right]\right\} \tag{45}$$

The form of C_x in this situation is now quite different from the quasi-steady case. We can break this expression up into two terms:

$$C_x = \underbrace{-\frac{iK}{\omega}e^{i\omega t}}_{(I)} + \underbrace{\frac{iK}{\omega}e^{i\omega t}\sqrt{\frac{R}{r}}\exp\left[-(1+i)\sqrt{\frac{\omega}{2\nu}}(R-r)\right]}_{(II)} \tag{46}$$

The first term (I) is the response associated with the inertia of the fluid in the pipe and is essentially an inviscid effect. (This term could in fact be derived from our simple unsteady diffuser situation if we let $A_2/A_1 \rightarrow 1$). The velocity associated with this term is constant over the annulus and is $-\pi/2$ out of phase with the pressure gradient driving term. The second part (II) is associated with the viscous effects and has a very different character. It dies off exponentially with distance from the wall, so that the distance to which the viscous effects "penetrate" the flow (by diffusion) can be seen to be of order

$$\text{Diffusion distance} \sim \sqrt{\frac{\nu}{\omega}}$$

We can now see the physical significance of the parameter $(\sqrt{\omega/\nu})R$. It can be viewed as the ratio of the pipe radius to the depth to which the effects of viscosity penetrate into the flow. Small values of $(\sqrt{\omega/\nu})R$ mean that the effects of viscosity are felt throughout this flow. Large values mean that the influence of viscous effects would be confined to a thin region of order $\sqrt{\nu/\omega}$ near the pipe wall, with the rest of the flow responding in an inviscid manner.

There is a further effect that is inherent in the form of (46). This is a phase shift in the velocity as one moves inwards from the wall so that the velocity near the wall is not in phase with that near the center. This is also reflected in the phase of the shear stress at the wall.

If we calculate this quantity we find that it is given by

$$\text{wall shear stress} \rightarrow \rho\frac{KR}{2}e^{i\omega t} \qquad \text{(low frequency);}$$

$$\rightarrow \rho\sqrt{\frac{\nu}{2\omega}}(i-1)\,Ke^{i\omega t} \qquad \text{(high frequency)}$$

At low frequencies the wall shear stress is thus in phase with the pressure gradient driving force, whereas at high frequencies there is a phase difference of $3\pi/4$ between the shear stress and the pressure gradient. [These are the limiting cases, and the phase difference between shear stress and pressure gradient actually increases continuously from zero to $3\pi/4$ as the parameter $\sqrt{\omega/\nu}$ R is increased from a very low value].

We can thus summarize the main points that have been found from the solution to this model problem. First, an important parameter is $\sqrt{\omega/\nu}$ R, with the size of this quantity determing the character of the solution. For small values, the flow can be regarded as quasi-steady. For large values the flow is quite unsteady, with the viscous effects confined to a small region near the wall and a phase difference between shear stress and pressure gradient, as well as between the velocity fluctuations near the wall and those in the center of the pipe.

This simple problem has served to introduce some of the features of unsteady viscous flows. The real interest, however, is in unsteady boundary layers and wakes in turbomachines, and there is still a great deal that is unknown about these phenomena. In particular, whereas it appears that the problem of calculating attached unsteady laminar boundary layers is reasonably well resolved [18] the boundary layers of most interest are turbulent and sometimes nearly separating or separated. Significant questions are still to be answered in the modelling of such flows, let alone in their application to turbomachine geometries. We therefore only make some qualitative comments to relate the above analysis to more complex flows.

As a start let us examine a boundary layer with the unsteadiness considered to be a perturbation on the steady flow. We can draw a rough analogy between the boundary layer thickness, δ, for the flow over a blade and the pipe radius in the above problem. (It is emphasized that this is *not* meant to be exact, but only to motivate what follows). Thus, for this boundary layer a useful non-dimensional parameter to show "how unsteady" the flow in the boundary layer is might be $(\sqrt{\omega/\nu})\delta$. If we assume that the (steady state) boundary layer thickness can be scaled as for a flat plate boundary layer, we have

$$\delta \sim \sqrt{\frac{\nu x}{U_\infty}}$$

where U_∞ is a velocity magnitude that characterizes the free stream. This yields for the non-dimensional unsteady flow parameter:

$$\sqrt{\frac{\omega}{\nu}}\,\delta \sim \sqrt{\frac{\omega x}{U_\infty}}$$

The parameter $\sqrt{\omega x/U_\infty}$, or $\omega x/U_\infty$ as it is generally used, can thus be interpreted as a measure of the importance of unsteady effects in a boundary layer with an impressed periodic disturbance, as is the situation in turbomachines. Although the analogy has only been drawn for the laminar case, this parameter $\omega x/U_\infty$ is also used to characterize turbulent unsteady boundary layers as well, and many of the results of calculations and experiments on unsteady boundary layers are plotted with this as the independent variable, although it should be remembered that the boundary layer behaviour will depend on Reynolds number as well.

As one might suspect the response of an unsteady boundary layer is considerably more complex than of the simple pipe flow, A useful introductory description of this is given by McCroskey [21] for the case of boundary layers in an external flow which has the form:

$$U_\infty = U_0(1 + A \sin \omega t).$$

where U_∞ is the free stream velocity. McCrosky discusses the phase relationships between the inner part of the boundary layer and the freestream flow as well as presents some results of calculations and experiments on unsteady boundary layers. (In this connection see also [18].)

Although an in-depth treatment of unsteady boundary layers is beyond the scope of this lecture, we can present some results that illustrate some of the overall features that are of interest in turbomachinery. These are taken from a recent paper [22] which describes a simple integral method for calculating unsteady turbulent boundary layers. The particular application is to non-steady flows in diffusers, but calculations are also carried out for other types of flows.

Figure 16, which is taken from [22], shows calculations of the boundary layer displacement thickness for the basic problem mentioned previously; a flat plate in an oscillating stream. The (normalized) fluctuations in this quantity are plotted as a function of $\omega x/U_0$, and it can be seen that there is a considerable difference in the displacement thickness perturbation as this parameter varies.

This method has also been applied to the fluctuating flow in a diffuser, which was the introductory problem that we looked at previously using a simple inviscid analysis. The inclusion of a boundary layer implies a different core velocity behaviour than that obtained in the inviscid flow. The computed core velocity is given in Figure 17 which also shows data from a 6 degree diffuser, with reduced frequency of unity. The parameter$(\Delta U_\infty/U_0)/A$ is the variation in core flow velocity divided by the variation in core flow velocity at inlet. The quantity ϕ_{U_∞} is the phase of the core velocity. Note that the inclusion of an unsteady boundary layer has caused a phase shift (a lag in this case) in the core flow velocity in addition to the change in the magnitude, and that, as shown in [22], these will vary with reduced frequency.

At present there is considerable activity on unsteady turbulent flows. Relevant survey references are the books by Telionis [18] Cebeci and Bradshaw [23], and Bradshaw, Cebeci and Whitelaw [24]. (The latter two, although not focussed specifically on unsteady flows, do have sections dealing with calculation procedures for non-steady boundary layers). Periodic unsteady turbulent boundary layers in adverse pressure gradients have been investigated by Covert and Lorber [25]. They measured ensemble averaged velocities and Reynolds stresses in the boundary layer on a NACA 0012 airfoil in an oscillating flow. The range of reduced frequencies was up to 6.4, which is a value characteristic of that encountered in turbomachinery. One important finding was that the influence of unsteadiness on the mean velocity profile in a strong adverse pressure gradient resulted in an alteration in the profile so that it appeared less like a separation profile. This general point will be commented on further below.

To end this section we can also mention three other related topics. The first concerns the type of unsteadiness that the boundary layer is subjected to. Many of the calculation procedures are based on having a free stream which undergoes oscillations of the form $e^{i\omega t}$. In many turbomachine applications, however, the unsteadiness more nearly resembles a travelling wave type of behaviour. This might be the case, for, example of a wake sweeping past an airfoil. In this situation, perturbations that are imposed on the boundary layer will be of the form:

$$\Delta U \sim e^{i(kx - \omega t)}$$

where k, the wave number is:

$$k = 2\pi/(\text{disturbance wavelength})$$

The phase velocity of this type of disturbance is ω/k and another nondimensional parameter that must be considered is the ratio of this phase velocity to a representative convection velocity U_0 ; i.e.

$$\frac{\text{phase velocity}}{\text{convection velocity}} = \omega/kU_0$$

For a given value of $\omega x/U_0$, changes in ω/kU_0 can alter the unsteady response of the boundary layer considerably [26], [27]. In particular, for certain values of ω/kU_0, the amplitude of the fluctuations in the boundary layer velocities can become considerably larger than those in the free stream [26]. An explanation for this is that these perturbations are near the natural eigenmodes of the boundary layer, and hence (very loosely) the latter is essentially being driven at a "near resonance" condition.

The second topic concerns the question of separation and stall in an unsteady flow. The articles by Telionis [18], [19], [28] should be referred to here, but one point should be noted. This is, as already alluded to, that generally separation appears to be delayed by unsteady effects. In an unsteady flow, therefore, a turbomachinery blade (or even a whole component) can operate transiently without stall in a regime which would result in stall if it were encountered in steady state operation. We will see an example of this in the section on non-uniform inlet flows.

The last point is that there are many situations in turbomachinery application where the phenomenon of (laminar-turbulent) transition is important. One example of this is in axial flow turbines, as will be described in more detail below. It appears that transition can be significantly affected by flow unsteadiness [29], although understanding of the process is far from complete.

6. SOME EXAMPLES OF UNSTEADY EFFECTS IN TURBOMACHINES

The foregoing has served to introduce some of the important unsteady flow concepts. In the remainder of this lecture we will describe several specific applications in which unsteady effects are of importance. It is to be remarked, however, that these examples give only a glimpse into the many features of flow in turbomachines in which unsteadiness occurs, and that no attempt is being made to provide a complete coverage of all the areas in which unsteady effects are important.

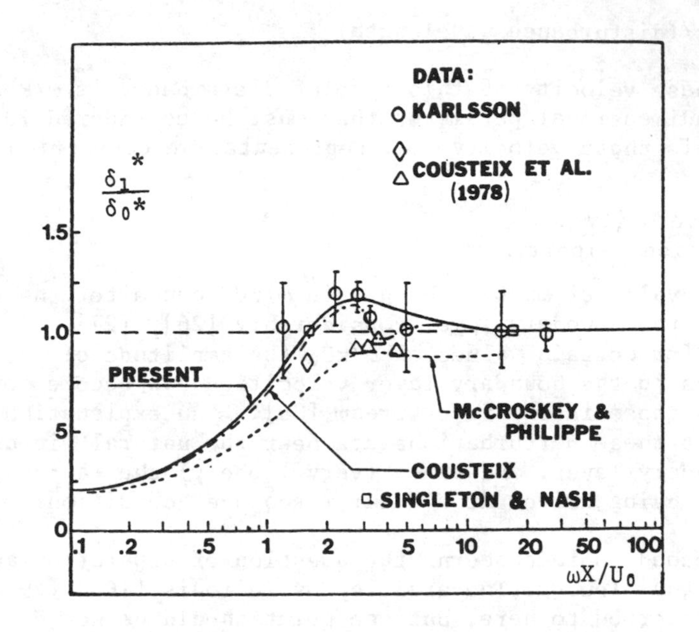

Figure 16. Amplitude of the boundary-layer displacement thickness fluctuations for a flat plate in an oscillating stream. The velocity is $U_0(1 + 0.125 \sin\omega t)$; $\delta^* = \delta_0^* + 0.125\ \delta_1^* \sin(\omega t + \pi + \phi)$ [22]

Unsteady Effects with Uniform Inlet Flows

We start by considering some examples of unsteady effects in situations in which the turbomachine is performing with a "uniform" inlet flow (as opposed to with an inlet distortion). If one considers a compressor blade row in a multi-row machine, it is apparent that there are sources of unsteadiness both upstream and downstream, since the neighboring rows are moving relative to the row under consideration. Two main types of unsteady effects therefore exist: those due to the upstream influence of the potential field of the downstream row, and those due to wakes from the upstream row. A schematic of the latter is shown in Figure 18 for a compressor rotor moving through a stator wake. As the blade passes through the stationary wake the velocity vector changes in both direction and magnitude so that the flow incident on the rotor is unsteady.

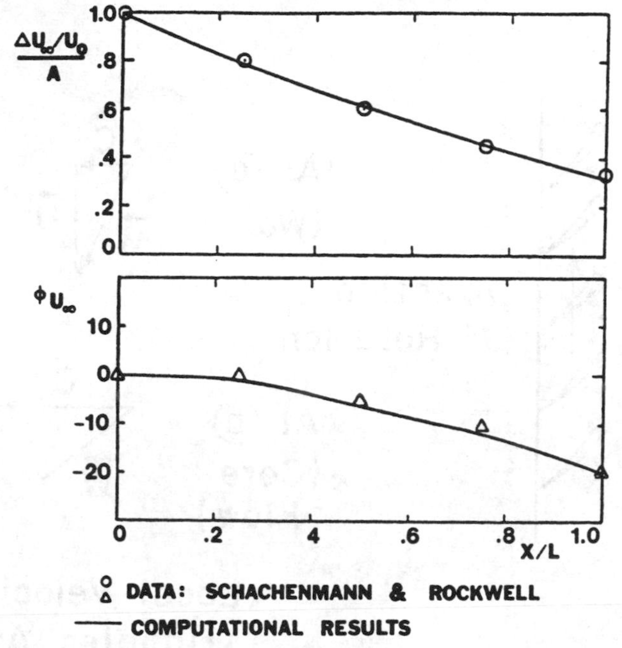

Figure 17 The velocity fluctuation amplitude (upper curve) and phase (lower curve) for the 6-deg unsteady conical diffuser of Schachenmann and Rockwell with 7% inlet fluctuation and $\Omega = 1$ [22].

The flow non-uniformities due to these effects can be large. In addition, the period of the fluctuations is roughly comparable to the time for a particle to be convected through the row so that unsteady reduced frequencies are unity or larger. Unsteady effects can therefore be important, and, as noted by Mikolajczak [30], in this situation there is no coordinate system that we can use to make the flow steady. The result is that not only can there be significant pressure fluctuations on the blades and vanes, but also that the (relative) stagnation temperature of a given fluid particle changes as the particle convects through the row. Although we will examine pressure and lift fluctuations on turbomachinery blades, we will take as the principle topic of this section the influence of unsteady effects on the aerodynamic performance of the turbomachine. In other words we will be concerned with the possibility of improving the design point effiency of the compressor or turbine through an improved understanding of the internal unsteady aerodynamics.

Unsteady Blade Forces Due to Blade Row Interactions

Let us consider the unsteady blade forces first. Some feeling for the magnitude of the effects that are to be expected is shown in Figure 19 which presents the fluctuating lift on a stator blade

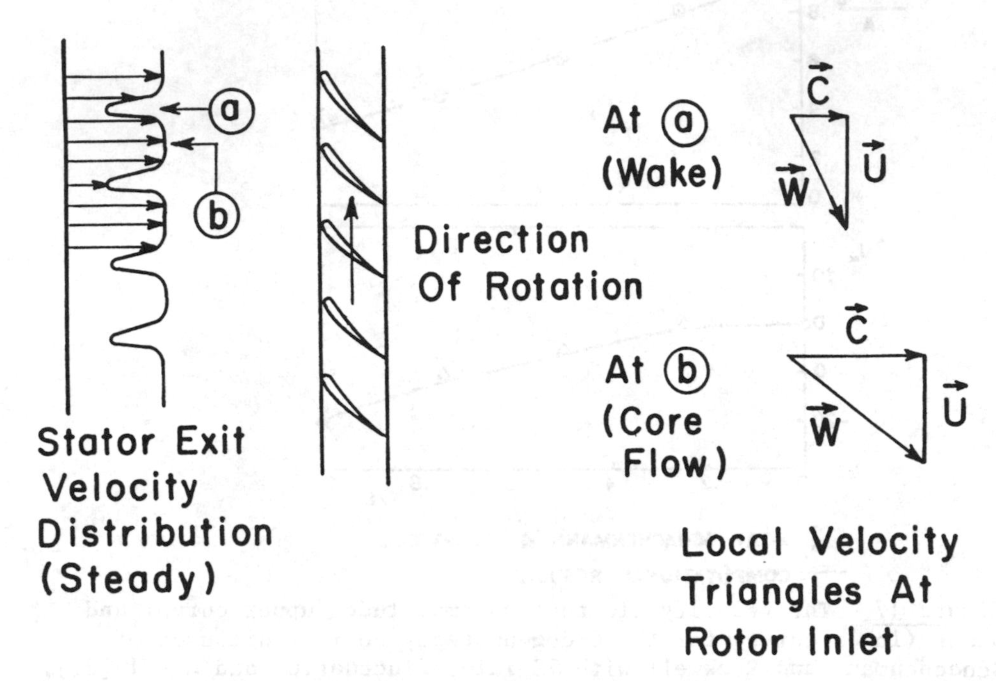

Figure 18 Unsteady Flow in a rotor due to stator wakes

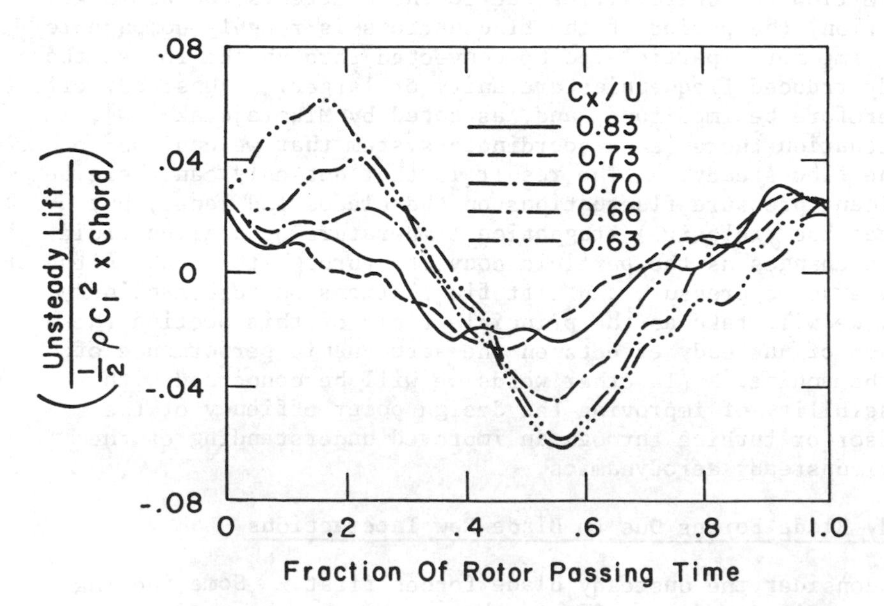

Figure 19 Measured periodic fluctuations of stator lift coefficient (after Gallus et al [31]).

in a single stage subsonic compressor [31]. The lift has been calculated from measurements of the fluctuating pressures on the stator at the midspan. The vertical axis is the non-dimensional fluctuating lift coefficient and the horizontal axis is a non-dimensional time based on one rotor passing period. The different curves correspond to different values of axial velocity parameter (C_x/U) with the smallest value (0.63) corresponding to the condition of highest incidence. The lift fluctuations are largest in this case. Note that at these low values of axial velocity parameter not only is there a high loading on the rotor and stator but also that the rotor wakes would be expected to be larger so that the stator incidence variations would be expected to be more severe.

Fluctuating forces on blades and vanes are perhaps the most obvious unsteady effects in multi-row turbomachines, and they are clearly of great import from an aeroelastic standpoint. As stated, however, our main concern here is the influence of unsteadiness on loss and efficiency, and the studies of fluctuating lift have not really provided any direct procedure for estimating how these quantities are affected by unsteadiness. We will therefore not discuss these further, but will rather concentrate on those investigations which are concerned with this topic.

Intra-Blade Wake Transport in Axial Turbomachines

A basic step towards obtaining an understanding of one of these problems, namely the effect of the upstream wakes on the flow inside blade rows, is a kinematic analysis of the rotor-stator interaction problem [32]. The flow model is illustrated in Figure 20a for a compressor stator. From consideration of the velocity triangles, it can be seen that when rotor wake fluid enters the stator passage it has a "slip velocity" towards the pressure side of the stator. Thus, as the rotor wakes pass through the stator they are transported towards the pressure side of the stator passage. The presence of the stator pressure surface will interrupt this transport with the result that the rotor wakes will be collected by the pressure side of the stator blades and the rotor wake fluid will tend to appear in the stator wakes. Since the rotor wake has an excess of stagnation temperature over that of the inviscid flow, the time-averaged temperature profile at the exit from the stator will be as shown in Figure 20b. Wake transport thus leads to a significant redistribution of stagnation enthalpy across the stator blade rows. The analysis, which agrees well with experimental results, has been used successfully to interpret compressor measurements and to estimate rotor losses [32], [33], [34].

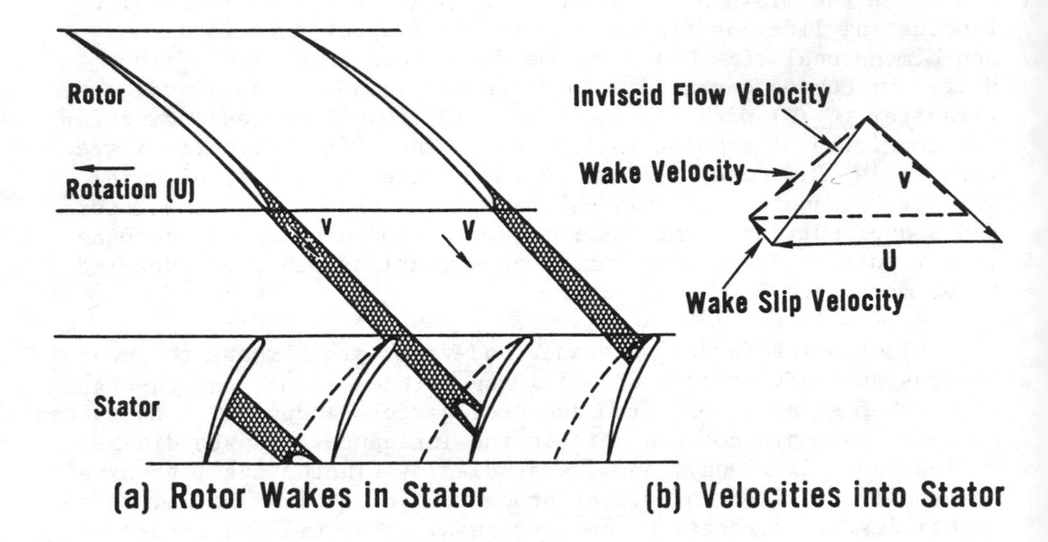

Figure 20a Schematic of a stage showing rotor wakes passing through stators; wakes have transport velocity towards the stator pressure surface [30].

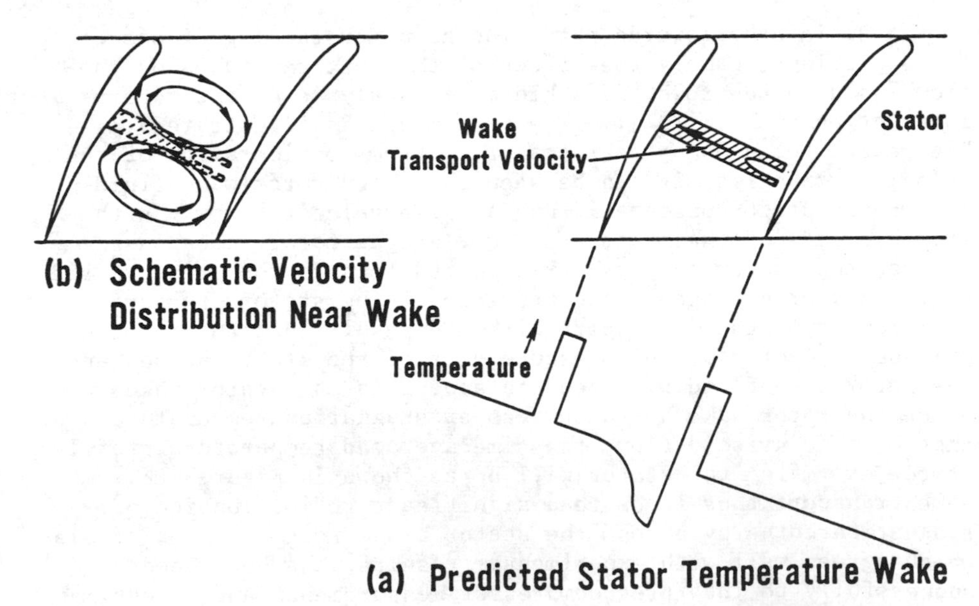

Figure 20b Wake transport leads to a time averaged stagnation temperature profile at stator exit, and flow circulation inside stator passages [30].

The kinematic model considers convection only and does not account for the effect of fluctuating static pressure on stagnation enthalpy (and pressure) redistribution inside a blade row. In order to satisfy continuity the flow near a transported wake is expected to be somewhat as shown in the figure, with sharp flow turning occurring near the bounding airfoil surfaces. The large streamline curvature associated with this turning would be expected to be associated with large fluctuations in static pressure with time and hence significant stagnation enthalpy (and stagnation pressure) changes. In addition to this, the convection velocity inside the blade row is itself not uniform over the passage, since particles near the suction side have a higher velocity than do those on the pressure side. This results in a reorientation of the wake as it is transported through the blade row. There have thus been attempts to include these aspects of wake transport in a more complete description of the overall phenomenon.

The quantities that the analysis of Kerrebrock and Mikolajczak focussed on were the stagnation temperature and stagnation pressure. Another feature of wakes is the high turbulence intensity that generally exists in them. In view of this, measurement of the turbulence intensity within a stator passage, say, can be very useful as an indicator of wake position. An example of this is given in Figure 21 which shows contours of "turbulence intensity parallel to the local main flow direction" [35] in a stator of a transonic axial compressor stage. The rotor wakes are delineated as regions of high turbulence. Only one passage is shown, and it should be noted that the circulation round the blades implies that the wakes will emerge from the stator in a "sawtooth" fashion.

The discussion of wake transport has been confined so far to compressors. Clearly, the same phenomenon occurs in turbine blade passages as well, although the drift of the wake is toward the suction surface of the turbine blade.

An examination of wake generated unsteadiness in axial flow turbines has been carried out by Hodson [36], [37]. He made measurements of the unsteady flow at the midspan section of a free vortex axial turbine rotor. As an example of these, Figure 22a shows the measured mean *periodic* unsteady "free stream" velocities on the pressure and suction surfaces of the turbine rotor, derived from phase locked averages. The nondimensionalizing parameter used is the rotor exit relative velocity. It can be seen that the level of these fluctuations on the suction surface is over twice that on the pressure surface. For comparison, Figure 22b shows the mean (actually the rms) *random* unsteadiness, i.e., the difference between instantaneous and phase locked average. The magnitude of the periodic and random components are comparable near the pressure side, but near the suction surface the periodic unsteadiness is considerably larger than the random component.

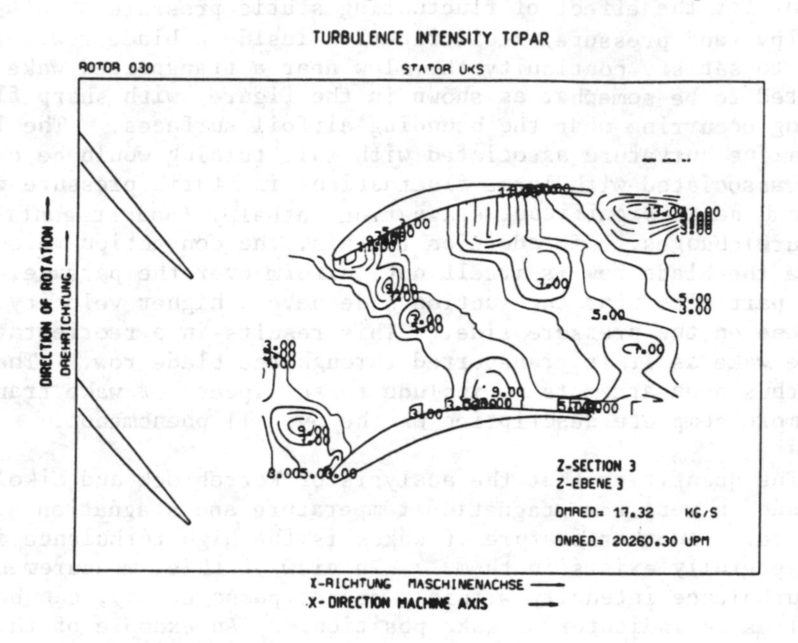

Figure 21 Distribution of turbulence intensity parallel to local main flow direction (TCPAR) throughout one stator blade channel [35].

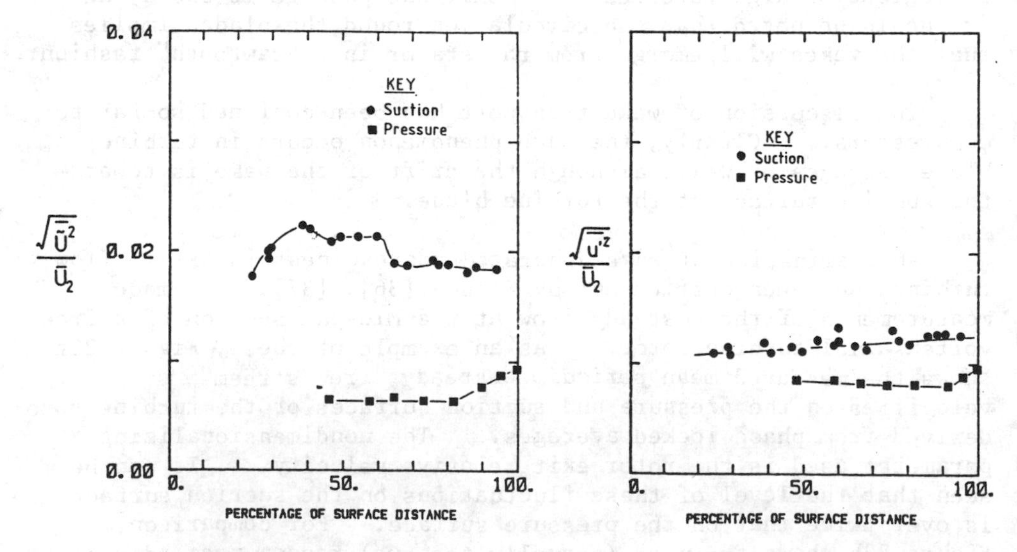

Figure 22a Mean Periodic Unsteadiness [36]

Figure 22b Mean Random Unsteadiness [36]

To track the stator wakes through the rotor, Hodson also employed the turbulence intensity as a marker. In [36], he presents plots of the unsteady velocity vectors, on which are superimposed contours of turbulence intensity, so that the wake shape can be seen.

To complement the experimental studies, calculations of wake transport have also been carried out using a (two dimensional) time marching procedure. The results of these are presented in Figure 23, which shows the computed unsteady velocity field in the rotor passage. The transport of the wakes across the passage, as well as the wake deformation, can be clearly discerned. Detailed discussion and further interpretation of this process, as well as some comparison with the experiments is discussed in the paper.

Effect of Wake Generated Unsteadiness on Boundary Layer Transition

The impingement of the wakes on the suction surface of the blades has yet another effect on the flow in the stator passage. This has to do with the laminar-turbulent transition in the blade surface boundary layers. The situation appears to be that the transition due to this type of unsteady flow can occur at a substantially lower Reynolds number than that found in boundary layers subject to oscillating free stream flows [38]. The mechanism by which this occurs is not clear, although it is hypothesized in [38] that it is the periodic variations of "free stream" turbulence due to the wakes that are responsible. In other words, as the wakes pass over the blades, the boundary layer at a given location varies periodically from laminar to turbulent.

A schematic representation of this process is given in [29] as illustrated in Figure 24. This shows sketches of quasi-steady laminar and turbulent velocity profiles along with hot wire traces during two cycles of the wake passage past a given location on the blade. It is to be emphasized, however, that this is only a schematic and that the actual situation is different since the instantaneous boundary layer profiles do not really resemble quasi-steady ones.

The importance of this unsteady transition process lies in its effect on the boundary layer and hence on the overall loss characteristics of the blade row. Comparison of the losses for the same blade section tested in a linear cascade (which had a steady uniform inlet flow) and in a turbine rotor give quite different results. Due to the (unsteady) transition in the latter situation, the loss was found to be approximately fifty per cent higher than in the former.

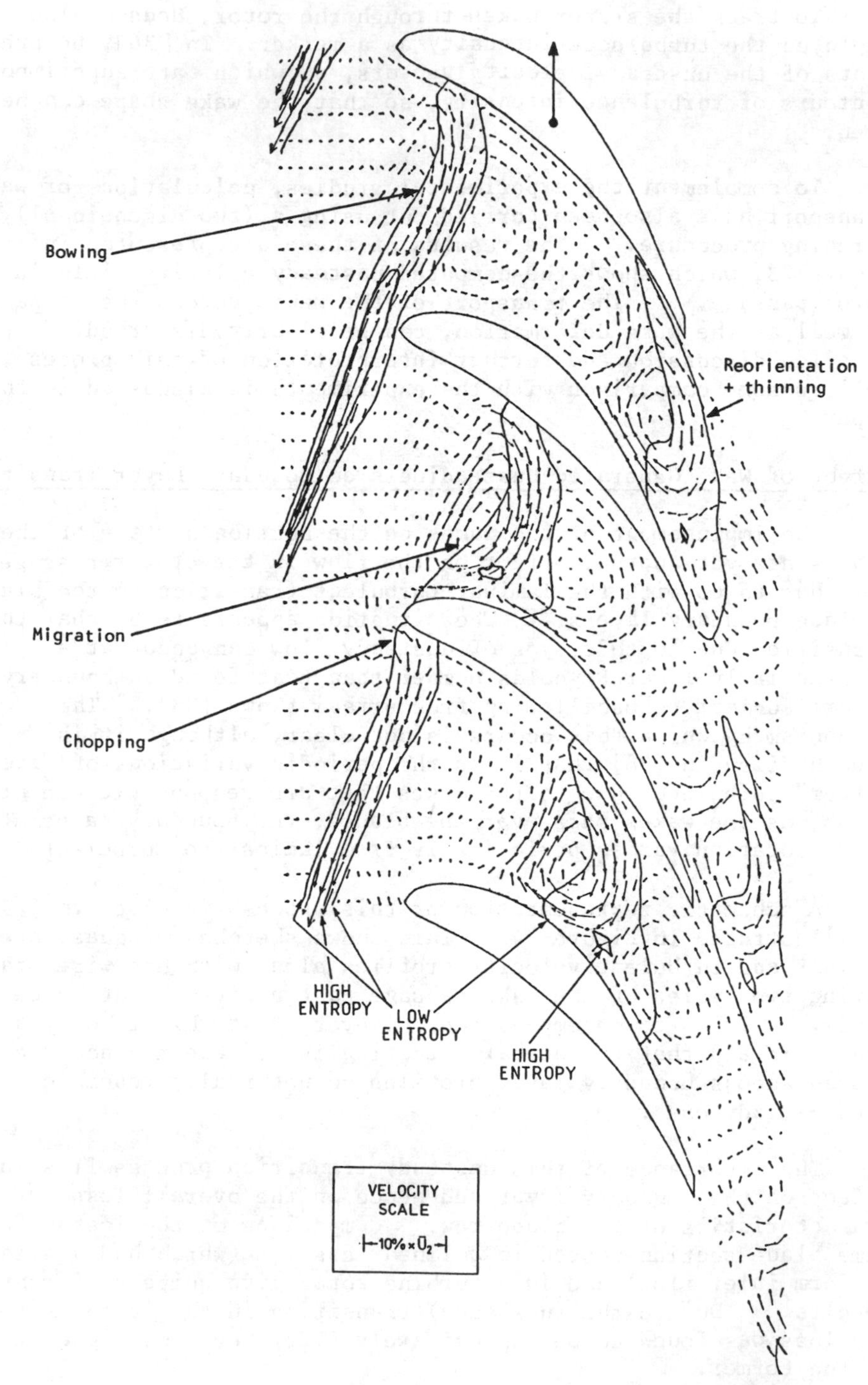

Figure 23 Unsteady Velocity Vectors and Entropy Contours [37]

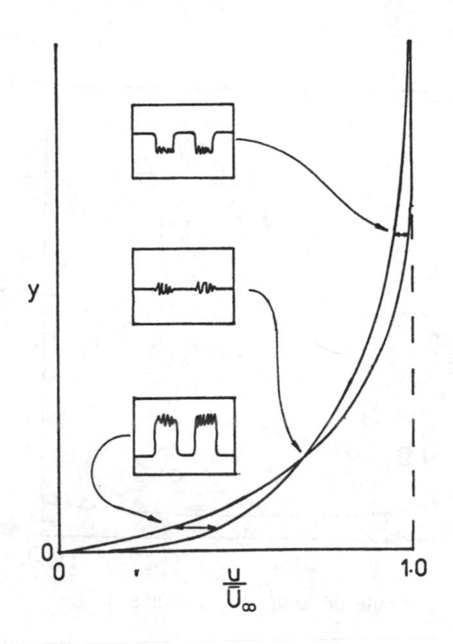

Figure 24 Schematic Representation of Unsteady Transition [38] (Pfeil & Herbst, 1979)

A somewhat more detailed look at the boundary layer behaviour is presented in Figure 25, which shows the time averaged shape factor, H, for the suction surface boundary layer. Data is shown from the turbine rotor and from the linear cascade (which was run at the same Reynolds number). The results of a turbulent boundary layer calculation are also plotted. It can be seen that over much of the blade surface the time averaged shape factor has a value which is midway between the limits of the laminar and fully turbulent boundary layers.

As a general comment on this work, it is perhaps useful to quote the recommendations given by Smith [39] in his Evaluation Summary of the 1983 AGARD Conference on Viscous Flows in Turbomachines:

"This work (i.e., that of [38]) brings into question whether boundary layer methods that ignore free stream unsteadiness, (or at best respond only to isotropic turbulence) are adequate for use in optimizing blade designs in multi-blade-row turbomachines. The type of research facility employed by Hodson, i.e. a large scale low-speed machine running at Reynolds numbers typical of those encountered in aircraft engine low pressure turbines, is ideally suited for this kind of research. Additional work should include experiments with more two-or three-stage configurations".

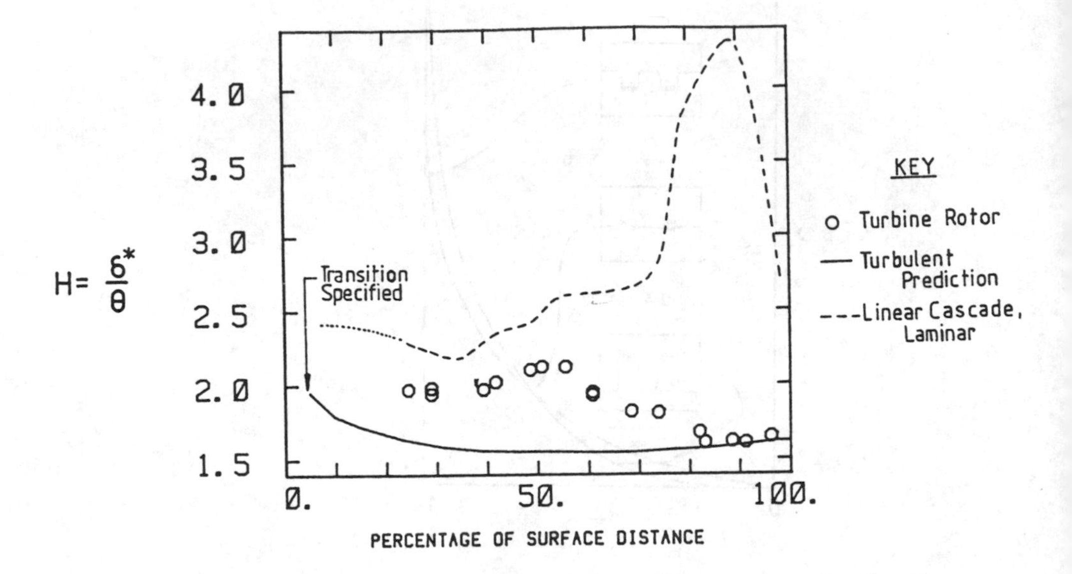

Figure 25 Shape Factor of the Time-Averaged Suction Surface Boundary Layer on the Rotor of an Axial Flow Turbine [38]

Detailed examinations of unsteady flows in turbines have also been carried out by Sharma et al [40], who examined the flow in a large scale 1 - 1/2 stage research turbine. Time resolved measurements were taken at the nozzle vane exit and at the rotor exit. Although the flow at the former station was "found to be essentially steady..." the flow downstream of the rotor was highly unsteady. As an example of the type of results that were obtained, Figure 26 shows contours of the total pressure (measured in the absolute system) at the maximum and minimum time. The contours give levels of total pressure coefficient Cp_T:

$$Cp_T = \frac{(P_{Tinlet} - P_T)}{\frac{1}{2}\rho U^2}$$

The location of the rotor trailing edges are indicated by the arrow marks at the top of the contour plots.

A central conclusion of the authors was that two regimes in the rotor exit flow field could be identified, corresponding to maximum and minimum interaction of the rotor leading edges with the upstream vane wakes. As stated in [40]:

"At the maximum interaction time location, the upstream vane wakes merge with the rotor airfoil wakes, and the passage between rotor airfoils shows relatively....uniform flow, while the suction side of the wake shows two large secondary-flow vortices. The overall flowfield at this location is...similar to those measured in low aspect ratio, high turning plane cascades tests. This indicates that the boundary layers do not significantly alter the structure of the flowfield.

At the minimum interaction time location, the upstream vane wakes pass through the rotor passage virtually without interacting with the rotor airfoils and the passage between rotor airfoils shows high turbulence, non-uniform flow in the exit plane. The flowfield at this time location is very different from those observed in stationary cascade tests. In particular, there is no evidence of the existence of strong secondary flow vortices. In their place a low static pressure, high loss region is present near the midspan location of the wake. The loss generation mechanisms in the flow at this time location should be very different from those in plane stationary cascades."

Other Illustrations of Unsteady Flows

A further way in which unsteady (potential field) interactions have been found to be important is in the context of turbomachine performance measurements. Probes placed behind a rotor tend to cause a flow blockage and hence a locally reduced velocity. This results in a local increase in rotor work output, so that the probe may not be reading an accurate indication of the "true" performance. It has also been found that when a compressor is operating near stall, introduction of these measuring probes can cause it to stall prematurely. Addition of large probes in closely spaced compressors may therefore modify the flow field that the probes are trying to measure.

It can also be commented that in a real turbomachine, the relative influence of potential flow effects and effects due to wakes may be difficult to separate. As an example, it has been reported that a very close spacing of compressor blade rows led to a 1% increase in efficiency compared to the nominal spacing [41]. Further evidence for this is presented by Mikolajczak [30], who mentions experiments conducted with a low speed compressor in which the axial gap between rows was varied from 55% to 10% of the chord. As shown in Fig. 27, a 1% increase in peak efficiency was obtained. It is stated that this improved performance could not be explained by incidence changes or by wake mixing between rows. It is clear that the unsteady effects due to *both* potential interactions and upstream wakes will be stronger than in the closely space case, although at present there is no method to quantitatively assess this difference. Thus, an improved understanding of the relative importance of unsteady wake transport and mixing, and of potential pressure field interaction, is needed to predict accurately the behavior of closely spaced compressors.

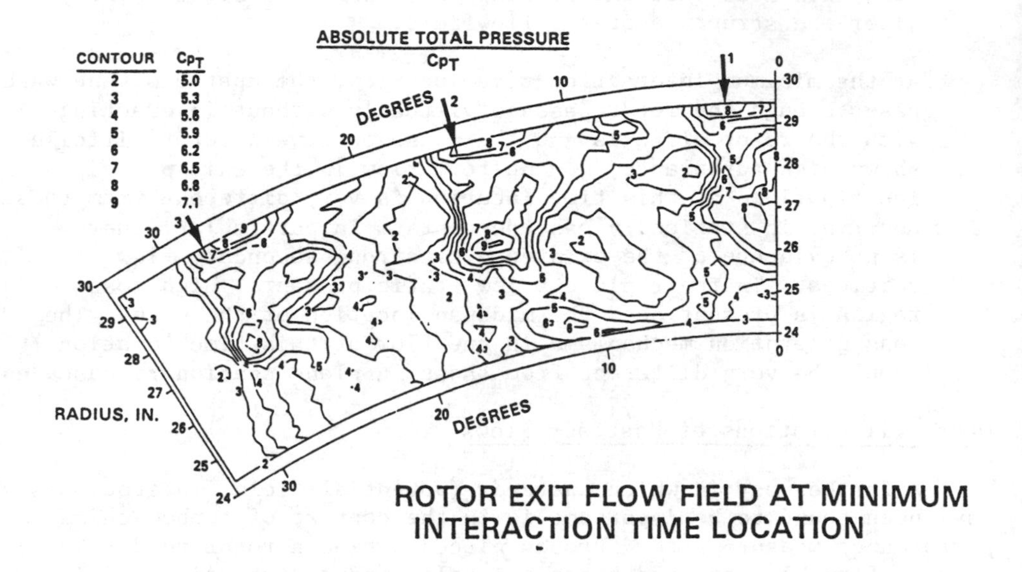

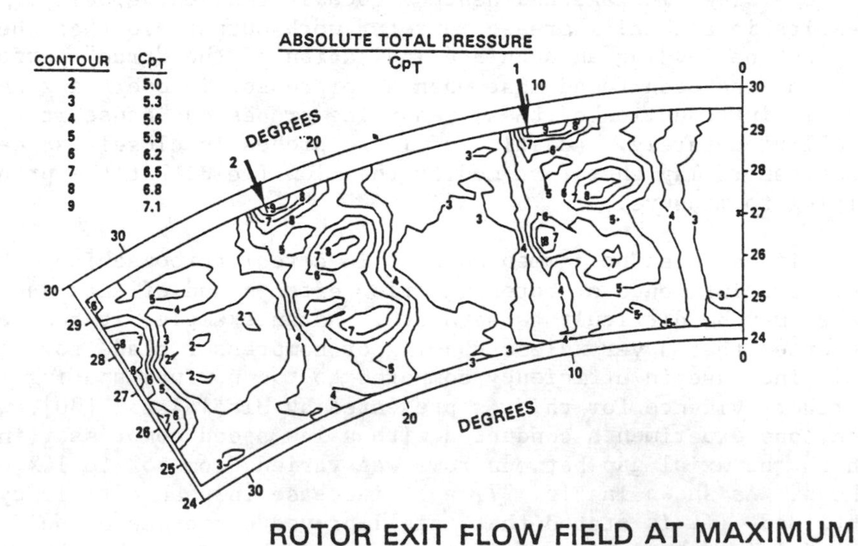

Figure 26. Instantaneous Turbine Rotor Exit Total Pressure [40].

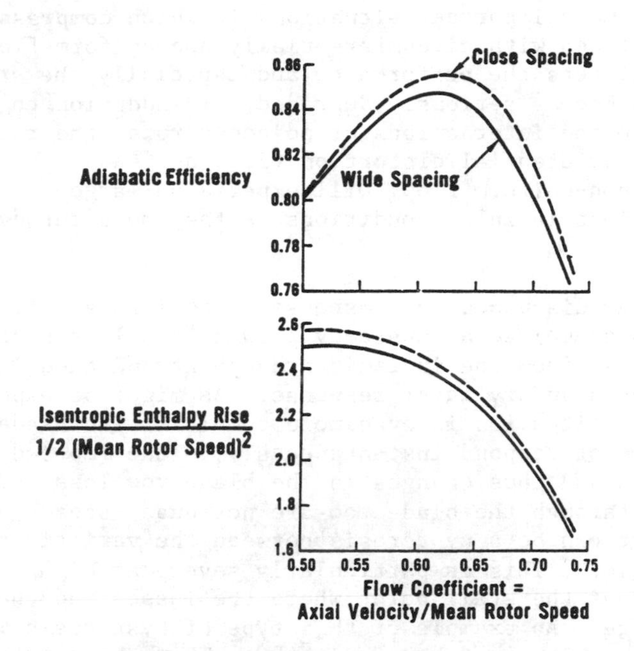

Figure 27. Effect of rotor and stator spacing on efficiency and pressure rise of a highly loaded compressor (average static pressure rise/dynamic head = 0.5) [30].

Finally, there is another aspect which deserves mention. Recent time-resolved measurements in transonic compressors indicate that the flow in an *isolated rotor* is not steady [42]. In other words, contrary to what one might expect, the flow in the rotor blade passages is not steady when viewed in the rotor relative frame. This mechanism associated with this unsteadiness is still being investigated, but it has been hypothesized that if the flow *is* unsteady, the shock location will oscillate. The entropy rise associated with this shock motion would be higher than that in a steady flow and therefore, if this picture is a representative one, could be part of the loss mechanism for this type of compressor.

The discussion in this section has been mainly on a qualitative basis. In part this has been due to a desire to keep the discussion at an introductory level. However, it also, to a large part, reflects the idea that an in-depth understanding of the basic fluid mechanics of many of these effects in a real compressor or turbine is far from complete.

Unsteady Effects with Non-Uniform Inlet Flows

The discussion so far has been of turbomachines in which the inlet flow is at least nominally uniform around the circumference.

However, there are many important situations in which compressors in particular are faced with circumferentially non-uniform flows. Under such circumstances the performance and especially the *stability* of the machine can become seriously degraded. In addition to the unsteadiness due to the interactions of adjacent rows, the rotors operating in a circumferential distortion (i.e. a steady circumferentially non-uniform flow) will experience large amplitude fluctuations in inlet conditions as they move through the distortion.

We have already discussed the response of a diffuser (i.e. compressor blade passage) to an unsteady flow at two levels of sophistication, first from the inviscid standpoint and then in the context of unsteady boundary layer response. As might be expected, however, the actual situation is even more complex. The blade boundary layers do not respond instantaneously to the imposed angle of attack variation and thus changes in the blade row loss and in the fluid turning through the blade row are not quasi-steady. As a consequence there can be a hysteresis between the variations in loss and inlet angle. This is particularly severe at high incidence angles near the stall point where the losses and the deviations are large. An example of this type of hysteresis is shown in Figure 28, which is taken from [43]. This shows total pressure loss, due to viscous effects, across a rotor operating in a 180° circumferential distortion. (Note that lower inlet angles imply larger angles of attack on the blades.)

Several features can be seen from this figure. First of all the instantaneous loss forms a loop around the quasi-steady data, i.e. the loss data taken on this rotor when operating with a uniform flow. Second it can be seen that the instantaneous excursions of the inlet angles are to conditions *below the uniform flow stall point*. This seems to indicate that large angles of attack on the blades, which would lead to stall in a uniform flow condition, can be tolerated transiently in the unsteady flow associated with the rotor motion through the distortion. This phenomena will be discussed further in the lectures on stability, but we can comment that inclusion of this behaviour is definitely not just a refinement of a steady flow approach, but is *vital* in developing models for predicting compressor response to inlet distortion.

Because of this, there has been a substantial amount of work on the phenomenon of *dynamic stall*, both analytically and experimentally (although much of this has been aimed at isolated airfoils, rather than blade rows). However, perhaps because of the complexity of the situation, little of this has been utilized as yet in *multistage* environments, which is the arena of greatest practical interest. Investigators of the unsteady response

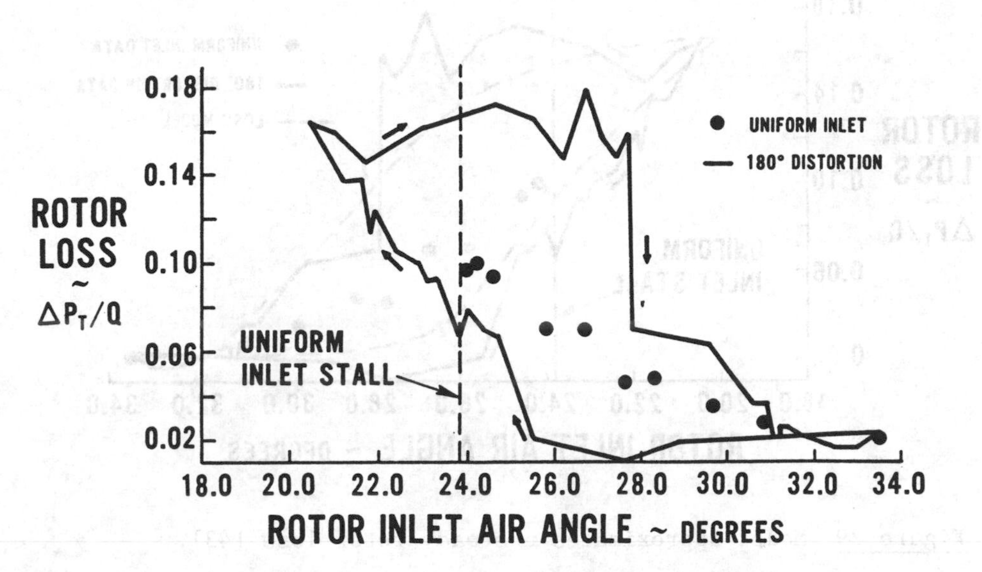

Figure 28. Circumferential distortion causes unsteady rotor loss [42].

of these types of devices have thus resorted to much more simplified approaches. One of these, which was suggested originally by Emmons [44] several decades ago, has been used with some success in a number of multistage compressor applications. In one form or another, in fact, it still provides a key part of many procedures for calculating compressor performance in distorted flow. The basic idea is to model the loss (or turning) characteristics of a blade row in an unsteady flow as a process which takes the form:

$$\tau \frac{\partial \text{ Loss}}{\partial t} = (\text{Quasi-steady Loss} - \text{Loss})$$

The term τ is an empirically determined time constant which has been found to be approximately equal to $\left(\frac{\text{axial chord}}{\text{axial velocity}}\right)$ in a wide variety of situations [45], [46]. This formulation has the feature that it reduces to the quasi-steady loss if the unsteadiness is small (i.e. if the fluctuations are "slow enough") and will yield a hysteresis loop of the sort shown above for reduced frequencies of interest. In order to apply it to flows below the stall flow, however one must have some model for the quasi-steady loss curve.

The simple first order model has been applied to the prediction of the unsteady loss and some reasonable agreement with data has been obtained as shown in Figure 29. However, the

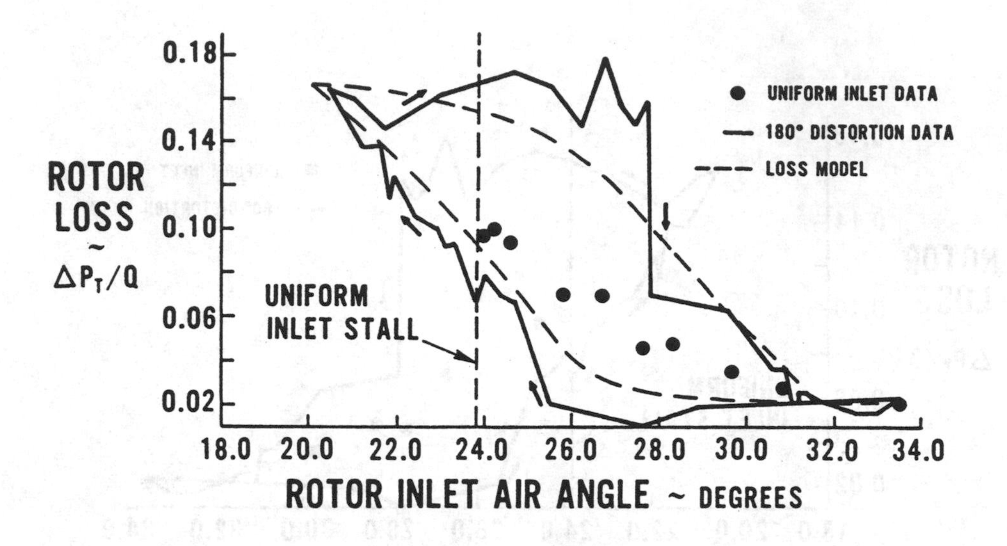

Figure 29 Model approximates unsteady rotor loss [43].

application of this type of model to prediction of deviation is less clear, and in general the area of the response of *highly loaded blades* to unsteady flow is an area in which there is a great deal of scope for obtaining further understanding and better prediction methods.

We close this section with two examples in which an understanding of unsteady phenomena would appear to be crucial in the development of improved prediction methods. These have to do with the transient behaviour of compression systems, particularly with respect to the behaviour subsequent to the initial onset of flow instability (rotating stall onset). The first concerns the evolution of the flow field in a compressor from unstalled condition to the (fully developed) rotating stall. In this context it is useful to distinguish the various time scales that characterize this evolution compared to those characterizing the departure of the individual *blades* (or blade passages) from quasi-steady behaviour.

The time scale for the latter is on the order of b/W where b is the blade chord and W is the relative velocity. However, let us consider the time scale associated with the transformation of an axisymmetric flow to the severely non-uniform one of rotating stall. A representative time scale might now be the disturbance wavelength (e.g. the mean circumference of the machine) divided by the through flow velocity. This is much longer than the time scale based on blade unsteady response. A qualitative physical argument for this scaling can be made by noting that the change from unstalled to stalled flow involves the shedding of blade circulation of the same sign over a significant circumferential extent of the flow annulus. The flow will only approach a "fully developed" state when this shed vorticity has been convected downstream some distance on the order of the disturbance wavelength.

These rough considerations imply that for a given machine the stall cell growth time might scale with axial velocity (or since C_x/U is approximately constant at stall, with rotor speed) and circumference. This has been found to be the case in one of the few instances where data is available, as shown in Figure 30 [47]. This response of the overall flow field is therefore likely to occur on a time scale an order of magnitude (or more) longer than that associated with the individual blades. The compressor performance (pressure rise, torque) during a stall transient may therefore differ substantially from the steady state performance. This is a facet of unsteady response which may be quite important in the application of overall compression system models [48], but is at present not well understood, and on which there is little data.

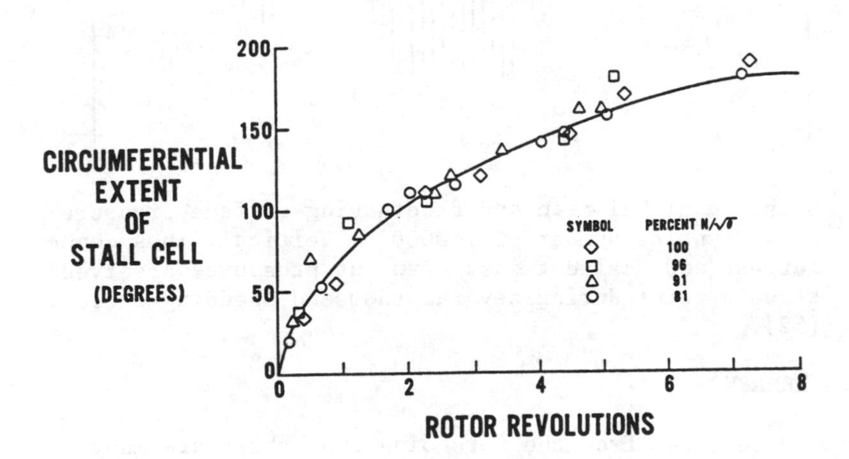

Figure 30 Circumferential growth of stall cells in a 3-stage compressor.

The second example is fully developed rotating stall. As will be described in the lectures on compressor stability, there has been some recent work on predicting the flowfield in this regime of compressor operation, initially using a semi-empirical approach [49] and subsequently on a more fundamental level [50]. A central feature pointed out by both these treatments was that the unsteady effects have important dynamical consequences. Their inclusion is thus fundamental in formulating descriptions of the flow which incorporate the relevant physical mechanisms of the phenomenon.

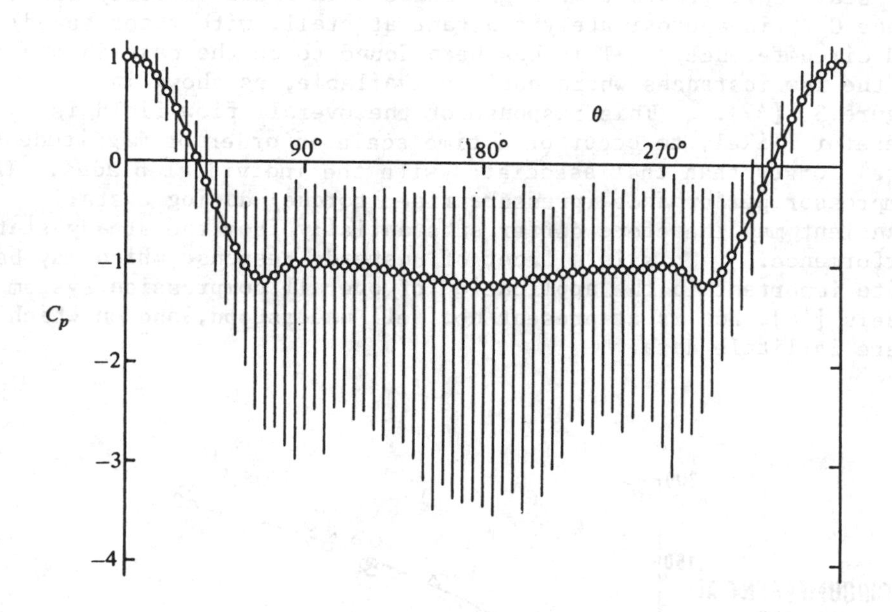

Figure 31 Measured global mean and fluctuating surface pressure at a Reynolds number of 140000. Vertical lines connect largest and smallest instantaneous pressures observed at each point during several thousand shedding cycles. [52].

7. CLOSING REMARKS

It should be clear from the foregoing that there are many situations in turbomachines in which unsteady effects are of import. (For further examples see [51]). To emphasize the general theme, however, we present one final figure. This does not arise in a turbomachinery application, but it is basic enough to be of relevance to the present lecture. The specific situation is that of flow round a cylinder; the experimental data cited is from the paper by Cantwell and Coles [52].

They carried out a detailed survey of the coherent (vortical) structure behind a two-dimensional cylinder. In the course of this, both time mean and unsteady measurements were made of the static pressure field around the cylinder, and these are presented as Figure 31. The static pressure coefficient, C_p,

$$(Cp = \frac{P - P_0}{\frac{1}{2}\rho U_0^2})$$

is plotted versus θ, the angular position round the cylinder, with $\theta = 0$ corresponding to the location of the stagnation line on the front of the cylinder.

The mean static pressure coefficient is denoted by the open symbols. The vertical lines connect the largest and smallest instantaneous pressures observed. It can be seen that it is not at all stretching a point to call the variations in static pressure sizeable. The time mean base pressure coefficent was - 1.21, however the *instantaneous* base pressure coefficient was -1.21, as high as -0.2 or as low as 3.5. It is clear that taking steady state data *only* in situations such as this can mask important information and can result, in the words of Cantwell and Coles, in a "ruinous loss of detail" about the flow.

8. OVERALL SUMMARY

The main ideas presented in this lecture are:

1. Unsteady flows are inherent in turbomachinery operation.

2. Flow unsteadiness can have an impact on different aspects of this operation, ranging from peak efficiency with uniform inlet flows, to performance and stability with circumferential inlet distortion, to post stall compressor transients.

3. Effects due to flow unsteadiness become important when the reduced frequency is not small, or equivalently when the time scale of the flow fluctuations is not large compared to the particle transport time.

4. The potential for unsteady interactions in turbomachines increases as the blade loading is increased and as the effects of compressibility become significant.

5. Flows with significant unsteadiness exhibit phenomena that are quite different from those associated with steady flows.

9. ACKNOWLEDGMENTS

Much of this lecture was written at the Whittle Laboratory while the author was a Visiting Fellow in the Department of Engineering. The author wishes to thank the Science and Engineering Council and the Royal Society for their support. The efforts of N.A. Cumpsty, Sir William Hawthorne and D.S. Whitehead in arranging this stay at Cambridge are also gratefully acknowledged.

Helpful suggestions and comments were rendered by Drs. H.P. Hodson and T.P. Hynes, and the manuscript was rapidly (and cheerfully) typed by Mrs. B. Roe. The efforts of Mr. R. Chue in performing some of the plotting are also appreciated.

10. NOMENCLATURE

a speed of sound
A flow-through area
$\vec{C}$ velocity in fixed coordinate system
C_x axial velocity
C_u circumferential velocity in fixed coordinate system
H Boundary layer shape factor, $H = \delta^*/\theta$
h used for enthalpy as well as to denote blade gap or vortex spacing
h_T stagnation enthalpy
k wavenumber; $k = 2\pi$/disturbance wavelength
ℓ distance along streamline
L length of fluid mechanic device
$\mathcal{L}$ effective length of fluid mechanic device
M Mach number
P static pressure
P_T stagnation pressure (total pressure)
ΔP pressure rise
r radial coordinate
R radius
s entropy
t time
T temperature
U mean rotor speed
U_0, U_∞ free stream velocity
W velocity in moving (rotor) coordinate system
x,y coordinate directions
Γ circulation
δ boundary layer thickness
δ^* displacement thickness
$\delta()$ perturbation quantity

ρ density
ϕ velocity potential, also used for phase shift in Fig. 16
θ momentum thickness
ν kinematic viscosity
ω radian frequency
Ω reduced frequency

Subscripts

1 inlet station
2 exit station

Superscripts

$(\bar{\ })$ mean flow quantity
$()'$ perturbation quantity; also used to denote moving coordinate system

11. REFERENCES

1. Dean, R.C., "On the Necessity of Unsteady Flow in Fluid Machines", ASME J. Basic Eng. March 1959, pp 24-28
2. Preston, J.H., "The Non-Steady Irrotational Flow of an Inviscid Incompressible Fluid, with Special Reference to Changes in Total Pressure Through Flow Machines", Aeronautical Quarterly Nov. 1961, pp 353-360.
3. Horlock, J.H., and H. Daneshyar, "Stagnation Pressure Changes in Unsteady Flow", Aeronautical Quarterly, XXII May 1970, pp 207-224.
4. Hetherington, R., and R.R. Moritz, "Influence of Unsteady Flow Phenomena on the Design and Operation of Aero Engines", Unsteady Phenomena in Turbomachinery, AGARD CP-177, 1975.
5. Sears, W.R., "Some Aspects of Non-Stationary Airfoil Theory and Its Practical Applications", J. Aero. Sci., 8, 1941.
6. Smith, S.N., "Discrete Frequency Sound Generation in Axial Flow Turbomachines", U. of Cambridge, Dept. of Eng. Rep. #CUED/A-TURBO/TR29, 1971
7. Verdon, J.M., "The Unsteady Aerodynamics of a Finite Supersonic Cascade with Subsonic Axial Flow", J. of Appl. Mech. Trans. ASME Series E, 40, Sept. 1973, pp 667-671.
8. Verdon, J.M., "Further Developments in the Aerodynamic Analysis of Unsteady Supersonic Cascades", ASME J. Eng. Power Oct. 1977, pp. 509-524.
9. Ni, R.H., "A Rational Analysis of Periodic Flow Perturbations in Supersonic Two-Dimensional Cascades", ASME Paper #78-GT-176 (1978).
10. Verdon, J.M., and Caspar, J.R., "Development of a Linear Aerodynamic Analysis for Unsteady Transonic Cascades", NASA CR 168038, October 1982.

11. Lighthill, M.J., Part II of Chapter I, Laminar Boundary Layers ed by L. Rosenhead, Oxford University Press, 1967.
12. Batchelor, G.K., An Introduction to Fluid Dynamics (section 3.6), Cambridge University Press, 1967.
13. Lamb, H., Hydrodynamics Section 156, Dover Publications, 1945.
14. Nikkanen, J.P., Private Communication.
15. Garrick, I.E., "Nonsteady Wing Characteristics", article in Aerodynamic Components of Aircraft at High Speeds, Princton, U. Press, 1957.
16. Kerrebrock, J.L., Aircraft Engines and Gas Turbines, Chapt. 9, "Aircraft Engine Noise", MIT Press, 1977.
17. Weyer, H.B., and H.G. Hungenberg, "Analysis of Unsteady Flow in a Transonic Compressor by Means of High-Response Pressure Measuring Techniques", from Unsteady Phenomena in Turbomachinery AGARD CP-177, 1975.
18. Telionis, D.P., Unsteady Viscous Flow, Springer Verlag 1982.
19. AGARD Report No. 679, "Special Course on Unsteady Aerodynamics June 1980.
20. Schlicting, H., Boundary Layer Theory, Seventh Edition, Chapter XV, Section e, McGraw-Hill Book Company, 1979.
21. McCroskey, W.J., "Some Current Research in Unsteady Fluid Dynamics", ASME J. Fluids Eng. 99, March 1977, pp. 8-38.
22. Lyrio, A.A., and Ferziger, J.H., "A Method of Predicting Unsteady Turbulent Flows and Its Application to Diffusers with Unsteady Inlet Conditions", AIAA Journal, Vol. 21, No. 4 pp. 534-540, April 1983.
23. Cebeci, T., and Bradshaw, P., Momentum Transfer in Boundary Layers, Chapter 10, Hemisphere Publishing Company, 1977.
24. Bradshaw, P., Cebeci, T., and Whitelaw, J.H., Engineering Calculation Methods for Turbulent Flow, Academic Press 1981.
25. Covert, E.E., and Lorber, P.F., "Unsteady Turbulent Boundary Layers in Adverse Pressure Gradients" AIAA Journal Vol. 22, No. 1, January 1984, pp 22-28.
26. Patel, M.H., "On Turbulent Boundary Layers in Oscillatory Flow" Proc. Roy. Soc., Series A, Vol. 353, Feb. 1977, pp. 121-144.
27. Patel, M.H., "An Integral Method for the Oscillating Turbulent Boundary Layer", Aeronautical Quarterly, 31, pp 271-298, November 1981.
28. Telionis, D.P., "Review-Unsteady Boundary Layers, Separated and Attached" ASME J. Fluids Eng. 101, March 1979, pp. 29-43.
29. Pfeil, H., Herbst, R., and Schnader, T., "Investigation of the Laminar-Turbulent Transition of Boundary Layers Disturbed by Wakes", ASME Paper 82-GT-124, 1982.
30. Mikolajczak, A.A., "The Practical Importance of Unsteady Flow", Unsteady Phenomena in Turbomachinery, AGARD CP-177, 1975.
31. Gallus, H.E., J. Lambertz, and Th. Wallmann, "Blade-Row Interaction in an Axial-Flow Subsonic Compressor Stage, ASME J. Eng. for Power, Oct. 1979, pp 359-370.

32. Kerrebrock, J.L., and A.A. Mikolajczak, "Intra-Stator Transport of Rotor Wakes and Its Effect on Compressor Performance", ASME J. Eng. for Power, Oct. 1970, pp 359-370.
33. Wagner, J.H., T.H. Okiishi, G.J. Holbrook, "Periodically Unsteady Flow in an Imbedded Stage of a Multistage, Axial-Flow Turbomachine", ASME J. Eng. Power, Vol. 101, Jan 1979, pp 42-51.
34. Zierke, W.C., and Okiishi, T.H., "Measurement and Analysis of Total-Pressure Unsteadiness Data from an Axial Compressor Stage", ASME J. Eng. Power, Vol. 104, April 1982, pp 479-482.
35. R. Dunker, et.al. "Redesign and Performance Analysis of a Transonic Axial Compressor Stator and Equivalent Plane Cascades with Subsonic Controlled Diffusion Airfoils", ASME J. Eng. Power, Vol. 106, April 1984, pp 279-288.
36. Hodson, H.P., "Measurements of Wake Generated Unsteadiness in the Rotor Passages of Axial Flow Turbines", ASME Paper 84-GT-189, 1984.
37. Hodson, H.P., "An Inviscid Blade-to-Blade Prediction of a Wake-Generated Unsteady Flow", ASME Paper 84-GT-43, 1984.
38. Hodson, H.P., "The Development of Unsteady Boundary Layers in the Rotor of an Axial-Flow Turbine" in AGARD Conference Proceedings No. 351, Viscous Effects in Turbomachines, 1983.
39. Smith, L.H., Jr. "Technical Evaluation Report" in AGARD Conference Proceedings, No. 351, Viscous Effects in Turbomachines, 1983.
40. Sharma, O.P. et.al., "An Experimental Investigation of the Three-Dimensional Unsteady Flow in an Axial Flow Turbine", AIAA Paper 83-1170, 1983.
41. Koch, C.C., and L.H. Smith, Jr., "Loss Sources and Magnitudes in Axial Flow Compressors", ASME J. Eng. for Power, July 1976, pp 441-424.
42. Ng, W.F., and Epstein, A.H., "Unsteady Losses in Transonic Compressors", ASME Paper 84-GT-183, 1984.
43. Mazzawy, R.S., "Multiple Segment Parallel Compressor Model for Circumferential Flow Distortion", ASME J. Eng. for Power, April 1977, pp 288-296.
44. Emmons, H.W., C.E. Pearson, and H.P. Grant, "Compressor Surge and Stall Propagation", Trans. ASME, May 1955.
45. Nagano, S., Y. Machida, and H. Takata, "Dynamic Performance of Stalled Blade Rows", Japan Soc. of Mech. Eng. Paper #JSME 11, presented at Tokyo Joint Int. Gas Turbine Conf., Tokyo Japan, Oct. 1971.
46. Takata, H., and S. Nagano, "Nonlinear Analysis of Rotating Stall", ASME Paper 72-GT-5.
47. Greitzer, E.M., "Review-Axial Compressor Stall Phenomena" ASME, J. Fluids Eng. Vol. 102, June 1980, pp 134-151.
48. Moore, F.K., and Greitzer, E.M., "A Theory of Post-Stall Transients in Multistage Compression Systems", NASA Contractor Report, to appear 1984.

49. Cumpsty, N.A., and Greitzer, E.M., "A Simple Model for Compressor Stall Cell Propagation", ASME J. Eng. Power Vol. 104, January 1982, pp 170-176.
50. Moore, F.K., "A Theory of Rotating Stall of Multistage Axial Compressors", Parts I-III, ASME J. Eng. Power, Vol. 106, April 1984, pp 313-336.
51. Platzer, M.F., "Unsteady Flows in Turbomachines - A Review of Current Developments" in AGARD Conference Proceedings 227, Unsteady Aerodynamics, 1977.
52. Cantwell, B., and Coles, D., "An Experimental Study of Entrainment and Transport in the Turbulent Near Wake of a Circular Cylinder", J. Fluid Mech., Vol. 136, 1983, pp 321-374.

12. ACKNOWLEDGMENTS FOR FIGURES

Figures 16, 17 and 26 are copyrighted by the American Institute of Aeronautics and Astronautics and reprinted with permission.

Figures 19, 21, 28, and 29 are copyrighted by the American Society of Mechanical Engineers.

Figure 31 is reproduced with permission of the Journal of Fluid Mechanics, Cambridge University Press.

FLOW INSTABILITIES IN TURBOMACHINES

E. M. Greitzer

Department of Aeronautics and Astronautics
Massachusetts Institute of Technology

1. INTRODUCTION: STATIC AND DYNAMIC STABILITY OF A BASIC PUMPING SYSTEM

In this lecture we discuss the instabilities that can occur in systems in which one moves a fluid through pipes, ducts, etc. by means of some type of turbomachine. An attempt has been made to make the lecture a self-contained survey of the subject. It should be noted, however, that the author has recently published a detailed review of this topic (1), which can be consulted for supplementary information.

The term "stability" can be defined in a general manner with respect to the equilibrium operating point of such a system. An operating point is stable if, when the system is disturbed slightly, it tends to return to the equilibrium point or at least does not keep moving further away from it. For the most part only overall guidelines can be given as to when instability is to be expected as well as for methods to enhance stability, since, in many cases, predictive methods have not yet been developed for the complex situations typical of engineering practice.

To illustrate the basic ideas, consider the simple (but quite relevant) pumping system, shown in Figure 1. An incompressible fluid is pumped from a large, constant pressure reservoir through a closed tank which contains a compressible gas (air) and then through a throttle valve into another large reservoir. The second reservoir does not necessarily have to be at the same pressure level as the first; however, in this example it *is* taken as such. If end effects are neglected the inlet of the pump (the "Inlet" station) and the exit of throttle (the "Exit" station) can

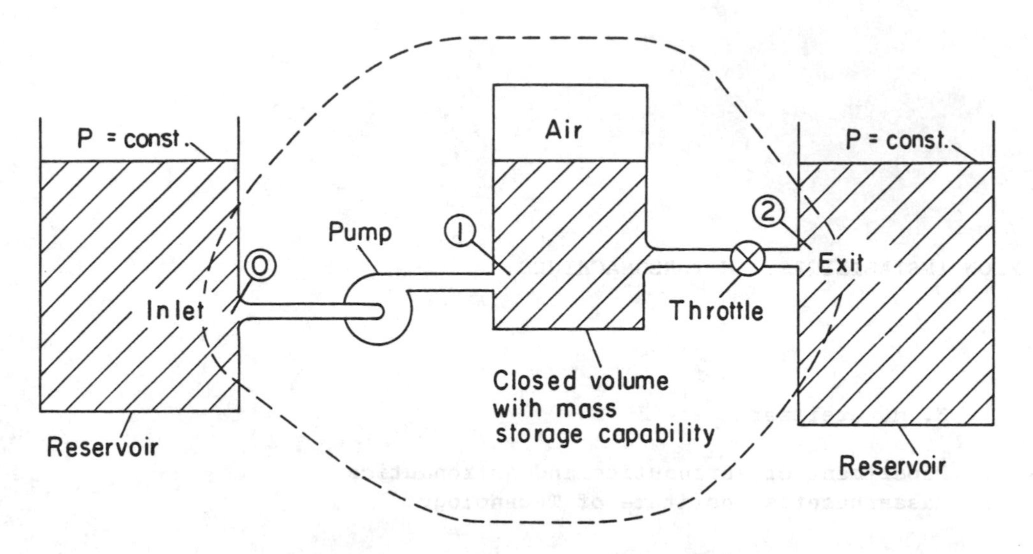

Figure 1 A basic pumping system (1)

also be assumed to be at this pressure, so that only the parts of the system within the dotted control surface need be discussed. The essential elements of the system are therefore the pump, the compliance (or mass storage capability) of the closed volume, the throttle (which controls the system flow rate) and the inertance of the fluid in the inlet and exit lines.

The steady-state system operating point is set by two conditions, namely that the flow through the pump and the flow through the throttle are the same and that the pressure rise through the pump is equal to the pressure drop due to the system resistance. These conditions imply that the steady-state operating point is at the intersection of the pumping characteristic and the throttle (or system resistance) curves.

Consider the stability of an arbitrary steady-state operating point. With reference to the lefthand side of Figure 2, the effect of a small perturbation in mass flow (a decrease, say) at operating point A is to cause the pressure drop across the throttle to be smaller than that produced by the pump. The resulting pressure imbalance causes fluid accelerations that return the system to operation at the initial point, so that point A is stable. This is true for all points to the right of point A or for points between A and B. At point B, however, where the throttle line is tangent to the pumping characteristic, the pressure forces that arise (due to a small decrease in mass flow) will cause the system to depart further from the initial operating point, so point B is an unstable operating point. This is the most basic of pumping system stability criteria--the system will become unstable if the slope of the pump (or compressor) pressure rise curve is steeper than the slope of the throttle curve.

This criterion is, however, *too simple* to describe many of the real phenomena that are observed in pumping systems since it only considers the *static stability* of the system, (static instability is associated with pure divergence from the initial operating point). In fact, it is often the criteria for *dynamic stability* (dynamic instability leads to growing *oscillatory* motion about the initial point) which are violated first.[1)]

Static instabilities can be inferred from viewing the transient performance of the system as a sequence of (quasi-) steady states. Hence knowledge of the steady-state pump characteristics and throttle lines, is enough to define the stability. In the prediction of dynamic instability, on the other hand, parameters such as system inertances and capacitances must be included since they play an essential role in determining the transient response of the system to disturbances. Thus knowledge of steady-state performance curves alone is not sufficient for prediction of dynamic instablity, and additional information about quantities such as volumes, duct lengths, etc., must also be included. As shown on the righthand side of Figure 2, the point to be emphasized is that a pumping system can be statically stable and still exhibit (dynamic) instability.

Lumped Parameter Analysis of Pumping System Instability

The dynamic stability of the system shown in Figure 1 can be simply analyzed using a lumped parameter model. In this, all the

1. The terms dynamic and static instability can be made more quantitative by the following illustration. Consider a simple second order system described by the equation

$$\frac{d^2x}{dt^2} + 2\alpha\frac{dx}{dt} + \beta x = 0$$

where α and β are constants of the system. The transient response of the system to an initial perturbation is given by

$$x = A \exp\left(\{-\alpha + \sqrt{\alpha^2 - \beta}\}\ t\right) + B \exp\left(\{-\alpha - \sqrt{\alpha^2 - \beta}\}\ t\right)$$

where the constants A and B are determined by the initial conditions. If $\beta > \alpha^2$ the condition for instability is simply $\alpha < 0$, which corresponds to oscillations of exponentially growing amplitude. Instability will also occur if $\beta < 0$, independent of the value of α; however in this case the exponential growth is non-oscillatory. It is usual to denote these two types of instability as dynamic and static respectively. Static stability is a necessary but not sufficient condition for dynamic stability.

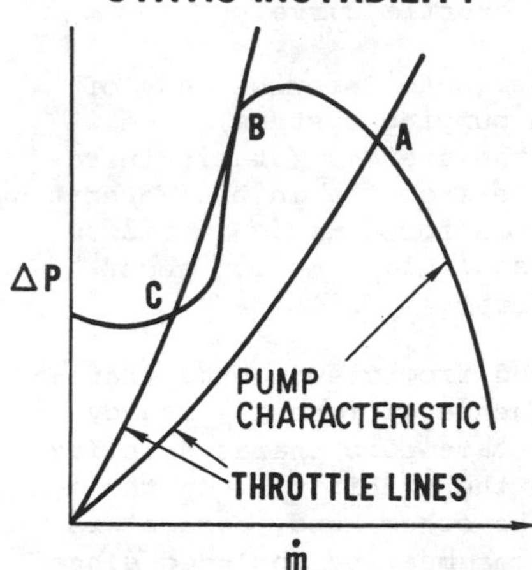

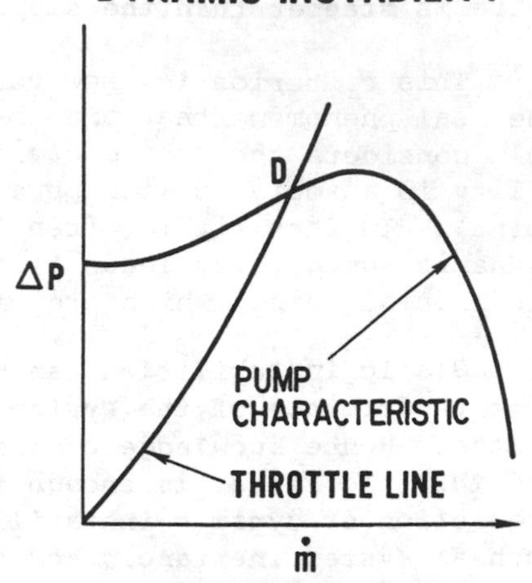

Figure 2 Static and dynamic system instabilities (1)

kinetic energy of the unsteady flow in the system is associated with the flow in the pump and throttle lines, and all the potential energy associated with the system transients is taken to arise from the expansion and compression of the fluid in the storage volume. (The system is viewed essentially as a Helmholtz Resonator.) The mass-spring-damper mechanical analogue of such a model is illustrated schematically in Figure 3 where the components of the simple pumping system are shown; these are a pump, a volume with capacity for mass storage, and a throttle. As is indicated, a key feature due to the pump is the ability to provide a negative damping (i.e. a net input of mechanical energy) to the system transients.

In the analysis, the liquid that is pumped is taken to be incompressible and the mass of gas (air) in the volume to behave isentropically. The transient pump performance is also taken to be quasi-steady. More elaborate allowances for unsteady behavior can be used but this approximation will suffice here. Assuming that the pump output is significantly larger than the dynamic pressure based on the through-flow velocity (as is generally true in a centrifugal pump), the first integral of the momentum equation between station 0 and station 1 can be written

$$P_0 - P_1 = \left[\int_0^1 \rho \frac{\partial C_x}{\partial t} dx\right] - \Delta P_p \tag{1}$$

where C_x is the axial velocity, ΔP_p is the pump pressure rise and is a function of pump mass flow, $\dot{m}_1$, and x is in the streamwise direction. Using continuity we can define a length L such that

$$L = A_{in} \int_0^1 \frac{dx}{A(x)} \tag{2}$$

where A(x) is the area at different values of x, and A_{in} is a convenient reference area, say at the inlet station. Thus:

$$P_0 - P_1 = \frac{L}{A_{in}} \frac{d\dot{m}_1}{dt} - \Delta P_p \tag{3}$$

since the mass flow is the same at all stations from 0 to 1.

A similar equation could be written for the throttle. However, it is found in practice that in many throttling elements by far the largest part of the pressure drop occurs due to the quasi-steady throttling characteristics of the device. The inertance in the throttle can therefore be neglected and we can write

$$P_1 - P_2 = \Delta P_T \quad , \tag{4}$$

where ΔP_T is the throttle pressure drop and is a function of throttle mass flow $\dot{m}_2$, which is not necessarily the same as $\dot{m}_1$.

In the volume we assume that fluid accelerations are negligible and the pressure is spatially uniform, although varying in

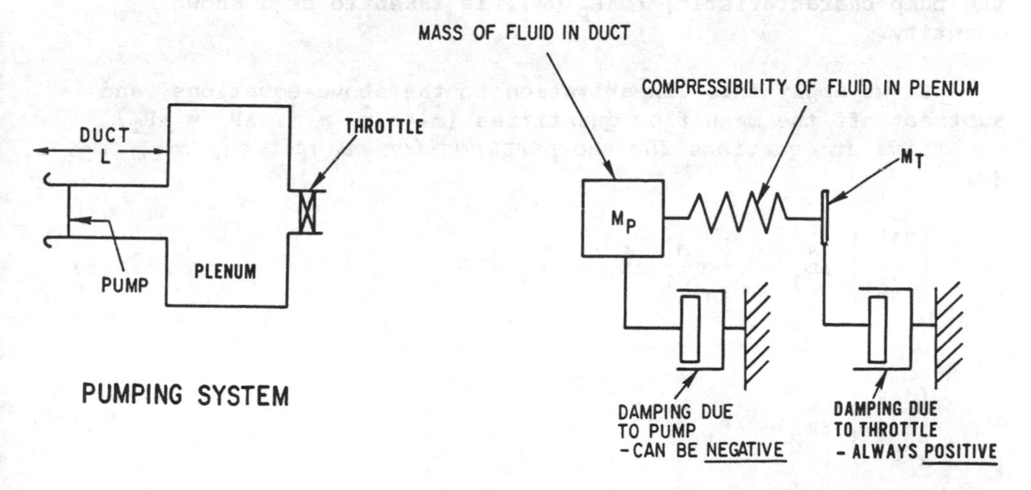

Figure 3 Mechanical analogue of simple pumping system (1)

time. The mass conservation equation for the volume is

$$\dot{m}_1 - \dot{m}_2 = -\rho \frac{dV_a}{dt} \tag{5}$$

where V_a is the volume of the air in the plenum. Mass conservation for the air in the volume is:

$$\frac{d}{dt}(\rho_a V_a) = 0 \tag{6}$$

These equations, plus the isentropic relation ($\ln P_a + \gamma \ln V_a$ = constant) for the gas in the plenum, describe the system behavior.

To find whether the system is stable or unstable we examine the response to small perturbations for a system operating at a given mean operating condition. To do this we will represent the flow quantities as being composed of a mean (time-independent) part, denoted by an overbar ( ), and a small perturbation, denoted by δ, and only retain quantities of the first order in δ. Under these conditions we can write the mass flow as

$$\dot{m} = \overline{\dot{m}} + \delta\dot{m}$$

the pump performance as

$$\Delta P_p = \overline{\Delta P_p} + \delta \Delta P_p = \overline{\Delta P} + \overline{\left(\frac{d\Delta P_p}{d\dot{m}}\right)} \delta\dot{m} \tag{7}$$

etc. In equation (7) we have used a Taylor series expansion about the mean operating point, ΔP_p, and the (steady-state) slope of the pump characteristic, $(d\Delta P_p/d\dot{m})$, is taken to be a known quantity.

If we apply this linearization to the above equations, and subtract off the mean flow quantities ($\dot{m}_1 = \dot{m}_2 = \dot{m}$; $\Delta P_p = \Delta P_T$), we arrive at equations for the *perturbation quantities*, $\delta\dot{m}_1$, $\delta\dot{m}_2$, δP_1:

$$\left(\frac{d\overline{\Delta P}_p}{d\dot{m}}\right) \delta\dot{m}_1 - \frac{L}{A_{in}} \frac{d}{dt} \delta\dot{m}_1 - \delta P_1 = 0 \tag{8}$$

$$\left(\frac{d\overline{\Delta P}_T}{d\dot{m}}\right) \delta\dot{m}_2 - \delta P_1 = 0 \tag{9}$$

$$\delta\dot{m}_1 - \delta\dot{m}_2 - \frac{\rho}{\gamma P_1} V_a \frac{d}{dt}(\delta P_1) = 0 \tag{10}$$

where we have used the conditions that P_0 and P_2 are constant.

Equations (8) to (10) are all linear with constant coefficients, and hence the solutions are of the form e^{st}. For stability the real part of s must be negative.

Substituting expressions of the form e^{st} for the perturbations we find that if there are to be non-trivial solutions to the resulting system of linear algebraic equations, the growth rate, s, must satisfy the equation:

$$s^2 + s\left\{\left[\frac{\gamma P_1}{\rho V_a}\,\frac{L}{A_{in}}\,\frac{1}{\left(\dfrac{d\overline{\Delta P_T}}{d\dot{m}}\right)} - \left(\frac{d\overline{\Delta P_p}}{d\dot{m}}\right)\right]_{(a)}\frac{A_{in}}{L}\right\}$$

$$+ \left\{1 - \frac{\left(\dfrac{d\overline{\Delta P_p}}{d\dot{m}}\right)}{\left(\dfrac{d\overline{\Delta P_T}}{d\dot{m}}\right)}\right\}_{(b)}\frac{A_{in}}{LV_a}\,\frac{\gamma P_1}{\rho} = 0 \,. \tag{11}$$

For *instability* either quantity in brackets can be negative. Condition (b) is the static stability criterion, while condition (a) is that for dynamic stability. These imply restrictions on the slope of the pump characteristic. In general the condition for dynamic stability, bracket (a), is the more critical, and since throttle slopes are often steep and volumes large, this may occur very near the peak of the pump characteristic curve. Bracket (b) represents the static stability criterion discussed previously.

Note that we have based this derivation on the liquid pumping system shown in Figure 1. We can also consider a compressor (whose pressure rise is small compared to the ambient level) operating in a gas. In this case the capacity for mass flow storage in the volume is due to the compressibility of the flowing medium. The analysis is not changed materially. The only change in equation (11) is that the density that appears is the air density, so that $\gamma P/\rho$ is now the square of the sound speed. We can thus write the equation for s as

$$s^2 + s\left\{\left[\frac{a^2}{V_a}\frac{L}{A_{in}}\frac{1}{\left(\overline{\frac{d\Delta P_T}{d\dot{m}}}\right)} - \left(\overline{\frac{d\Delta P_p}{d\dot{m}}}\right)\right]\frac{A_{in}}{L}\right\}$$

$$+ \omega^2\left\{1 - \frac{\left(\overline{\frac{d\Delta P_p}{d\dot{m}}}\right)}{\left(\overline{\frac{d\Delta P_T}{d\dot{m}}}\right)}\right\} = 0 \qquad (12)$$

where $\omega = a\sqrt{A/LV_a}$, the Helmholtz resonator frequency for the system. This is the natural frequency of the system and can be seen to be independent of compressor speed.

Finally note that if one carries out the analysis using non-dimensional quantities [denoted by tildes ($\tilde{\ }$)] which are defined by

$$\phi = \frac{\dot{m}}{\rho A_{in} U},$$

$$\delta\tilde{P}, \Delta\tilde{P}_p, \Delta\tilde{P}_T = \frac{\delta P}{\frac{1}{2}\rho U^2}, \frac{\Delta P_P}{\frac{1}{2}\rho U^2}, \frac{\Delta P_T}{\frac{1}{2}\rho U^2},$$

$$\tilde{t} = \omega t$$

it is found that equation (12) can be written as

$$\tilde{s}^2 + \tilde{s}\left\{\frac{1}{B\left(\overline{\frac{d\Delta\tilde{P}_T}{d\phi}}\right)} - B\left(\overline{\frac{d\Delta\tilde{P}_P}{d\phi}}\right)\right\} + \left\{1 - \frac{\left(\overline{\frac{d\Delta\tilde{P}_P}{d\phi}}\right)}{\left(\overline{\frac{d\Delta\tilde{P}_T}{d\phi}}\right)}\right\} = 0 \qquad (13)$$

where B is the parameter

$$B = \frac{U}{2\omega L}. \qquad (14)$$

As will be seen below, for given compressor (or pump) and throttle characteristics, changing the value of B can considerably alter the damping and hence the system dynamic response.

Physical Mechanism for Dynamic Instability

The physical mechanism associated with dynamic instability can be understood by considering the system undergoing oscillations about a mean operating point. Since the flow through the throttle is dissipative, there must be energy put into the system to maintain the oscillation (or to increase its amplitude in the case of instability). The only source of this energy is the pump. The mass flow and pressure rise perturbations through the pump are shown in Figure 4 which presents the *perturbations* in mass flow ($\delta\dot{m}$) and pressure rise ($\delta\Delta p$) through the pump, plotted versus time over a period of one cycle (assuming that the pump responds quasi-steadily to fluctuations in mass flow), as well as the product of the two, $\delta\dot{m} \times \delta\Delta p$. The integral of the last quantity over a cycle is equal to the net *excess* (over the steady-state value) of the rate of production of mechanical energy.

In the case of a positive slope it is seen that a favorable condition for energy addition occurs, since high mass flow rate and high rate of mechanical energy addition (in the form of pressure rise) go together. The net amount of mechanical energy that the pump puts into the flow will thus be higher than if the system were in steady operation at the mean flow rate. (In a

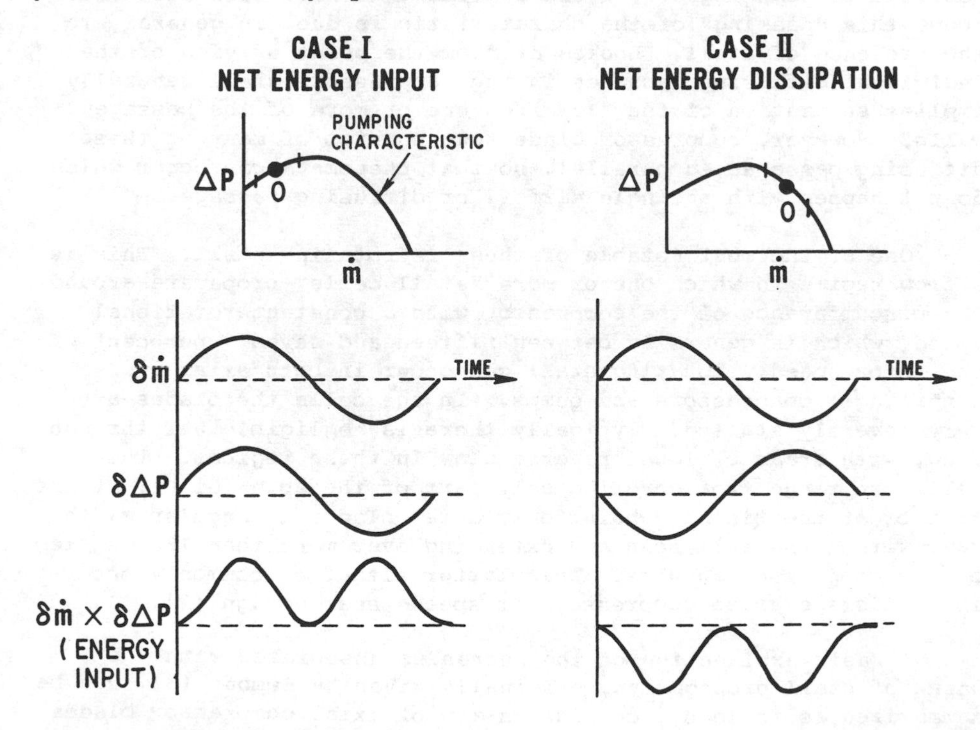

Figure 4 Physical mechanism for dynamic instability (1)

similar fashion the net dissipation due to the throttle will also be higher than if the system were in steady operation). When the net energy input over a cycle balances the dissipation, a periodic oscillation can be maintained; this corresponds to the boundary between stability and instability. For an operating point on the negatively sloped region, as shown in the right-hand side of the figure the pump adds less mechanical energy over a cycle than in steady operation (since high mass flow is associated with low energy input); perturbations will therefore decay and the operating point will be stable. Dynamic instability for this system can thus occur when the mechanical energy input from the pump is greater than during a mean (steady) flow (i.e. when there is the required amount of negative damping), and this will be true only if the pump characteristic is positively sloped so that high mass flow and high mechanical energy input per unit mass flow go together.

2. COMPRESSOR INSTABILITY: ROTATING STALL

The system instabilities that have been described will occur when the slope of the compressor or pump characteristic becomes positive to some degree. In the systems that have been discussed above this "peaking" of the charateristic is due, in general, to the presence of stall. Looked at from the point of view of the individual diffusing passages in the compressor, stall generally implies separation of the flow from one or more of the passage walls. However, compressor blade rows consist of many of these diffusing passages in parallel, so that phenomena can occur which do not happen with a single airfoil or diffusing passage.

One of the most notable of these is rotating stall. This is a flow regime in which one or more "stall cells" propagate around the circumference of the compressor with a constant rotational speed, which is generally between fifteen and seventy per cent of the rotor speed. Rotating stall can occur in both axial and centrifugal compressors and pumps. In the cells the blades are very severely stalled. Typically there is negligible net through-flow, with areas of local reverse flow in these regions. The cells can range from covering only part of the span (either at the root or at the tip) and being only a few blades in angular width, to covering the full span and extending over more than 180 degrees of the compressor annulus. This latter situation commonly occurs in multistage axial compressors at speeds near design (2).

A basic explanation of the mechanism associated with the onset of stall propagation, originally given by Emmons (3), can be summarized as follows: consider a row of axial compressor blades operating at a high angle of attack, such as is shown on the left hand side of Figure 5. Suppose that there is a non-uniformity

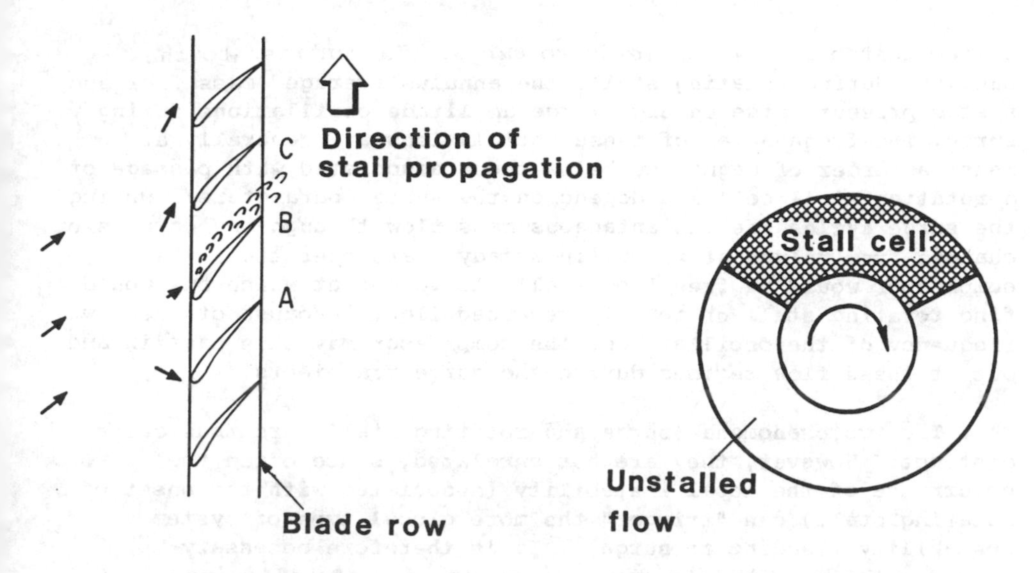

Figure 5 Rotating stall

in the inlet flow such that a locally higher angle of attack is produced on blade B which is enough to stall it. If this happens, the flow can separate from the suction surface of the blade so that a substantial flow blockage occurs in the channel between B and C. This blockage causes a diversion of the inlet flow away from blade B and towards C and A to occur (as shown by the arrows), resulting in an increased angle of attack on C and a reduced angle of attack on A. Since C was on the verge of stall before, it will now tend to stall, whereas the reduced angle of attack on A will inhibit its tendency to stall. The stall will thus propagate along the blade row in the direction shown, and under suitable conditions it can grow to a fully developed cell covering half the flow annulus or more, as shown on the righthand side of Figure 5. In this fully developed regime the flow at any local position is quite unsteady although the annulus averaged mass flow is steady, the stall cells serving only to redistribute this flow.

The onset of rotating stall is associated with an instability which arises due to the stall of the compressor blade passages.[2] As far as the overall system is concerned, this can be regarded as a localized instability. However, there is also a more global

---

2. The onset of stall can, however, be affected by other closely coupled components in the system as described below.

system instability which leads to *surge*. In contrast to the behavior during rotating stall, the annulus averaged mass flow and system pressure rise undergo large amplitude oscillations during surge. The frequencies of these oscillations are generally at least an order of magnitude below those associated with passage of a rotating stall cell and depend on the system parameters. During the surge cycles the instantaneous mass flow through the compressor changes from values at which (in steady state operation) the compressor would be free from stall, to values at which one would find rotating stall or totally reversed flow. Because of the low frequency of the oscillations, the compressor may thus pass in and out of these flow regimes during the surge transients.

The two phenomena (surge and rotating stall) are thus quite distinct. However, they are not unrelated, since often the occurrence of the local instability (associated with the onset of rotating stall) can "trigger" the more global type of system instability (leading to surge). It is therefore necessary to consider the possibility for *both* types of instability and develop methods for their prediction.

Prediction of Stall Onset in Axial Compressors

The prediction of the point at which stall occurs has been attacked by many investigators at quite different levels of approach. The most empirical are the correlations that have been developed for stall onset. The basic concept is to find a parameter (or parameters) which correlates the "stall point" (defined here as the condition at which the steady axisymmetric flow becomes unstable) for a number of different blade geometries, compressor designs, etc. In a design procedure for a low hub-tip ratio fan, for example, the parameter could be applied at different span locations along the blading, using the local flow conditions generated by use of one of the many axisymmetric compressor flow field calculations, to see whether any section would be operating under too adverse a condition, while for a multistage compressor the parameter might be applied only on a meanline or averaged basis.

One well-known example of this type of approach, which is still much in use, is the work of Leiblein (4). He developed a parameter which he called the diffusion factor, or D-factor, defined as:

$$D = 1 - W_2/W_1 + \Delta W_\theta / 2\sigma W_1$$

where W_1 is the inlet relative velocity, W_2 is the exit relative velocity, ΔW_θ is the change in circumferential velocity component, and σ is the solidity. This parameter is related to the adverse pressure gradient on the suction surface of the airfoils.

It is found that the total pressure loss correlates quite well with D, and, based on Leiblein's cascade results, one can see a rather sharp rise in loss occur as D is increased past a value of roughly 0.6. This can therefore be taken as an approximate criterion for the onset of *airfoil* stall in a cascade. Although much of the work done by Leiblein was based on two-dimensional cascades, the use of the D-factor has been carried over to axial as well as to centrifugal compressors (5). Features such as the differences between the flows in a cascade and the flows at the tips of axial compressor rotors, for example, are "recognized" by noting that different limiting values of the D-factor are used for the rotor tips than for other sections.

The diffusion factor, however, is really a measure of the tendency of the boundary layers on the compressor *airfoil* to separate and does not account for the fact that it is very often the flow in the annulus *endwall* regions which is the central cause of rotating stall onset. In order to approach this it has been found useful to view the blade passages in another way, namely as a diffuser. A method of doing this is described by Koch (6) who has examined data from over fifty compressor builds to obtain a correlation for the peak (stalling) pressure rise capability of axial compressors. His data include tests to show the effect of Reynolds numbers, tip clearance, and blade geometry and velocity triangles.

The results of the correlation are presented as an adjusted stage average peak pressure rise coefficient ($\Delta P/1/2\rho W_1^2$) versus the nondimensional passage length, L/g_2. This latter parameter is essentially the chord length divided by the "staggered pitch" (the pitch x cos β_2 where β_2 is the exit metal angle of the blade). Adjustments in the pressure rise coefficient are made due to Reynolds number effects, tip clearance effects, blade stagger angle (essentially inlet boundary layer skew), and axial spacing. To account for the third of these, an effective dynamic pressure F at blade inlet is defined. This is shown in Figure 6 which is for a *stator*; similar considerations are used for rotors. Reynolds number effects on the peak pressure rise are illustrated in Figure 7 and tip clearance effects are shown in Figure 8.

Although this correlation has been derived from multistage machines with near "equilibrium" endwall layers, it has also been applied with good success to single stages. The basic correlation is shown in Figure 9 which presents the adjusted peak pressure rise achievable in an axial compressor row plotted versus the non-dimensional length parameter L/g_2 (defined in Figure 6). Also shown in the Figure is a curve from a two dimensional diffuser correlation by Sovran and Klomp (7). All data has been corrected to a Reynolds number of 130,000, a tip clearance/staggered gap ratio of 5.5%, and an axial gap of 38% of stator chord. The main

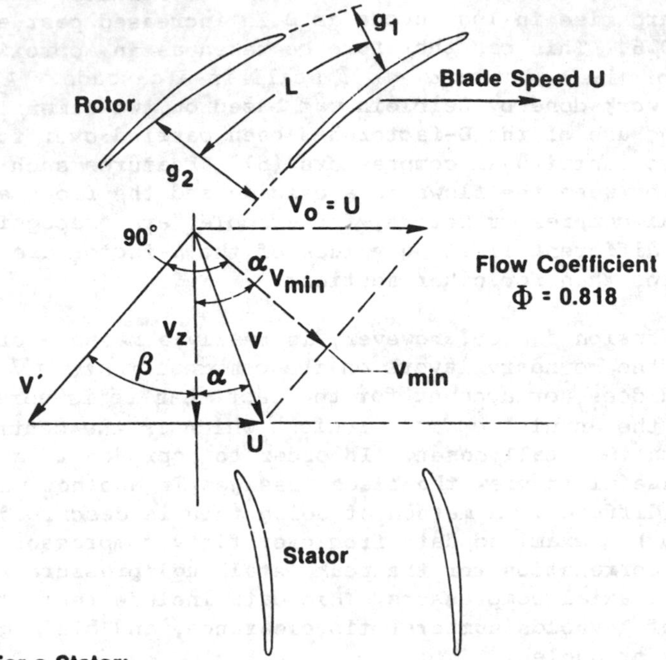

For a Stator:

$$\mathcal{F}_{ef} = \frac{V^2_{ef}}{V^2} = \left(V^2 + 2.5\ V^2_{min} + 0.5\ V_o^2\right) \Big/ 4.0\,V^2$$

$$\frac{V^2_{min}}{V^2} = \sin^2\left(\alpha + \beta\right), \text{ if } \left(\alpha + \beta\right) \leq 90^\circ \text{ and } \beta \geq 0^\circ$$

$$\frac{V^2_{min}}{V^2} = 1.0\ , \quad \text{if } \left(\alpha + \beta\right) > 90^\circ$$

$$\frac{V^2_{min}}{V^2} = \frac{V_o^2}{V^2}, \text{ if } \beta < 0^\circ$$

Figure 6 Effective dynamic pressure (6)

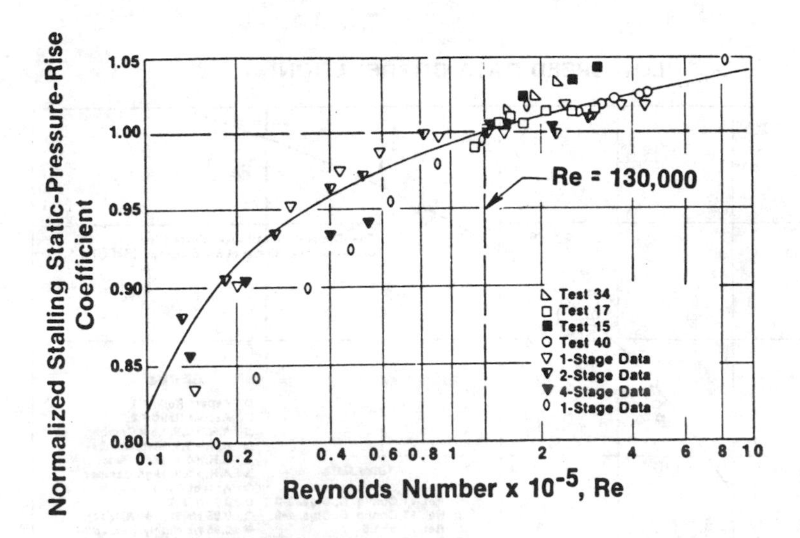

Figure 7 Effect of Reynolds number on stalling pressure rise coefficient (6)

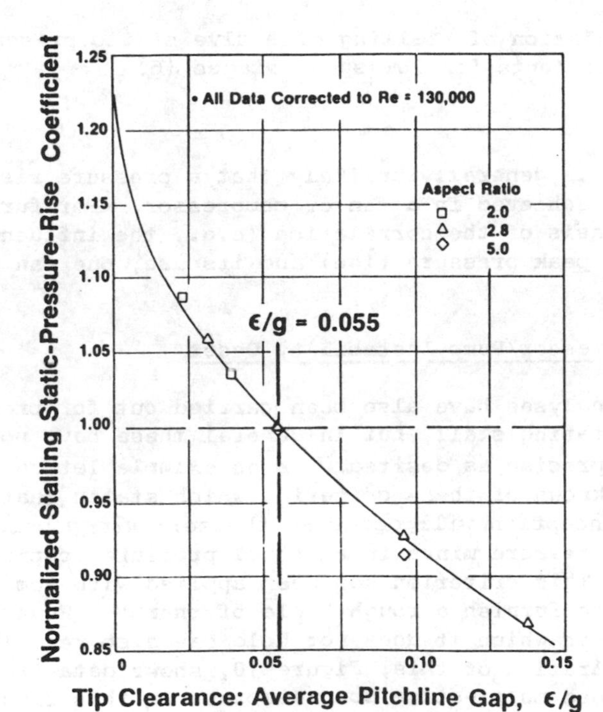

Figure 8 Effect of tip clearance on stalling pressure rise coefficient (6)

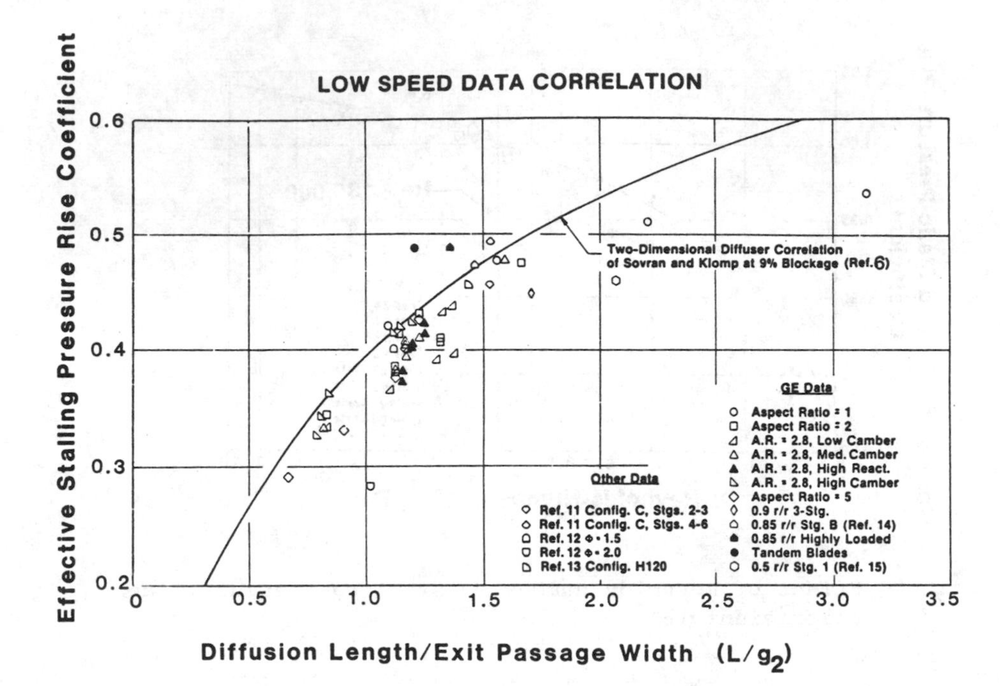

Figure 9 Correlation of stalling effective static pressure rise coefficients for low speed stages (6)

point is that it is generally unlikely that a pressure rise above the curve can be achieved in a fan or compressor. For further details on the basis of the correlation (e.g., the influence of tip clearance on peak pressure rise) and its use, one can refer to Reference (6).

Analyses of Compressor/Pump Instability Onset

Stability analyses have also been carried out for prediction of compressor rotating stall, but in general these have not been as quantatively precise as desired. As an example let us consider one of the best known of these criteria, which states that rotating stall inception will occur at the zero *slope* point of the exit static pressure minus inlet total pressure compressor characteristic. This criterion has been applied with some success and does appear to furnish a rough "rule of thumb". However, counter examples in which it does not hold can also readily be found. An illustration of this, Figure 10, shows data from a sample of low-speed multi-stage compressors. In this figure the horizontal axis is axial velocity parameter, ϕ ($= C_x/U$) and the vertical axis is ψ_{TS}, the non-dimensional total-to-static pressure rise. [$\psi = (P_{exit} - P_{T\ inlet})/(\rho U^2)$].

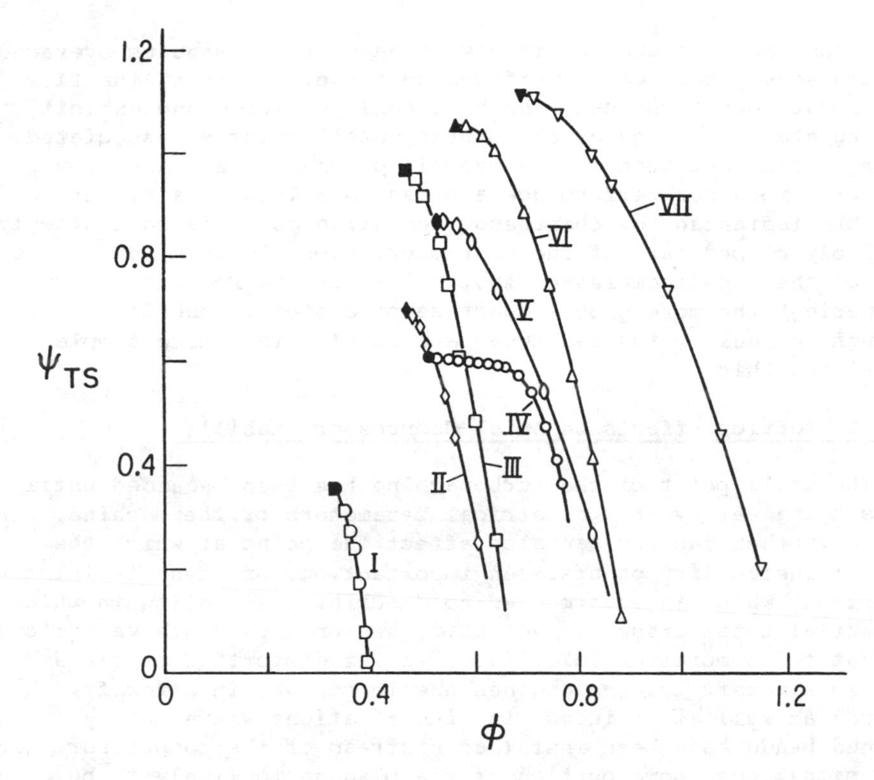

Figure 10 Multistage compressor performance (low speed rigs) (1)

Curves V, VI, and VII do appear to show approximately zero slope (within the accuracy of the data), but curves I-IV have a negative slope right up to the stall points. This situation, where there does not appear to be a zero slope region of the compressor characteristic, is even more apparent in high-speed multistage compressor data. The use of this criteria in any quantitative manner thus should be treated with some caution.

Analytical methods also exist for predicting the onset of the more global type of instability leading to surge. This is a system phenomenon and one in which, in contrast to rotating stall, the compressor appears to participate in a roughly one-dimensional manner. The calculation of the system stability point is based on using analyses of the type described in the introduction, and leads to a situation where the critical parameters are the slopes of the steady-state (uniform flow) compressor speed lines at the stall point. In general, however, these are not known a priori with the desired accuracy.

One connection between the two types of instability can be viewed as follows: (system) instability is a basically one-dimensional phenomenon, involving an overall, annulus averaged (in some sense) compressor performance curve. However, the flow through the compressor can also be locally unstable and exhibit rotating stall. The onset of rotating stall is often associated with a precipitous drop in the overall pressure rise mass flow curve of compressor performance and can thus lead to a situation where the instantaneous compressor operating point is on a steeply positively sloped part of the characteristic. In this sense, the onset of the local compressor instability can be regarded as "triggering" the more global compression system instability, although in reality the two types are coupled in a more complex manner than this.

Inlet Distortion Effects on Axial Compressor Stability

The stall point of the turbomachine has been regarded until now as being set by the geometrical parameters of the machine. In practice, other factors can also affect the point at which the onset of instability occurs. An important one of these is inlet distortion, which is a term used to describe a situation in which substantial total pressure, velocity, and/or flow angle variations exist at the compressor inlet face. Inlet distortion problems can occur in aircraft turbine engines due to changes in aircraft attitude as well as in industrial installations where poorly designed bends have been installed upstream of the compressor. In these situations, some portion of the blading is likely to be operating under more unfavorable conditions than would occur with a uniform flow at the same mass flow rate, leading to a decrease in the useful range of operation of the machine.

An illustration of the effect of inlet distortion on compressor stability and performance is given in Fig. 11, which shows data from a nine stage axial flow compressor (1). The horizontal axis shows corrected mass flow, and the vertical axis is the total pressure ratio. The dashed lines indicate the measured performance with a uniform inlet, while the solid curves give the measured performance with a circumferential distortion, i.e., a circumferentially nonuniform inlet flow. It can be seen that there is a substantial degradation in performance and, far more importantly, a large drop in the position of the stall line (the stability boundary) due to the distortion. While the consequence may not be this severe with all compressors, there is generally a reduction in the stability of the compression system associated with the presence of an inlet distortion.

Because of the widespread occurrence of inlet distortion and its adverse effects on system stability, there has been a large amount of work on the problem of predicting compressor response to flow distortion, ranging from correlations to more basic analyses (2),(8),(9). In these investigations the flow non-uniformities

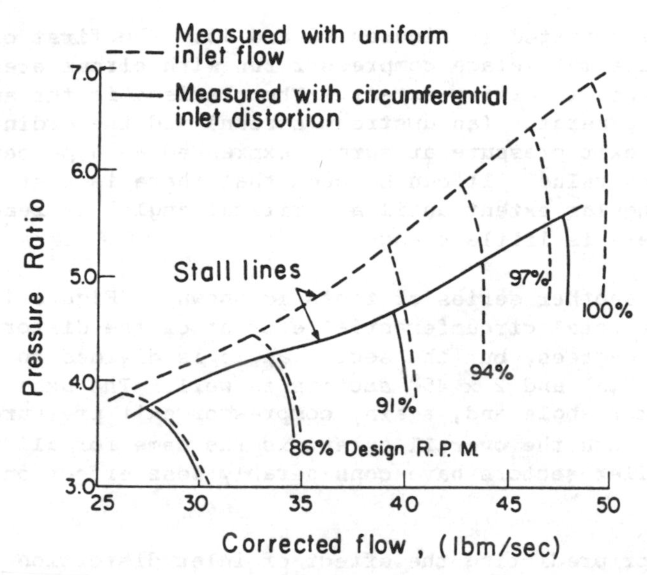

Figure 11 Effect of inlet distortion on axial compressor performance and stability (nine-stage compressor) (1)

are commonly divided into radially varying steady-state, circumferentially varying steady-state, and unsteady distortions. In reality the distortions encountered are combinations of two or possibly all three of these types, but significant progress has been made using the above simplifications.

An important point that emerges from these investigations is that there is a strong interaction between the compressor and the distorted flow field. Put another way, the compressor does not passively accept the distortion, but plays an active role in determining the velocity distribution that will occur at the compressor face, which is what the individual compressor airfoils actually respond to.

A further key result is that for transient distortions, or for a steady circumferentially non-uniform flow (which the moving compressor rotor sees as an unsteady flow), the effect of unsteady blade row response is to mitigate the degradation of stall point by the distortion. This effect increases as the reduced frequency (based on blade length, throughflow velocity, and a frequency which characterizes the time scale of the distortion) increases. Hence, for the circumferential distortions, it is the distortions with low harmonic content, i.e. a small number of "lobes" around the circumference that are most serious and, in general, distortions with numbers of lobes greater than two or three, say, or of small circumferential extent, will not affect the stall point to an appreciable extent.

This is illustrated in Figures 12 and 13. The first of these shows data from a multistage compressor run with circumferential inlet distortions of varying extent. The abscissa is the angle of the distortion generator (an upstream screen) and the ordinate is the compressor exit pressure at surge, expressed as a percent of the uniform flow value. It can be seen that there is a strong influence of angular extent until a "critical angle" is reached, after which there is little change.

Data from another series of tests is shown in Figure 13. In this figure the total circumferential extent of the distortion is kept at ninety degrees, but the sector angle is divided so that there are 4 x 22.5° and 2 x 45° sectors as well. The axes are individual sector angle and, again, compressor exit pressure at surge. Even though the overall extent is the same for all three tests, the smaller sectors have considerably less effect on the stability.

Methods for predicting the effect of inlet distortion on loss in surge margin are described in the references mentioned. The current status of these methods can perhaps be inferred from Figure 14 which shows predicted and measured loss in stall pressure ratio for a J85 with several different combinations of stagnation temperature and pressure. Reasonable agreement is seen, but predictive capability becomes less certain when one examines combined radial and circumferential distortion. Although the predictive procedures have been very useful for obtaining insight into the basic physical concepts (8)(10), it thus seems fair to state that current techniques for assessing the engine stall line loss due to inlet distortion have a strong empirical input (8),(9).

<u>Effect of Downstream Components on Compressor Stability</u>

The stability boundary for a compressor i.e., the inception point of rotating stall, can also be affected by other components in a compression system. This should be apparent for components upstream of the compressor, since they affect the compressor inlet conditions. Although perhaps not so apparent, however, the stability boundary can also be influenced by components downstream of the compressor, since they alter the downstream boundary conditions on flow perturbations. The extent of the change in stability boundary is dependent on the circumferential length scale of the flow perturbation, since this determines how close is the coupling between compressor and downstream component. For many situations of practical interest the predominant mode of instability occurs with a one-lobed (single cell rotating stall) type of disturbance, so that the relevant axial distance within which there can be a strong interaction is on the order of the machine diameter. This generally means that in terms of the stability boundary the compressor is *not* isolated from the

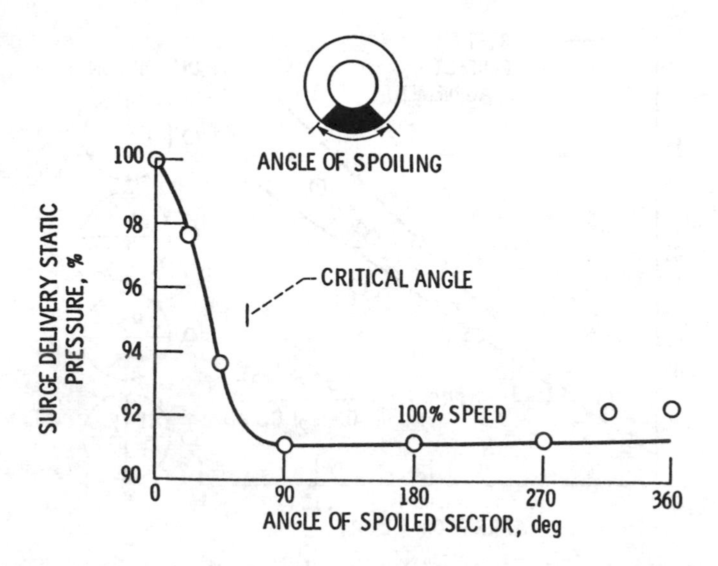

Figure 12 Effect of varying distortion sector angle on compressor stall margin (9)

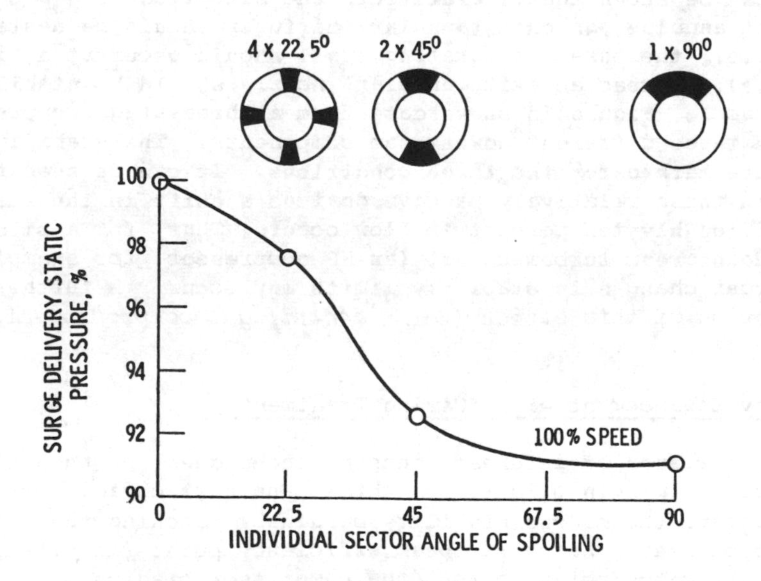

Figure 13 Effect of dividing distortion sector angle on compressor stall margin (9)

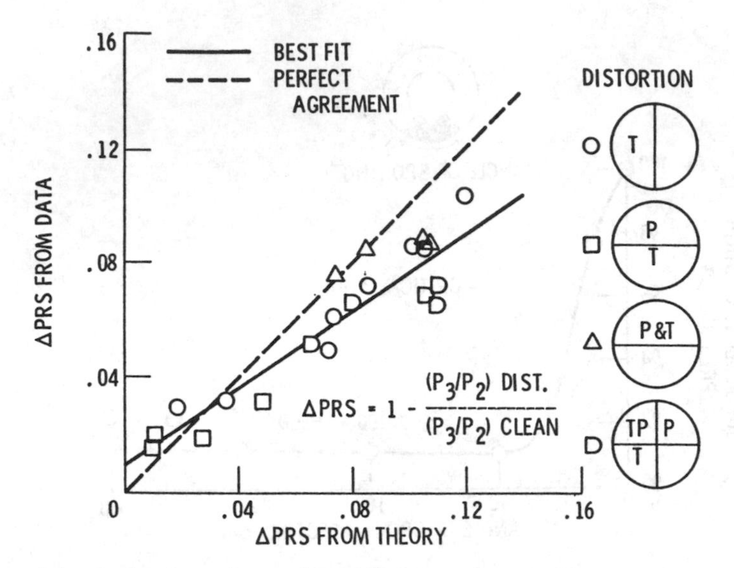

Figure 14 Actual and predicted loss in stall pressure ratio (9)

influence of downsteam components (1).

It can be shown that, relative to the situation with a constant area exit annulus, an exit (annular) diffuser should be destabilizing (i.e., the onset of rotating stall should occur at a higher flow rate), whereas an exit (annular) nozzle should be stabilizing. As an example, Figure 15 shows data from a three-stage compressor run with three different downstream components. The stability limits are marked for the three conditions. It can be seen that even with these relatively passive devices a shift in the stall point of roughly ten percent in flow occurs. Thus, for a situation with a downstream turbomachine, (an HP compressor, for example) even larger changes in stability limits may occur. A further illustration of this effect (for a *centrifugal* compressor) is given in (11).

Stability Enhancement with "Casing Treatment"

Another area of interest concerns techniques for enhancing the stability margin of a turbomachine. The most obvious of these is to achieve the needed stability margin by matching the compression system below its peak efficiency point (in effect setting the match point so that the compressor blading has, incidences and pressure rises far below the maximum). Although this provides an increase in airfoil incidence range between the operating line and the stability limit, it leads to decreased efficiency on the (down-rated) operating line, and this is generally unacceptable.

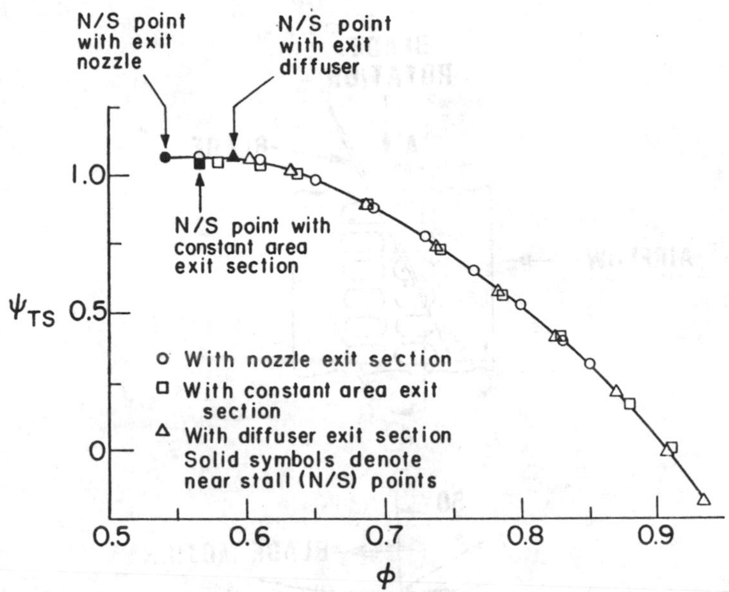

Figure 15 Effect of exit conditions on compressor stability (1)

A different solution to the problem is the use of so-called "rotor casing treatment" to improve the stability of compressors. This casing treatment consists of grooves or perforations, over the tips of the rotors in an axial compressor, or located on the (outer) shroud in a centrifugal machine. Numerous investigations of these types of configuration have been carried out under widely varying flow conditions, and these have demonstrated that the range of usefulness of these casing configurations extends from compressor operation in basically incompressible flow (relative Mach numbers of $\sim$ 0.15) to the supersonic flow regime (relative Mach numbers $\sim$ 1.5) (2). A sketch of one of the more successful of these casing configurations, known as axial skewed grooves, is shown in Figure 16, and a typical improvement in stall line brought about by use of these grooves is shown in Figure 17 for a transonic axial fan. It is to be noted that far larger improvements have been seen, that casing treatment has also been used to inhibit instability in centrifugal compressors (as will be discussed in a subsequent section), and that stator *hub* treatment has also been used to increase stall range.

In connection with this last point we can examine data from some experiments with a compressor stage with a heavily loaded stator hub (12). The configuration used was a cantilevered stator (from the OD wall) and a rotating hub surface. Tests were carried out with a smooth hub and with an axial skewed groove configuration. In the experiments, care was taken to insure that the rotor was matched well away from stall so that it only acted as a flow generator to the stator.

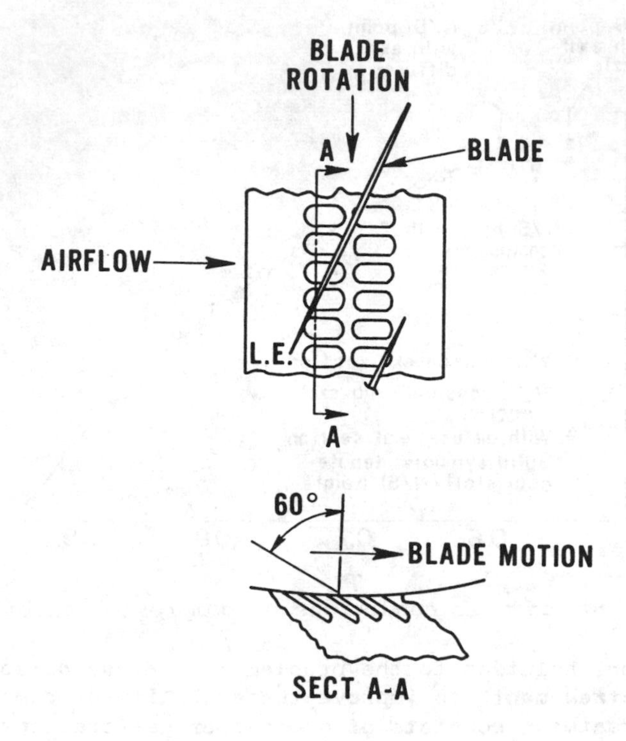

Figure 16 Axial skewed groove casing treatment (1)

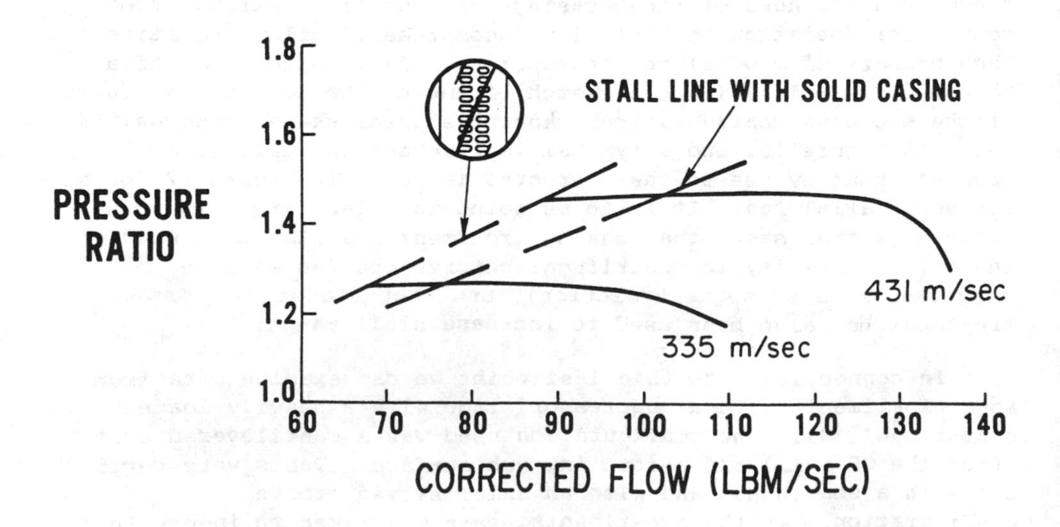

Figure 17 Axial compressor stall margin increase due to casing treatment (1)

The central results of these tests are shown in Figure 18, which shows the static pressure rise across the stator, for smooth hub (solid wall, SW) and for hub treatment (HT). It can be seen that the application of hub treatment has resulted in a substantial change in the pressure rise curve. The flow coefficient at stall onset has been decreased by 10% and the peak stator static pressure rise increased by over 50%. It is also shown in (12) that with the hub treatment the stagnation pressure actually increases across the hub section of the stator. This implies that there is a substantial amount of work done on this fluid by the rotating hub, due to the greatly increased momentum transfer associated with the hub treatment.

The basic mechanism of operation of casing treatment has not yet been fully elucidated; however, some important points for its use have emerged. First, it is clear that for casing treatment to work it must be the rotor tip section that is setting the stall point. In addition, as shown in (12) and (13), it appears that the type of stall must be an endwall related stall rather than an airfoil stall. Thus, although casing treatment does have the potential for large increases in stable flow range, one should be sure that its application is warranted before it is used.

Finally it should be noted that although casing treatment is a potent remedy in increasing stall range, it is not a panacea, since there is generally some penalty in efficiency for those treatments which have the most success in improving stall range.

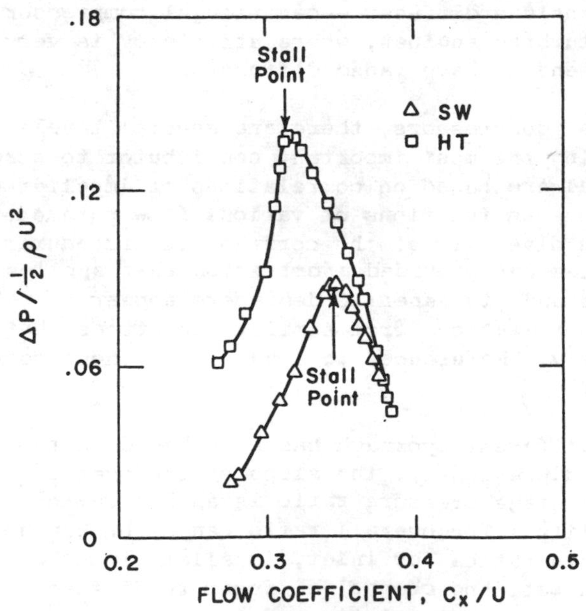

Figure 18 Stator static pressure rise characteristic with solid wall (SW) and hub treatment (HT) (12)

3. STABILITY OF CENTRIFUGAL COMPRESSOR AND PUMPS: STAGE STALL IN A CENTRIFUGAL COMPRESSOR

Although many of the same phenomena occur with centrifugal compressors as with axial turbomachines (especially when viewed on an overall system basis) there seem to be, especially in high pressure ratio centrifugal compressors, differences in the fluid mechanic features associated with the onset of instability. At the present level of understanding of compression system instability, one can say that the basic ideas concerning overall *system* behavior should apply to the centrifugal as well as axial compression systems. When one examines the specifics of the local phenomena that can "trigger" this overall instability, however, there do appear to be differences between the two types of machine. Because of this, it is useful to discuss the centrifugal compressor separately and, to focus on the phenomena that are associated with the stall of the centrifugal compressor *stage*, since this lies at the root of the system instability.

Centrifugal compressor stages consist of an impeller and a diffuser, with the latter of either vaned or vaneless design. The type of diffuser used can have a considerable effect on the stable flow range of the stage. As an illustration of this, Table I, taken from (14), divides diffusers into two types according to whether the principal objective is high efficiency or wide flow range, with the choice of design depending strongly on the application. Process compressors, which can require very broad range, may use vaneless diffusers--centrifugal compressor stages in aircraft gas turbine engines, where efficiency is very important, will tend to have vaned diffusers.

As with axial compressors, there are several levels of approach to finding the most important contributor to stage stall. The most empirical are based on correlations of impeller or diffuser flow range as functions of various flow parameters. However, from the diversity of the correlative procedures one can infer that none has yet provided a criterion that applies in all cases of interest and, in general, designers appear to rely heavily on data correlations from similar geometries that have been run previously. References to a number of these correlations are given in (1).

A somewhat different approach has been based on the idea that, as discussed previously, the slope of the overall (total-to-static) stage pressure ratio is an indication of the onset of instability. The overall ratio can be broken up into the product of separate ratios for inlet, impeller, impeller exit to vaned diffuser throat, and channel diffuser to diffuser exit. The first of these is a total-to-static and the rest are all static-to-static pressure ratios. Writing this overall ratio as a

TABLE I: DIFFUSER TYPES (from (14))

Wide Range	High Efficiency
Dump Collector	Vaneless plus cascade (1-3 row)
Vaneless diffuser with: dump collector or scroll or scroll plus discharge diffuser	Thin vane: single row multiple row Contoured vane single row multiple row
	Vane-island with: curved straight centerline passages
	Pipe" diffuser (UACL patent) with: circular non-circular "pipe" cross-section
	Rotating-wall vaneless plus vaned diffuser

product, an expression can be obtained for the normalized slope of the overall stage pressure ratio $PR_{overall}$, in terms of the different element pressure ratios:

$$\left(\frac{1}{PR_{overall}}\right) \frac{\partial PR_{overall}}{\partial \dot{m}} = \sum_i \frac{1}{(PR)_i} \frac{\partial PR_i}{\partial \dot{m}}$$

where the sum over i indicates the different elements of the stage. This normalized slope may be regarded as a stability slope parameter. An illustration of the behavior of this stability parameter is plotted in Figure 19, taken from (14). The values of the stability slope parameter are presented for the individual elements as well as for the overall stage, plotted as functions of the mass flow rate. Positive values indicate a tendency toward instability (positive slopes); negative values indicate stability. It can be seen that, in this example at least, the channel diffuser appears to be the element that has the largest destabilizing effect and that as the flow is reduced, the counteracting influence of other elements decreases so that the overall value moves towards positive.

The conclusion from (14) appears to be that in medium to high pressure ratio centrifugal compressors it is very often the vaned diffuser that sets the stall point, although this is not true in all cases. For example, a variable geometry diffuser, which could

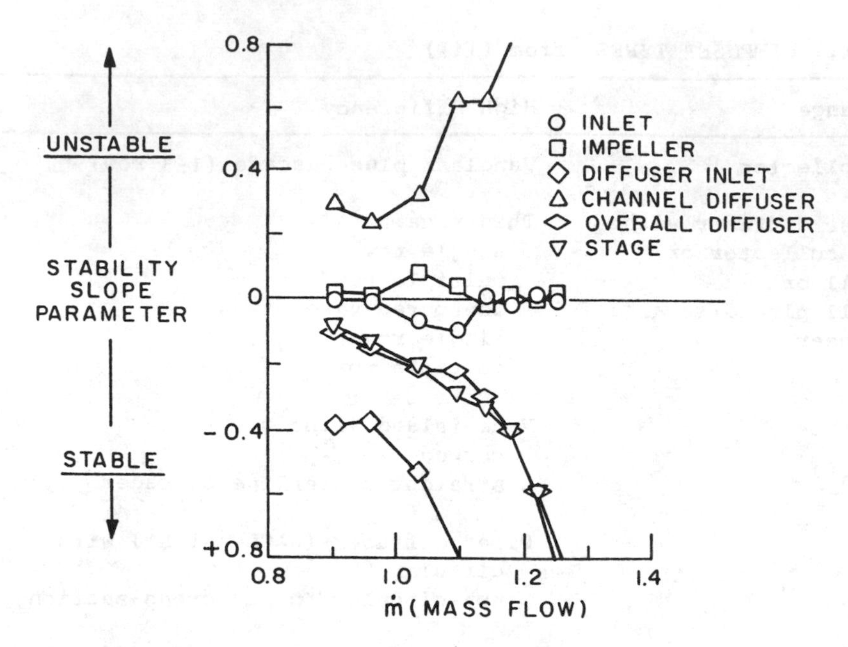

Figure 19 Influence of stage elements on centrifugal compressor stage stability (14)

be set close to choke over a wide flow range, could well be used as the stabilizing element in some stages; an example of this (11) has already been mentioned. (*Rotating* vaned diffusers have also been used as a stabilizing influence to increase flow range compared to the situation with a stationary vaneless diffuser). It should also be emphasized that, as with the axial compressor, the usefulness of the overall total to static pressure ratio is primarily as a *guide* to the initiation of stage stall, rather than as a direct quantitative criterion.

The situation is thus that the stalling element cannot *always* be regarded as being the diffuser, but must be found from examination of the aerodynamics of each component. In addition, the choice of parameters to correlate the onset of stall is still under considerable debate as are the unsteady physical phenomena that characterize the stage stall process. At present, therefore, one can say only that stage stall in a moderate to high pressure ratio centrifugal compressor with vaned diffuser is very often, but not always, set by the diffuser at speeds near design, that there can be substantial fluctuations in mass flow rate and pressure before the onset of surge (reverse flow), and that the presence of rotating stall in the diffuser may not be necessary for the initiation of overall system instability.

Stability Enhancement in Centrifugal Compressors

There are many applications of centrifugal compressors in which considerable flow range is required even if some efficiency must be sacrificed. As with the axial compressor, there are several techniques that have been developed to do this. One of these is to use inlet guide vanes to impart a "prewhirl" to the inducer, shifting the pumping characteristics on the map and altering the flow rate at which one encounters instability. The effects of using this technique are well documented in the literature, and a discussion of this method, with references, can be found in (14). Other means for increasing stability are the use of backward impeller leaning blades to create a negatively sloped pumping curve, and/or the use of vaneless diffusers (as mentioned above) rather than vaned.

Apart from these methods, however, there are other approaches that are perhaps less well known. The first is the use of a closely coupled system resistance to extend the stable operating range. The basic idea can be illustrated with respect to Figure 20. If operation on the positively sloped characteristic leads to instability, we change the characteristic slope by closely coupling another element to the stage to make the combined slope negative. Thus, curve C is the pumping characteristic and curves A and B are two examples of resistance curves (exit pressure dropping as flow is increased). If operation at a low flow rate such as A' is needed, then resistance A can be closely coupled to

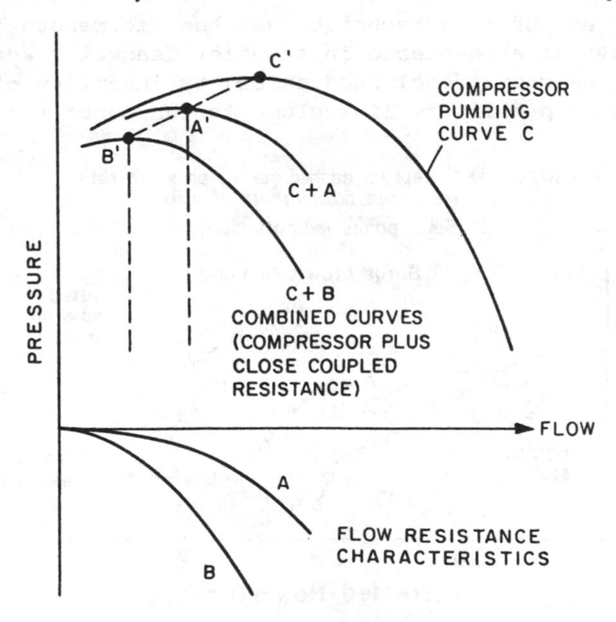

Figure 20 Surge control using close coupled resistance (15)

the pump; if further reduction to point B' is needed, then resistance B can be used. The combined curves in the two cases are shown as having basically zero slope at the desired operating points; negative slopes could be obtained with larger resistances.

This remedy has been applied to a compressor that had exhibited large amplitude surge cycles (including severe reverse flow) as a result of system instabilities (15). The closely coupled downstream throttling was achieved by means of overlapping plates with slots cut in them so that the open area, i.e. the downstream resistance curve, could be varied. The results are shown in Figure 21, which gives the measured compressor characteristics for one configuration tested. The data shown are from the baseline (no downstream resistance) run and from several other runs with varying resistance. The possibility of extending the stable flow range is definitely exhibited.

Note that the instability in the region of steep positive slope was not suppressed with the more open configurations, although it was much less severe than with no resistance. However, this instability was suppressed with the more closed (higher resistance) configurations. The downstream resistance did not completely inhibit the onset of rotating stall, although this was associated with a gradual decrease in performance so that the overall "stage" curve was still rising and the system behavior was stable.

A related use of this technique has been to reduce to pressure pulsations encountered in a boiler feedwater system (due to the multistage centrifugal feed pumps) by insertion of an orifice in the pump delivery line close to the pump (see (1)).

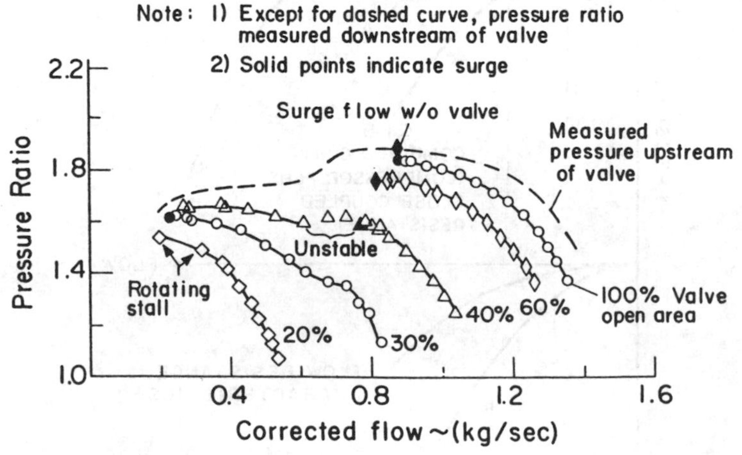

Figure 21 Centrifugal compressor performance with close coupled surge control (15)

Another approach is the use of casing treatment. If the stage stall line is controlled by the inducer (impeller) at low speed and by the vaned diffuser at high speed, one might expect to improve stability over a considerable speed range by the use of treatment on both the impeller and the diffuser. This has been done using an essentially axial, skewed, groove over the tip of the impeller at inlet, as well as with the impeller hub wall extended under the inlet of the vaned diffuser so that a treated (grooved) wall moved under the diffuser inlet (16). The locations of the casing treatment are indicated in Figure 22. The (rotating) treatment used under the diffuser inlet was a radial skewed groove.

Tests carried out with various centrifugal compressor stages showed significant improvements in stall range with grooves at the impeller inlet for speeds from 0.70 to 1.05 of design speed, although there was some small loss in efficiency. The "hub treatment" under the vaned diffuser also gave an improvement in stall flow margin, as well as an increase in choke flow.

Instability in Radial Vaneless Diffusers

Vaneless diffusers (as well as vaned diffusers) can also exhibit a type of propagating disturbance at high inlet swirl angles. This has also come to be called "rotating stall" although there does not appear to be as direct a connection with stall as there is in the case of the bladed cascades.

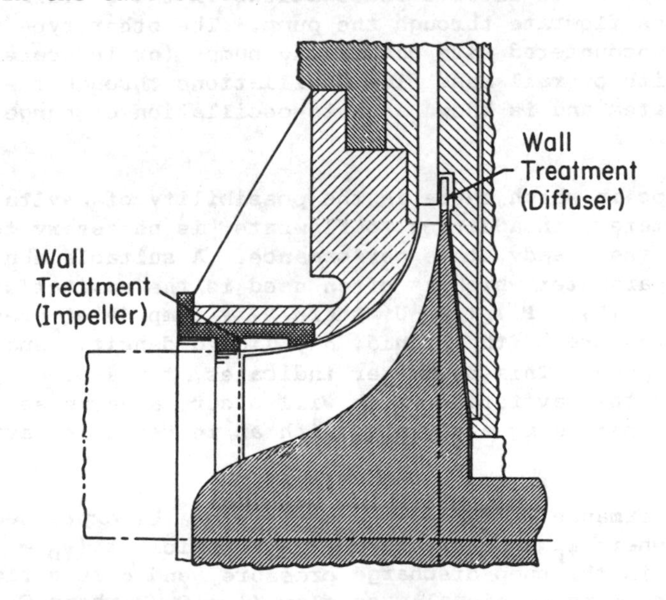

Figure 22 Casing treatment locations in centrifugal compressor stage (16)

Experiments have shown that there can be (at least) two different types of oscillatory behavior, one occurring at a speed of roughly one-fourth to one-third of the impeller velocity and one which occurs at less than ten per cent. These oscillations are local, in the sense that they merely redistribute the flow around the diffuser and that the overall mass flow is *not* affected, and are not truly a system phenomena, since the *system* behavior is *stable*. Studies of this phenomenon have been carried out; references to these can be found in (1), (17), (18).

4. OSCILLATIONS IN SYSTEMS WITH CAVITATING TURBOPUMPS

The instabilities occurring in the systems discussed above have been associated with stall. However, there are other classes of instabilities in pumping systems in which stall is not the primary cause, and these can occur even when the turbomachine is operating at its design flow rate. Such types of instability occur in systems with cavitating turbopumps or in pumps with two-phase flow.

The instabilities associated with cavitating turbopumps fall into two general categories. One of these affects only the local flow in the inducer and inlet, and is known as rotating cavitation. It is manifested as an unsteady cavitation pattern at the inducer inlet that rotates with respect to the inducer blades, and it does not appear to involve fluctuations in the overall (annulus averaged) mass flowrate through the pump. The other type of instability encountered with cavitating pumps (or inducers) is associated with overall mass flow oscillations through the entire hydraulic system and is known as auto-oscillation or surge (19), (20).

For pumps in which there is the possibility of cavitation, another parameter, in addition to flowrate, is necessary to characterize the steady state performance. A suitable nondimensional parameter which is often used is the cavitation number, defined by $\sigma = (P_1 - P_v)/1/2\ \rho U^2$ where P_1 = pump inlet pressure, P_v = vapor pressure of the liquid, ρ = liquid density, and U = the rotor (tip) speed. This parameter indicates, for a given geometry, the extent of the cavitation which will occur, a decrease in cavitation number being associated with an increase in cavitation extent.

The performance of the pump can therefore be expressed as $\psi_p = f(\phi,\sigma)$ where ψ_p is a pressure rise coefficient ($\psi_p = (P_2-P_1)/1/2\ \rho U^2$), P_2 is the pump discharge pressure, and ϕ is a flow coefficient or nondimensional mass flow ($\phi = C_x/U$ where C_x is based on the volumetric flow rate and the inlet area). Under non-cavitating conditions the performance is only dependent on ϕ.

As cavitation becomes important, however, the pressure rise becomes a function of cavitation number as well.

Representative performance curves for a rocket pump impeller are given in Figure 23 for a range of cavitation numbers (20). In this figure the horizontal axis is cavitation number and the vertical axis is the pressure rise coefficient, ψ_p, with the different curves corresponding to different non-dimensional mass flows. The design value of ϕ is 0.07. Indicated on the figure by the stars are the points at which the instability known as auto-oscillation was encountered with this system.

It can be seen clearly that this instability occurred at design (and higher) flow on the negatively sloped part of the performance curve. This is in direct contrast to the behavior that is found in the single phase systems, and it is evident that stall is not involved in this instance. It is apparent, however, that cavitation is connected with the onset of instability.

A *simplified* analysis that shows the qualitative features of these instabilities can be carried out by modelling the pump performance as a quasi-steady function of two variables, the inlet pressure and mass flow (1). For small perturbations therefore:

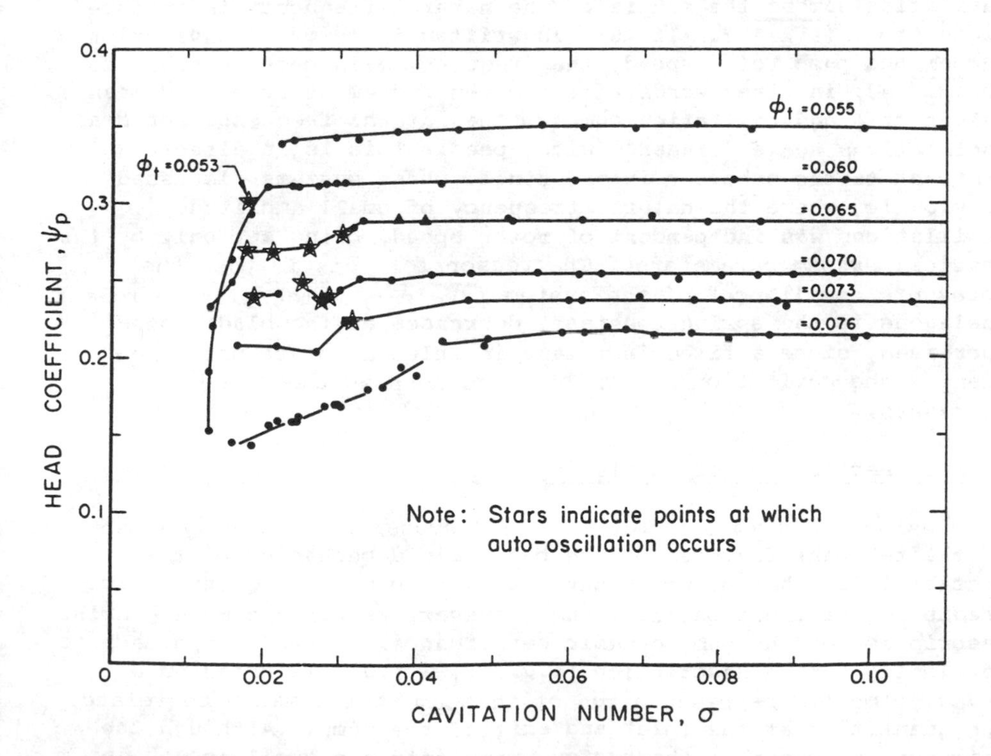

Figure 23 Cavitation performance of impeller (20)

$$\delta\Delta P_p = \delta P_2 - \delta P_1 = \left(\overline{\frac{\partial\Delta P_p}{\partial P_1}}\right)\delta P_1 + \left(\overline{\frac{\partial\Delta P_p}{\partial \dot{m}_1}}\right)\delta\dot{m}_1$$

$$\delta\dot{m}_2 - \delta\dot{m}_1 = \rho\left(\left(\overline{\frac{\partial V_c}{\partial P}}\right)\frac{d\delta P_1}{dt} + \left(\overline{\frac{\partial V_c}{\partial \dot{m}}}\right)\frac{d\delta\dot{m}_1}{dt}\right)$$

In these relations ΔP_p is the pump pressure rise, $\dot{m}$ is the mass flow, V_c is now the volume occupied *by the cavitation*, and the subscripts 1 and 2 denote pump inlet and outlet.

Using this representation of the transient pump behavior, if one analyzes the stability of a system with a cavitating turbopump, it is found that for pumps operating near the design value of ϕ, the system will become unstable when $(\partial V_c/\partial\dot{m})$, which is referred to as the *mass flow gain factor*, becomes sufficiently negative.

A further result that can also be extracted from the stability analysis based on this simple model is the determination of the natural frequency of the system. It is found that for a given cavitation number the square of the natural frequency is proportional to $1/(\overline{\partial V_c/\partial P})$. If this is written in terms of cavitation number and pump rotor speed, the frequency ω is seen to scale as $(U/\overline{\partial V_c/\partial\sigma})$; in other words, for a given system at specified mean values of ϕ and cavitation number, the natural frequence for small oscillations scales linearly with speed. This is in direct contrast to the behavior in the single phase systems discussed previously, where the natural frequency of small amplitude oscillations was independent of rotor speed, being set only by the physical *system* parameters. The reason for this is that the "pressure compliance" of the system $(\overline{\partial V_c/\partial P})$, which plays a role analagous to the spring constant, decreases as the blade speed increases, since a fixed increment in inlet absolute pressure changes the cavitation number less and less as the speed increases.

Dynamic Performance of Cavitating Pumps

Using the simple model one can discuss *qualitatively* some of the relevant features of the basic fluid mechanics of the instabilities that occur in cavitating turbopumps. Quantitative prediction of these oscillations, however, requires a more precise description of the pump dynamic performance. A useful approach to developing such a description (e.g., (19),(20)) is based on a *transfer matrix* representation of the pump performance to relate the quantities at the inlet and exit of the pump. Although the approach assumes that the system transients are small enough in amplitude so that a linearized description of the motion can be adopted, it has proved of value in clarifying the nature of dynamic performance of cavitating pumps.

The forms of these transfer matrices have been examined theoretically and experimentally for a cavitating pump (a rocket pump impeller) and have been used in a detailed system stability calculation. The approach used is to focus on the net flux of mechanical energy out of the various elements in the system. This can be found for the pump by using the experimentally measured transfer matrices, and Figure 24a presents (a cross plot of) results from (20) which shows the *activity parameter* (essentially the non-dimensional flux of mechanical energy out of the pump), as a function of cavitation number. The flow rate is $\phi = 0.07$, the design value. The vertical scale in this figure indicates the non-dimensionalized value of this net flux. Positive values indicate a dynamically active element, i.e. one that is feeding energy into the oscillations. The activity parameter is strongly dependent on oscillation frequency as well as cavitation number and mass flow. The quantity shown in the figure is the maximum occurring at a particular cavitation number, which would be the frequency that characterizes the largest outflux of mechanical energy.

A positive value of the dynamic activity of the pump is a necessary, but not sufficient condition for system instability, since there can be dissipation occurring in the rest of the system to offset the pump behavior. To find the overall stability of the system, one must couple the dynamic pump behavior to the rest of the system dynamics; this can readily be done using the transfer matrix techniques. Figure 24b shows the results of this overall energy analysis and again gives net flux of mechanical energy as a function of cavitation number. As before, the convention used here is that a positive number indicates a trend toward instability whereas negative values indicate stability. Note that it is the value of this quantity, rather than that shown in Figure 24a, that is directly tied to whether the system is unstable or not, and (again) the bars indicate the maximum value (as a function of frequency) for the particular cavitation numbers shown.

The important point is that there are large differences in the values of these activity parameters as the cavitation number is reduced. These differences appear to be strongly associated with the mass flow gain factor which becomes very large in magnitude over certain ranges in cavitation number. To analyze the system stability, it is thus basic to be able to understand how the mass flow gain factor behaves over the different regimes of pump performance.

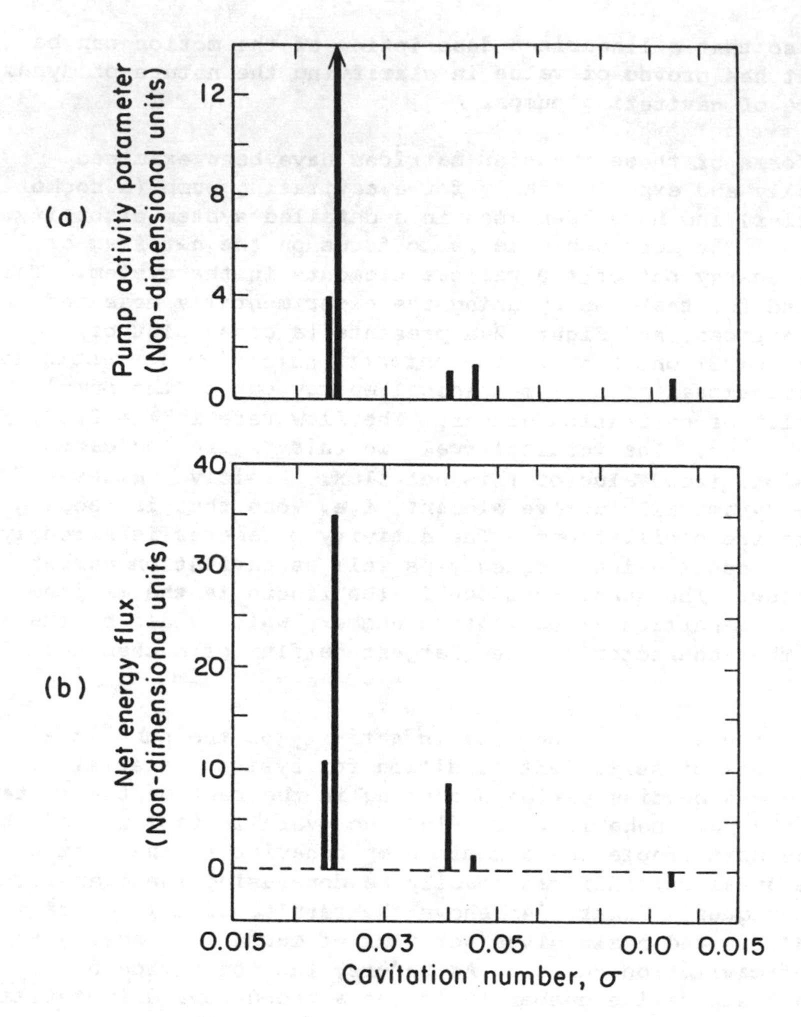

Figure 24 **Effect of cavitation number on pump activity parameter and net system energy flux (20)**

Cavitation Induced Instabilities at Off-Design Conditions

Although the discussion has been focussed so far on the flow regimes near design, instability is also seen during operation of cavitating pumps at "throttled", i.e. low flow, operation. In many instances it appears that system oscillations can in fact become more severe as the flow is reduced from the design. The occurrence of severe system oscillations due to cavitation is not limited to axial pumps described above and they can also **occur with centrifugal pumps, for example, boiler feed pumps.**

Observations of the flow fields in both types of pumps (axial and centrifugal) reveal that during operation at low flows there is a strong backflow in the tip region of the pump. This can give rise to a strong tangential velocity component in the fluid upstream of the inducer. The inlet flow can also contain a significant cavitation volume *upstream* of the pump and can thus contribute to the mass flow gain factor. Flow visualization has shown that the extent of the upstream pre-rotation region and the cavitation volume can vary substantially during the surge cycle. In this connection it has been found that augmentation of the "prewhirl" upstream of the pump, by injecting high pressure fluid from the pump discharge in a tangential direction can have a signficant stabilizing effect.

5. PUMP SURGE DUE TO TWO-PHASE FLOW

Although not directly tied to cavitation, a related type of instability has also been encountered in two-phase flow systems. This has been termed "pump surge" (21) in view of the apparent similarity to the types of instability observed with single phase flow in pumps and compressors. Large amplitude oscillations have been found in two-phase flow systems with both centrifugal and axial pumps.

At present no in-depth quantitative analysis of this phenomenon appears to have been carried out, although it appears that the techniques mentioned in the previous two sections could be usefully applied here as well. However, some qualitative observations relating to the system stability can be made. The steady state behavior of the headrise (pressure rise) coefficient of a centrifugal pump in a two-phase flow as the inlet void fraction (the volumetric concentration of the gas phase) is increased as shown in Figure 25a. The vertical axis is the head coefficient (pump head rise divided by impeller tip wheel speed squared) normalized by the head coefficient for single phase flow, and the horizontal axis is the inlet void fraction. The different symbols represent different total flow coefficients (defined as total volumetric flow at inlet divided by the product of impeller tip wheel speed and impeller discharge area). As the void fraction is increased a significant falloff in head coefficient can be seen. In addition, at constant void fraction the falloff in head coefficient increases as the flow coefficient is decreased.

With this as background, it is pertinent to plot this data in the usual headrise versus flow coefficient format, as is done in Figure 25b. The single phase curve (the circles) is negatively sloped except for a small region at quite low flow rates. The curves for higher inlet void fractions, however, have a considerable positive slope over much of the range shown. Thus,

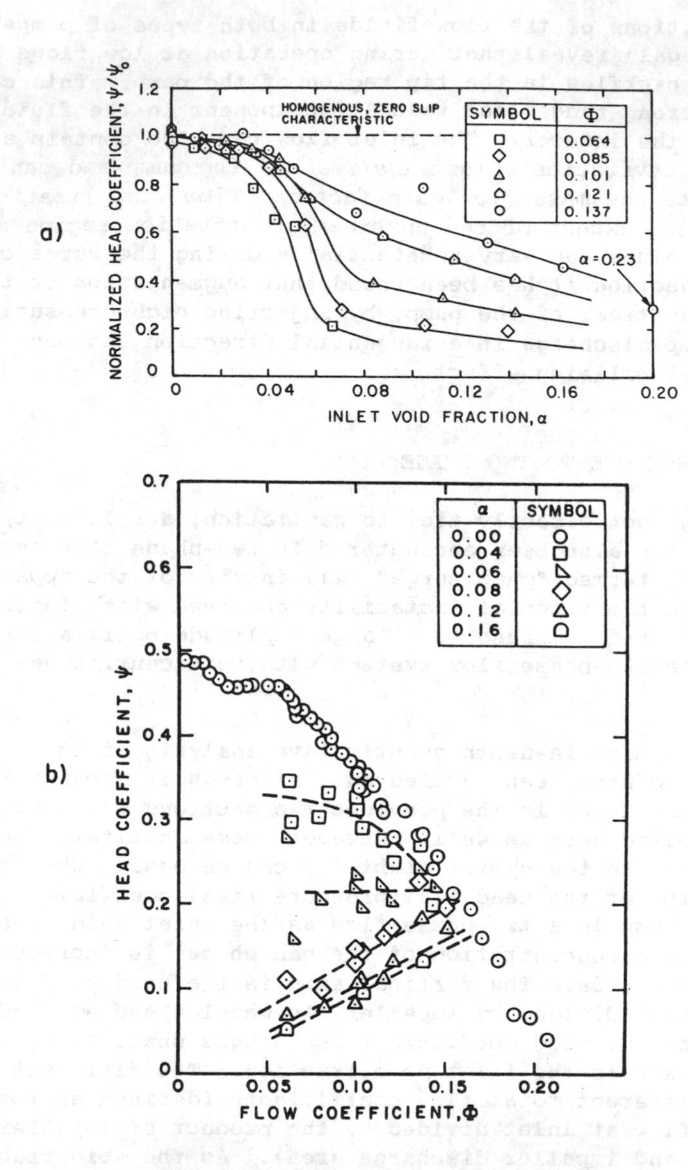

Figure 25 Steady state pump performance in two-phase flow (21)
a) Pressure rise coefficient versus inlet void fraction for different flows
b) Pressure rise coefficient versus flow for different inlet void fraction (crossplot of a)

if one considered small perturbations which occurred in a system with a pumping characteristic of this type, there would be a definite possibility for the promotion of an instability as described previously.

For the general case of an arbitrary perturbation in system operating point there is no need for the inlet void fraction to remain constant and even *if* one considers the pump transient performance to be approximately quasi-steady, the headrise will then be a function of both the instantaneous inlet void fraction and the flow coefficient. The pump operating point could thus traverse a path that cut across the curves in Figure 25b and the effective slope could be more or less steeply sloped than shown. Nevertheless, it does appear that the falloff in head coefficient as the result of a two phase inlet flow can be a potentially important factor in promoting instability, *in addition to* the effects associated with the mass flow gain factor, for these types of pumps operating in two phase flow.

6. SELF-EXCITED OSCILLATIONS IN HYDRAULIC SYSTEMS

There can be many other instabilities encountered in the general area of hydraulic pumping systems (1),(22), although space precludes a detailed description here. Severe pulsations have been known to arise due not only to system oscillations, but also to local instability (i.e. rotating stall) in the pumps. For example, self-excited oscillations of check valves with spring dampers can also occur. Experimental studies of a simplified model of this phenomenon have been carried out to clarify the cause of instability which was associated with high rates of change of discharge in the last few degrees of closing. Instabilities have been encountered with hydraulic gate seals and hydraulic turbine penstock valves and these have also been analyzed in terms of a negative damping which was responsible for the growth of the oscillations.

7. FLOW TRANSIENTS AND INSTABILITIES IN COMPOUND PUMPING SYSTEMS (SYSTEMS WITH PUMPS IN PARALLEL)

In many practical situations two or more pumping devices operate in parallel. Under these conditions not only are there possibilities for instabilities of the types that we have already discussed, but new problems, associated with the compound system, can also appear. In particular there can be difficulties associated with transients, such as startups, in these kinds of systems.

To examine some of these problems consider the model pumping system pictured in Figure 26. Suppose that the two pumps have different pressure rise and mass flow capabilities. The steady state operating points of such a configuration might be as given in Figure 27. Curves A and B are the individual curves for each machine; the dash-dot curve C is the combined. For system pressure requirement curve 1, the two machines do better than one, while for the system resistance curve 3 the capacity of the two machines is actually smaller than that of the larger machine by itself. As it is drawn the smaller machine would be operating with reverse flow. The overall system operation , however, is stable in that the pump performance changes smoothly and continuously as the throttle area is decreased.

An often encountered situation is that one or the other of the pumps has been shut off and is brought on line. To be specific, let us suppose that pump A is operating with pump B off, and B is now brought on line. If the throttle curve is considered fixed at 3, A's operating point would undergo a large transient from a to a', with operation of B at point b'. If the demand (flow volume) is considered to be fixed, then the transient would be to a". In either case the result could be that a pump that was running near peak efficiency undergoes a transient to a region of much less benign operation.

The difficulties can be compounded if the pump curves have regions with positive slope, as one can see using arguments similar to those presented in the first section. In general it is therefore found that for pumps or fans to operate satisfactorily in parallel they should have negatively sloped pressure rise versus flow curves, and they should have roughly the same percentage reduction in capacity over the operating range. Some design rules for different specific combinations of fans and compressors have been worked out by Eck (23).

The transients just described can be large and can lead to considerable difficulty. In addition to this excursive behavior, however, there are also situations in which compound systems give rise to an oscillatory type of transient. An example of this is

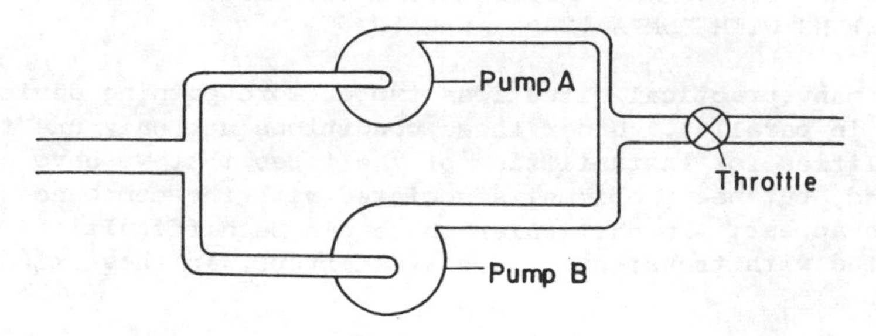

Figure 26 Compound pumping system with pumps in parallel (1)

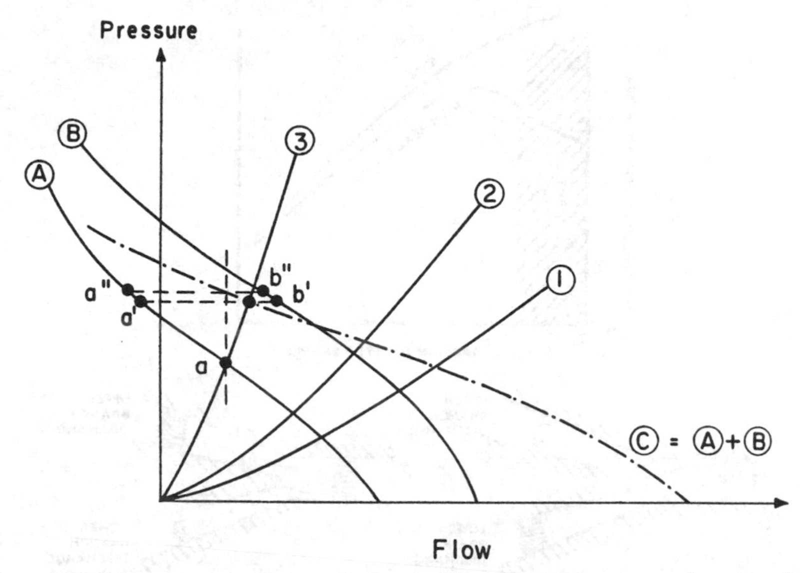

Figure 27 Pressure-flow characteristics of pumps in parallel (1)

described by Ehrich (24), who examined the basic compound system associated with a branched diffuser configuration. This is a situation encountered in gas turbine annular combustors. Each branch of the system has a diffuser, a volume with a mass storage capability, and a downstream resistance similar to the configuration shown in Fig. 1 for the simple system. The "pumping characteristics" of one branch of the system are shown in Fig. 28. A key feature is that the static pressure recovery in either diffuser branch is a function of the flow fraction entering that branch. A small flow fraction will result in low velocity at the branch discharge and, ideally, high static pressure recovery. However, this would imply a high diffusion which may not actually be possible, so that separation could occur. If so, the pressure recovery characteristic will have its peak at some intermediate flow fraction as shown by the curve marked "with real losses" in the figure.

The analysis of the branched diffuser system showed that operation to the left of the maximum is unstable and that, for the conditions studied, a relaxation type of oscillation would be set up. For symmetric diffusers this results in the flow through the combustor shells, i.e., the downstream resistance varying periodically in an approximately triangular fashion while the system pressure drop will vary periodically (approximately as a sawtooth wave) with twice the frequency of the other parameters. Results of the analysis are shown in Fig. 29 which presents the oscillations in branch inlet and exit flow and in overall system

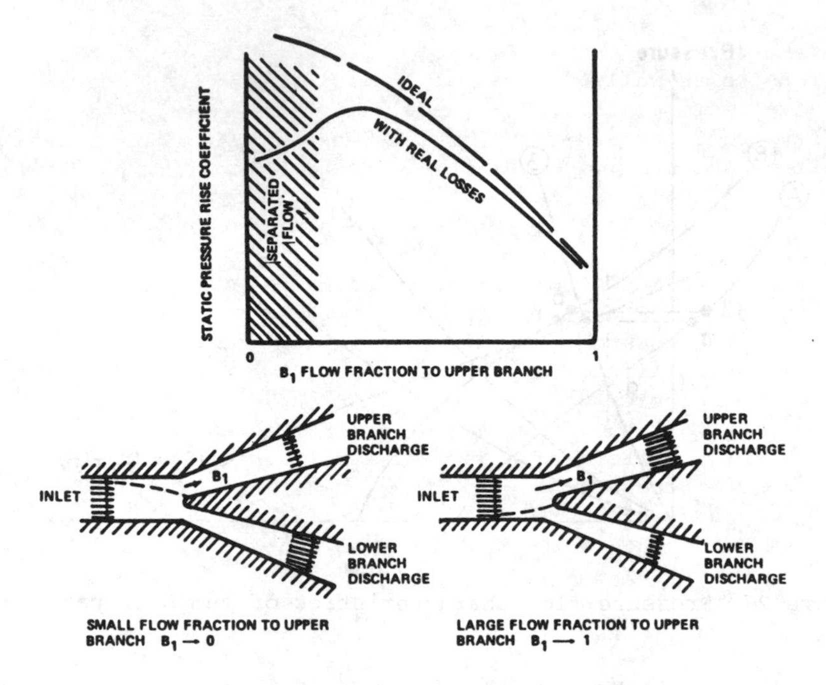

Figure 28 Static pressure recovery characteristic of a branched diffuser system (24)

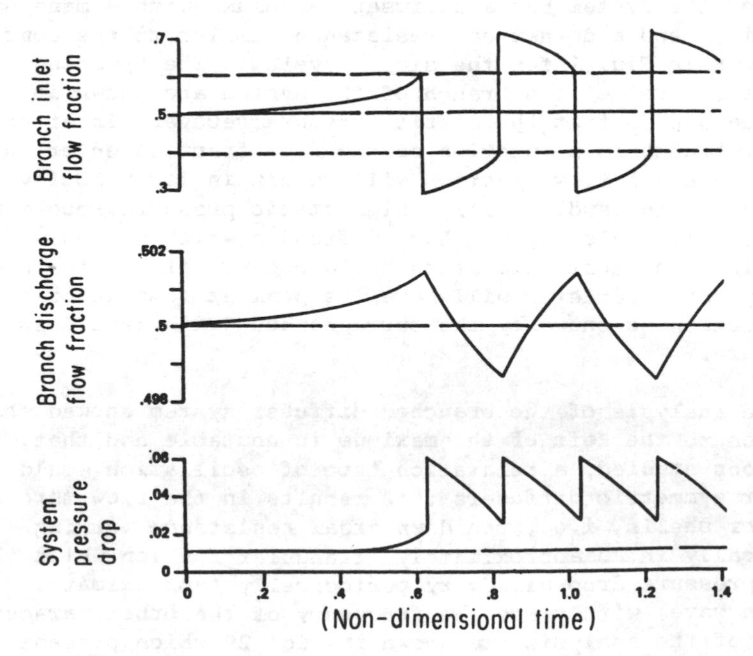

Figure 29 Computed flow oscillations in branched diffuser system(24)

pressure drop. One point that should be noted in interpreting these results as well as others in (24) is that there are no inertial effects considered. These will: (1) modify the stability criterion somewhat, moving the point of instability to the left of the peak, and (2) change the shape of the oscillations shown for the nonlinear calculations, with the tendency being to make the waveforms less like relaxation oscillations.

A further example of flow transients that can arise in pumping systems in parallel has been documented in (25). This was a phenomenon that occurred in a dual entry centrifugal compressor, where a large amplitude, sudden flow shift occurred, accompanied by a substantial pressure change. The phenomenon is illustrated schematically in Fig. 30, where the two sides of the impeller, having slightly different characteristics, are shown. In position A_1 and T_1 (for Accessory and Turbine sides, respectively) stable operation exists, although with some asymmetry. As the overall flow is reduced, the flow in the turbine side of the impeller (point T_1) reaches the stall line first and ends up at point T_2. The operating point of the accessory side then falls back to A_2, with an attendant change in flow and pressure rise. The machine is thus operating in rotating stall on the turbine side and at a high flow rate on the accessory side. Reference (25) also cites instances in which the flow at T_2 might not be stable, in which case this side encounters mild surging.

As an illustration of the actual performance of the individual sides of the impellers, Fig. 31 shows the curves of pressure ratio versus flow at different speeds for each side of the machine. The circles are the accessory (A) side and the triangles

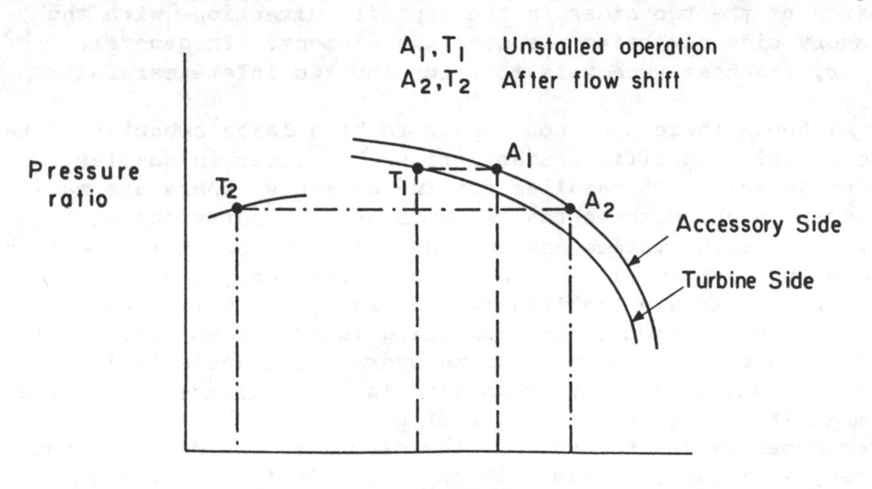

Figure 30 Flow transient in dual entry impeller (25)

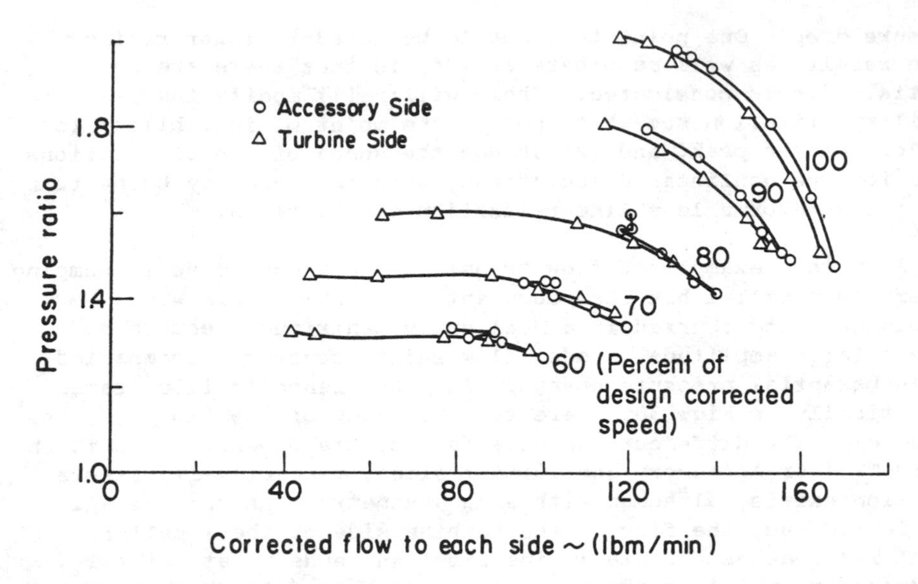

Figure 31 Measured pressure-flow characteristics of each side of a dual entry impeller (25)

the turbine (T) side. It can be seen that in the lower speed ranges there is the possibility for a large "flow shift" even when the accessory side is operating far from stall. One cure for this type of instability is to reduce the pressure rise in the stronger side by using an inlet screen until the two sides are matched. This was tried successfully, although one must be careful to tailor the screens properly since too much blockage can cause a mismatch of the two sides in the opposite direction, with the accessory side now being the stalling element. In general, however, the best remedy is to match the two inlets carefully.

Although there does not appear to be a large amount of literature covering specific instabilities that occur in complex pumping systems with parallel pumping elements, there are many instances in which these can be important. To carry out an analysis of such systems one must dynamically couple the various components. For small oscillatory perturbations, one method of doing this is to use the transfer matrix approach that was described in the section on cavitating inducer stability. This technique has been used to analyze hydraulic transients in complex piping systems (22). For predicting large amplitude transients, one must integrate the equations of motion for the flow in the system numerically; however, if the stability boundary is what is desired, a linear analysis will suffice. As far as overall results are concerned, one cannot make a general statement about the stability of the complicated pumping systems that exist in practice, execpt to note that a very conservative criterion would be that if none of the components in the system are dynamically active, i.e., if none of them have a net outflux of mechanical energy over a complete period, then the system will be dynamically stable.

8. SYSTEM TRANSIENTS SUBSEQUENT TO INSTABILITY INITIATION

We have, until this point, been concerned with phenomena associated with the *onset* of instability. There are, however, important problems associated with turbomachinery and compression system behavior *subsequent* to this onset, and we will briefly describe some of the relevant issues; these are discussed in greater detail in Refs. (1) and (2).

For an axial or centrifugal compression system, there are two modes of system behavior, surge and rotating stall, that can result from the initial instability. As stated previously, the former consists of large amplitude oscillations in pressure rise and flow during which the compressor undergoes periodic excursion from a stalled to an unstalled condition. Operation in rotating stall, on the other hand, is characterized by a situation in which, after an initial transient, the overall (annulus averaged) mass flow and pressure ratio are steady. In addition, these quantities can be greatly reduced from the values that existed prior to the onset of instability.

For an aircraft gas turbine engine, a key requirement is recovery from stalled conditions once instability occurs (26). In this connection, therefore, it is important to note that even though surge may be associated with severe transient stresses due to the large amplitude flow variations, the engine does operate in an unstalled condition over part of the surge cycle. Because of this, one is able to open bleed valves or make other changes that will have an effect on bringing the system out of surge. In contrast to this, rotating stall, and in particular the full-span, large extent, rotating stall which is characteristic of a multi-stage compressor, can be very difficult to recover from. Therefore, it is the surge mode that is much more favorable for recovery in systems such as aircraft gas turbine engines and is hence the more desirable one. On the other hand, for an industrial centrifugal compressor or pump, it may be that operation in rotating stall can be tolerated, and that throttle movement is wide enough so that recovery is readily achieved. In this situation it may be that the often violent oscillations that occur during surge make this the less desirable of the two modes.

Whatever the situation, however, it is apparent that an important question is whether a given pumping system will exhibit large amplitude oscillations of mass flow and pressure ratio (surge), or whether the system will operate in rotating stall where the annulus average mass flow and pressure ratio are essentially steady, but are greatly reduced from the pre-stall values.

To answer this, one must analyze the nonlinear system behavior. This was done in (27) using the Helmholtz resonator type of compression system model introduced by Emmons (3) (for the linear case). The analysis shows that for a given compression system, i.e., specified compressor characteristic, plenum volume, compressor length, etc., there is an important non-dimensional parameter on which the system response depends. This parameter is denoted as B:

$$B = \frac{U}{2\omega L_c}$$

where ω is the Helmholtz resonator frequency of the system, L_c is an "effective length" of the compressor duct, and U is the rotor speed. If we insert the definition of ω we can define B in terms of geometric and physical system parameters as

$$B = \frac{U}{2a}\sqrt{\frac{V}{A_c L_c}}$$

where a is the speed of sound, A_c is the compressor flow-through area and V is the plenum volume.[3] *For a given compressor characteristic*, there is a critical value of B which determines whether the mode of instability will be surge or rotating stall. Thus if we examine the system transient response as represented on a compressor map format (pressure rise versus flow) we find the following situation, as shown schematically in Fig. 32. The steady-state pumping characteristic is shown for reference. Systems with B above the critical value, as shown on the right (e.g., speeds above a critical value) will exhibit surge oscillations, while those having B lower than the critical (speeds below the critical value) will undergo an initial transient to the (annulus averaged) steady flow and pressure rise associated with operation in rotating stall.

Experimental evaluation of this concept is shown in Fig. 33, in which data is presented for a three-stage axial compressor that was run with several different downstream volumes (27). The horizontal axis is corrected speed and the vertical axis is the B parameter. The open symbols mark the experimental values of B at which the changeover from rotating stall to surge occurred for the different volumes tested. (The solid point indicates the highest value of B that could be obtained with the smallest volume used.)

---

3. For the system of Fig. 1, V should be the volume of the compressible gas in the closed volume.

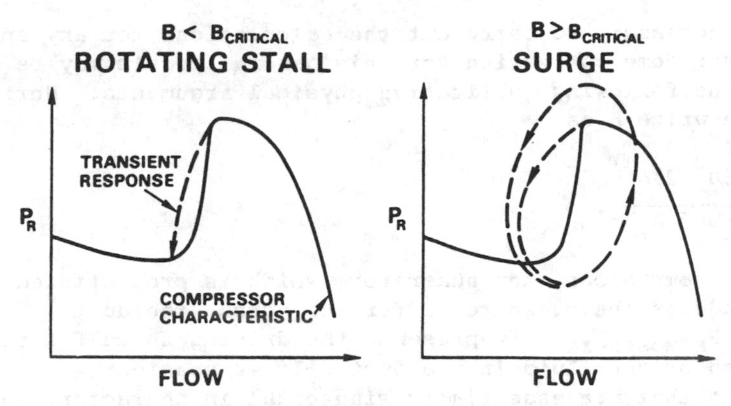

Figure 32 Schematic of transient system response subsequent to initial instability (1)

The regions in which one encounters surge (B above the critical value) or rotating stall (B below critical) as the mode of instability are also indicated. It can be seen that the prediction of a constant B as the boundary between surge and rotating stall is well borne out, although the value at which the change occurs is somewhat above that predicted by the theory. (It should again be emphasized that the value of 0.8 is for this particular compressor.)

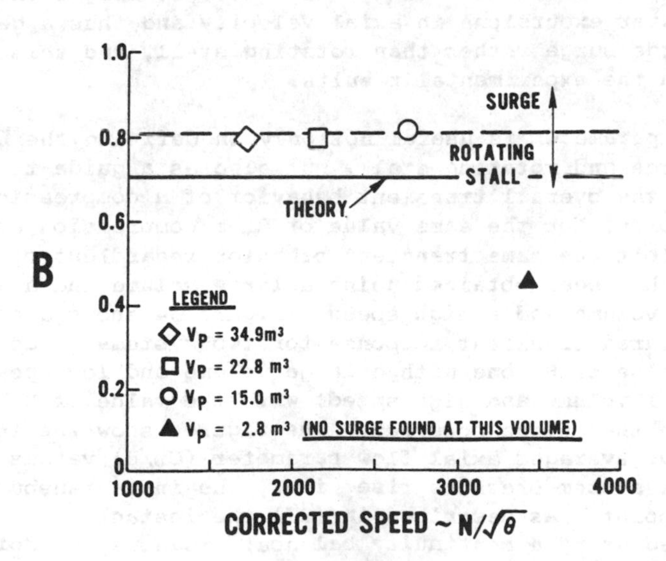

Figure 33 Surge/rotating stall boundary for different plenum volumes (27)

It is necessary to carry out the calculations for any specific case, however some motivation for this general result may be gained by the following qualitative physical arguments. Notice that we can write B as

$$B = \frac{(\rho U^2/2)A_c}{\rho U \omega L_c A_c} .$$

For a given compressor, the numerator, which is proportioned to the magnitude of the pressure difference across the duct ($P_{plenum} - P_{atmospheric}$), represents the driving force for the acceleration of the fluid in the duct. If we consider oscillations that are essentially sinusoidal in character, the order of magnitude of the inertial forces that arise because of these local fluid accelerations will be given by the product of the fluid density, ρ, the frequency of the oscillations, ω, the amplitude of the axial velocity fluctuation, and the volume of the fluid in the duct, $L_c A_c$. Hence, if we assume that the axial velocity fluctuation is a specified fraction of the mean wheel speed, the magnitude of these inertial forces will be proportional to the denominator. The ratio of the two forces (pressure to inertial) is therefore proportional to B. Thus as B is increased, for example by increasing the rotor speed, a fixed percent amplitude of the compressor axial velocity oscillations will result in inertial forces that are relatively smaller and smaller compared to the driving force due to the pressure differential. The capability to accelerate the fluid in the duct is thus increased as B increases. Hence, as B becomes larger one would expect greater excursions in axial velocity and thus a general trend towards surge rather than rotating stall, and this is in accord with the experimental results.

The B parameter is useful not only in defining the boundary between surge and rotating stall, but also as a guide to the scaling of the overall transient behavior of a compression system. In other words, for the same value of B, a compression system should exhibit the same transient behavior regardless of whether this value has been obtained using a large volume and a low speed or a small volume and a high speed. Figure 34 shows a comparison of the measured transient response for two systems at approximately the same value of B, one with a large volume and low speed and one with a small volume and high speed, with the value of B high enough such that surge occurred. The figures show the instantaneous annulus averaged axial flow parameter (C_x/U) versus non-dimensional system pressure rise, i.e., the instantaneous system operating point. As described in (27) the instantaneous mass flow was obtained using a continuity balance; reasons for doing this and difficulties of alternative ways of measuring this quantity are also discussed. The steady-state compressor curves are also shown for reference. As can be seen, the two surge cycles show extremely similar behavior, emphasizing the influence of B as the relevant non-dimensional parameter for the phenomena under study.

The surge cycles of Fig. 34 do not show periods of overall flow reversal. However, with high speed multistage compressors, surge cycles having periods of reversed flow are, in fact, the more common situation. In addition, although the mass flow and pressure rise oscillations associated with surge cycles of the type shown in Fig. 34 are relatively sinusoidal (as a function of time), the variations in mass flow and pressure rise that occur in compressors of high pressure ratio are quite non-sinusoidal and are more nearly a relaxation type of oscillation. The two types of oscillation were referred to as "classic surge" and "deep surge", respectively, in (27), where it was shown that this deep surge mode could also be seen with a low speed compressor as one reduced the throttle area somewhat from that value associated with the initial instability.

An illustration of the form of this type of oscillation is presented in Fig. 35, which shows data taken during deep surge of a compressor. Note that these two figures show several seconds of

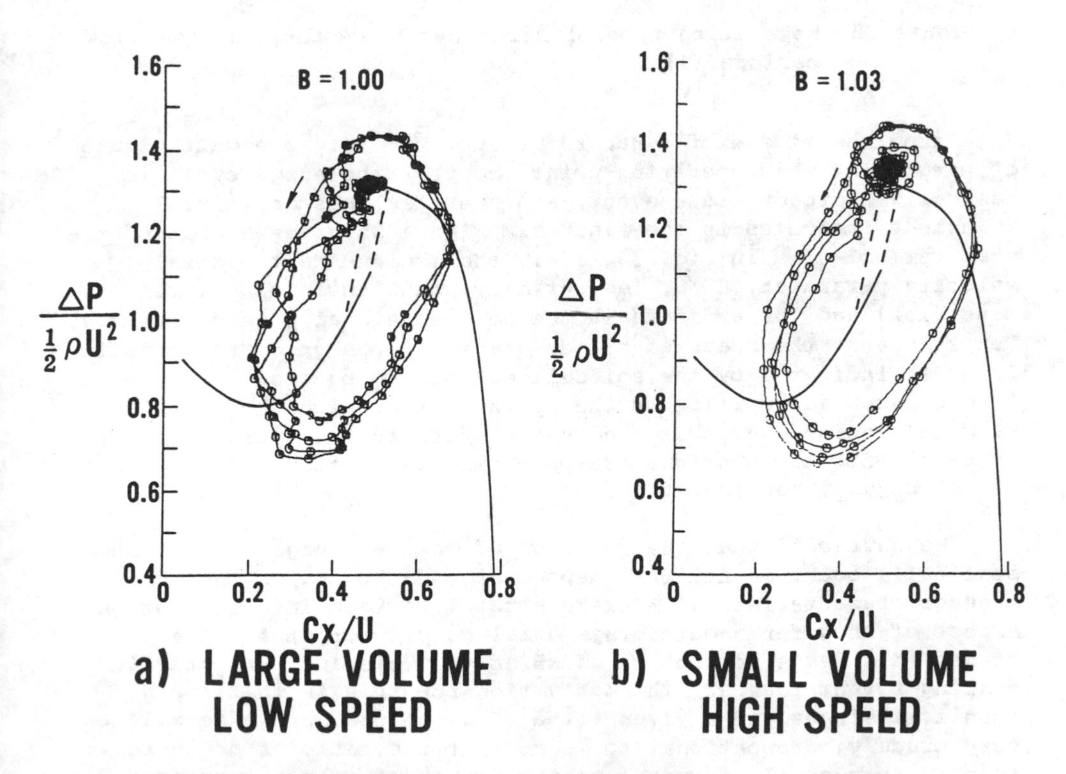

Figure 34 Dependence of system response on B (1)

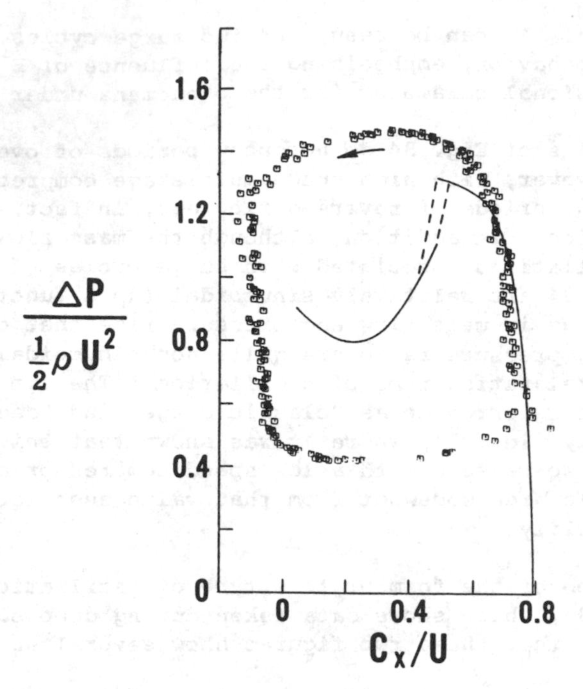

Figure 35 Relaxation type of surge cycle showing reverse flow periods (1)

data from the middle of a record of approximately one half minute of surging at this operating point, so that the surge cycle is basically a steady-state event, and there is thus no initial transient indicated in the figures. Figure 35 gives a plot of the same variables as in Fig. 34, i.e., the abcissa is the axial velocity parameter, C_x/U, (essentially nondimensional compressor mass flow) and the ordinate is the nondimensional pressure rise. For reference the measured steady-state compressor characteristic is again indicated by the solid line. It can be seen that there is a substantial portion of the cycle during which the overall flow is reversed and that the general form of the cycle is roughly "square" with two sections of approximately constant pressure and two of roughly constant flow.

We have said that the question of whether surge or rotating stall will occur is not only dependent on B, but upon the compressor characteristic. A basic attempt to take into account one aspect of this for a multistage axial compressor, the number of stages (N), leads to the use of NB as the relevant parameter for scaling rather than B. The motivation for this is that for N identical stages and a given value of B the pressure rise will be, very crudely, proportional to N, so that the ratio of pressure to inertia forces will increase as the number of stages increases.

There is thus a strong trend toward lower values of B needed to encounter surge, rather than rotating stall, as the number of stages and overall pressure rise of the machine is increased. Further discussion of this point is given in (28).

Additional comments on the use of these types of lumped parameter models (and the utility of this B-parameter) for high pressure ratio compressors and core engines can be found in (26) which gives calculations showing the effects of combustor heat release and number of stages. Application to centrifugal compressors is described in (29).

We have described the necessity for certain system parameters to be of a specified magnitude in order to encounter surge and we can also comment here on the physical mechanism for maintaining the surge cycles. These are large amplitude, nonlinear limit cycle phenomena; however, the overall mechanism is similar to that described with reference to dynamic stability. This is an increased mechanical energy input from the compressor during the cycles compared to during a mean steady flow (which results from a positively sloped compressor characteristic). However, it is now not just a local value of the slope that matters, but rather the instantaneous values over a finite range of flows.

Another important aspect of stall recovery concerns the performance of the compressor in rotating stall. A critical issue is that of hysteresis in the stall/unstall process (26). A compressor which encounters rotating stall at a given throttle setting will generally require considerable opening of the throttle from this setting in order for the rotating stall to cease. The problem can be particularly severe for multistage compressors, and can necessitate engine shutdown and restart.

As a result of this, there has been a substantial amount of recent work aimed at predicting the performance of the compressor in rotating stall and, in particular, the stall cessation point. A basic correlation for stall recovery is described in (1) which includes data from over thirty (low speed) multistage compressors. Another examination of the topic from a more fundamental viewpoint is given in (30). The latter presents calculations of the stall flow performance and recovery point for several different three-stage compressors.

As a final note on this topic, we may mention a forthcoming NASA contractor report (31), which presents an approximate theory for general post-stall transients in compression systems. This treatment, which encompasses both surge and rotating stall as special cases, gives a conceptual framework from which to view the complex unsteady behavior that occurs during these types of strongly nonlinear, nonsteady and severely non-axisymmetric flows.

9. ACKNOWLEDGMENTS

Much of this lecture was written at the Whittle Laboratory, Cambridge University, while the author was a Visiting Fellow in the Department of Engineering. The author wishes to thank the Science and Engineering Council and the Royal Society for their support. The efforts of N.A. Cumpsty, Sir William Hawthorne, and D.S. Whitehead in arranging this stay at Cambridge are also gratefully acknowledged. Partial support during the preparation of the manuscript was also provided by Air Force Office of Scientific Research under contract F49620-82-K0002.

10. NOMENCLATURE

a	speed of sound
A	flow-through area
A_c	compressor flow-through area
B	B parameter; $B = (U/2a)\sqrt{V/AL_c}$
C_x	axial velocity
L	length of compressor (or pump) ducting
$\mathcal{L}$	effective length
$\mathcal{L}_c$	compressor effective length
$\dot{m}$	mass flow
P	pressure
P_v	vapor pressure
U	blade speed
V	volume
V_a	air volume
V_c	cavitation volume
γ	specific heat ratio
α	absolute flow angle
β	relative flow angle
δP	pressure perturbation
δm	mass flow perturbation
ΔP	pressure difference
ΔP_p	pressure difference across pump or compressor
ΔP_T	pressure difference across throttle
W	velocity
ρ	density
σ	cavitation number (also used for solidity in definition of D-factor)
ϕ	axial velocity parameter; $\phi = C_x/U$
ψ	pressure rise coefficient, $\psi = \Delta P/ 1/2\rho U^2$
ψ_{TS}	inlet total to exit static pressure rise coefficient, $\psi_{TS} = (P_{out} - P_{T_{in}})/\rho U^2$

Subscripts and Superscripts

$(\bar{\ })$	mean flow quantity
$\delta(\ )$	perturbation quantity
1	inlet
2	exit

11. REFERENCES

1. Greitzer, E.M., The Stability of Pumping Systems--The 1980 Freeman Scholar Lecture. ASME J. of Fluids Engineering, 103 (June 1981) 193-242.
2. Greitzer, E.M., Review--Axial Compressor Stall Phenomena. ASME J. of Fluids Engineering, 102 (June 1980) 134-151.
3. Emmons, A.W., C.E. Pearson, and H.P. Grant, Compressor Surge and Stall Propagation. Trans. ASME, 79, (April 1955) 455-469.
4. Leiblein, S., Experimental Flow in Two-Dimensional Cascades, Aerodynamic Design of Axial Flow Compressors, Chapter VI (NASA SP-36, 1965).
5. Rodgers, C., A Diffusion Factor Correlation for Centrifugal Impeller Stalling. ASME J. of Engineering Power, 100, (October 1978) 592-603.
6. Koch, C.C., Stalling Pressure Rise Capability of Axial Flow Compressor Stages. ASME J. of Engineering Power, 103 (October 1981) 645-656.
7. Sovran, G. and E.D. Klomp, Experimentally Determined Optimum Geometries for Rectilinear Diffusers with Rectangular Conical, or Annular Cross-Section, in Fluid Mechanics of Internal Flow (G. Sovran, ed., Elsevier Publishing Co., 1967).
8. Hercock, R.G., Effects of Intake Flow Distortion on Engine Stability, Paper 20 in Engine Handling (AGARD CP324, October 1982).
9. Bowditch, D.N. and R.E. Coltrin, A Survey of Inlet Engine Distortion Capability, NASA TM 83421 (prepared for AIAA Joint Propulsion Specialists Conference, June 1983).
10. Mazzawy, R.S., Multiple Segment Parallel Compressor Model for Circumferential Flow Distortion. ASME J. Eng. for Power, (April 1977) 288-296.
11. Railly, J.W. and H. Ekeral, Influence of a Closely-Coupled Throttle on the Behavior of a Radial Compressor Stage, ASME Paper 84-GT-190 (1984).
12. P. Cheng, et al., Effects of Compressor Hub Treatment on Stator Stall and Pressure Rise. J. Aircraft, 21 (July 1984) 469-476.
13. Smith, G.D.J. and N.A. Cumpsty, Flow Phenomena in Compressor

Casing Treatment, Cambridge University Engineering Department Report CUED/A-Turbo TR-112 (1982).

14. Dean, R.C., Jr., The Fluid Dynamic Design of Advanced Centrifugal Compressors, Creare Technical Note TN-185 (July 1974).
15. Dussourd, J.L., G.W. Pfannebecker, and S.K. Singhania, An Experimental Investigation of the Control of Surge in Radial Compresssors Using Close Coupled Resistances. ASME J. of Fluids Engineering, 99, (1977) 64-76.
16. Jansen, W., A.F. Carter, and M.C. Swarden, Improvements in Surge Margin for Centrifugal Compressors, presented at AGARD 55th Specialists' Meeting, Centrifugal Compressors, Flow Phenomena and Performance, Brussels, Belgium (1980).
17. Kinoshita, Y. and Y. Senoo, Rotating Stall Induced in Vaneless Diffusers of Very Low Specific Speed Centrifugal Blowers, ASME Paper 84-GT-203 (1984).
18. Frigne, P. and Van der Braembussche, A Theoretical Model for Rotating Stall in the Vaneless Diffuser of a Centrifugal Compressor, ASME Paper 84-GT-204 (1984).
19. Ng, S.L. and C.E. Brennen, Experiments on the Dynamic Behavior of Cavitating Pumps. ASME J. of Fluids Engineering, 100 (June 1978) 166-176.
20. Braisted, D.M. and C.E. Brennen, Auto-Oscillation of Cavitating Inducers, Proc. of Conference on Polyphase Flows (ASME, July 1980).
21. Rothe, P.H., P.W. Runstadler, Jr. and F.X. Dolan, Pump Surge Due to Two-Phase Flow, in Polyphase Flow in Turbomachinery, ASME Winter Annual Meeting (ASME, 1978).
22. Wylie, E.B. and V.L. Streeter, Fluid Transients (McGraw Hill Publishing Co., New York, 1978).
23. Eck, B., Fans, Design and Operation of Centrifugal, Axial-Flow, and Cross-Flow Fans, First English Edition (Pergamon Press, 1973).
24. Ehrich, F.E., Aerodynamic Stability of Branched Diffusers. ASME J. of Engineering for Power, 92 (July 1970) 330-334.
25. Dussourd, J.L. and W.C. Putnam, Instability and Surge in Dual Entry Centrifugal Compressors, in Symposium on Compressor Stall, Surge and System Response (ASME, 1960).
26. Stetson, H.D., Designing for Stability in Advanced Turbine Engines, Engine Handling (AGARD CP324, October 1982).
27. Greitzer, E.M., Surge and Rotating Stall in Axial Flow Compressors, Parts I and II. ASME J. Eng. for Power, 98, (April 1976) 190-217.
28. Mani, R., Compressor Post Stall Operation, Lecture Notes from AIAA Professional Study Seminar on Airbreathing Propulsion, Gordon C. Oates, course director (June 1982).
29. Hansen, K.E., P. Jorgensen, and P.S. Larsen, Experimental and Theoretical Study of Surge in a Small Centrifugal Compressor. ASME J. Fluids Engineering, 103, (September 1981) 391-395.

30. Moore, F.K., A Theory of Rotating Stall of Multistage Axial Compressors, Parts I-III. ASME J. Eng. Power, 106 (April 1984) 313-337.
31. Moore, F.K. and E.M. Greitzer, A Theory of Post-Stall Transients in Multistage Compression Systems, NASA Contractor Report, to appear.

12. ACKNOWLEDGMENTS FOR FIGURES

Figures 1-4, 6-11, 15-17, and 19-35 are copyrighted by the American Society of Mechanical Engineers.

Figure 18 is copyrighted by the American Institute of Aeronautics and Astronautics and is reprinted here with permission.

PANEL DISCUSSION

SUGGESTIONS FOR FUTURE WORK AND DEVELOPMENT

The panel consisted of:

Dr. P. Stow (Chairman)
Dr. J. Denton
Professor E. Greitzer
Professor R. Van den Braembussche

Dr. Stow opened the panel discussion by saying that from the lectures and course notes one should have a very good appreciation of the state of the art in turbomachinery on both the computational and experimental sides and that the idea behind the panel discussion was to talk about future work that people believed was needed. He said it was planned to cover four main topics with the subjects being introduced by a panel member namely

Axial Compressors - Professor Greitzer
Turbines - Dr. Denton
Radial Machines - Professor Van den Braebussche
Computational Work - Dr. Stow

1. AXIAL COMPRESSORS

Professor Greitzer started by saying that he would divide his presentation into two areas. First design point operation, where methods are based on getting the design point operation as efficient as possible, and second off-design operation where considerations often compete with the former. He said both need to be optimised simultaneously if one wanted to achieve increased performance.

1.1 Design Point Operation

Professor Greitzer started by asking the question "Do we already do well enough from correlations". He said that the answer was no, one reason being that a lot of designs tended to be evalutionary in that they evolve from existing designs by making small parameter changes but it may well be that some new and different designs exist that could not be obtained from this approach using correlations. Calculation methods or understanding of the flow were needed to allow these jumps to be made.

He presented a number of specific problems and phenomena for future work.

1.1.1 End-wall flow. Professor Greitzer said that the most important multi-stage problem was the end-wall flow in terms of the fluid dynamics and thermodynamics. He said that it was not really a boundary layer phenomenon as there were substantial pressure differences through the end-wall region. It was a three-dimensional viscous flow; tip clearance effects were also important. End-bends were being used but designed very much on a trial and error basis. He said that the flows were not well understood.

1.1.2 Mixing in turbomachines. He said that Smith and Adkins posed the problem "Why don't the end-walls heat up". He said that they had made a significant case for spanwise transport of stagnation enthalpy. They focussed on one mechanism for radial transport, namely, large scale secondary flow. A student of Dr. Cumpsty at Cambridge was looking at the problem from another aspect and Professor Greitzer said that it was not clear that the question was totally resolved so mixing remains another problem area.

1.1.3 Effects of unsteadiness on performance. Professor Greitzer posed two questions here: "How does unsteadiness impact on the steady design process?" and the most important question "Can we take advantage of it?". He presented a slide first shown by Mikolajczak some years ago demonstrating the effect of axial spacing on Compressor performance, see Figure. He said that close spacing brings in all unsteady effects - potential effects, wake effects etc., and it was clear that it was a better thing to do. The question was, he said, how close do the rows have to be? Was this something to do with unsteadiness and could we take advantage of it more? He said that this was still an open question.

1.1.4 Multi-row effects. He said that machines of interest were not single rows and we needed to take account of iteractions in a more correct manner. For example wakes, pseudo Reynolds stresses due to flow non-uniformity and may be most importantly the boundary conditions; rows were not isolated and could be fairly strongly coupled.

1.1.5 Tip section losses in transonic flows. Professor Greitzer posed the problem first raised by Kerrebrock in his AIAA Dryden lecture "Why are the tip section losses in transonic flows substantially higher than one might calculate?" This was still an important unanswered question he said.

1.1.6 Non gas path fluid mechanics. He said this was an important area possibly not viewed as being as glamorous as blade aerodynamics. Very important were the non gas path fluid dynamics of seals and parasitic losses under the blade. He posed the question of their size relative to end-wall losses and whether we should be working on the end-wall loss problem or were these other losses bigger. He said that there was not a good understanding of how to design efficient seals and how to calculate these parasitic losses.

1.1.7 Aero-elastic problems. He put forward a number of topics not covered in the course, flutter and forced response. He asked the question "What are the sources of excitation?" Some strange frequencies could appear, he said, as a result of wake-wake coupling in compressors and turbines and this was not really well understood.

He said that self-excited oscillations could occur due to tip clearance variations in turbines; Also oscillations associated with the flow in seals. Neither of these were well understood.

1.2 Off-Design Operation

Professor Greitzer said that off-design operation including stall offered all the problems of design except that they were generally more difficult.

On stall prediction he said that there was no such thing and that analysis relied heavily on correlations. Consequently we needed to understand the stall inception process from a more fundamental aspect in order to carry out prediction before building the machine.

On the effects of inlet distortion he said again this was by correlation. One would like to be able to assess the effects of parametric changes such as clearance, stagger etc computationally or analytically.

On casing treatment he said that it was not clear how it worked even years after being utilised in actual engines. Understanding was still lacking and guidelines for optimisation were needed.

As regards unsteadiness he said that as the flow moved towards stall the flow got more and more unsteady. If this could be damped in some way he wondered whether there would be more stall potential.

Stall recovery in multi-stage compressors he said was another area requiring substantial work.

1.3 Discussion

Dr. Hall asked whether we shouldn't be aiming to collect design rules; for example, he said, compressors used diffusion factor. Should we look for other rules he asked. Professor Greitzer replied that part and parcel of the understanding of phenomena should lead to much better criteria than diffusion factors.

Dr. Eralp commented on the effects of axial gap on performance and asked whether there was any work published on the effect on the onset of the stability point. Professor Greitzer replied that the Cooke correlation had one graph showing this effect on stall onset.

Dr. Casey commented on parasitic losses saying that they were much more important in radial machines and that quite a lot of work had been done.

On the subject of parasitic losses Dr. Stow said that work in disc cavities was an important area from the loss and heat transfer points of view.

2. TURBINE WORK

Dr. Denton started by saying that with turbines the problems were rather less than with Compressors as they tended to operate close to their design conditions so the interest was in performance.

He said we wanted high efficiency and high work per stage. High efficiency could be achieved at low stage loading and low Mach number; for example one could achieve an efficiency of 93-94% with no problems. However, in practise, certainly in gas turbines, we choose to operate at lower levels of efficiency because we wanted to increase the work per stage in the interest of reducing weight and cost of the machine. He said we could increase the work per stage either by a high loading coefficient or high Mach number, i.e. high rotational speed. The problem was to increase the loading and Mach number without decreasing the efficiency too much. Gas turbine designers had been striving to do this for many years; efficiencies had not increased but loadings had and the number of stages had been reduced. The problems were he said that as we increased stage loadings we increased the secondary losses of the blade quite dramatically and that reduced efficiency. As we increased the Mach number for transonic points we get greatly increased trailing edge losses particularly on cooled blades with large trailing edges. Areas where we needed to continue to work were in the reduction of secondary losses at high stage loadings by the increasing use of three-dimensional design methods. Geometries with highly curved trailing edges have been investigated by many Companies and institutions and shown to give substantial increases in efficiency. He said there were equally plausible arguments for curving the blades so that the pressure surface moved towards the end-walls or away, although only the former had been tried but on an ad hoc trial and error basis. He said that there was not much understanding of why the secondary losses were reduced and we needed further investigation in order to fully exploit the freedoms given.

Dr. Denton said that another way of reducing secondary losses, again tried on a trial and error basis, was end-wall contouring. The Russians were the first to show that it increased stage efficiency. Since then it had been tried by others, sometimes with success and other times without and the reason was not understood. He said we needed to understand the three-dimensional influences of contouring on the secondary flows and losses to fully exloit freedoms.

Dr. Denton raised the problem of trailing edge losses. He said that looking at profile loss against Mach number it shot up as sonic conditions were approached, probably doubling, and beyond there it came down again, but only sometimes; this was what prevented us going to higher Mach numbers and hence higher enthalpy drops per stage. He said that in fact over recent years the Mach numbers levels in HP gas turbines had been deliberately reduced in the interests of getting higher efficiency. He said that we didn't really know what caused this and that there had been a lot of measuring and correlating base pressures but not much on the origin of the loss itself. Further work was needed to understand what was going on in this area and then we needed to see if we could operate in the transonic region with losses not significantly higher than for subsonic flow.

Another question he said was the state of the blade surface boundary layers in a multi-stage turbine. Hodson's work at Cambridge had shown that the state of the boundary layer in a single stage turbine was very complex and was somewhere between laminar and turbulent. The question was whether a laminar boundary layer was possible after the first stage and consequently was there any point in designing blade surface velocity distributions on the assumption that there was laminar flow. He said that this was important when deciding whether fore or aft-loaded blades would give best performance and also from a heat transfer point of view. Once we knew the state of the boundary layer we would be in a much better position to optimise the blade profile.

He said that this was still optimisation on the basis of profile loss only, the mid-span loss of the blades, and that we did not know at all how to optimise low aspect ratio blades to achieve minimum secondary loss. Throughout the machine as a whole secondary loss was probably at least as important as profile loss yet we always optimised parameters like pitchchord ratio, surface velocity distributions etc on the basis of profile loss alone as tested in cascade; we needed to consider secondary losses. Dr. Denton said that the earlier statements on section stacking were applicable only to nozzles, for rotor blades there were less freedoms.

Dr. Denton said that another area for speculation was contra rotating blades in a stage; this would give much higher stage loading and work per stage but at the expense of mechanical complexities of two shafts rotating in opposite directions. He said one would get reduced weight and improved efficiency since there would be much less turning for the same stage loading if both blades rotated.

He said that radial inflow turbines may also become of interest as cycle pressure ratios become higher and the volume flow going into the HP turbine got less. They had several advantages, e.g. they could have a very high pressure ratio per stage, 4:1, without running into any problems of supersonic flow and the temperatures dropped very quickly minimising cooling problems.

Like Professor Greitzer he said there was a need to turn attention to parasitic losses, e.g. windage, inter-stage annulus boundary layers, mixing of the coolant with mainstream flows etc. He said there were lots of minor losses that become relatively more important as the main sources of irreversibility got worked on and reduced.

2.1 Discussion

Dr. Hourmouziadis gave answers to a number of questions raised by Dr. Denton. Regarding three-dimensional blade shapes, he said that MTU up to now had been able to correlate changes in efficiency with changes in the profile loading across the span. On the question of the existence of laminar boundary layers beyond the first blade row, he said that they had identified laminar separations in all four rows of a two stage turbine.

Dr. Denton replied saying that he felt that rather than considering surface pressure distributions people should be looking at more fundamental features, for example shed vorticity, to see whether it was driving the secondary flows towards the end walls, or should consider radial pressure gradients. He said he felt that the problem was not understood. Dr. Denton said that he was interested to hear about laminar boundary layers on the two stage machine and questioned whether the bubbles were leading edge separations. In reply Dr. Hourmouziadis said that some were beyond 50% axial chord.

Dr. Hall referred to statements made about reducing secondary losses rather than reducing secondary flows and asked whether this was deliberate. Dr. Denton replied that it was; he said that secondary flows did not necessarily produce any secondary losses. The secondary flow can be predicted reasonably well using inviscid methods. He continued that secondary losses were related to the flow but one could get losses which were incidental to the secondary flow; not all the kinetic energy of the secondary vortex would be recovered but some of it would. He believed that most of the secondary loss occured because the end-wall boundary layer was thinned as a result of the flow and exposed to high mainstream flow giving much higher shear stresses and greater rates of dissipation. He said that the secondary flow was a way of sucking off the existing boundary layer and exposing the surface to the high velocity free stream flow. The loss was a consequence of the secondary flow.

Dr. Casey commented on the fact that GE and NASA had a joint project, seen at the Farnborough Air Show, for an engine with contra-rotating turbine blades driving contra-rotating propellers.

Dr. Gregory-Smith raised another aspect of secondary flow namely the effects of the secondary flow of one row on subsequent rows. He said that whereas secondary flows may not produce any loss their effects may be such as to cause deterioration in the performance of the next blade row. He asked whether there were any comments on the importance. Dr. Denton said that he could not comment on the relative importance but agreed that it was an area that needed to be looked at. Dr. Hourmouziadis added the comment that the fluctuatios of exit angle moving from a relative to absolute system or vice-versa could amplify from say $\pm 3^o$ to $\pm 10^o$ which would affect the losses.

Professor Sieverding raised a subject that he said had not been addressed, namely impulse turbines with small aspect ratios and supersonic inlet and the match between supersonic outlet of the nozzle and supersonic inlet to the rotor. He asked Dr. Denton whether he felt it was still of interest. In reply Dr. Denton said that this was an area of very great current interest in the UK and also to Dr. Hall of Volvo Flygmotor. He said these were usually small turbines that didn't have the same economical importance as large gas turbines but there were problems where some people needed to work; he said that the efficiencies of these small supersonic turbines were in the order 50-60% and there was scope for improvement.

Professor Macchi said that in the last twenty years there had been a big effort in producing better computational work and also a lot of experimental work had been performed. He asked whether it was felt that designers could now produce more efficient turbines than say ten years ago. Dr. Denton replied by quoting engine designers as saying that they could not necessarily produce more efficient turbines but could do it in a much shorter time. Dr. Hourmouziadis added the comment that designers were now much surer of achieving efficiency goals than ten years ago but all the improvements in efficiency had been put into reducing size and stages and hence reducing weight.

Professor Sieverding raised two further points. He said that we knew much more about the radial distribution of angles and probably much less about losses. He asked whether there was any chance of taking the former into account in blade design. His second question was whether it would be possible to put spanwise mixing into through flow calculations in a better form than today. Replying on spanwise mixing Dr. Denton said that he had this feature in his through flow program for ten years but had only one piece of data on which to base a mixing coefficient. He suggested that it could be done more easily than Adkins and Smith if only more data were available for the model. Professor Greitzer added that Mr. Galimore at the Whittle Laboratory, Cambridge, was working on mixing in multi-stage compresors and looking at adaptations of mixing in a through-flow calculation to see what effects it had.

3. RADIAL MACHINES

Presentation by Professor Van Den Braembussche: Professor Van den Braembussche started by saying that he felt that there was a need for more realistic calculations. He commented on the fact that very good comparisons between three-dimensional calculations and experimental results had been shown for axial turbines but non yet for radial compressors. He said that there had been important improvements in numerical methods in the last decade and this had resulted in much better methods for inviscid flow. It was time now he said to put more effort into three-dimensional viscous-inviscid coupling and also in Navier-Stokes solutions. This would be more profitable than any further work on three-dimensional inviscid calculations. He also added that methods must be three-dimensional since two-dimensional calculations were not sufficient - the flow was very three-dimensional.

He said that the problem with flow in the impeller and diffuser was that not many detailed measurements had been made up to now. Those that existed were at design conditions and were mainly to verify calculation methods. However, at off-design the measurements would be much more difficult as the flow may well be unsteady. Also local reverse flow was not easily handled by hot wires or laser systems. Measurements were needed, he said, to indicate what the flow was doing at off-design. Only then could we work on better descriptions of it.

He continued that he was often surprised by how many centrifugal compressors were working with unsteady flow in terms of impeller rotating stall and diffuser rotating stall. He said there were some famous cases where compressors had been operated at different inlet conditions and where the unsteady flow phenomena had resulted in drastic mechanical failures. In general many centrifugal compressors were much more strongly built than axial ones; a lot were shrouded with thicker blades. Also they had three-dimensional blade shapes that may make the blade more resistant to vibration. He continued by saying that one thing that happened was impeller or diffuser rotating stall which created unsteadiness. Then there was the impeller-diffuser interaction; if one had a vaned diffuser then there was a strong interaction between the diffuser an impeller. He mentioned measurements made by Eckert at DFVLR but said it was still not understood what was going on.

Another area where unsteadiness was important was, he said, where there was non-axisymmetric outlet flow conditions. Compressors continuously operated with unsteady flow in the rotor and not much work had been done in this area.

A further point made was inlet distortion which was he said the same for other machines but usually there was a radial to axial bend before entry to the compressor. The effect of the unsteadiness was very important. He said that we needed to do a lot of work to get a better understanding of where losses came from and how performance was influenced by unsteadiness.

Presentation by Dr. Casey: At the request of the panel Dr. Casey made a short presentation. He mentioned a number of areas where he felt that more work was needed.

He said that more should be done on performance prediction and quoted problems with work input or slip factors. If one was trying to calculate the performance of an impeller then one didn't know what work input was going to come. He said that most of the correlations used were on the safe side. One usually got more pressure ratio using the correlations available and this was accepted as a safety factor in the design. He said that effectively there were even fewer loss correlations to work with in radial compressors than in axial ones and in turbines. Every Company had its own correlations but these were only usually valid for the impellers used to provide the correlations. There was he said need for a lot more work on efficiency prediction.

He commented on operating range and said that there was no real method of predicting surge or rotating stall. We could predict choke, to a certain extent, but there were fudge factors that came into that too.

He said that one aspect of industrial compressors that was probably different from turbo-chargers and radial compressors in small gas turbines were the problems of scaling and real gas effects. For example, for the effect of the diameter on performance, one may have a scale factor of four between a model and prototype and this changed the diameter and roughness effects. Also he said there were problems with different gases and this was two-fold. Firstly there were the problems of predicting these effects on impeller performance and secondly with gases used in industrial compressors the values of K and R, were often not known.

He continued by saying that there was a need for better flow models and that probably it would be much more satisfactory to move to three-dimensional viscous models since it really was a three-dimensional problem in radial compressors. He reiterated Professor Braembussche's point about the need for more measurements to calibrate methods.

He said that component interaction was another important aspect that little was known about. Changing to a vaneless diffuser he said caused a change in the slip factor and changing the diffuser can also cause a change. There were also the effects of disturbances caused by the scroll on the diffuser and on the impeller.

One other point he said that may become more important in the future, as more and more often we found variable geometry in industrial compressors, inlet guide vanes or diffuser angle variations, was the need to have better models of the matching of the various components at design and off-design. Perhaps he said this could be performed with a simple one-dimensional analysis at least to start with; there was nothing in the literature.

4. COMPUTATIONAL WORK

Dr. Stow listed a number of aspects of computational work that he felt were impotant for the future.

4.1 Method Reliability

He said that from an industrial point of view method reliability was very important; methods quickly became discarded by users if they were unreliable. A lot of effort was needed to produce reliable user-proof methods and what companies tended to do was develop their methods to get the most out of them rather than moving over to new methods.

Another important aspect of development he said was evaluation. This was more and more essential these days as methods become more sophisticated. He said there had been some good examples shown during the week and he encouraged people to do more in the future. He said that there needed to be good communications between people developing methods and those doing experiments and often they needed to work as teams.

4.2 Speed of Methods

He said that the faster the methods could be made the better, as this would allow additional optimisation in design times available and he advocated looking at the development of methods for supercomputers and parallel processing machines. He said that there was going to be a computing revolution in terms of speed and core of machines available and that there was need to be able to take maximum advantage of this.

4.3 Optimum Computational Grids

The question of what was the optimum computational grid was raised. It was clear he said that around the leading edge of blades the full geometry should be used but it was not yet clear what should be used around the trailing edge especially with cusp models in inviscid calculations. Also it was not yet clear what level of sophistication was needed in the type of grid to be used.

4.4 Method/Algorithm Versatility

He said that the situation was that blading methods were very advanced but they were often difficult to adapt easily to other flow configurations, e.g. intakes, exhaust systems etc. If they could be written in a more general form, for example working in a finite element sense in terms of connectivity tables, so that all one had to do for a new problem was to generate the grid and the method would solve the equations, we would be able to make progress in a lot of areas very quickly. This, he said, was one of the main reasons for the success of finite elements in solid mechanics. The same sort of ideas could be applied to finite volume methods.

4.5 Boundary Condition Treatment

Often, he said, one needed a "settling" distance with blading methods in order to match up with through-flow conditions at inlet and outlet. More needed to be done on reducing these distances since as well as reducing the number of grid points this would also be of advantage when moving on to stage calculations.

4.6 Multi-grid Techniques

He said that over the next few years there would be much more work emerging on Navier-Stokes solutions in two and three-dimensions. There was room for work on multi-grid techniques for viscous flow; whereas inviscid flows had received attention little had yet been done for viscous flows.

4.7 Three-dimensional Design Calculations

He introduced this topic but questioned how such a method would be used. May be, he said, a mixed design and analysis mode calculation would be of use here.

4.8 Analytical Solutions

Prompted earlier by Dr. Hourmouziadis he suggested it was worth looking for analytical solutions for checking computational work especially for three-dimensional flow and unsteady flow.

In conclusion Dr. Stow said that he expected to see more work on Navier-Stokes calculations in two and three-dimensions, more on semi-inverse boundary layer coupling with inviscid methods and also emerging unsteady Navier-Stokes solutions. It was still early days in these areas and there was a lot of room for contributions to be made.

4.9 Discussion

Dr. Gregory-Smith said that one area not mentioned was turbulence modelling in Navier-Stokes solutions. He asked how big a problem people saw this and how it might be tackled. In reply Dr. Stow said that he felt that it was a formidable problem. The signs were that one was starting to get the numerics under control and at that stage the quality of the answers was governed by transition and turbulence models.

Transition modelling he said was a very difficult problem, especially in three-dimensions, where there was little data and few clues as to what to do apart from simple three-dimensional extensions to two-dimensional correlation models. On the turbulence modelling side he said that satisfactory answers could be obtained in many cases using fairly simple mixing length models but that higher order models would most probably be needed for complex flow configurations.

Dr. Hall said that of the methods seen during the course time-marching appeared to be dominant. As we were interested in the main in steady flows he questioned whether this was the way to achieve the fastest steady state solution. He suggested one should look at the mathematics involved. Dr. Stow pointed out that a number of non time-marching methods had been presented e.g. pressure-correction, finite element methods, velocity potential approaches etc. He said that Applied Mathematicians looking at flow problems often advocated a Newton-Raphson type of approach; this was very similar to implicit time-marching methods used for viscous flow. One thing worth saying in favour of explicit time-marching was he said that there was a lot of the physics encompassed in this type of approach.

Professor Üçer said that we should aim to have standard experimental test cases on which to check numerical solutions. He suggested that this was lacking. Dr. Stow suported the use of test cases saying that in industry it was usual to test methods against experiment quite extensively before general release.

Dr. Hourmouziadis said that as the computational procedures became more detailed like Navier-Stokes solutions, then one should not forget to look at the boundary conditions which are getting more complicated. When going over to a stage calculation with a Navier-Stokes calculation one should look at the cross section of a real turbine, for example the gaps between the hub of the rotor and hub of the stator and the secondary flow in the rows. He continued by appealing to the teaching community in saying that he felt that today young engineers tended to compute for the sake of it and may be didn't have such a good feeling for the physics that they might. In replying to the latter point Dr. Denton said that with a good numerical method one could actually learn and get physical insight much more quickly than by doing experiments. He said that this was his experience, for example changing blade geometries, running the program and looking at the results and he advocated this approach. In support Dr. Stow said that this was his experience also but as Dr. Denton had suggested you needed to know that you could rely on the results of the method and that extensive evaluations were needed before this stage was reached.

Dr. Van den Braembussche advocated care on the use of test cases. He quoted an example where using a non-conservative difference scheme better comparisons with experiment were achieved than with a conservative scheme because the numerical errors in the differencing were of the same order as the differences between the numerical solution and experiments for the grid used.

Professor Sieverding suggested that nomenclature in computational work was a problem. He complained about the non-standard situation especially in turbines and advocated that an international committee should consider the problem. Dr. Stow said he had seen enough problems with trying to standardise within companies to wonder whether this would be possible.

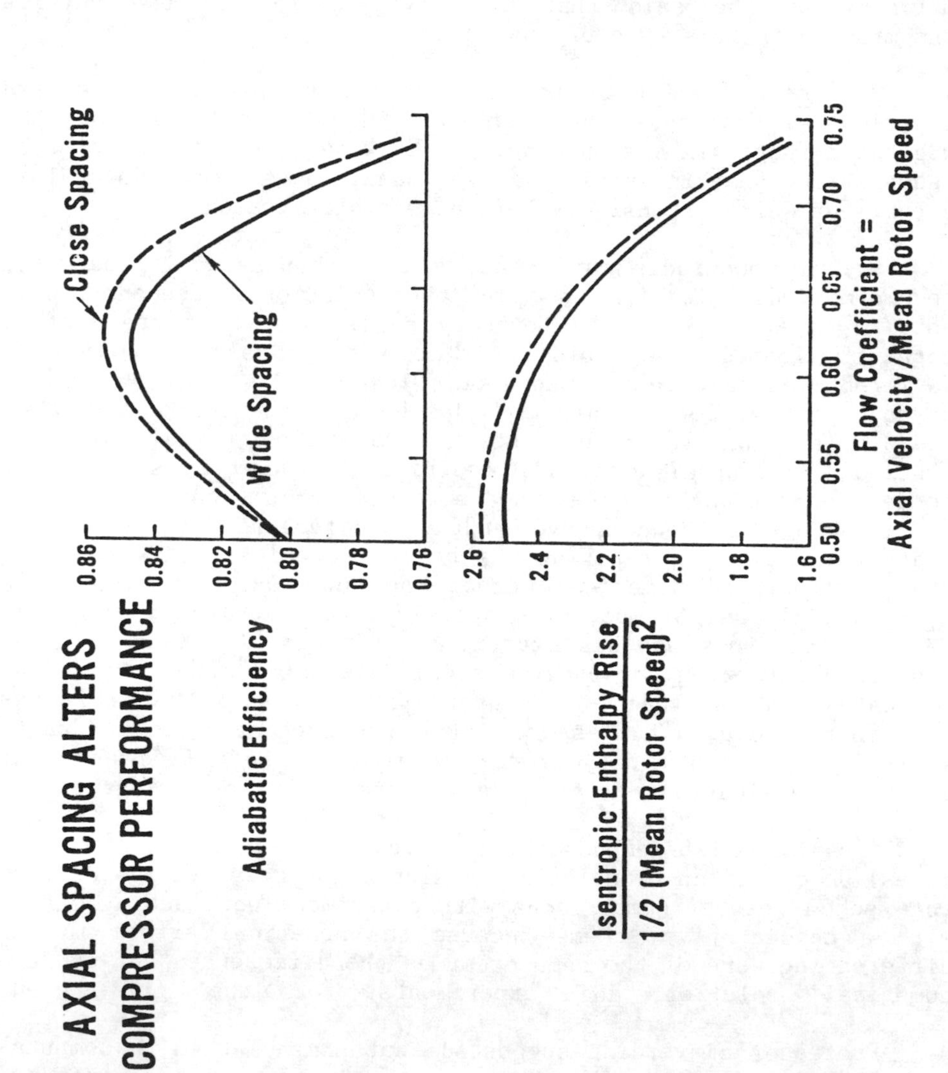
AXIAL SPACING ALTERS
COMPRESSOR PERFORMANCE
Adiabatic Efficiency
0.86
0.84
0.82
0.80
0.78
0.76
Close Spacing
Wide Spacing
Isentropic Enthalpy Rise
1/2 (Mean Rotor Speed)2
2.6
2.4
2.2
2.0
1.8
1.6
0.50
0.55
0.60
0.65
0.70
0.75
Flow Coefficient =
Axial Velocity/Mean Rotor Speed

INDEX